Meet the Faculty and Post-Doc Candidates Poster Session – Sponsored by the Education Division 2019

Held at the 2019 AIChE Annual Meeting

Orlando, Florida, USA
10 - 15 November 2019

ISBN: 978-1-7138-0551-9

Printed from e-media with permission by:

Curran Associates, Inc.
57 Morehouse Lane
Red Hook, NY 12571

CURRAN ASSOCIATES INC.
proceedings
.com

Some format issues inherent in the e-media version may also appear in this print version.

AIChE
120 Wall Street, FL 23
New York, NY 10005-4020

Phone: (800) 242-4363
Fax: (203) 775-5177

www.aiche.org

Additional copies of this publication are available from:

Curran Associates, Inc.
57 Morehouse Lane
Red Hook, NY 12571 USA
Phone: 845-758-0400
Fax: 845-758-2633
Email: curran@proceedings.com
Web: www.proceedings.com

TABLE OF CONTENTS

(6B) ENGINEERING THE FUTURE OF AGING, HEALTHSPAN AND LYMPHOMA TIME 1
Jude M. Phillip

(6D) DESIGNER BIOMATERIALS AND INTEGRATED BIOSENSORS TOWARD PRECISION MEDICINE 3
Jouha Min, Ralph Weissleder

(6E) BIOMATERIAL INVESTIGATION AND DEVELOPMENT FOR BIOMEDICAL ADVANCEMENT 4
Antonio C. F. Dos Santos

(6F) MOLECULAR SIMULATIONS OF BIOLOGICAL SELF-ASSEMBLY 6
Gul H. Zerze

(6H) PEPTIDE- AND PROTEIN-BASED FUNCTIONAL BIOMATERIALS 7
Jugal Kishore Sahoo

(6I) CONVERSION OF WASTE BIOMASS INTO BIOPRODUCTS (BIOENERGY, BIOMATERIALS, BIOCHEMICALS) 9
Ezinne Achinivu

(6K) ENGINEERING BIOLOGY SOLUTIONS FOR A SUSTAINABLE FUTURE 11
Niju Narayanan

(6N) EXPLORING THE DESIGN SPACE OF LIVING SYSTEMS: EXPERIMENTAL AND THEORETICAL TOOL-KITS FOR MULTI-FUNCTIONAL MATERIALS 15
Symone Alexander

(6R) ULTRASMALL C' DOTS ACTIVATE PRO-INFLAMMATORY ANTI-TUMOR RESPONSES IN THE MICROENVIRONMENT OF PDGF-B DRIVEN HIGH GRADE GLIOMAS 17
Steven Zanganeh

(6S) METABOLIC SUGAR LABELING FOR CANCER TARGETING AND IMMUNOENGINEERING 19
Hua Wang, David J. Mooney

(6U) MULTISCALE DYNAMICS IN BIOLOGICAL SOFT MATTER AND POLYMERIC FLUIDS 21
Amir Saadat

(6V) MOLECULAR TENSION SENSORS IN PROTEIN HYDROGELS 24
Joshua Baccile, David Tirrell

(6W) 3D-ENGINEERING OF FUNCTIONAL LIVING COMPOSITE MATERIALS 25
Seunghyun Sim

(6AB) MOLECULAR DESIGN PRINCIPLES FOR CHEMICALLY TUNABLE BIOMATERIAL PLATFORM TECHNOLOGIES 27
Crystal K. Chu

(6AC) SHINING LIGHT ON THE NERVOUS SYSTEM: FROM BIOMATERIALS TO BIOELECTRONICS 29
Jing Tang

(6AE) 3D PRINTED BIOELECTRONICS FOR TISSUE ENGINEERING AND REGENERATIVE MEDICINE 33
Alexandra Rutz

(6AG) LIPID NANOPARTICLES FOR HIGHLY EFFICIENT NON-VIRAL GENE EDITING 36
Jie Li, Niren Murthy

(6JZ) MULTISCALE MODELING FOR THE DESIGN OF FUNCTIONAL SOFT MATERIALS 37
Nicholas Jackson

(6G) FABRICATION OF HIGHLY SENSITIVE LSPR BIOSENSOR FOR THE C-REACTIVE PROTEIN (CRP) DETECTION BASED ON THE IMMUNOCOLLOIDAL GOLD NANOPARTICLES 39
Seo Yeong Oh, Tea-Joon Jeon, Eunseon Lee, Mi-Wha Oh, Yun Suk Huh

(6Y) MULTISCALE BIOMECHANICS OF PLATELET-DRIVEN BLOOD CLOT CONTRACTION AND INTRACELLULAR MECHANISMS OF ITS TERMINATION 40
Oleg Kim

(6KK) HIERARCHICAL SELF-ASSEMBLY AND BIOLOGICAL INTERACTIONS OF FUNCTIONAL SYNTHETIC AND NATURAL SUPRAMOLECULAR SYSTEMS 42
Herdeline Ann M. Ardoña

(6KQ) GENETICALLY ENGINEERED PROBIOTICS DESIGNED AT THE INTERCONNECT OF SYNTHETIC BIOLOGY AND METABOLIC ENGINEERING 44
Amin Zargar
(6KR) QUANTITATIVE SINGLE-CELL ANALYSIS OF RNA REGULATION AT THE SINGLE MOLECULE LEVEL 46
Fangyuan Ding
(6KS) THEORY AND MODELING OF BIOPOLYMERS AND BIO-INSPIRED SOFT MATERIALS 49
Kai Huang
(6L) UNDERSTANDING PHYSICALLY CROSSLINKED POLYMER NETWORKS TO RATIONALLY DESIGN HYDROGEL BIOMATERIALS 51
Hector Lopez Hernandez, Eric A. Appel
(6M) BIOCOMPATIBLE ZWITTERIONIC POLYMERS FOR MEDICAL APPLICATIONS 54
Xiaojie Lin
(6P) ELECTROCHEMISTRY ENABLING NEW FRONTIERS IN BIOMATERIALS ENGINEERING 56
Sina Jamali
(6AA) ENGINEERING MICROORGANISMS AND THEIR ENVIRONMENT FOR INCREASED PERFORMANCE IN BIORENEWABLE APPLICATIONS 59
Kirsten Davis
(6A) NANO-OPTICAL AND -ELECTRONIC DEVICES WITH MACHINE LEARNING FOR BIOMARKER DISCOVERY AND DIAGNOSTICS IN PERSONALIZED MEDICINE 60
Lee Korshoj
(6C) BIOSAFE, ECO-FRIENDLY LEVAN POLYSACCHARIDE TOWARD TRANSIENT ELECTRONICS 63
Kiyoon Kwon, Tae-Il Kim
(6J) CONSTITUTIVE MODELING OF COMPLEX BIOMATERIALS 64
Jeffrey S. Horner
(6O) INTEGRATING MOLECULAR MODELING WITH EXPERIMENTAL WORK TO PROPEL A BIOPRODUCTS PIPELINE 66
Emma C. Brace
(6AF) ADVANCING TECHNOLOGIES FOR PRENATAL AND WOMEN'S HEALTH 68
Christina M. Bailey-Hytholt
(210A) BIOFILM PREVENTION BY UVC SIDE EMITTING OPTICAL FIBERS 70
Mariana Lanzarini-Lopes, Sergi Garcia-Segura, Paul Westerhoff
(6KT) FORCE RESPONSE DEFINES BOTH SUBCELLULAR ARCHITECTURES AND DYNAMIC PROTEIN BINDING 72
Peter Chung
(6KW) ENGINEERING BIOMIMETIC MATERIALS AND CELLS FOR DIAGNOSTIC AND THERAPEUTIC APPLICATIONS 75
Zongmin Zhao
(6KZ) MULTI-MODAL BIOFABRICATION APPROACHES FOR BIOMATERIALS DEVELOPMENT AND TISSUE ENGINEERING 77
Jenna M. Shapiro
(6LE) AN UNSTEADY STATE REACTOR ENGINEERING MODEL FOR SPATIALLY HETEROGENEOUS WINE FERMENTATIONS 81
Konrad Miller, David E. Block
(6LG) OPTOELECTRONIC DEVICES AND SMART BIOMATERIALS 82
Dena Shahriari
(6LN) MOLECULAR ENGINEERING APPROACHES TOWARDS PLATFORM IMMUNO-BIOMATERIALS 84
Owen S. Fenton, Robert Langer
(6LO) DEFINING CAPTURE AND RELEASE MECHANISMS OF BIOCHEMICAL LIGANDS IN MULTIFUNCTIONAL BIOMATERIALS TO CONTROL CELL FUNCTION 86
Linqing Li, Christopher Chen
(6AI) LIVING BIODEVICES FOR PRECISION MEDICINE:FROM MORPHING ELECTRONICS TO MEDICAL NANOROBOTS 88
Jinxing Li
(6AJ) SOFT, FLEXIBLE TISSUE-INTEGRATED CHEMICAL SENSORS: FROM WEARABLE TO IMPLANTABLE NEURAL SYSTEMS 92
Amay J. Bandodkar, John A. Rogers

(6AL) DEVELOPING TOOLS FOR EMERGING LIQUID BIOPSY APPLICATIONS 95
Jose C. Contreras-Naranjo
(6AM) INTERFACING CELL/TISSUE ENGINEERING WITH GENE EDITING TOOLS AND SEQUENCING TECHNOLOGIES FOR REGENERATIVE MEDICINE 97
Halil Tekin
(6AN) ENGINEERING MICROSYSTEMS FOR REGULATING CELLULAR BEHAVIOR: FROM IMPLANTABLE DRUG FACTORIES TO NOVEL PLATFORMS FOR SINGLE-CELL GENOMICS 99
Suman Bose
(6AO) BIOLOGICALLY INSPIRED, ELECTRICALLY ACTIVE MEMBRANES FOR SUSTAINABILITY AND MEDICINE 101
Thomas B. H. Schroeder
(6AP) COMPUTATIONAL MODELING IN CANCER SYSTEMS BIOLOGY: STOCHASTICITY, COMPLEXITY, AND MULTISCALE DYNAMICS IN DISEASE PROGRESSION AND DRUG RESPONSE 104
Leonard Alfredo Harris
(6AR) MULTIDIMENSIONAL SINGLE CELL ANALYSIS: DEVICES AND TECHNOLOGY FOR CANCER BIOLOGY 106
Alex Xu
(6AT) THE ROLE OF TOPOGRAPHICAL CUES IN CANCER CELL MIGRATION AND METASTASIS 108
Colin D. Paul
(6AU) REVERSE PERFLUOROCARBON EMULSIONS FOR PULMONARY DRUG DELIVERY 110
Diane L. Nelson
(6AV) ENGINEERING DNA-POLYMER ASSEMBLIES 111
Alexander E. Marras
(6AW) SYNERGIZING ENGINEERING, CHEMICAL, AND IMMUNOLOGICAL CONCEPTS TO DESIGN THE NEXT GENERATION OF THERAPEUTICS FOR UNMET CLINICAL NEEDS 113
Benjamin Umlauf
(6AX) ENGINEERING MATERIALS TO RECAPITULATE THE STEM CELL MICROENVIRONMENT 115
Christopher M. Madl
(6AY) IONIC LIQUIDS TO OVERCOME OBSTACLES IN NANOPARTICLE DRUG DELIVERY 116
Eden E L Tanner
(6BA) UNDERSTANDING NUCLEATION AND CRYSTAL GROWTH OF ORGANIC MOLECULAR MATERIALS WITH APPLICATION TO PHARMACEUTICAL MANUFACTURING 118
Gerard Capellades
(6AS) A POLYMERIC REACTOR FOR THE SYNTHESIS OF SUPERPARAMAGNETIC-THERMAL TREATMENT OF BREAST CANCER 119
Roa' Fardous
(6LJ) DEVELOPING METHODS TO ACCELERATE THE DESIGN, DEVELOPMENT, AND IMPLEMENTATION OF THERAPEUTICS, MATERIALS, AND EQUIPMENT FOR DISEASES WITH URGENT UNMET CLINICAL NEEDS 120
Donald Belcher
(6LT) HIGH PERFORMANCE MATERIALS FOR WEARABLE ELECTRONICS AND SENSORS 122
Lauren Taylor
(6LU) BIONANOTECHNOLOGY: DNA NANOTECHNOLOGY-BASED MOLECULAR DEVICES, BIO-SENSORS, AND HYBRID MATERIALS FORBIOLOGICAL AND NANOPHOTONIC APPLICATIONS 124
Palash Dutta
(6BC) PROCESS INTENSIFICATION FOR SUSTAINABLE FUELS AND ENERGY PRODUCTION 126
Cornelius Mduduzi Masuku
(6BD) MOLECULAR MODELING AND MACHINE LEARNING FOR CATALYSIS AND SEPARATIONS 129
Tyler R. Josephson
(6BG) SCALING UP FIRST PRINCIPLE SIMULATION IN REALISTIC ENVIRONMENT: SOLVENT EFFECTS AND EXCITED STATE PROPERTIES IN COMPUTATIONAL CATALYSIS 132
Fang Liu

(6BH) ELUCIDATING ATOMIC DANCES THROUGH REACTION LANDSCAPES 133
Arthur J. Shih
(6BI) FROM 1- AND 2-DIMENSIONAL MATERIALS TO ARCHITECTURAL PROPERTIES IN CATALYSIS: RATIONALIZING, PREDICTING AND DESIGNING THROUGH FIRST-PRINCIPLES METHODS 135
Roberto Schimmenti
(6BK) DYNAMIC HETEROGENEOUS CATALYSIS TO ENHANCE TURNOVER FREQUENCY VIA SURFACE RESONANCE 137
M. Alexander Ardagh
(6BM) SUSTAINABLE FUEL AND CHEMICAL SYNTHESIS VIA CATALYTIC VALORIZATION OF ABUNDANT AND RENEWABLE RESOURCES 139
Nathaniel Eagan
(6BN) DISCOVERY AND OPTIMIZATION OF PROCESSES AND CATALYSTS FOR THE ORGANIC ELECTROSYNTHESIS OF CARBON-NEUTRAL FUELS AND CHEMICALS 140
Ezra L. Clark
(6BP) ENABLING NEW CHEMISTRIES THROUGH CATALYST DESIGN 142
Marcella Lusardi, Mark E. Davis
(6BQ) PRECISION SYNTHESIS TO CONTROL CATALYST SURFACE STRUCTURES FOR IMPROVED REACTIVITY AND PERFORMANCE 144
Madelyn R. Ball
(6BS) RESHAPING THE CARBON CYCLE WITH CATALYSIS: SELECTIVE ACTIVATION OF CHEMICAL BONDS FOR PRODUCING CARBON NEUTRAL FUEL AND CHEMICALS 146
Alyssa Hensley
(6BU) DEVELOPMENT OF THIN FILM DEPOSITION & ETCHING PROCESSES FOR CHALLENGING MATERIALS 149
David Barlaz
(6BV) UNIFYING PRINCIPLES IN THERMALLY AND ELECTROCHEMICALLY DRIVEN CATALYTIC REACTIONS 151
Joaquin Resasco
(6BZ) FIXED FEED TEMPERATURE PROGRAM MODULATION (FFTPM) - A VERSATILE METHOD FOR EXTRACTING ADSORPTION THERMODYNAMICS FOR WEAKLY ADSORBING SYSTEMS OVER A RANGE OF ADSORBATES 153
William T. Gibbons
(6CA) MECHANISTIC STUDY AND ENGINEERING OF VAPOR/SOLID INTERFACE IN-OPERANDO: APPLICATIONS IN HETEROGENEOUS CATALYSIS FOR ENERGY UPGRADING REACTIONS 154
Xueqiang Zhang
(6CF) UNDERSTANDING AND CONTROLLING MULTIELECTRON TRANSFER CHEMISTRY FOR SUSTAINABLE ENERGY TECHNOLOGIES 155
Adam Nielander
(6CH) COMPUTATIONAL-ACCELERATED MATERIALS DESIGN FOR NEGATIVE EMISSION TECHNOLOGIES 158
Paul Meza-Morales
(6CI) CATALYST DESIGN THROUGH SIMULATIONS AND MACHINE LEARNING 161
Seoin Back
(6CJ) ACCELERATING NET-ZERO CARBON EMISSIONS BY ELECTROCHEMICAL CATALYSIS: UNDERSTANDING AND CONTROLLING THE REACTIONS AT INTERFACES 163
Lei Wang
(6KG) DESIGN OF DYNAMIC AND SOLVATED REACTION ENVIRONMENTS FOR ENHANCED CONVERSION AND SELECTIVITY WITH HETEROGENOUS CATALYSTS 166
Jennifer Jocz
(6KN) SUSTAINABILITY TOWARDS THE FUTURE:BRIDGING THE GAP BETWEEN CATALYSIS SCIENCE AND REACTION ENGINEERING 168
Yang Xiao
(6CD) RATIONAL CATALYSTS AND PROCESS DEVELOPMENT FOR LIGHT HYDROCARBONS UPGRADING FROM NATURAL GAS/SHALE GAS 171
Weijian Diao
(6BB) RATIONAL DESIGN OF ALLOY CATALYSTS BY BUILDING ATOMIC-SCALE STRUCTURE-PROPERTY-ACTIVITY RELATIONSHIPS 174
Liang Cao

(6BJ) COMBINING FIRST-PRINCIPLES MODELING AND NANOMATERIAL SYNTHESIS TO UNDERSTAND AND IMPROVE ENVIRONMENTAL CATALYSIS 175
Hanyu Ma
(6CK) PROCESS-INFORMED DESIGN OF ELECTROCATALYSTS FOR SUSTAINABLE CHEMICAL TRANSFORMATIONS 177
Gastón O. Larrazábal
(6CG) DEVELOPMENT OF N-DOPED OXIDES AS AN ELECTRON RICH SUPPORT: BRIDGING HOMOGENEOUS AND HETEROGENEOUS CATALYSIS 179
Juan Jimenez
(6BF) FUNDAMENTAL DESIGN OF SUSTAINABLE CATALYTIC REACTIONS AT THE WATER-ENERGY-FOOD NEXUS 181
Lea Winter
(6BO) ELECTROSYNTHESIS FOR SUSTAINABLE CHEMICAL PRODUCTION 183
Matthew Jouny
(6BR) STUDY OF CATALYTIC REACTIONS FOR THE PRODUCTION OF FUELS AND CHEMICALS FROM RENEWABLE FEEDSTOCKS 185
Gabriel Seufitelli
(6BT) DIRECT AND NON-OXIDATIVE CONVERSION OF METHANE TO VALUE ADDED PRODUCTS 187
Vaidheeshwar Ramasubramanian
(6BW) ONE-STEP SYNTHESIS OF OXIDE CATALYSTS USING AEROSOL REACTORS FOR ENVIRONMENTAL APPLICATIONS: THEORETICAL AND EXPERIMENTAL STUDY 189
Sungyoon Jung
(6CB) ENGINEERING THE CATALYTIC ENVIRONMENT: SYNTHETIC, MECHANISTIC, AND SPECTROSCOPIC APPROACHES FOR DEVELOPING DESIGN PRINCIPLES 190
Daniel T. Bregante
(6CC) FIRST-PRINCIPLES APPROACHES FOR CATALYST DESIGN: NOVEL DESCRIPTORS AND STRATEGIES FOR MATERIALS DISCOVERY 193
Joseph Gauthier
(6KU) NOVEL APPROACHES TO BIOMASS UPGRADING FOR A MORE SUSTAINABLE CHEMICAL INDUSTRY 194
Andrew W. Tricker
(6LC) CATALYST DESIGN WITH ATOMIC PRECISION FOR FUEL GAS PROCESSING AND POLLUTION PURIFICATION REACTIONS 195
Ming Yang
(6CL) MOLECULAR SIMULATIONS OF INTERFACIAL DYNAMICS IN BIOLOGICAL SYSTEMS 198
Viviana Monje-Galvan
(6CO) REPRODUCE - REMEDIATION AND PRODUCTION USING COMPUTATIONAL AND ELECTROCHEMICAL APPROACHES 200
Damilola A. Daramola
(6CZ) FROM BIOSYNTHESIS TO HUMAN HEALTH: HARNESSING BIOMOLECULES WITH MULTISCALE QUANTUM MECHANICS–MOLECULAR MECHANICS SIMULATIONS 203
Zhongyue Yang, Heather J. Kulik
(6JX) DESIGNING THE STRUCTURE AND FUNCTION OF SOFT, COMPLEX MATERIALS WITH COMPUTATIONAL MODELING 205
Michael P. Howard
(6KI) COMPUTATIONAL PROTEIN DESIGN USING OPTIMIZATION PROGRAMS AND FORCE-FIELD CALCULATIONS 207
Ratul Chowdhury, Costas Maranas, Manish Kumar
(6CP) DATA-DRIVEN MODELING AND CONTROL OF BATCH AND BATCH-LIKE PROCESSES 209
Abhinav Garg
(6CQ) OPTIMIZATION AND CONTROL OF CHEMICAL PROCESS SYSTEMS UNDER UNCERTAINTY 211
Rohit Kannan
(6CR) COMPUTATIONAL DESIGN AND DEVELOPMENT OF ADVANCED CATALYTIC MATERIALS 214
Mingjie Liu
(6CS) STUDYING SOFT MATERIALS IN- AND OUT-OF-EQUILIBRIUM USING ANALYTICAL AND NUMERICAL FIELD THEORIES 215
Douglas Grzetic

(6CT) QM-BASED MULTISCALE SIMULATIONS FOR APPLICATIONS IN ELECTROCATALYSIS, INTERFACIAL CHEMISTRY, AND ENERGETIC MATERIALS 216
Saber Naserifar
(6CU) COMPUTATIONAL HETEROGENEOUS CATALYSIS FOR ENERGY STORAGE AND CONVERSION 220
Zhenghang Zhao
(6CV) COMPUTATIONAL CRYSTALLIZATION AND ASSEMBLY OF POLYMERS AND SOFT MATTER 222
Wenlin Zhang
(6CW) COMPUTATIONAL MODELING OF CHEMICAL INTERACTIONS AT INTERFACES FOR ENVIRONMENTAL APPLICATIONS 224
Zhizhang Shen
(6CY) MODELING AND OPTIMIZATION FOR PROCESS SYSTEM ENGINEERING PROBLEMS 226
Yachao Dong
(6DB) PREDICTIVE DESCRIPTORS FOR THE TARGETED SYNTHESIS OF SOLID-STATE MATERIALS 228
Christopher J. Bartel
(6CN) UNDERSTANDING MULTIMODAL INTERACTIONS THROUGH DEEP NEURAL NETWORKS AND STATISTICAL MECHANICS 229
Camille Bilodeau
(6DD) MECHANISTIC ELECTROCHEMISTRY DRIVING ENERGY SCIENCE: FROM CATALYSTS TO BATTERIES 231
Srini Ramakrishnan, Byungchun Park, Robert M. Waymouth, Christopher E. D. Chidsey, Bryan D. McCloskey
(6DI) HETEROGENEOUS ELECTROCATALYSIS: DEVELOPING STRATEGIES TO ENGINEER INDUSTRIALLY RELEVANT CATALYSTS FROM FUNDAMENTAL ACTIVITY TRENDS 234
Michaela Burke Stevens
(6DN) CORROSION MITIGATION AND WASTEWATER TREATMENT 237
Zineb Belarbi
(6DO) NANOSCALE DESIGN OF ELECTROCATALYSTS FOR CO_2-TO-FUELS POWERED USING RENEWABLE ELECTRICITY 240
Yuguang (Chris) Li
(6DP) REINFORCED ANION EXCHANGE MEMBRANE (AEM) SEPARATORS BASED ON TRIBLOCK COPOLYMERS FOR ELECTRODE-DECOUPLED REDOX FLOW BATTERIES (RFBS) 241
Shrihari Sankarasubramanian
(6DF) DEVELOPING ELECTROCHEMICAL-BASED PREDICTIVE BATTERY HEALTH MONITORING FOR FUTURE BATTERY MANAGEMENT SYSTEMS 244
Saeed Khaleghi Rahimian
(6DE) CHARGE STORAGE AND TRANSPORT IN ELECTROCHEMICAL SCIENCE: FROM BULK TO INTERFACES 247
Chia-Chin Chen
(6DG) ELECTROCATALYSTS AND GAS DIFFUSION ELECTRODES FOR VAPOR-PHASE CARBON DIOXIDE ELECTROLYZERS AND HYDROGEN FUEL CELLS 249
Dong Un (Daniel) Lee, David Koshy, Kabir Abiose, Daniel Corral, Christopher Hahn, Thomas F. Jaramillo
(6DH) COMBINING ADVANCED MATERIALS AND MANUFACTURING AND ELECTROCHEMICAL ENGINEERING FOR BUILDING A SUSTAINABLE FUTURE 251
Jeremy T. Feaster
(6JY) DESIGN OF ELECTROCHEMICAL BIOSENSING STRATEGIES TOWARD HIGH-SENSITIVITY, NON-FOULING, MULTIPLEX POINT-OF-CARE SYSTEMS 252
Yifan Dai, Xiaowei Wu, Jiwei Yao, Yinuo Chen, Yuan Wang, Wei Xu, Le Luo, Liang-Yuan Chiu, Jianzhi Huang, Huichun Zhang, Blanton Tolbert, Chung-Chiun Liu
(6MC) HIGH-ENERGY AQUEOUS LI-ION BATTERY 253
Chongyin Yang
(6DQ) ADVANCED MATERIALS FOR EFFICIENT ENERGY CONVERSION BASED ON SPECTROSCOPIC AND MECHANISTIC STUDY 256
Xuan Yang, Yushan Yan, Bingjun Xu
(6DX) FABRICATION OF CONDUCTING POLYMERS AND SUPERCONDUCTORS USING CHEMICAL VAPOR DEPOSITION METHODS FOR ENERGY AND ELECTRONIC APPLICATION DEVICES 258
Meysam Heydari Gharahcheshmeh

(6DY) SUSTAINABILITY IN PROCESS SYSTEMS ENGINEERING 260
Styliani Avraamidou
(6EA) NANOSTRUCTURED MATERIALS FOR NEXT-GENERATION LITHIUM-SULFUR BATTERIES 261
Somayeh Zamani, Yong Lak Joo
(6EB) KINETICS AND CHEMISTRY OF FAST PYROLYSIS OF CARBON-BASED MATERIAL USING NOVELS REACTORS APPROACHES 264
Ali Zolghadr
(6EC) ENERGY STORAGE AND CONVERSION WITH ORGANIC MOLECULES AND ADVANCED POROUS ELECTRODES 268
Michael R. Gerhardt
(6ED) MATERIALS FOR ENERGY STORAGE APPLICATIONS: FUNDAMENTAL INSIGHT FOR RATIONAL DESIGN AND DEVELOPMENT 271
Jeffrey Lopez
(6EH) REALIZING CO_2 UTILIZATION AND CARBON-NEUTRAL ENERGY CONVERSION IN PRACTICAL SCENARIOS --- MINING THE AIR, EXTRACTING INDUSTRIAL WASTES, ONE-PASS CONVERSION, AND SCALE-OUT 273
Xu Lu
(6EI) ADVANCING THE DEVELOPMENT OF FUEL FLEXIBLE COMBUSTION, COMPACT ENERGY SYSTEMS, AND THE SUSTAINABILITY ANALYSIS METHODS 275
Sampath Gunukula
(6EN) ELECTROCHEMICAL ENGINEERING FOR ENERGY, ENVIRONMENT AND SEPARATION APPLICATIONS 277
Tao Gao
(6KJ) ILLUMINATING SOLID-WATER INTERFACIAL PROCESSES AT NANOSCALE FOR SUSTAINABLE ENVIRONMENTAL REMEDIATION AND RESOURCE RECOVERY 278
Haesung Jung
(6DR) WINNING THE ENERGY DILEMMA: CARBON-FREE ENERGY AND ENVIRONMENTAL SUSTAINABILITY 281
Simona Liguori
(6DU) ORGANIC MOLECULAR ELECTROCATALYSTS FOR ENERGY-WATER NEXUS AND BEYOND 285
Xi Yin
(6EJ) ENERGY CONVERSION AND STORAGE 290
Yijin Kang
(6EK) EFFICIENT CARBON MODIFICATION FOR SUSTAINABLE FOOD/ENERGY/WATER NEXUS 292
Baharak Sajjadi, Wei Yin Chen, Daniell Mattern
(6EO) MULTISCALE APPROACH TO FUTURE ENERGY SCIENCE AND ENGINEERING 293
Yingda Lu
(6EL) BIOPOLYMER ENCAPSULATED LIPASE FOR BIODIESEL PRODUCTION 295
Ravindra Pogaku
(6KP) FUNCTIONALIZED MATERIALS FOR SUSTAINABLE ENERGY APPLICATIONS: STRUCTURE AND REACTIVITY 299
Gengnan Li
(6DT) SUSTAINABLE CONVERSION PROCESSES AND REACTOR DESIGN TO PRODUCE FUELS AND CHEMICALS 301
Muhammad Siddiquee
(6EE) SUSTAINABLE ENERGY PRODUCTION FROM RENEWABLE AND FOSSIL FUELS 304
Saikat Das
(6EF) A STRUCTURED APPROACH TO THE DESIGN AND OPTIMIZATION OF SUSTAINABLE ENERGY SYSTEMS 307
Nathanial J. Cooper
(6DS) A NOVEL ROUTE FOR THE FLEXIBLE PREPARATION OF HYDROCARBON JET FUELS AND VALUABLE CHEMICALS FROM BIOMASS-BASED PLATFORM CHEMICALS: A CASE OF USING 2, 3-BUTANEDIOL AS A FEEDSTOCK 310
Yuchen Bai, Xuebing Zhao, Dehua Liu
(6LF) ENGINEERING MULTIFUNCTIONAL NANOMATERIALS FOR ENERGY AND THE ENVIRONMENT 311
Michael Bozlar
(6LZ) PROBING HYDROGEN ACTIVATION TO SOLVE CURRENT ENERGY CHALLENGES 312
K. B. Sravan Kumar, Todd Whittaker, Bert D. Chandler, Lars C. Grabow

(6MA) UTILIZING BIOENERGY FOR PRODUCTION OF ENERGY AND MITIGATION OF CLIMATE CHANGE ... 314
Sonal Thengane
(6MB) PRODUCING CLEANER FUELS USING WASTE AND RENEWABLE RESOURCES ... 317
Efstratios Svinterikos
(6EP) ACTIVE SOFT MATERIALS: DATA DRIVEN STUDY TO UNDERSTAND, CONTROL, AND DESIGN BIO-SOFT-MATERIALS ... 320
Mehdi Molaei
(6EQ) DYNAMICS, TRANSPORT, AND SELF-ASSEMBLY IN FLOWING POLYMERIC LIQUIDS ... 322
Sarit Dutta
(6ET) CONSTITUTIVE MODELLING FOR COMPLEX FLUIDS IN COMPLEX FLOWS ... 323
Joseph Peterson
(6EV) STRUCTURE, RHEOLOGY AND PROCESSING OF COMPLEX FLUIDS TOWARDS SCALABLE MANUFACTURING OF FUNCTIONAL DEVICES ... 325
Sunilkumar Khandavalli
(6EW) COMPLEX AND BIOLOGICAL FLUIDS ... 327
Debasish Das
(6KL) MICROSCALE TRANSPORT PHENOMENA: DATA ANALYTICS AND FLUID DYNAMICS ... 329
Shiyan Wang
(6EU) NUMERICAL SIMULATION BASED DESIGN OF POINT OF CARE DIAGNOSTIC DEVICES ... 331
Debayan Das
(6ES) DYNAMICS OF SOFT MATERIAL SYSTEMS UNDER ELECTRIC FIELDS ... 332
Rajarshi Sengupta
(6KX) INVESTIGATING YIELDING DYNAMICS AND FRACTURE MECHANICS IN SOFT MATERIALS ... 334
Alan Ranjit Jacob
(6LS) UNDERSTANDING THE HIERARCHY OF STRUCTURE, DYNAMICS, AND INTERACTIONS IN SOFT MATTER ... 335
Ryan Poling-Skutvik
(6LV) ELECTROKINETICALLY ENHANCED PARTICLE FOCUSING IN NON-LINEAR MICROFLUIDIC FLOWS ... 337
A. Choudhary, Thiruvengadam Renganathan, Subramanium Pushpavanam
(6LY) RECYCLE, MICROMIXING AND PARTICLE MIGRATION IN A MICROCHANNEL ... 339
T. Krishnaveni, T. Renganathan, Subramanium Pushpavanam
(6EX) UNDERSTANDING AND DESIGNING NON-EQUILIBRIUM STATES AT SOFT/HARD MATERIAL INTERFACES ... 340
Yanhao Yu
(6EY) QUANTIFYING INTERFACIAL TRANSPORT PHENOMENA FOR ENVIRONMENTAL AND BIOCHEMICAL SYSTEMS ... 343
Henry C. W. Chu
(6EZ) MICROBIAL BIOFILM PROCESSES: MULTISCALE MODELING, SIMULATION, AND VISUALIZATION ... 344
George E. Kapellos
(6FB) DESIGN OF FUNCTIONAL SOFT MATERIALS FOR A SUSTAINABLE FUTURE ... 346
Karthik Nayani
(6FD) INTELLIGENT SYSTEMS, ACTIVE COLLOIDS, AND ADVANCED RESPONSIVE MATERIALS ... 348
Nicholas G. Chisholm
(6FF) DESIGNER SOFT MATTER: FROM PASSIVE TO ACTIVE TO BIOMIMETICS ... 350
Hanumantha Rao Vutukuri
(6FG) COUPLED ION TRANSPORT AND FLUID FLOW IN ENERGY AND ENVIRONMENTAL SCIENCES ... 352
Mohammad Mirzadeh
(6FH) DILATIONAL RHEOLOGY OF LUNG SURFACTANT INHIBITORS AND ITS EFFECTS ON ACUTE RESPIRATORY DISTRESS SYNDROME ... 353
Sourav Barman
(6FI) IONIC AND ELECTRONIC TRANSPORT PROPERTIES IN COVALENT ORGANIC FRAMEWORK AND NANOMATERIAL ... 355
Ankit Agrawal

(6FE) FLOW OF SOFT MATTER IN COMPLEX GEOMETRIES 356
Yu-Jiun Lin
(6FC) ROLE OF SPECIES TRANSPORT ON THE STABILITY OF INTERFACES 358
Vineeth Chandran Suja
(6FK) SILK-BASED MATERIALS: A SUSTAINABLE ROUTE TOWARDS MULTIFUNCTIONAL MATERIAL APPLICATIONS 360
Jaewon Choi
(6FM) INTERFACIAL MATERIALS FOR ELECTROCHEMICAL AND BIOMEDICAL DEVICES 362
Jie Zhao, Yi Cui, John A. Rogers
(6FN) MACHINE LEARNING AND DATA-ENABLED DESIGN AND DISCOVERY OF SOFT AND NANO MATERIALS 363
Tarak Patra
(6FQ) RATIONAL DESIGN OF SMART SOFT MATERIALS 365
Yimin Luo
(6FR) TAILORING ORGANIC MATERIALS FOR ELECTRONICS AND ENERGY: FROM MOLECULAR DESIGN TO TWO-DIMENSIONAL POLYMERIZATION 367
Yu Zhong
(6FS) LEARNING FROM NATURE: RATIONAL DESIGN OF MULTIFUNCTIONAL HYBRID MATERIALS 369
Hyosung An
(6FU) RATIONAL DESIGN OF ADVANCED MATERIALS AND MEMBRANES FOR ENERGY, SEPARATIONS AND CATALYTIC APPLICATIONS 373
Sameh Elsaidi
(6FT) COMPUTATIONAL ASSEMBLY ENGINEERING FOR BIO-INSPIRED NANOMATERIALS 376
Trung Nguyen
(6KD) HIERARCHICALLY ENGINEERED STRUCTURES USING SELF- AND DIRECTED-ASSEMBLY FOR UNIQUE ELECTRONIC, OPTICAL AND MECHANICAL PROPERTIES 377
Yuyin Xi
(6FJ) 3D PRINTING OF POLYLACTIC ACID MIXED WITH WOOD 380
Samarthya Bhagia
(6FV) ADVANCE MANUFACTURING OF MATERIALS FOR FLEXIBLE AND STRETCHABLE DEVICES 381
Kunal Mondal
(6FW) BIO-INSPIRED SOFT ELECTRONICS 382
Alex Chortos
(6FO) SCALABLE SYNTHESIS OF NANO- AND MICRO-SCALE MATERIALS FOR ENERGY APPLICATIONS 383
Clayton Kacica
(6KC) MOLECULAR ENGINEERING APPROACH FOR THE DEVELOPMENT OF ADVANCED FUNCTIONAL MATERIALS: ENGAGING STUDENTS THROUGH IMPACTFUL RESEARCH 385
Anthony Engler
(6LB) POLYMER SCIENCE IN DEVELOPING FIBROUS MATERIALS FOR ADVANCED TECHNICAL APPLICATIONS 387
Behzad Nazari
(6LI) ADVANCED MATERIALS AND MANUFACTURING FOR ELECTROCHEMICAL METHODS IN WATER-ENERGY NEXUS 389
Arkadeep Kumar
(6LL) MULTISCALE FRAMEWORK TO ENGINEER NON-EQUILIBRIUM RESPONSES AT COMPLEX MATERIALS INTERFACES 390
Alexander J. Pak
(6FZ) ADVANCING NANOMATERIALS FOR ENERGY AND WATER APPLICATIONS USING ATOMISTIC AND QUANTUM CHEMICAL SIMULATIONS 393
Ananth Govind Rajan, Michael Strano, Daniel Blankschtein, Emily A. Carter
(6GA) CATALYTIC POROUS MATERIALS WITH TAILORED NANOSTRUCTURES FOR SELECTIVE AND SUSTAINABLE CHEMICAL PROCESSES 394
Hong Je Cho
(6GB) BUILDING AUTONOMOUS NANOMACHINES AT THE INTERFACE OF COLLOIDS AND ELECTRONICS 396
Volodymyr Koman

(6GD) CONTROLLING ORGANIZATION AND EFFECTS OF NANOPARTICLES ON POLYMER-NANOPARTICLE COMPOSITES 398
Hamed Emamy, Oleg Gang, Francis W. Starr, Sanat K. Kumar
(6GE) ENGINEERING REDOX-ACTIVE MATERIALS FROM ELECTROCATALYSIS TO PSEUDOCAPACITANCE 400
Xueli Zheng
(6GF) ENGINEERING NANOMATERIALS FOR BIOMEDICAL APPLICATIONS 401
Devleena Samanta
(6GI) AN ACTIVE APPROACH TO COLLOIDAL SELF-ASSEMBLY 403
Stewart Mallory
(6GJ) LOW-COST, SCALABLE, AND RAPID PLATFORMS IN ULTRATHIN MATERIALS SYNTHESIS FOR WATER AND ENERGY TECHNOLOGIES 404
David S. Bergsman
(6GN) NANOSTRUCTURED ANODIC OXIDES OF METALS: FROM CORROSION PROTECTION TO NANOTECHNOLOGY AND EMERGING APPLICATIONS 407
Wojciech Stepniowski, Wojciech Misiolek, Kuo-Kuang Wang
(6GO) ELECTROCHEMICAL AND OPTOELECTRONIC TRANSFORMATIONS IN DYNAMIC SEMICONDUCTOR NANOMATERIALS 410
Clayton J. Dahlman
(6GP) APPLICATIONS OF NANOFILLERS IN ADVANCED COMPOSITES 412
Aniruddh Vashisth
(6GQ) CHARGE TRANSPORT IN SELF-ASSEMBLED BIOMOLECULAR SYSTEMS 413
Bo Li, Charles M. Schroeder
(6GY) RATIONAL SYNTHESIS AND ASSEMBLY OF DOPED NANOMATERIALS FOR QUANTUM AND OPTOELECTRONIC APPLICATIONS 414
Matthew Crane
(6GZ) DESIGNING NANOMATERIALS AND ELECTRONIC DEVICES FOR MULTI-LEVEL NEURAL INTERFACING 415
Huiliang Wang
(6HB) LIQUID METAL ENCAPSULATION: TOWARDS NEXT GENERATION FLEXIBLE ELECTRONICS 417
Megan A. Creighton
(6KB) DESIGN OF FLOW REACTORS AND FUNCTIONAL NANOMATERIALS FOR ENERGY & PHOTOCATALYTIC APPLICATIONS 418
Ioannis Lignos
(6GG) MULTI-SENSOR SYSTEM FOR COMPLEX CHEMICAL ENVIRONMENT DETECTION 419
Yixin Liu
(6GT) EVALUATION OF THE INFLUENCE OF CE4+/CE3+ REDOX-COUPLE ON THE CYCLYC REGENERATION OF NIO-PDO/CEO2 NANOPARTICLES FOR ASPHALTENE STEAM DECOMPOSITION 420
Oscar Medina
(6GK) BEYOND LITHIUM AQUEOUS ELECTROCHEMICAL ENERGY STORAGE 421
Daniel S. Charles
(6GL) INTERFACIAL ENGINEERING OF 2D NANOMATERIALS FOR BIOMEDICAL AND ENERGY APPLICATION 423
Dorsa Parviz
(6HA) UNDERSTANDING OF NANOPARTICLE SELF-ASSEMBLY MECHANISMS AND ITS APPLICATIONS TO ENERGY STORAGE AND BIO-MEDICAL APPLICATIONS 425
Jaewon Lee
(6GC) COLLOIDAL ELECTRONIC CELLS AS BUILDING-BLOCKS FOR SYNTHETIC, TISSUE-LIKE SCAFFOLDS 427
Albert Tianxiang Liu, Michael Strano
(6GH) ENHANCEMENT ROUTES OF CORROSION RESISTANCE IN STEEL-REINFORCED CONCRETE BY USING NANOMATERIALS 429
Arash Etemad
(6GR) ECO FRIENDLY STRATEGIES FOR NANOCELLULOSE PRODUCTION USING A NON COMMERCIAL ENZYMATIC COCKTAIL 430
Paula Squinca, Stanley Bilatto, A. C. Badino, Cristiane Sanchez Farinas
(6GS) INTERFACIAL MOLECULAR ENGINEERING OF BIO-INORGANIC SYSTEMS FROM AQUEOUS MOLECULES TO ASSEMBLED HIGH ORDER FUNCTIONAL MATERIALS 431
Tyler D. Jorgenson

(6GV) MEET THE FACULTY: INVESTIGATING THE PROPERTIES OF NANOPARTICLES CONTAINING FULLERENE-LIKE MOLECULES 433
Kimberly Bowal, Jacob W Martin, Laura Pascazio, Markus Kraft
(6GW) RECOMBINANT PEPTIDE-TEMPLATED PALLADIUM NANOPARTICLE SYNTHESIS AND THEIR CATALYTIC ACTIVITY IN COUPLING REACTIONS 434
Imann Mosleh, M. Hassan Beyzavi, Robert Beitle
(174AE) PLANT GENOME ENGINEERING WITH NANOTECHNOLOGY FOR AGRICULTURAL APPLICATIONS 435
Gozde Sultan Demirer, Huan Zhang, Juliana Matos, Abhishek Aditham, Brian Staskawicz, Markita Landry
(6KV) HARNESSING THE POWER OF NANOMATERIALS 438
Atefe Hadi, Matthew G. Panthani
(6KY) NEXT-GENERATION FLUORESCENCE BIOIMAGING: INTEGRATING CHEMICAL AND OPTICAL TOOLS 440
Yang Zhang
(6LD) SCALABLE ELECTRONIC SEPARATION OF SINGLE-WALLED CARBON NANOTUBES (SWCNTS) VIA NOVEL CHROMATOGRAPHIC METHODS 442
Payam Rezaei, Lisa Pfefferle, Douglas D. Frey
(6LP) DEEP CHARACTERIZATION BOOSTED NEXT-GENERATION HIGH-PERFORMANCE ENERGY STORAGE SYSTEMS 443
Chengcheng Fang
(6LR) NOVEL NANOSTRUCTURED TRANSITION METAL OXIDES FOR CATALYSIS. 445
Yulian He, Shu Hu, Victor S. Batista, Lisa Pfefferle
(6HD) DYNAMICS OF CARBONACEOUS NANOPARTICLES: CLIMATE IMPACT & FIRE DETECTION 446
Georgios A. Kelesidis
(6HE) PARTICLE SIMULATION IN ROTATING REACTORS 449
Dan Sun
(6HF) DEVELOPMENT OF STRATEGIES TO COMBAT GENOMIC INSTABILITY IN CELL CULTURE ENGINEERING, BIOPHARMACEUTICAL PRODUCTION, DISEASE AND AGEING 450
Philipp Spahn
(6HI) ENGINEERING LIPID AND POLYMER-BASED NANOMATERIALS FOR DRUG, PROTEIN AND GENE DELIVERY 452
Yamin Li
(6HJ) PRIMING THE INNATE IMMUNITY USING NANOMEDICINE 453
Fan Zhang
(6HG) ZWITTERIONIC POLYMERIC PLATFORMS FOR BIOLOGIC DRUG DELIVERY AND TISSUE ENGINEERING 455
Caroline Tsao, Shaoyi Jiang
(6LX) HIGH DIFFUSIVE PEPTIDES FOR DRUG CARRIERS IN TUMOR EXTRACELLULAR MATRIX. 457
Rashmi Mohanty, Debadyuti Ghosh
(6HL) MECHANISMS OF DIFFUSION IN ASSOCIATIVE POLYMER NETWORKS: EVIDENCE FOR CHAIN HOPPING 459
Peter Rapp, Ahmad Omar, Bradley Silverman, Zhen-Gang Wang, David A. Tirrell
(6HM) PROGRAMMABLE DEFORMATION AND ASSEMBLY OF SOFT ACTIVE MATERIALS 460
Ji-Hwan Kang
(6HN) MOLECULAR-BASED MODELING OF POLYMER DYNAMICS FOR MATERIAL DESIGN AND PROCESSING 463
Marat Andreev
(6HP) MOLECULAR–SCALE ENGINEERING OF CHARGE–CONTAINING POLYMERS AND ELECTROCHEMICAL INTERFACES 465
Snehashis Choudhury
(6HR) OPTIMIZING THE PROPERTIES OF ADVANCED MATERIALS THROUGH MOLECULAR DESIGN 467
Bassil El-Zaatari
(6HS) POLYMER MECHANOCHEMISTRY IN MOLECULAR MACHINES AND MEDICAL TREATMENTS 468
Qiong Wu
(6HU) MICROFLUIDIC PROCESSES TO ENGINEER HYDROGEL PARTICLES AND THEIR APPLICATIONS IN BIOMEDICAL ENGINEERING 471
Jae Jung Kim

(6HV) FUNCTIONAL POLYMERS FOR MOLECULAR AND MATERIALS DESIGN 473
Alice B. Chang
(6HZ) TOWARDS SUSTAINABLE POLYMERS THROUGH INTEGRATION OF DYNAMIC COVALENT CHEMISTRY AND ADVANCED PROCESSING 476
Kailong Jin
(6IC) SPECIALIZED POLYMERS FOR INTEGRATING ADVANCED SYNTHETIC AND BIOLOGICAL MATERIALS 478
Jeffrey M. Ting
(6ID) 3D PRINTED RECONFIGURABLE LIQUID CRYSTAL ELASTOMERS VIA DYNAMIC-COVALENT BONDS 481
Emily Davidson
(6IE) GRANULAR COMPOSITE-ENABLED MULTI-SCALE DYNAMIC RESPONSIVE MATERIALS 482
Yin Fang, Endao Han, Yuanwen Jiang, Yiliang Lin, Xianghui Xiao, Jin Wang, Heinrich M. Jaeger, Bozhi Tian
(6IF) MULTISCALE MODELING AND ENHANCED SAMPLING TO PROBE PEPTIDE AND PEPTIDOMIMETIC ASSEMBLIES 483
Janani Sampath, Jim Pfaendtner
(6KO) EXPLORING THE STRUCTURE OF GRADIENT DOUBLE NETWORK GELS 485
Pandiyarajan Chinnayan Kannan
(6HW) ENGINEERING THE MOLECULAR PACKING AUTONOMOUSLY 487
Jie Xu
(6HO) SUSTAINABLE APPROACHES TO POLYMER RECYCLING 490
Michael Miranda
(6HX) FUNCTIONAL MATERIALS DESIGN GUIDED BY POLYMER PHYSICS 492
Renxuan Xie, Michael L. Chabinyc, Ralph H. Colby, Enrique D. Gomez
(6HY) MECHANICS AND PROCESSING OF POLYMERIC MATERIALS FOR SOFT ELECTRONICS AND ROBOTICS 494
Samuel E. Root
(6LH) DESIGN OF PROTEIN-POLYELECTROLYTE COMPLEXES 495
Rituparna Samanta
(6LQ) RATIONALE DESIGN OF POLYMERIC MATERIALS FOR BIOLOGICAL ANDENERGY APPLICATIONS USING MULTISCALE MODELING 496
Vaidyanathan Sethuraman, Venkat Ganesan, David Morse, Kevin D. Dorfman
(6IG) ACCELERATING CHEMICAL DISCOVERY AND PROCESS DEVELOPMENT WITH THEORETICAL MODELS AND MACHINE LEARNING 498
Hanyu Gao
(6II) SYSTEMS APPROACH TO ADVANCED DECISION-MAKING IN CHEMICAL ENGINEERING, BIOMANUFACTURING, AND SOCIETY 500
Yu Luo
(6IK) LEVERAGING SCALE-APPROPRIATE PRINCIPLES OF METAL-OXIDE REACTION ENGINEERING AND PARTICLE TECHNOLOGY SCIENCE FOR ENERGY CONVERSION SOLUTIONS 501
Mandar Kathe
(6IL) ONLINE PROCESS OPTIMIZATION OF COMPLEX CYBER-PHYSICAL SYSTEMS 505
Dinesh Krishnamoorthy
(6IN) PROCESS CONTROL AND DEVELOPPEMENT FOR ENERGY-EFFICIENT APPLICATIONS 507
Esma Ines Achouri
(6KF) UNCERTAINTY QUANTIFICATION AND RISK-BASED DECISION SUPPORT METHODOLOGIES FOR HEALTHCARE AND ADVANCED MANUFACTURING 508
Francesco Rossi, Flavio Manenti, Guido Buzzi-Ferraris, G. V. Rex Reklaitis
(6IJ) REVIEW OF RECENT TRENDS IN SULFUR UNIT CAPACITY EXPANSION PROJECTS 511
Avinashkumar Karre
(6IO) MODELLING OF GAS RECOVERY FROM SOUTH AFRICAN SHALE RESERVOIRS 512
Diakanua Nkazi, Thembinkosi Qwabe Sr.
(6IP) MODELING AND SIMULATION OF MODULAR REFINERY FOR PRODUCTION OF FUELS WITH LOW ENVIRONMENTAL POLLUTION 513
Diakanua Nkazi, Mbinzi Kita Deddy Ngwanza Sr.
(6IS) 20-PLUS YEARS AFTER THE PIONEERING WORK AT THE UNIVERSITY OF OKLAHOMA-CHEMICAL ENGINEERING: CAPILLARY CONDENSATION OF LIGHT HYDROCARBONS IN MCM-41 TYPE NANO-POROUS MEDIA 514
Mariana Ioneva

(6IT) DETAILED STRESS ANALYSIS OF FLOATING PLATFORMS ... 515
David Etemad
(6IM) PROCESS SYSTEMS ENGINEERING AND DYNAMIC PROCESS CONTROL OF LARGE-SCALE SYSTEMS ... 516
Khalid Rashid
(6IH) CO_2 UTILIZATION THROUGH DRY REFORMING OF METHANE REACTION ... 518
Shaik Afzal, Nimir Elbashir, Mahmoud El-Halwagi
(6LA) PROCESS SYSTEMS ENGINEERING (PSE) ... 519
Ravendra Singh
(6IV) GAS HYDRATES RESEARCH: FROM FUNDAMENTAL SCIENCE TO ENGINEERING APPLICATIONS ... 522
Ahmad Abdul Majid
(6IW) THE DESIGN, SYNTHESIS AND TESTING OF ADVANCED HIGH-PERFORMANCE IONENE POLYMERS AS GAS SEPARATION MEMBRANES ... 524
Irshad Kammakakam, Jason E. Bara
(6IZ) HYBRID MEMBRANES FOR CHALLENGING ENERGY SEPARATIONS ... 525
Yang Liu
(6JA) ADVANCED MEMBRANES FOR SUSTAINABLE SEPARATIONS AT THE WATER-ENERGY NEXUS ... 527
Oishi Sanyal
(6JB) RATIONAL DESIGN OF NANOFILTRATION MEMBRANES FOR THE REMOVAL OF TRACE ORGANIC CONTAMINANTS FROM WASTE WATER BASED ON SINGLE-MOLECULE DYNAMICS AND HETEROGENEITY ... 529
Daniel Kienle, James S. Weltz, Andres Chaparro Sosa, Rebecca Falatach, Joel L. Kaar, Daniel K. Schwartz
(6JC) MEMBRANE MATERIALS AND TRANSPORT STUDIES FOR SUSTAINABLE WATER, ENERGY AND LIFE SCIENCES: FROM FUNDAMENTALS TO APPLICATIONS ... 531
Ngoc Bui
(6JN) BIOMIMETIC AND POLYMERIC MEMBRANES FOR ENHANCED SELECTIVITY IN DESALINATION AND IONIC SEPARATIONS ... 533
Jay R. Werber
(6JI) BIOSORPTION OF METHYLENE BLUE USING CITRIC ACID MODIFIED ARJUN BARK POWDER ... 535
Dilip Mondal, Shyamal Roy
(6JJ) NEXT-GENERATION MOLECULARLY SELECTIVE MATERIALS AND PROCESSES FOR SCALABLE ENERGY EFFICIENT SEPARATIONS ... 536
Canghai Ma
(6JK) ENHANCEMENT OF CO_2 ABSORPTION FROM GAS STREAMS USING NANOFLUIDS IN HOLLOW FIBER GAS–LIQUID MEMBRANE CONTACTOR ... 538
Seyed Mojtaba Mirfendereski
(6JL) CO_2 REMOVAL USING HIGHLY PERMEABLE, ULTRA THIN AND WELL ORIENTED SAPO-34 ZEOLITE MEMBRANE SYNTHESIZED BY A NOVEL FOUR-STEP METHOD ... 539
Seyed Mojtaba Mirfendereski
(6JM) ACOUSTIC CAVITATION INDUCED CHEMICAL FUNCTIONALIZATION OF BIOCHAR: A FEASIBLE STRATEGY FOR EFFECTIVE REMOVAL OF HEAVY METALS ... 540
Baharak Sajjadi, Wei Yin Chen, Daniell Mattern
(6IX) SYNTHESIS OF ENVIRONMENTALLY FRIENDLY AND SUSTAINABLE MULTI-FUNCTIONAL SURFACES AND INTERFACES ... 541
Sebastian Hernandez
(6JD) MICROPOROUS MOLECULAR SIEVES FOR MEMBRANE AND ADSORPTIVE SEPARATIONS ... 544
Shaowei Yang
(6JQ) BIOMIMETIC MICROELECTRONIC SYSTEMS FOR DECIPHERING NEUROTRANSMISSION AND HORMONE RELEASE ... 546
Sathish Ramakrishnan
(6JR) FSP-DDF COUPLING MODEL OF LBM FOR THE FLUID FLOW AND HEAT TRANSFER IN POROUS MEDIA ... 549
Shuyan Wang, Shuren Yang, Ruichao Tian, Baosli Shao, Yujia Chen, Qiji Sun
(6JP) NOVEL APPROACHES FOR PREBIOTIC DETECTION AND CONTROL OF MICROBIAL COMMUNITIES ... 550
Fatima Enam
(6JW) PROTEIN CONDENSATION IN NATURE: FUNCTIONAL OR PATHOLOGICAL? ... 551
Mohammad Safari

(6JU) THERMODYNAMIC BEHAVIOR AND THERMOPHYSICAL PROPERTIES OF UNUSUAL FLUIDS: HEAVY OILS 552
Francisco Ramos-Pallares
(6LM) ENHANCING NANO-BIOMATERIALS AND CHEMICAL METHODS FOR DETECTION, ISOLATION AND SEPARATION OF CHEMICAL AND BIOLOGICAL CONTAMINANTS FROM INFECTED SAMPLES 554
Mohammadali Masigol
(6LW) THERMOCATALYTIC DEPOLYMERIZATION OF LIGNIN AND HYDRODEOXYGENATION OF LIGNIN-DERIVED MONOMERS 557
Deepak Raikwar, Debaprasad Shee, Saptarshi Majumdar
(14A) CUSTOM-BUILT POLYMERS PROMOTE STABILIZATION, DELIVERY, AND BIOAVAILABILITY IN PRECISION DRUG FORMULATION STRATEGIES 559
Jeffrey M. Ting, Frank S. Bates, Theresa M. Reineke, Matthew V. Tirrell
(14B) ENGINEERING BIOMATERIALS TO RECAPITULATE THE STEM CELL MICROENVIRONMENT 560
Christopher M. Madl
(14C) LIQUID INFUSED-ELASTOMERS AS A MULTI-FUNCTIONAL MATERIAL IN IMPLANTABLE BIOELECTRONICS 561
Alexandra Rutz, George Malliaras
(14D) RAPID ACTUATION AND TUNABLE CONTROL OF DNA-BASED MECHANISMS 562
Alexander E. Marras, Carlos E. Castro
(14E) BIOMIMETIC PRODUCTION OF POLYMER FIBERS FROM REVERSIBLY CROSS-LINKED POLYSACCHARIDE NETWORKS IN WATER 563
Crystal K. Chu, Alby Joseph, Suman Bose, Robert Langer, Daniel G. Anderson
(14F) DESIGNER BIOMATERIALS AND INTEGRATED PLATFORMS TOWARD PRECISION MEDICINE 564
Jouha Min
(14G) MECHANOPHORE-BASED BIOMATERIALS FOR IN VIVO ULTRASOUND-TRIGGERED LIGHT GENERATION 565
Qiong Wu, Gun Kim, Abigail Halmes, Shang Ning, Clair Lundberg, Michael Oelze, King Li, Jeffrey Moore
(14H) PRESENT AND FUTURE OF BIOINSPIRED PHENOLIC COATINGS FOR SURFACE AND INTERFACIAL ENGINEERING 566
Kyueui Lee
Author Index

581912 Engineering the Future of Aging, Healthspan and Lymphoma Time

Research Interests:

With an educational background rooted in chemical engineering and medicine, my research agenda focuses on applying fundamental engineering principles to answering key questions in aging and cancer biology. To this end, my long term goal is to translate the discoveries and disruptive technologies from my laboratory into the medical arena to improve the healthspan and wellbeing of individuals. I envision leading a diverse team of interdisciplinary researchers working collaboratively on the central theme of aging. In the early stages of my lab, I intend to study two key aspects of aging, mainly: 1) cellular biomarkers of aging in health and disease; and 2) the role of the aging tumor immune-microenvironment (TiME) in lymphoma, and its impact on disease progression and treatment responses. I believe that studying the aging process at this juncture is important, since the doubling of the aging population above the age of 60 (2 billion people) by the year 2050 is projected to have significant socio-economic implications.

The process of aging is inevitable, and although we gain one chronological year with the passing of each birthday, the phenotypic outlook among individuals is quite variable. These inter-individual differences stemming from a combination of genetic and environmental factors, influence our biological age. This concept of biological age, which is associated with chronological age but not necessarily equal, is now known to be a key risk factor for the onset of chronic disease states such as cancer and frailty. However, biological age, typically measured at the anatomical level is often difficult to interpret, and rely on prior manifestation of physiological dysfunction. Because of this, during my PhD, I postulated and demonstrated, that cells represent a robust window into the aging process, and provide a means to evaluate one's biological age. Employing high throughput single-cell techniques to assess features such as cell morphology and motility, together with modern data science approaches and AI, I intend to improve and develop new cell-based predictors of biological age. I foresee this as being an important challenge, since by understanding these aging trajectories we can potentially intervene and course correct. As such, these discoveries could have significant implications for maintaining healthy aging trajectories, and extending human healthspan.

In the context of cancer, there is an intimate relationship between cancer incidences and age, however it is still ambiguous as to why. During my postdoctoral studies at Weill Cornell Medicine, I pioneered a study geared towards elucidating the understudied role of the tumor microenvironment in orchestrating a pro-lymphoma phenotype. Here, probing the phenotypic effects of the lymphoma microenvironment, regarding tumor mechanics, extracellular matrix content and organization, as well as stromal and immune compositions, I identified a critical molecular regulator that facilitates the cellular reprogramming. Building along these lines, I intend to characterize the cellular and non-cellular compartments of the lymphoma microenvironment with age, with the goal of identifying key vulnerabilities to prevent the co-opting and reprogramming of stromal, and immune cells by malignant B-cells. Further, the age-dependent increase of senescent cell populations, both senescent fibroblasts and senescent immune cells have been linked to the promotion of malignant transformations through the secretion of pro-survival signals and immune exhaustion. By studying the effect of senescent populations in lymphomas, we will seek to identify novel methods to mitigate these effects, and find way to reprogram the microenvironment into an immuno-permissive microenvironment. Leveraging my previous research experiences, together with my vision to innovate, I aim to become a leader and pioneer in the fields of mechanobiology of aging, and geriatric immuno-oncology.

Teaching Interests:

I am filled with excitement by the possibility of training the next generation of engineers that will solve the pressing problems facing our world today. I was fortunate to have creative and dedicated teachers whose encouragement and adaptive teaching strategies enabled me to succeed. From them I have learned the importance of advocating for the success of my students, while promoting an inclusive environment where diversity can shine, and where the strengths of each student is highlighted.

I believe that collaborative learning is essential to the field of engineering and is an area of great interest as I reflect on my teaching plan. To foster this collaborative environment both in my lab and in the classroom, I intend to use modern pedagogical techniques, which would include peer-based team learning and application-based projects. Within the classroom, I appreciate that student learns differently, and to address this I would utilize various education forms to appeal to different students, such as using real life examples that tangibly demonstrate engineering concepts to students, and the use of visual aids for visual learners. Investing the time to help students realize their learning styles will be key. For the graduate students and trainees in my lab I intend to organize seminars that will expose them to resources that will prepare them for this changing engineering climate, such as entrepreneurial training and startup models, as well as developing business models. I would also encourage them to engage in some type out teaching outreach with neighboring schools to help them cement their own knowledge and develop their own teaching styles, as this was a key aspect of my own development.

As I prepare to transition to faculty positions in chemical engineering, I took some time to think about the course that had the most impact on me as an undergraduate student, which is the course I would love to teach. I believe that Introduction to chemical engineering principles and practices present students with the breadth of chemical engineering possibilities, and inspires them to push the envelope to go beyond the norm, which is a key aspect of my philosophy. In addition to the introductory class, I am interested in designing courses at the undergraduate and graduate levels on the implementation of engineering principles for real life applications in healthcare, with a focus on cell-based approaches to combat disease. In this course, we will not only address the scientific aspects, but also the epidemiological and societal implications.

I believe that engineering provides limitless potential to address the various “problems” of society but the field is only as good as those who commit to it. During my time as an undergraduate I recall meeting eager and passionate students who had the potential to make a meaningful impact on the field but were deterred by difficulty with introductory courses. In order to support students with the desire to pursue careers in science, technology, engineering and math (STEM) a pipeline program must exist to foster this desire as early as possible. One of my goals as a professor would be to collaborate afterschool programs to foster and promote success in STEM for disadvantaged and underperforming high school and middle school students.

Teaching is more than a passion to me, I believe it is a calling. My goal is to reach learners at different level and not only provide them with information but enable them to be the critical and creative thinkers that will boast the field of engineering reach its full potential.

581998 Designer Biomaterials and Integrated Biosensors Toward Precision Medicine

Research Interests: Precision or personalized medicine is an emerging approach that matches patients with the most appropriate treatment based on the precise molecular features of an individual patient's disease. Despite huge advances in translational research, diagnosis and treatment continue to pose profound challenges in addressing significant health problems, including cancer and infectious diseases. In the MTFC presentation, I will showcase three research projects that demonstrate molecular engineering efforts toward precision medicine. Two focus on new developments in diagnosis: (i) bioanalytical sensing of plasma markers predictive of impending sepsis based on a magneto-electrochemical approach, and (ii) molecular profiling of breast cancer assisted by deep-learning algorithms for global health and point-of-care diagnostics. The third utilizes a novel programmable multi-therapy release strategy for orthopedic implants, enabling both bacterial eradication at the bone-hardware interface and accelerated bone tissue regeneration. These studies highlight the potential of integrating materials science, molecular engineering, and clinical science approaches to develop advanced molecular diagnostics and innovative therapies for precision medicine.

I received my BS from Cornell University, and PhD in Chemical Engineering from MIT (Advisors: Paula T. Hammond and Richard D. Braatz). I then joined Dr. Ralph Weissleder's lab as a postdoctoral fellow at Massachusetts General Hospital (MGH) / Harvard Medical School in Boston, where I have made some exciting developments in translational research. My research interests include the development of advanced therapeutic biomaterials, new tools for early disease detection and molecular profiling, and modeling methods for systems analysis. As faculty I would like to continue developing materials-based molecular biosensing technologies for precise non-invasive imaging and detection of human disease for early diagnosis, and to ultimately improve the quality of global healthcare and serve humanity in the long-term.

Complete List of Published Work in My Bibliography:
https://scholar.google.com/citations?user=LEpfhOQAAAAJ&hl=en&oi=ao

Teaching Interests: I am interested in teaching students the basic chemical engineering concepts, as well as recent advancements and perspectives in chemical and biomedical engineering. Specifically I am looking forward to teaching the transport courses and polymer courses. In addition, I am excited to create new courses such as an undergraduate-level polymer laboratory and a graduate-level special topic course in biomedical and biotechnology that showcases various approaches to translate traditional engineering concepts into clinical reality.

582008 Biomaterial Investigation and Development for Biomedical Advancement

Research Interests:

Synthesis and development of biomaterials for health applications are the main focus of my research. Advancement of the biomanufacturing technologies will impact drug development, delivery and diagnostics. Protein characteristics, transport, and interaction with complex biopolymers introduce numerous challenges to these processes. They are of particular interest due to their ability to be catalysts, drugs or biomarkers. Harnessing their potential is crucial to developing innovative technologies. This requires understanding function and structure, and as necessary, engineering improved versions towards translatable results. This can lead to valuable bioproducts (such as drugs, excipients and cofactors), development platforms for future studies, and screening techniques.

Throughout my experience and training in pharmaceutical sciences and biochemical engineering, I have acquired a wide range of skills in systems biology, catalysis, reactor design, bioprocessing and prototyping. Being part of multiple projects that touched on a variety of topics has taught me the value of distinct perspectives. As a faculty member, I will strive to maintain a diverse and multi-disciplinary group that can investigate the multiple facets of biotechnology development. My students and I will employ experimental and computational methods to: (1) explore protein-biomaterial interactions; (2) synthetize and modify biomaterials of interest; (3) develop and validate applications for the health industry.

PhD dissertation: "Enzymatic Inhibition by Lignin during Second Generation Ethanol Production", advised by Michael Ladisch, Agricultural and Biological Engineering, Purdue University

Postdoctoral Projects:

- "Analytical Modeling of Biomass Transport and Feeding Systems", part of successful grant DE-FOA-0001689 "Integrated Biorefinery Optimization".
- "Development of Improved Delivery of Injectable Medicines"

Both under the supervision of Michael Ladisch, Purdue University

Teaching Interests:

My teaching philosophy is based on my experiences working in industry, conducting collaborative research and mentoring students. I aim to help undergraduate and graduate students to (i) understand concepts as they relate to other topic areas, (ii) identify and address the gaps in knowledge they may have, (iii) discover, create and develop new ideas that bridge academics and the real world. My studies in both pharmaceutical sciences and biological engineering give the tools to teach a wide range courses included in chemical and biological engineering curricula. Topics include catalysis, kinetics, (bio)reactor design, (bio)process engineering, among others. As part of my doctoral training at Purdue, I had the opportunity to be a teaching assistant for a bioprocess engineering course. I prepared and taught lectures on enzymatic processes and modelling of their activity, in addition to preparing and grading assignments. In the laboratory, I have mentored multiple undergraduate students, guiding them as they develop their own projects. Besides experimental techniques, I have invested in teaching theoretical concepts to foster critical thinking and independence.

Selected publications:

dos Santos, A. C. F., Ximenes, E., Kim, Y., & Ladisch, M. R. (2019). Lignin–enzyme interactions in the hydrolysis of lignocellulosic biomass. *Trends in biotechnology*, 37, 5, 518-531. doi: https://doi.org/10.1016/j.tibtech.2018.10.010

Zanchetta, A., **dos Santos, A. C. F.**, Ximenes, E., Nunes, C. D. C. C., Boscolo, M., Gomes, E., & Ladisch,

M. R. (2018). Temperature dependent cellulase adsorption on lignin from sugarcane bagasse. *Bioresource technology*, 252, 143-149. doi: https://doi.org/10.1016/j.biortech.2017.12.061

Zhang, L., **dos Santos, A. C. F.**, Ximenes, E., & Ladisch, M. (2017). Proteins at heterogeneous (lignocellulose) interfaces. *Current Opinion in Chemical Engineering*, 18, 45-54. doi: https://doi.org/10.1016/j.coche.2017.09.003

582010 Molecular Simulations of Biological Self-Assembly

Research Interests: The overarching goal in my research program is the design and engineering of biologically related materials, ranging from protein self-assemblies to bio-nano sensing devices. Along the way of achieving this goal, my main contribution is understanding the physics behind the biological phenomena including but not limited to protein folding/non-folding/misfolding, DNA hybridization, aqueous solvation and bio-nano interfaces. The methods that I use to execute the research mostly relies on molecular simulations, computational modeling, and programming. A key aspect of my research is the continuous contribution to the improvement of computational methods.

This poster will include my experience in atomistic simulations of biologically related materials including protein folding/unfolding/non-folding/misfolding, aqueous solvation and bio-nano interfaces.

Teaching Interests: I am extensively experienced in and ready to teach all core subjects of Chemical Engineering at the undergraduate level, including Physical Chemistry, Classical and Statistical Thermodynamics, Phase and Reaction Equilibria, Chemical Reaction Engineering, Transport Phenomena, Applied Mathematics and Numerical Methods. Additionally, I have an advanced knowledge on Molecular Simulation Techniques, Physical Chemistry, Statistical Thermodynamics, Numerical Methods, Polymer Physics that will help me design and teach advanced classes of the mentioned subjects. I am also keen on designing online-friendly teaching materials and demonstrative tutorials as well as online repositories of the teaching material.

582055 Peptide- and Protein-Based Functional Biomaterials

Peptide- and Protein-based Functional Biomaterials

Abstract

Research Interests: There continues to be strong interest over the last few decades in the design and engineering of peptide- and protein-based functional nanomaterials. These materials are important in applications for energy, nanotechnology and biomedicine. Thus, the rational design of these materials and their incorporation into emerging applications remains both a challenge and a goal. Some of the challenges include: (a)How to rationally design the biomaterials and predict their self-assembly patterns towards targeted functions, and (b) How to predict translational potential in physiologically-relevant environments,

With the above goals in mind, my research laboratory aims to work on the interface between soft nanotechnology, supramolecular chemistry and biomaterials. My plan is to use nanostructures to generate functional materials *via* molecular and nanoscale self-assembly via rationally designed peptide-based materials. This will include the study responsive and adaptive self-assembly properties. Such peptide-based assembly can be used to control and direct nanoparticle self-assembly towards specific functions. Special emphasis will be given to peptide-based adaptive, autonomous biomolecular systems in a non-equilibrium environment. In addition, the lab will be working on silk protein-based biomaterials and studies of their crosslinking by enzymatic and non-enzymatic routes for hydrogel formation and for application in different biomedical fields.

For initial investigations, I will leverage my broad research skillset towards three major aims:

- Design, discovery and development of multiple short peptides, their stimuli responsiveness, hydrogel formation and their potential applications as a drug delivery vehicle.
- Peptide- and protein-based materials as soft templates for nanoparticles synthesis and self-assembly
- Investigative multiple non-enzymatic (covalent and non-covalent) routes to crosslink silk-based biomaterials for utility in different fields of biomedicine e.g. drug delivery and scaffolds for tissue engineering.

Prior Research Training: My research training includes chemistry, materials science, peptide- and protein-based biomaterials and supramolecular chemistry. I have led several multidisciplinary research projects to completion that resulted in more than 25 peer reviewed publications to position me to initiate and lead many fundamental and applied research projects in the areas of peptide and protein-based materials.

Doctoral Thesis: Surface functionalization of metal chalcogenide (MQ_2, M= Mo, W, Re, Ti, Zr: Q=S, O) nanoparticles (Advisor: Prof. Wolfgang Tremel)

Postdoctoral Trainings: (Advisors: Prof. Rein Ulijn, Prof. Matthew Webber, Prof. David Kaplan)

1) Biocatalytic self-assembly of aromatic peptide amphiphiles and their stimuli responsiveness

2) Pathway-dependent biocatalytic self-assembly of short peptides and their application in nanomaterials synthesis

3) Synthesis of enzyme responsive dynamic peptide surfaces for mesenchymal stem cell culture

4) Chemical functionalization of silk-based biomaterials, enzymatic and non-enzymatic crosslinking for hydrogel formation and biomedical applications

Teaching Interests: My teaching interests include courses related to materials (e.g. Materials science and engineering) including topical elective courses like Biomaterials, Principles of biomolecular engineering or Nanoscience and nanotechnology, Materials surface engineering for undergraduate and graduate students. I am also comfortable teaching core courses in chemical engineering like Thermodynamics.

I strongly believe teaching is a medium to promote and facilitate learning and to effectively transfer knowledge to establish the next generation of academics as well as informed students. Hence, my teaching philosophy focuses on creating and promoting a dynamic and interactive learning environment where the students are active participants in the classroom and can be mentored to become independent leaders. Learning objectives of the courses, assignments and examinations will be designed to equip students with creative thinking in early stages of their career. In this way, I strive to be a catalyst in different ways to guide their overall education in my classroom while consolidating their strengths and helping improve their areas of weakness, while valuing their diversity.

During my Ph.D., I was a teaching assistant in two undergraduate inorganic laboratory courses (first and second semester students). This was an exciting opportunity for me as a teacher and I was involved in their laboratory experiments, trouble shooting and evaluating their performance. As a research associate, I mentor undergraduate and graduate students to tap their full potential in laboratory research.

Public Outreach and Diversity: My research group will give utmost importance to public outreach and diversity. Students from diverse background will be encouraged to take up science and engineering. Special emphasis will be given to build a diverse group with representation from women, people of different races, disabilities and LGBTQ community, while not compromising merit.

Selected Publications:

1. J. K. Sahoo, M. A. Vandenberg, M. J. Webber, 'Injectable network biomaterials via molecular and colloidal self-assembly.' ***Adv. Drug. Deliv. Rev***., **2018**, *127*, 185-207
2. J. K. Sahoo, C. G. Pappas, I. R. Sasselli, Y. M. Abul-Haija, R. V. Ulijn, 'Biocatalytic self-assembly cascades.' ***Angew. Chem. Int. Ed***., **2017**, *56*, 6828.
3. J. K. Sahoo, S. Roy, N. Javid, K. Duncan, L. A. Aitken, R. V. Ulijn,' Pathway dependent gold nanoparticle formation by biocatalytic self-assembly.' ***Nanoscale***, **2017**, *9*, 12330.
4. J. K. Sahoo, C. Nazareth, M. A. Vandenberg, M. J. Webber, 'Self-assembly of amphiphilic tripeptides with sequence dependent nanostructure.' ***Biomaterial Science***, **2017**, *5*, 1526.
5. J. K. Sahoo, N. M. S. Sirimuthu, A. Canning, M. Zelzer, D. Graham, R. V. Ulijn 'Analysis of enzyme responsive peptide surface by Raman spectroscopy.' ***Chem. Commun***., **2016**, *52*, 4698.
6. J. K. Sahoo, S. K. M. Nalluri, N. Javid, H. Webb, R. V. Ulijn, 'Biocatalytic amide condensation and gelation controlled by light.' ***Chem. Commun***., **2014**, *50*, 5462.
7. J. K. Sahoo, M. N. Tahir, F. Hoshyargar, B. Nakhjavan, R. Branscheid, U. Kolb, W. Tremel, 'Molecular camouflage: Making use of protection group chemistry to control the self-assembly of janus particles onto metal chalcogenide nanotubes by pearson hardness.' ***Angew. Chem. Int. Ed***., **2011**, *50,* 12271.
8. J. K. Sahoo, M. N. Tahir, A. Yella, T. D. Schladt, S. Pfeifer, B. Nakhjavan, E. Mugnaioli, U. Kolb, W. Tremel, 'From single molecule to nanoscopically structured materials: self-assembly of metal chalcogenide/metal oxide nanostructures based on degree of pearson hardness.' ***Chem. Mater***., **2011**, *23 (15)*, 3907.
9. J. K. Sahoo, M. N. Tahir, A. Yella, T. D. Schladt, E. Mugnaioli, U. Kolb, W. Tremel, 'Reversible self-assembly of metal chalcogenide/metal oxide nanostructures based on pearson hardness.' ***Angew. Chem. Int. Ed***., **2010**, *49*, 7578.

582077 Conversion of Waste Biomass into Bioproducts (Bioenergy, Biomaterials, Biochemicals)

Research Interests:

Effectively partitioning waste biomass into its various fractions and converting these fractions into high value products is of high importance for the valorization and commercialization of bioproducts from bioresources from waste biomass. Waste biomass is released at over 2 billion tons per year in the U. S and has the potential to serve as a source of bioproducts with high commercial value. My research interests involve advancing green engineering and green materials, particularly with regard to their link to new production methods for a sustainable energy infrastructure. This typically involves the application of green chemistry/engineering principles, towards the design of sustainable products and processes that minimize the use and generation of hazardous substances. The core of my research strategies fall within the chemical and biomolecular engineering specialty, however, I often draw on key techniques in biology/microbiology, organic/analytical/computational chemistry and materials processing/characterization. In particular my expertise encompasses the development of effective practices for the valorization of waste bioresources into commercially viable bioproducts (biofuels/bioenergy/biomaterials). My unique contribution is to apply the idea of sustainable development (i.e. the triple bottom line—economics, environment and society) towards establishing long lasting research innovations in my chosen field.

Proposed Research Thrusts

- Isolation, Purification and Characterization of Compounds of Significance from Waste Biomass (Biopolymers and Bioactives)
- Application Development (Chemical Functionalization/Deriviatization, Material Formation and Product Formulations)
- Process Development (Life Cycle Analyses, Techno-Economic Analyses and Process Modeling/Scale up)

Relevant Experience

- Postdoctoral Fellow: Biomass Pretreatment and Process Development; Supervisor: John Gladden and Blake Simmons, Joint BioEnergy Institute and Sandia National Labs
- Postdoctoral Research (Fulbright Postdoctoral Fellowship): Extraction and Recovery of Sinapic Acid from Oleaginous Biomass (Mustard Bran): A Sustainable Access to a Valuable Phenolic Platform Chemical; Supervisor: Professor Florent Allais, Chaire ABI Agro-Biotechnologies Industrielles-AgroParisTech
- Ph.D. Dissertation: Isolation and Recovery of Lignin from Lignocellulosic Biomass Using Recyclable Protic Ionic Liquids (PILs) for a Cost-Effective Biomass Processing Technique; Ph.D. Advisor: Dr. Wesley Henderson, Chemical and Biomolecular Engineering; North Carolina State University

Selected Publications

- **Achinivu E**. Protic Ionic Liquids for Lignin Extraction—A Lignin Characterization Study. ***International Journal of Molecular Sciences***, **2018**, *19*, 428; DOI: 3390/ijms19020428
- **Achinivu E. C.** et. al. Lignin extraction from biomass with protic ionic liquids. ***Green Chemistry***, **2014**, *16*, 1114-1119; DOI: 10.1039/C3GC42306A

Teaching Interests:

In addition to my research, I have developed my capacity for engineering principles through hands-on instruction. My major teaching experience is related to the Chemical Engineering curriculum, although, I have also taught Chemistry and Mathematics courses in the past. I have also acquired formal training/experience in teaching lectures through the "Certificate of Accomplishment in Teaching (CoAT)" program at North Carolina State University. As part of this program, I was required to lead at least forty-five hours of lecture, attend six training seminars on fundamentals in teaching and develop a teaching portfolio. During this program, I taught Transport Processes and notably, I co-developed a course titled: Engineering Challenges at the Energy Frontier.

582162 Engineering Biology Solutions for a Sustainable Future

Title: Engineering biology solutions for a sustainable future

Author: Niju Narayanan, Postdoctoral Associate, Los Alamos National Laboratory, Los Alamos, New Mexico

Research Interests: Depleting petroleum reserves, energy security concerns and global climatic changes have driven our search for alternate technologies in order to shift to a more bio-based sustainable society. Living cells are adept at creating all these molecules that are currently sourced from petroleum. However, the success of a biosynthetic pathway is highly context dependent - microbial host, pathway genes and their expression levels, promoters, enzyme activities, product and intermediate metabolite titers, cofactor availability, and often their delicate balance is offset owing to the complexities of biology. The cell's physiological goals and the engineering objectives are often in a war resulting in poor cell growth, toxicity and low product yields. Decoupling the cell growth from the catalyst synthesis opens a new paradigm for synthetic biology. Advances in synthetic biology methods has enabled the construction and rapid generation of billions of varying metabolic design. However, the evaluation of such designs by traditional metabolite measurement techniques currently available, lag by several orders of magnitude. Genetically encoded biosensors that act as screens for productive designs can solve these problems, identify optimal combination of genetic components, and enhance product titers and yields.

My research interest is in the engineering of biosynthetic pathways in living cells & cell free systems and their evolution using biosensors.

Research Experience: My research background has provided me with a wide range of experience in synthetic biology, high throughput technology, metabolic engineering, protein engineering, recombineering, molecular biology, biochemistry, fermentation technology, downstream processing, and analytical methods, some are described below:

Postdoctoral Study:

1. "Genetic biosensors for small molecules to engineer biosynthetic pathways in *Pseudomonas putida*" Advisors: Dr. Ramesh Jha and Dr. Taraka Dale, Los Alamos National Laboratory

This project focusses on using transcription factor based whole cell biosensors with fluorescence readouts for enhancing product titers and yields, evolving their promoters for transition to alternate production hosts, enzyme evolution, strain performance and selection of best producers, adaptive evolution for better growth of engineered host strains, utility of renewable non-conventional feedstocks etc.

2. Biosynthesis of Isopentenol in *Escherichia coli* via the Methyl Erythritol Phosphate Pathway

Advisor: Prof. Greg Stephanopoulos, Massachusetts Institute of Technology

The project was focused on improving isopentenol titers produced by the MEP pathway by soluble and functional expression of dephosphorylase and the pathway enzymes, maturation of the Fe-S cluster proteins, investigation of possible iron donors, NADP regeneration routes for cofactor supply, development of product and pathway metabolite analysis methods, culture media and isopentenol extraction procedures.

3. Pathways for synthesis of Advanced biofuels

Advisor: Prof. Greg Stephanopoulos, Massachusetts Institute of Technology

This project was focused on isovaleryl coA pathway for synthesis of C5 compounds, identification, cloning, characterization and variants of putative pivalyl coA mutases, investigation of prenol synthesis from isovaleryl coA pathway, development of analytical methods for pivalic acid and isovaleric acid determination.

PhD Dissertation: "Molecular and genetic strategies to enhance the functional expression of recombinant proteins in *Escherichia coli*"

Advisors: Prof. C. Perry Chou and Prof. Bill Anderson, University of Waterloo

The research work during graduate study was focused on the exploration of 'strategies for functional expression of recombinant proteins in *E. coli* using model proteins (lipase, penicillin acylase, yellow fluorescent protein, β-galactosidase, hCD83ext). They included posttranslational modifications, cytoplasmic and periplasmic expression, surface display, extracellular secretion, molecular chaperones, site directed and error-prone mutagenesis, stress regulatory pathways, host strains, and optimization of culture conditions.

Industrial collaboration project with Argos Therapeutics at University of Waterloo

"Development of 'bioprocess for production of recombinant human therapeutic protein hCD83ext"

Advisor: Prof. Murray Moo-Young and Prof. C. Perry Chou

The project included optimization of culture media and bioreactor conditions for high level expression of hCD83ext, Development and optimization of purification protocol for hCD83ext: GST affinity column, thrombin cleavage, ion-exchange chromatography, ultrafiltration, analytical methods for GST assay and endotoxin assay, construction of hCD83 mutants for improved protein stability

Funding/Awards:

NSERC Postdoctoral Fellowship, 2013

Murray Moo-Young Biotechnology Scholarship, 2008

NSERC Graduate Studentship, 2009

International Doctoral Student Award, 2006-2009

Teaching Interests: My background in Chemical Engineering, coupled with my research training in Biological Engineering has equipped me to teach a variety of courses in Chemical Engineering, Biotechnology, Bioengineering, Biochemistry, Protein Engineering, Metabolic Engineering and Synthetic Biology. However, I would be particularly interested in teaching 1. Biochemical Engineering and 2. Synthetic Biology. The coursework would be designed to encompass theoretical and computational approaches, practical laboratory classes, experimental model systems, and current research & classic literature.

Teaching Experience: I have had the opportunity to be a teaching assistant for various courses at the University of Waterloo and at Indian Institute of Technology Delhi. Some of them include Bioprocess Engineering, Advanced Biochemical Engineering, Engineering Biology, Introduction to Biotechnology, Chemical Engineering Concepts, and Physical Chemistry. This has given me teaching experience for 5-6 years at undergraduate level and around 8 months at the graduate level. Teaching helped me to hone my technical knowledge in the curriculum subjects and develop my interpersonal and communication skills and I was able to enjoy a good rapport with my students and colleagues.

Supervisory experience: I have had the chance to mentor and work with 7 graduate students, 4 visiting scholars, 7 undergraduate co-op students and 3 technical staff during the course of my research career of graduate studies at UWaterloo, the Industrial collaboration project for therapeutic protein production at UWaterloo, and postdoctoral study at MIT and LANL. Most students who started without any background in biotechnology and bioengineering, developed the essential skills within a short period of time through research training. This is also demonstrated by their co-authorship on our publications.

Teaching Certificate Program: I have completed a teaching certificate program at MIT which was based on seven workshops aimed at development of teaching skills, exposure to research on teaching and learning, and structure of future learning. The workshop was followed by two microteaching sessions that are videotaped, evaluated and feedback provided.

Manuscripts in preparation

1. Bentley, G., **Narayanan, N.**, et al. Engineering glucose metabolism for enhanced muconic acid production in *Pseudomonas putida* KT2440
2. **Narayanan, N.**, Henelly, S., Dale, T., and Jha, R. K. Engineering a *b*-ketoadipate biosensor in *Pseudomonas putida* and evolution of aromatic catabolism pathway for biomanufacturing from corn stover hydrolysate.
3. **Narayanan, N**., Edgar, S., Woolston, B., and Stephanopoulos, G. Maturation of Fe/S clusters of Methyl erythritol phosphate pathway enzymes in *Escherichia coli*.
4. **Narayanan, N**., Woolston, B., and Stephanopoulos, G. NADPH regeneration in *Escherichia coli* improves the biosynthesis of isopentenol via Methyl erythritol phosphate pathway.
5. **Narayanan, N**., Duggar, D., Pereira, B., and Stephanopoulos, G. Identifying and Engineering of putative pivalyl CoA mutase for the biosynthesis of C5 branched chain compounds.

Publications in refereed journals

1. Jha, R. K., **Narayanan, N.** et al. (2019) Sensor-enabled alleviation of product inhibition in chorismate pyruvate-lyase. ACS synthetic biology. 8, 775-786.
2. **Narayanan, N.,** Khan, M., and Chou, C. P. (2011) Enhancing functional expression of heterologous *Burkholderia* lipase in *Escherichia coli*. Molecular Biotechnology. 47: 130-143
3. Srivastava, P., Singh, P., **Narayanan, N.** and Deb, J.K. (2011) Physiological and biochemical consequences of host–plasmid interaction – A case study with *Corynebacterium renale*, a multiple cryptic plasmid containing strain. Plasmid. 65: 110-117
4. Zhang, L., **Narayanan, N.**, Brand, S. R., Nicolette, C. A., Baroja, M., Arp, J., Wang, H., Moo-Young, M., and Chou, C. P. (2010) Structural identification of recombinant human CD83 mutant variant as a potent therapeutic protein. Protein Expression and Purification. 73: 140-146
5. **Narayanan, N.**, Khan, M. and Chou, C. P. (2010) Enhancing functional expression of heterologous lipase B in *Escherichia coli* by extracellular secretion. Journal of Industrial Microbiology and Biotechnology. 37: 349-361
6. **Narayanan, N.**, and Chou, C. P. (2009) Alleviation of proteolytic sensitivity to enhance recombinant lipase production in *Escherichia coli*. Applied and Environmental Microbiology.75: 5424-5427
7. **Narayanan, N.**, and Chou, C. P. (2008) Physiological improvement to enhance *Escherichia coli* cell- surface display via reducing extracytoplasmic stress. Biotechnology Progress 24: 293-301
8. **Narayanan, N.**, and Chou, C. P. (2008) Periplasmic chaperone FkpA reduces extracytoplasmic stress response and improves cell-surface display on *Escherichia coli.* Enzyme and Microbial Technology. 42:506-513
9. **Narayanan, N.**, Follonier, S., and Chou, C. P. (2008) In vivo monitoring and alleviation of extracytoplasmic stress to recombinant protein overproduction in the periplasm of *Escherichia coli*. Biochemical Engineering Journal. 42: 13-19
10. **Narayanan, N.**, Xu, Y., and Chou, C. P. (2006) High-level gene expression for recombinant penicillin acylase production using the *ara*B promoter system in *Escherichia coli*. Biotechnology Progress. 22: 1518-1523
11. **Narayanan, N.**, Hsieh, M. Y., Xu, Y., and Chou, C. P. (2006) Arabinose-induction of *lac*-derived promoter systems for penicillin acylase production in *Escherichia coli*. Biotechnology Progress. 22: 617-625
12. Xu, Y., Hsieha, M. Y., **Narayanan, N.**, Anderson, W. A., Scharer, J. M., Moo-Young, M., and Chou, C. P. (2006) Cytoplasmic overexpression, folding, and processing of penicillin acylase precursor in *Escherichia coli*. Biotechnology Progress. 21: 1357-1365
13. Xu, Y., Weng, C. L., Hsieh, M. Y., **Narayanan, N.**, Anderson, W. A., Scharer, J. M., Moo-Young, M., and Chou, C. P. (2005) Chaperone-mediated folding and maturation of the penicillin acylase precursor in the cytoplasm of *Escherichia coli*. Applied and Environmental Microbiology. 71: 6247-6253

14. **Narayanan, N.**, Roychoudhury, P. K., and Srivastava, A. (2004) Isolation of *adh* mutant of *Lactobacillus rhamnosus*.sp for production of L(+) lactic acid. Electronic Journal of Biotechnology. 7: 72-84
15. **Narayanan, N.**, Roychoudhury, P. K. and Srivastava, A. (2004) L (+) lactic acid fermentation and its product polymerization. Electronic Journal of Biotechnology.7: 167-179

582185 Exploring the Design Space of Living Systems: Experimental and Theoretical Tool-Kits for Multi-Functional Materials

Symone Alexander, Ph.D.

Eckert Postdoctoral Research Fellow, Dept. of Chem. and Biomol. Engineering

Georgia Institute of Technology

Research Interests:

From an engineering perspective, nature creates motion through fundamental concepts like fluid transport, thermodynamics, and material properties. These fundamental design parameters can then be utilized to develop materials with tailorable transport, responsiveness and motion. My research focuses on identifying the key design concepts used by living systems to control the speed, direction, and magnitude of response to dynamic environments. I aim utilize both experimental and theoretical strategies to investigate the role of polymeric structures in storage, compartmentalization, and delivery in living systems, and aim to employ fundamental design concepts learned from nature to target toxins and tailor motion in complex environments via dynamic polymer materials.

Research Experience:

As a chemical engineer with a passion for molecular design and assembly, my research trajectory has included elements of chemistry, materials science and engineering, polymer physics, and biophysics. My experience within these research disciplines surrounds small molecule and macromolecular assemblies and interfaces and the use of hierarchical materials for rapid motion. My dissertation focused on two main concepts: influencing self-assembly by diversifying the fluid environments of molecules and translating those hierarchical assembly and interfacial interactions to solid-state functional materials. Through this work, I developed a research platform for hygromorphic bilayer composites and explored a new area in the field of molecular gels by utilizing their self-assembly in both the solution and solid states. One of the key findings of this fundamental investigation is the ability to achieve solid-state reversibility of small molecules by tailoring their solution state assembly. As a part of this project, I submitted a highly-ranked, successful beam time proposal to Argonne National Laboratory's Advanced Photon Source. My research project design was also instrumental to a successful ACS-PRF New Directions grant. I also utilized electrospun nanofibers to guide transport of fluids in hierarchical polymer nanocomposites, and was a Visiting Scholar at M.I.T., where my preliminary research findings on fiber-wrapped light guides were used to foster scientific collaboration between the Korley and Kolle research groups. Throughout my diverse investigations of self-assembling systems, I have mastered an assortment of characterization techniques, and translated the impact of molecular assembly across multiple length scales and scientific disciplines.

As a postdoc, I seek to uncover more material design parameters used in nature by studying the physics and engineering of living systems. I am currently exploring the following question: how can we push the utility of natural polymeric materials to achieve ultrafast motion? To do so, I am investigating a tiny arachnid native to the Peruvian Amazon Rainforest known as the "Slingshot Spider." This spider employs a conical 3-D web to achieve accelerations exceeding 1300 m/s^2 without sustaining physiological harm. In this research, I am examining how slingshot spiders achieve ultrafast motion and power amplification using polymeric materials, fundamental physics concepts, and engineering design, while sharing insight about the utility of this extraordinary prey capture strategy.

Teaching Interests:

My academic background and research experience have equipped me to teach core courses, laboratory courses, and electives that are inclusive of macromolecular science and engineering, chemical engineering, and chemistry departments. In the classroom, I want to foster a learning environment that encourages curiosity and engagement. By setting clear learning objectives that are supplemented with metacognitive activities and assignments, I want to empower students to reflect on their learning and studying processes and improve or reinforce their strategies for success. Additionally, I aim to instill the importance of individuality and inclusivity by introducing class content and projects that are relevant to a diverse range of cultural, societal, and economic challenges. As a mentor, I plan to be a launching pad for my students to their chosen career path. I want to be a resource for advice, opportunity, and connection with additional mentors. As a research mentor, my goal is to help my students achieve research excellence by building confidence, critical thinking skills, and safe, efficient research practices. In addition, I plan to train my students to excel in interpersonal and scientific communication, so that they are able to build a strong professional network while connecting with and helping educate our communities.

Additional Roles: Social Media Contributor, *Biophysical Journal*

Successful Proposals/Awards: 2018 MIT Rising Stars in Chemical Engineering, Eckert Postdoctoral Research Fellowship, NSF Graduate Research Fellowship (NSF-GRFP), Argonne National Laboratory General User Proposal (APS-GUP)

Featured Publications:

Symone L. M. Alexander & LaShanda T. J. Korley, Restricting molecular mobility in polymer nanocomposites with self-assembling low-molecular-weight gel additives, **ACS Appl. Mater. Interfaces**, 2018, *10 (49)*, 43040–43048

Symone L. M. Alexander and L. T. J. Korley, Tunable hygromorphism: structural implications of low molecular weight gels and electrospun nanofibers in bilayer composites, **Soft Matter**, 2017, ***13***, 283-291 (*Soft Matter* emerging investigators 2017)

Symone L. M. Alexander, Lindsay E. Matolyak and LaShanda T. J. Korley, Intelligent Nanofiber Composites: Dynamic Communication between Materials and Their Environment, **Macromol. Mater. Eng.**, 2017, 1700133 (Featured on Advanced Science News)

581386 Ultrasmall C' Dots Activate Pro-Inflammatory Anti-Tumor Responses in the Microenvironment of PDGF-B Driven High Grade Gliomas

Research Interests: My research goals are directed toward developing Clinical Translational nanoscale technologies with particular emphasis on developing nanoscale biomaterials and immunoengineering systems for immuno-oncology, immunoengineering, and molecular imaging. My main research interest lies in immuno-oncology of solid tumors, with a focus on myeloid cells, particularly tumor-associated macrophages, dendritic cells, and myeloid-derived suppressor cells.A major question my research is investigating is, "What role do nanoparticles play in the tumor microenvironment to regulate innate and adaptive immune systems?" Thus, I am studying the impact of proteins and signals that promote the activation and migration of pro-inflammatory cells into tumor microenvironment and also affect the phenotype of immune cells. In addition, I study human antigen presenting cells (APCs), chemokines, and myeloid cells in linking innate and adaptive immunity. My focus has been on dynamic interactions of macrophages and dendritic cells with T cells during antigen presentation and the possibility of using nanoparticles in immuno-oncology to prime T cells for immune checkpoint inhibitors and CAR-T Cell Therapy.

Teaching Interests: My teaching interests lie in the interface of Immunoengineering, cancer immunotherapy, and nanomedicine. I plan to develop highly engaging and critically important courses such as: (1) nanoparticles in cancer immunotherapy. (2) Advanced topics in biomaterials and cancer therapy with a special focus on the relationships between the tumor microenvironment, nanoparticles, and the immune system. I will also arrange journal clubs in order to: Improve competence and confidence in leading discussion and resolving conflicts and improve and refine presentation and communication skills. Through the journal club, I will share research findings or the latest literature in the field.

Introduction: Therapeutic approaches targeting high-grade glioma have largely failed. One alternative strategy is to regulate cells in the tumor microenvironment (TME), such as tumor-associated macrophages and microglia (TAMs). TAMs account for as much as 30% of the tumor mass in mouse models of high-grade glioma and in brain tumor patients; their accumulation is associated with a higher glioma grade and poor patient prognosis. TAMs closely resemble the M2-polarized macrophages and have been shown to contribute to tumor initiation and maintenance, as well as influence anti-tumor autoimmunity via cytokine release and inflammatory recruitment to the TME. Tumors, in turn, can promote the polarization of monocytes into M2 TAMs by releasing factors, such as TGF-beta and M-CSF. We have previously showed that systemically administered sub 8-nm targeted Cornell prime dots (C' dots) are preferentially internalized by cancer cells, rather than TAMs, and can induce ferroptosis under select conditions. However, the influence of these particle-based probes on immune cell populations (*e.g.,* TAMs) within the TME is unclear. In this work, we hypothesize that non-targeted particles (i.e., PEG-C' dots) may induce a pro-inflammatory immune response by promoting M1 macrophage polarization and depleting M2 macrophages.

Materials and Methods: Bone marrow derived macrophages (BMDMs) were isolated from nestin–Tv-a; Cdkn2a−/− transgenic mice (PDG) and treated over time with a range of PEG-C' dot concentrations. Gene signature profiles and cytokine secretion profiles were evaluated in vitro by quantitative real-time PCR and luminex assays, respectively. Cell phenotyping, T cell priming, and phagocytosis rates were evaluating by flow cytometry and immunofluorescence assays. In vivo, PDGF-B driven high grade glioma mice (mGBM) were i.v.-injected with a single-dose PEG-C' dots to assess for alterations in the TME, including (1) polarization of TAMs, (2) tumor cell phagocytosis, and (3) promotion of pro-inflammatory cytokine secretion profiles.

Results and Discussion: In vitro, exposure of macrophages to PEG-C' dots led to upregulated M1- and downregulated M2 signatures over a 24-hour period. Macrophages also demonstrated increased pro-inflammatory cytokines and phagocytosis rates in comparison with untreated controls. In addition, PEG-C' dot exposure led to proliferation and activation of T cells, as well as a significant increase in proliferation of antigen presenting cells. In vivo, PEG-C' dots significantly enhanced pro-inflammatory responses in the TME. Multicolor flow cytometry of brain tumor tissue specimens revealed prominence of infiltrating macrophages within PEG-C' dot-treated tumors when compared with untreated tumors and WT brains, accounting for about 50% of all live cells in the brains of treated animals. M1-like subtype (MHC-II high Ly6C-) expression levels were more prevalent in PEG-C' dot-treated tumors, unlike M2-like subtype (MHC-II- Ly6C-) expression levels, which were lower in treated tumors. Concurrent Luminex assay studies revealed significantly suppressed expression of pro-tumoral cytokines. In addition, anti-inflammatory cells and angiogenesis marker levels were significantly higher in untreated groups, confirmed by $CD206^+$ and $CD31^+$ immunohistochemistry staining.

Conclusion: In a first-of-its-kind study, a distinct pro-inflammatory mechanism was induced in non-functionalized C' dot-exposed immune cells (TAMs, T cells) and mGBM tumors. The results of this work suggest that anti-tumor responses may be augmented by induction of this pro-inflammatory phenotype, which might further be exploited as part of a combinatorial treatment regimen.

581304 Metabolic Sugar Labeling for Cancer Targeting and Immunoengineering

Title: Metabolic Sugar Labeling for Cancer Targeting and Immunoengineering

Hua Wang, David J. Mooney

Harvard John A. Paulson School of Engineering and Applied Sciences, Wyss Institute for Biologically Inspired Engineering, Harvard University, Cambridge, MA 02138, USA.

Abstract: Cancer is still one of the leading causes of human deaths worldwide, and the development of effective and safe cancer therapies remains a significant challenge. One of my general interest in the field of cancer research is to understand how cancer cells and immune cells can be manipulated or engineered to improve and innovate cancer therapies. For example, cancer cells can actively metabolize unnatural monosaccharides bearing functional groups and express them on the cell surface in the form of glycoproteins. The cell-surface chemical tags, coupled with efficient Click chemistries, allow for conjugation of molecules of interest to monitor or manipulate cancer cells. Similarly, unnatural sugars can be utilized to metabolically label and regulate immune cells. I have been exploring how metabolic labeling and targeted modulation of tumor cells and immune cells can be utilized to develop potent cancer therapies. Immune responses involved in tissue repair and biomedical implant has proven critical, and efforts have been made to steer these immune responses towards promoting tissue regeneration. I am interested in understanding the role of T cells, dendritic cells, and neutrophils in tissue regeneration, and utilizing biomaterials to modulate these immune cells in situ to regulate tissue repair.

Research Interests:

1. Metabolic Labeling and Targeting of Cancer Cells and Immune Cells
2. Cancer Immunotherapy
3. Immunoengineering for Regenerative Medicines
4. Polymer Chemistry and Macromolecular Materials

Teaching Interests: Core undergraduate and graduate courses in Bioengineering/Chemical engineering/Materials Science/Chemistry. Courses that I have experience in teaching include Polymer chemistry, Biomaterials, Immunoengineering, Cancer biology.

Representative Publications:

1. Wang, H., Sobral, M., Cartwright, A., Zhang, D., Li, A., Dellacherie, M., Koshy, S., Wucherpfennig, K., Mooney, D. In Vivo Metabolic Labeling and Targeted Modulation of Dendritic Cells. *Nat. Biotechnol*. 2019, under revision.
2. Wang, H., Liu, Y., Xu, M., Cai, K., Wang, R., Cheng, J. In Vivo Cancer Targeting via Glycopolyester Nanoparticle Mediated Metabolic Cell Labeling Followed by Click Reaction. *Biomaterials* 2019, under revision.
3. Wang, H., Najibi, A., Sobral, M., Li, A., Verbeke, C., Mooney, D. Antigen-Free Cancer Vaccine to Treat Poorly Immunogenic Tumors. *Nat. Commun*. 2019, under revision.
4. Wang, H., Mooney, D. Biomaterial-Assisted Targeted Modulation of Immune Cells in Cancer Treatment. *Nat. Mater.* **2018**, *17*, 761-772.
5. Wang, H., Wang, R., Cai K., He H., Liu, Y., Yen, J., Wang, Z., Xu, M., Sun, Y., Zhou, X., Yin, Q., Tang, L., Wang, Z., Dobrucka, I.T., Dobrucki, L.W., Chaney, E.J., Boppart, S.A., Fan, T.M., Yin, L., Lezmi, S., Chen, X., Yin, L., Cheng, J. Selective In Vivo Cell Labeling Mediated Cancer Targeting. *Nat. Chem. Biol.* **2017**, *13*, 415-424.
6. Wang, H., Tang, L., Liu, Y., Dobrucka, I.T., Dobrucki, L.W., Yin, L., Cheng, J. In Vivo Targeting of Metabolically Labeled Cancers with Ultra-Small Silica Nanoconjugates. *Theranostics* **2016**, *6*, 1467-1476.

7. Wang, H., Gauthier, M., Kelly, J. R., Miller, R.J., Xu, M., O'Brien, W.D., Cheng, J. Targeted Ultrasound Assisted Cancer-Selective Labeling and Imaging. *Angew. Chem. Int. Ed.* **2016**, *128*, 1-6.
8. Wang, H., Xu, M., Xiong, M., Cheng, J. Reduction-Responsive Dithiomaleimide-Based Nanomedicine with High Drug Loading and FRET-Indicated Drug Release. *Chem. Commun.* **2015**, *51*, 4807-4810.
9. Wang, H., Tang, L., Tu, C., Song, Z., Yin, Q., Yin, L., Zhang, Z., Cheng, J. Redox-Responsive, Core-Cross-Linked Micelles Capable of On-Demand, Concurrent Drug Release and Structure Disassembly. *Biomacromolecules* **2013**, *14*, 3706-3712.

579891 Multiscale Dynamics in Biological Soft Matter and Polymeric Fluids

Research Interests:

The overarching theme of my research is to utilize numerical simulations in conjunction with experiments and direct visualization of soft matter dynamics for a bevy of engineering and biological applications. These applications extend beyond simply academic interest and concern human health, bioengineering, and rational design of industrially critical materials.

The platforms that I have developed for my investigations include computational mechanics of complex non-Newtonian fluids and deformable objects (immersed boundary method) and molecular stochastic simulation of polymeric solutions. I exploit high-performance-computing (HPC) and massive large-scale parallelism to accelerate the numerical simulations which are otherwise intractable. Graphical processing units (GPUs) and message passing interface (MPI) are the two computational tools to facilitate parallel computing that I have been using extensively.

A major part of my graduate and postdoctoral research has involved experimental investigation, either directly or through collaboration. In my postdoctoral research, I have been conducting microfluidic experiments and automatic image analysis for quantification of red blood cell biomechanics. In my graduate research, I have studied the dynamics of single comb DNA macromolecules in cross-slot microfluidic hydrodynamic trap and also used linear and nonlinear viscoelastic measurements for characterizing rheological properties of polymer melts.

Research Experience:

Designing Microfluidic Medical Assays

(most recent Postdoctoral research advised by Eric S. G. Shaqfeh for the computational part and Juan Santiago for the experimental part. The project is in collaboration with Stanford Genome Technology Center)

My main objective has been to combine direct microfluidic experiments with computational modeling to design a multi-purpose microfluidic platform which can diagnose morphological and bio-mechanical changes to red blood cells (RBCs) to diagnose Myalgic Encephalomyelitis/Chronic Fatigue Syndrome (ME/CFS), a mysterious disease with no definitive cure which is estimated to affect millions of individuals around the world, and 2 out of 100 children. The platform is designed to accurately measure the cell size and flexibility to diagnose differences between healthy controls and ME/CFS patients. It will identify factors contributing to differences in deformability, namely, the shear resistance of the cytoskeletal spectrin network and cytoplasmic viscosity. In this regard, the simulations are performed for the following specific aims:
1. Optimizing the operational and geometrical features of the micro-channels.
2. Converting cell shape and deformation to the characteristic elastic and viscous properties.

Collective Dynamics of Red Blood Cells (RBCs) in Micro-vasculature

(Postdoctoral research advised by Eric S. G. Shaqfeh)

I studied the effect of RBC deformability and cytoplasmic viscosity on its migration and distribution in small-medium arterioles (20-35 μm). In particular, we addressed the following aims:
1. The effect of RBC deformability on its distribution and clear fluid layer in small arterioles,
2. The impact of cytoplasmic viscosity on cell shapes, migration, and clear fluid layer.
Cell deformability alters in certain pathologies and cytoplasmic viscosity is mostly determined by hemoglobin concentration. Under normal conditions, a red blood cell has an internal viscosity 5-6 times higher compared to the viscosity of the plasma. It was found that the distribution of RBCs in smaller

arterioles are subjected to a qualitatively different behavior compared to medium sized arterioles. I also found that the cytoplasmic viscosity significantly reduces the RBC-free layer, changes the concentration distribution of the cells normal to the walls, and slows down the formation of RBC accumulation at the center of the channel. The simulations indicated the formation of rich configurational diversity among the cells, from simple slipper-liked to trilobe and multilobe shapes which have been observed experimentally. Interestingly, the single cell hydrodynamic lift is significantly affected for higher viscosity contrast, whereas two cell collision remains almost unaffected..

Transport of Architecturally Complex Macromolecules in Dilute and Semi-dilute Polymeric Solutions

(PhD research advised by Bamin Khomami in collaboration with Charles M. Schroeder)

I have developed a simulation platform ("BDpack" http://amir-saadat.github.io/BDpack/) on the basis of Brownian dynamics to do high fidelity prediction of both transient and steady-state properties of semi-dilute polymer solutions while including full long range hydrodynamic interactions (HI) and excluded volume effects. The simulation platform is used to study the transport of interacting multiple chains under equilibrium condition and also in shear and planar elongational flow.
I used this platform to study the dynamics of comb polymers in dilute solutions, and the simulation results revealed the significant impact of branches on the unraveling stretching dynamics and also relaxation behavior of comb polymers with respect to the linear chains. Currently, I am continuing the project by investigating the dynamic behavior of architecturally complex macromolecules, in particular comb polymers, in a bath of semi-dilute chains.

Teaching Interests:

Teaching and mentoring younger students are part of my motive to continue working in academia. Throughout my graduate and postdoctoral research, I had the privilege to mentor/supervise four undergraduate students and help them develop the required research skills.

I am interested in teaching courses in the chemical and mechanical engineering syllabi in graduate and undergraduate levels. In particular, fluid and continuum mechanics as applied to medicine/biology, applied numerical methods applied in computational fluid dynamics of soft matter and deformable objects and stochastic simulation of polymers, all of which fit well with my background. I also enjoy teaching many of the undergrad level courses, namely, introductory or advanced fluid mechanics, and finite element method. I am open to accept other topics and not afraid of taking over courses that are not in my area of expertise.

Teaching Experience:

1. "Numerical Analysis and Linear Algebra" for the 2017 AHPCRC (army high-performance-computing research center) summer institute.
2. Co-thought graduate level core course "Transport Phenomena" (I also served as a teaching assistant for this course for two consecutive semesters).
3. "Plastic engineering and polymer processing" (I prepared undergraduate students for graduate level national entrance exam)
4. Meshing software "Fluent" and "Gambit" for two semesters (practical numerical skills as well as engineering problems)

Selected Awards and Successful Proposals:

1. July 2017 -- NVIDIA-Stanford ICME grant for leveraging GPU computing ($50,000)
2. Apr 2019-present -- NSF's XSEDE computational resources (Allocation of computational time equal to ~$40,000)
3. Aug 2018-present -- Open Medicine Foundation (OMF) Postdoctoral Scholarship
4. Feb 2017-Aug 2018 -- Army High-Performance Computing Research Center (AHPCRC) Postdoctoral Scholarship
5. Feb 2014-Aug 2016 -- Eastman graduate student fellowship from Eastman Chemical Company

Selected Publications:

1. **A. Saadat**, C. J. Guido, E. S. G. Shaqfeh, Effect of cytoplasmic viscosity on Red blood cell migration in small arteriole-level confinements, *bioarxiv preprint* (submitted to *Phys. Rev. Fluids Rapid Communications*)**, 2019**.
2. **A. Saadat***, C. J. Guido*, G. Iaccarino, E. S. G. Shaqfeh, Immersed-finite-element method for deformable particle suspensions in viscous and viscoelastic media**,** *Phys. Rev. E***, 2018**, 98, 063316.
3. J. Mai*, **A. Saadat***, B. Khomami, C. M. Schroeder, Stretching Dynamics of Single Comb Polymers in Extensional Flow, *Macromolecules*, **2018**, 51, 1507.
4. Y. Lin*, **A. Saadat***, A. Kushwaha, E. S. G. Shaqfeh, Effect of Length on the Dynamics of Wall Tethered Polymers in Shear Flow, *Macromolecules*, **2017**, 51, 254.
5. **A. Saadat**, B. Khomami, Letter to the Editor: BDpack, an Open Source Parallel Brownian Dynamics Simulation Package. *Journal of Rheology*, **2017**, *61*, 147-149.
6. **A. Saadat**, B. Khomami, A New Bead-Spring Model for Simulation of Semi-Flexible Macromolecules. *The Journal of Chemical Physics*, **2016**, 145, 204902.
7. **A. Saadat**, B. Khomami, Matrix-Free Brownian Dynamics Simulation Technique for Semidilute Polymeric Solutions. *Physical Review E,* **2015**, *92*, 033307.
8. **A. Saadat**, B. Khomami, Molecular Based Prediction of the Extensional Rheology of High Molecular Weight Polystyrene Dilute Solutions: A High-Fidelity Brownian Dynamics Approach. *Journal of Rheology,* **2015**, *59*, 1507-1525.
9. **A. Saadat**, B. Khomami, Computationally Efficient Algorithms for Incorporation of Hydrodynamic and Excluded Volume Interactions in Brownian Dynamics Simulations: A Comparative Study of the Krylov Subspace and Chebyshev Based Techniques. *The Journal of Chemical Physics*, **2014**, *140*, 184903.
10. **A. Saadat**, H. Nazockdast, F. Sepehr, M. Mehranpour, Linear and Nonlinear Melt Rheology and Extrudate Swell of Acrylonitrile-Butadiene-Styrene and Organoclay-Filled Acrylonitrile-Butadiene-Styrene Nanocomposite. *Polymer Engineering & Science,* **2010**, *50*, 2340-2349.
11. **A. Saadat**, H. Nazockdast, F. Sepehr, M. Mehranpour, Viscoelastic Modeling of Extrudate Swell of Acrylonitrile-Butadiene-Styrene/Clay Nanocomposite. *Applied Rheology,* **2013,** *23*, 12131.

579892 Molecular Tension Sensors in Protein Hydrogels

Soft materials with made-to-order chemical and physical properties have significant potential value in fields ranging from regenerative medicine and *in vitro* cell culture to soft robotics and drug delivery. Artificial protein hydrogels (APHs) are uniquely suited to be designer materials via exploitation of the near limitless possible combinations of functional protein domains that can be used to create novel polymer building blocks with tailored properties. Recent work in model systems has demonstrated APHs can be engineered to improve wound healing, for controlled drug delivery and for the creation of materials with programmed characteristic relaxation times. However, engineering APHs to have enhanced mechanical strength and toughness has so far not been possible and remains a critical obstacle for their clinical deployment and use in other applications. One potential approach to make tougher APHs is by creating them from protein structures encoded with a higher propensity to dissipate energy through unfolding in response to mechanical stress. This approach seems promising; however, it is not currently understood how features at the molecular level permeate to the bulk physical properties of APHs. I will present our efforts to address this gap through the development of a fluorescence microscopy method for real-time visualization of protein conformational changes that result from mechanical stress induced on an APH. The presentation will focus on two key points: (1) Design and synthesis of a tunable force-responsive FRET sensor that can be readily incorporated into APHs and other types of hydrogels; and (2) fundamental insights on the mechanism of energy dissipation in protein hydrogels gained from visualizing spatially resolved mechanical stress. Lastly, I will show an application of visualizing mechanical stress in APHs – quantifying cell traction forces in three-dimensional cell culture systems.

Research Interests:

Protein hydrogels, Bioconjugation, Cell material interactions, Nanocomposites, Biocatalysis, Biosynthetic pathway engineering, Metabolomics, Small-molecule structure elucidation.

Teaching Interests:

Biomaterials, Protein engineering, Metabolomics, Organic synthesis for engineers, Quantitative analysis for engineers, Bioprocess engineering, Polymer chemistry.

579894 3D-Engineering of Functional Living Composite Materials

Seunghyun Sim, Helen Hay Whitney Postdoctoral Fellow, California Institute of Technology

Ph.D. Research (Advisor: Takuzo Aida, The University of Tokyo):

Protein Nanostructure Engineering. My Ph. D. research focuses on the design of nanostructures using protein as a building block and their biomedical application. Combinations of genetic engineering, implementing rationally designed chemical moiety, and interdisciplinary solutions such as utilizing magnetic field made it possible to control protein self-assembly structure in multiple dimensions.

Postdoctoral Research (Advisor: David A. Tirrell, California Institute of Technology):

Engineered Living Materials. My current postdoctoral research in the Tirrell lab at Caltech is developing strategies to include "living" bacterial entities as an active ingredient of smart materials. Two novel living materials have developed: (1) artificial protein network produced by cells *in situ* and (2) 3D-printable material encasing engineered bacterial cells and spores. During the postdoctoral training, I have broadened my skill set in the areas of molecular biology and characterization of soft matter.

Research Interests:

3D-Engineering of Functional Living Composite Materials. The research program I propose to implement centers on developing smart 3D living composite materials. Bacterial cells possess ideal qualities for generating emergent materials: they can assemble, communicate, synthesize biomolecules, and adapt to the various conditions. The first outcome of this research will be a functional library of 3D-printable living composite materials using different combinations of chemical scaffolds and engineered microorganisms. The interaction of the chemical matrix and bacterial cells will be studied in various length scales– from the molecular level to macroscopic tensile testing –to engineer them for specific goals. Bacterial cells will be genetically engineered to produce artificial proteins or effector molecules for mechanical reinforcement, cellular communications, and therapeutic applications using a variety of tools in synthetic biology. In particular, we will aim to accomplish controlled mechanical actuation, chemical conversion, and biologics delivery with these materials.

Teaching Interests:

Majoring both chemistry and biology as an undergraduate and later earning Ph. D. from the school of engineering, I would be comfortable teaching a wide range of subjects in chemistry, chemical engineering, biology, bioengineering, and materials science. Also, I have a special interest in developing interdisciplinary courses with an emphasis on introducing the recent scientific trend.

Research Funding:

Merit Fellowship (2013–2015), JSPS Fellowship for Young Scientists (2015–2017), Helen Hay Whitney Fellowship (2019–current).

Publications:

S. Sim, Y. Kim, T. Kim, S. Lim, M. Lee, "Directional Assembly of a-Helical Peptides Induced by Cyclization", *J. Am. Chem. Soc.* **2012**, *134*, 20270-20272.

S. Sim, D. Miyajima, T. Niwa, H. Taguchi, T. Aida, "Tailoring Micrometer-Long High-Integrity 1D Array of Superparamagnetic Nanoparticles in a Nanotubular Protein Jacket and Its Lateral Magnetic Assembling Behavior", *J. Am. Chem. Soc.* **2015**, *137*, 4658–4661.

S. Sim, T. Niwa, H. Taguchi, T. Aida, "Supramolecular Nanotube of Chaperonin GroEL: Length Control for Cellular Uptake Using Single-Ring GroEL Mutant as End-Capper", *J. Am. Chem. Soc.* **2016**, *138*, 11152-11155.

S. Sim, T. Aida, "Swallowing a Surgeon: Toward Clinical Nanorobots", *Acc. Chem. Res.* **2017**, *50*, 492–497.

D. Kashiwagi, **S. Sim**, T. Niwa, H. Taguchi, T. Aida*, "Protein Nanotube Selectively Cleavable with DNA: Supramolecular Polymerization of "DNA-Appended Molecular Chaperones", *J. Am. Chem. Soc.* **2018**, *140*, 26-29.

- Publication as a corresponding author
- Featured as the journal's cover
- Featured in JACS Spotlights, "DNA, Proteins Combine to Yield Highly Stable Nanotubes"

S. Sim, D. Miyajima, N. Nitta, I. Aoki, T. Aida, "Novel Approach for High-Performance T1 MRI Contrast Agent: Effect of Adjacent Charged Groups to the Metal Oxide Nanoparticle Surface (tentative title)", *To be submitted.*

S. Sim, P. K. Hashim, K. Okuro, Y. Anraku, K. Kataoka, T. Aida, "Functional Design of de novo RNA Virus based on Chaperonin GroEL (tentative title)", *To be submitted.*

D. Kashiwagi, **S. Sim**, T. Niwa, H. Taguchi, T. Aida, "Supramolecular Lamellar Assembly of DNA-appended Janus GroEL via AND Logic Gate (tentative title)", *To be submitted.*

S. Sim, Y. Hui, D. A. Tirrell, "Engineered living film reinforced by artificial protein assembly in situ (tentative title)", *To be submitted.*

S. Sim, Y. Hui, D. A. Tirrell, "3D-printable living composite materials of B. subtilis for portable therapeutic protein delivery (tentative title)", *To be submitted.*

Selected Awards:

Award for Academic Excellence, Seoul National University, 2009, 2010, and 2011.

The Rising Star Award, CEMS ISSC&FM, 2013.

BASF PhD Rising Star Award, BASF, 2014.

PMSE Future Faculty Scholars, ACS PMSE Division, 2018.

Global Outstanding Student Award in Polymer Science and Engineering, ACS PMSE Division, 2019.

577784 Molecular Design Principles for Chemically Tunable Biomaterial Platform Technologies

Research Interests:

Organic chemistry is a central science for the development of functional materials. My research program will specifically address the molecular design of biomaterials to elucidate structure-property relationships that can be utilized for unexplored applications and production methods. During my graduate training, I established new reaction methodologies and design principles for catalyst development. In my postdoctoral research, I am applying these fundamental concepts from organic chemistry to establish rational design strategies of polymeric materials with tunable properties for drug delivery and tissue engineering. The translation of such technologies to the clinic has the potential to alter poor patient prognoses and improve the quality of life of patients on a global scale. Because of this, I will continue a molecular approach to establish structure-property relationships for organic materials design in the next stage of my career, and am excited to mentor students and postdocs in a program dedicated to research at the intersection of organic chemistry, materials science, and chemical engineering with the goal of addressing longstanding challenges in medicine.

Graduate Research at the California Institute of Technology, Division of Chemistry and Chemical Engineering. *Advisors: Professors Robert H. Grubbs and Gregory C. Fu*

Development of New Catalysts and Materials Applications of Olefin Metathesis. Olefin metathesis has proven to be a highly versatile reaction for the synthesis of both small molecules and polymers, and is utilized by organic, polymer, and materials chemists. Despite this, the production of ultra high molecular weight polymers with low dispersities remains challenging. We synthesized series of new ruthenium catalysts with aminophosphine ligands and performed systematic studies to determine design principles to better relate the effect of ligand structure on catalyst activity. Specifically, we incorporated P–N bonds in the phosphine ligand structure to investigate incongruent substituents on the phosphine architecture. Kinetic and computational studies disentangled contributions from ligand electronic, steric, conformational, and distortion effects on catalyst activity. We have also been able to utilize Grubbs catalysts to form brush block copolymers that rapidly self assemble into ordered nanostructures studied by small angle X-ray scattering. The resulting polymer films have been explored in applications such as pressure-responsive materials and photonic crystals.

Installation of Bioisosteres in Place of Carbon and Hydrogen. The replacement of specific functional groups of lead compounds with bioisosteres has become a common strategy in medicinal chemistry. This is exemplified by the incorporation of fluorine in place of hydrogen to tune metabolic stability, lipophilicity, and bioactivity. Following identification of lead compounds, site-selective fluorination remains a synthetic challenge. My graduate research addressed a major challenge in fluorination chemistry – the selective formation of beta-fluorinated carbonyl compounds. Previously discovered methods largely suffer from poor regioselectivity, difficult substrate synthesis, or harsh reaction conditions. The strategy I employed involves the reaction of readily accessible allylic fluorides using a robust Wacker oxidation exhibiting reversed regioselectivity to produce aldehydes. These aldehydes can then be easily derivatized to a variety of compounds, providing a mild route to diverse fluorinated building blocks that can be utilized by medicinal chemists toward drug targets.

In similar fashion to fluorination chemistry, silicon bioisosteres have been incorporated in place of carbon atoms to study the effect on bioactivity of target compounds. Hydrosilylation is sensitive to steric interactions, and many compounds that would be derived from tri- and tetrasubstituted olefins cannot be accessed in this way. Over the last decade, nickel-catalyzed cross-coupling reactions have been established as efficient, mild strategies toward the production of C–C bonds. I developed a nickel-catalyzed cross-coupling reaction to form C–Si bonds tolerant of sterically hindered alkyl electrophiles. We demonstrated broad functional group tolerance of the reaction, and accessed organosilanes that cannot be produced from hydrosilylation. These reaction methodologies provide efficient alternative strategies for the incorporation of hydrogen and carbon bioisosteres in drug discovery.

Postdoctoral Research at the Massachusetts Institute of Technology, Koch Institute for Integrative Cancer Research. *Advisors: Professors Robert S. Langer and Daniel G. Anderson*

Biomimetic Production of Polymer Fibers. Polymer fibers, ubiquitous in functional materials toward biomedical applications, are typically produced using protocols requiring organic solvents and high energy processes. I have recently discovered a new method for forming polysaccharide fibers from dynamic cross-linked networks without the need for electrospinning or extrusion. Like spider silks, these fibers can be formed by pultrusion and are drawn from a viscous liquid precursor. We have investigated the effects of the molecular structure, molecular weight, and degree of branching of the cross-linkers on the mechanical properties of the resulting materials. These studies allow for a deeper understanding of how organic chemistry can be used to rationally design fibrous materials with interesting properties.

Glucose-Responsive Materials for Insulin and Glucagon Delivery. Patients with type 1 diabetes regularly experience hyper- and hypoglycemic episodes. Glucose-responsive insulin delivery has been the focus of significant research efforts, but a major challenge in translating effective systems to clinic is glucose recognition. Common approaches to glucose recognition include the use of enzymes or lectins, which can be immunogenic and unstable, and synthetic phenylboronic acids, which suffer from poor selectivity. The focus of my research towards glucose-responsive insulin is to design and synthesize a lectin mimic that is stable and glucose-selective to be conjugated to insulin-loaded materials. In order to achieve tight glycemic control, I have developed and performed release studies on related materials to release glucagon, a hormone secreted by the pancreas of healthy individuals in response to hypoglycemia.

Teaching Interests:

From a diverse laboratory background requiring a broad skill set, I am excited to teach laboratory courses in addition to lecture courses. I am interested in designing undergraduate courses in Organic Chemistry, Polymer Science, and Thermodynamics as well as advanced courses in Physical Organic Chemistry, Homogeneous Catalysis, and Soft Materials. Students I have mentored, as well as myself during my education, have found that research-focused courses are very informative and beneficial to developing their career goals. Because of this, I am excited to design hands-on laboratory and proposal-based seminar courses in the subjects of Reaction Development as well as Pharmaceutical Products and Medical Devices. These topics will extend to my research program, and multidisciplinary training in a collaborative environment will provide the foundation for students and postdocs to conduct impactful research and excel in their career aspirations.

578767 Shining Light on the Nervous System: From Biomaterials to Bioelectronics

Personal Web Page: http://www.jingtang.org/

Research Interests

Biomedical Devices, Biosensors, Soft Materials, Nanomaterials, Nanomedicine, Bioinspired Materials and Devices, Neuromodulation, Biomaterials and Drug Delivery, Neuroengineering, Tissue Engineering, Electrochemical Interfaces, Biocatalysis, Sustainability.

Postdoctoral Research Summary at Stanford University: Bioinspired Materials and Devices for Healthcare and Sustainability

Advisor: Yi Cui (Materials Science and Engineering and Chemistry), in a close collaboration with Zhenan Bao (Chemical Engineering), Jeffrey A. Reimer (Chair of Chemical and Biomolecular Engineering at UC Berkeley) and Friedrich Prinz (Mechanical Engineering and Materials Science and Engineering), Stanford and Peter Ercius (National Center for Electron Microscopy, Lawrence Berkeley National Laboratory)

Postdoctoral Research Summary at Harvard Medical School and MIT: Shining Light on Pain Therapeutics: from Biomaterials to Smart Drug Delivery Systems

Advisor: Daniel Kohane (Boston Children's Hospital) and Robert Langer (Chemical Engineering and Biomedical Engineering)

PhD Dissertation: Nanowire Arrays Restore Vision in Blind Mice

Advisor: Gengfeng Zheng and Dongyuan Zhao (Department of Chemistry and Laboratory of Advanced Materials, Fudan) and in a close collaboration with Jiayi Zhang (Institutes of Brain Sciences at Fudan) and Wenlong Cheng (Chemical Engineering at Monash, Australia)

Research Experience

My academic training and research experience have provided me with a background in highly interdisciplinary area combing scientific disciplines including nanomaterial, bioelectronics, biomaterials, chemical engineering, tissue engineering, neuroengineering and biomedical engineering.

Dysfunctional neuronal signaling underlies a wide range of psychiatric and neurological disorders. Existing therapeutics often lack clinical efficacy to reverse signaling imbalance and are both non-specific and invasive. To improve treatments of brain-related diseases will require new tools and methods to map and to repair the brain with precision and biocompatibility. I am working at the interface of materials science, electronics and neuroscience with the goal of advancing understanding and treatment of disorders of the nervous system. I design, synthesize and fabricate pharmacological, optical, and electrical toolsets that manipulate and record neuronal activity and development. The dichotomy between advanced materials and brain has driven the curiosity of scientists to explore the wonders of the brain, as well as motivated the continued innovations of novel technologies based on advances in materials science and engineering to understand the brain. These results, along with a discussion of future neural interfaces, aim to improve our understanding of the nervous system and to inform new therapeutic approaches for bioelectronics.

- **I developed a light triggered smart drug release device system.** Current treatments of pain heavily rely on opioids, resulting in significant side effects such as addiction, tolerance, leading to the Opioid Overdose Crisis as we know of today. Smart drug delivery systems may provide an effective solution. Here I present the development of externally-triggerable drug delivery systems for on-demand, repeatable and adjustable local anesthesia using new polymer nanoparticles, where the timing, duration, and intensity of nerve block can be controlled through external energy triggers such as the optical tool. In addition to traditional pharmacological approaches, bioelectronic platforms to enhance our insights into the retina.

- **I developed an artificial retina biomedical device.** The restoration of light response with complex spatiotemporal features in retinal degenerative diseases towards retinal prosthesis has proven to be a considerable challenge over the past decades. Herein, inspired by the structure and function of photoreceptors in retinas, I develop artificial photoreceptors based on gold nanoparticle-decorated titania nanowire arrays, for restoration of visual responses in the blind mice with degenerated photoreceptors. Green, blue and near UV light responses in the retinal ganglion cells (RGCs) are restored with a spatial resolution better than 100 μm. ON responses in RGCs are blocked by glutamatergic antagonists, suggesting functional preservation of the remaining retinal circuits. Moreover, neurons in the primary visual cortex respond to light after subretinal implant of nanowire arrays. Improvement in pupillary light reflex suggests the behavioral recovery of light sensitivity. My study will shed light on the development of a new generation of optoelectronic toolkits for subretinal prosthetic devices. Through pharmacological, optical, and electrical toolsets, I aim to develop effective therapeutic solutions to neurological disease states. These results, along with a discussion of future neural interfaces, aim to improve our understanding of the nervous system and to inform new therapeutic approaches for biomaterials and bioelectronics.
- **I developed bioinspired single-atom-based devices for mental health.** Cortisol is a stress hormone from cholesterol in the two adrenal glands. It's significant for helping your body handle stressful situations, as your brain triggers its release in response to many different kinds of stress. however, when cortisol levels are too high for too long, this hormone can hurt you more than it helps. Here I generate atomically dispersed metal via photochemical reduction of metal acid solution using ultraviolet light. And then I emulate optimum natural systems following Murray's law using a top-down approach. Such bio-inspired materials and devices, whose pore sizes decrease across multiple scales and finally terminate in size-invariant units like lung systems provide hierarchical branching and precise diameter ratios for connecting multi-scale pores from macro to micro levels. My Murray devices mimics enable highly enhanced mass exchange and transfer in liquid-solid, gas-solid and electrochemical reactions and exhibit enhanced performance in electrochemical interfaces. I introduce the integration of an electrochemical biosensing and a tailor-made bioinspired device, which acts as a stable and selective probing of the human stress hormone cortisol.

Research Proposal Experience

1. **Shell Global, Shell Research and Engineering Grant (Yi Cui, Steven Chu), Bioinspired Materials and Devices for Sustainability, 2019-2021, Pending, Role: Written by me**
2. R01GM073626-11, NIGMS, Prolonged Duration Local Anesthesia (Daniel Kohane) 2015.8.1-2019.4.30 Role: Main Participant
3. R01 GM116920 NIGMS, Nanoparticle-driven systems for tunable local pain relief (Daniel Kohane), 2016.4.27-2020.4.26, Role: Main Participant
4. **Nanjing Leading Technology and Entrepreneurship Program, Photoelectrochemical Biosensor (Gengfeng Zheng), 2015-2019, Role: Written by me**
5. **Fudan, The Interdisciplinary Outstanding Doctoral Fellowship for PhD Thesis (Jing Tang), Carbon Quantum Materials for Drug Delivery and Bioimaging, 2012-2015, Role: PI**
6. **FDUROP Undergraduate Research Grant**, Hui-Chun Chin and Tsung-Dao Lee Chinese Undergraduate Research Endowment, Wang-Dao Undergraduate Research Grant, Fudan Undergraduate Academic Research and Innovation Program, and Student's Academic Science and Technology Innovation Action Support Program. Furthermore, they got National Grand Prize winner of "Challenge Cup" national undergraduate academic science and technology event in China, 2012-2015, **Role: Mentor**
7. **Dow Chemical Company, Dow chemical Innovation Fellowship for Sustainable Development** (Jing Tang, Kun Jiang, Yongcheng Wang, Tao Jiang, Fei Wang, Jingjing Zhang), Functional Nanomaterials for Clean Energy, Awarded to only 2 teams for scientific innovation in sustainable biology and energy, 2014-2015, **Role: MPI**
8. **BASF, BASF Outstanding Doctoral Fellowship (Jing Tang), Awarded to only one graduate student in Laboratory of Advanced Materials, Nanowire Nanosensor, 2013, Role: PI**
9. The National Natural Science Foundation of China, Nanowire Functional Materials (Gengfeng

Zheng), 2014.1-2016.12, Role: Main Participant

10. Ministry of Science and Technology, 973 Program, Rational Design of Functional Mesoporous Materials for Energy (Gengfeng Zheng, Dongyuan Zhao), 2013.1-2017.8, Role: Main Participant
11. Shanghai Municipal Education Commission, Nanowire Photoelectric Functional Materials (Gengfeng Zheng), 2012.1-2014.12, Role: Main Participant
12. Ministry of Education, Controllable synthesis of core-Shell Mesoporous Materials-Metal Oxide Nanowires and Their Applications in PEC Water Splitting (Gengfeng Zheng), 2013.1-2015.12, Role: Main Participant

Selective Awards & Honors

- **Young Investigator Award in Northern California Chapter American Association of Physicists Symposium (2019)**

- Winner of the 2018 NanoArt Image Contest, Lawrence Berkeley National Laboratory (2018)
- Singapore Challenge, Global Young Scientists Summit (GYSS) @one-north, Singapore (2016)
- Outstanding Poster Award of Vision Forum (2017)
- Outstanding Shanghai Graduate Students on Graduation, Shanghai, China (2016)

- Baosteel's Principal Scholarship, BAOWU STEEL GROUP (2015)

- **First Prize of Unilever-The Royal Society of Chemistry Presentation Competition, Royal Society of Chemistry (2014)**
- **Grand Prize Winner of Dow Sustainability Innovation Student Challenge Award, Dow Chemical Company (2014)**
- Kwang-Hua Scholarship, Kwang-Hua Education Foundation (2014)
- National Scholarship for Graduate Students, Chinese Government (2014)
- Representative of Universitas 21 (U21) Graduate Research Conference in New Zealand, Universitas 21 (2014)

- **Distinguished Award of BASF Scholarship, BASF Corporation (2013)**

Community Service and Leadship

- Safety Officer and Ordering, Yi Cui Group, Stanford University (2019-Present)
- Seminar Committee, Stanford Neurosciences Institute (2019-Present)
- Session Chair, Materials Research Society Fall Meeting (2018)
- Leader, Dow Sustainability Innovation Student Team (2014)
- Organizer and Host, Fudan Advanced Materials Laboratory Doctoral Forum (2014)
- Chair of the Academics, Research and Careers Committee, Fudan Advanced Materials Laboratory Students Council (2012-2015)
- Co-President, The Royal Society of Chemistry-Fudan University Students Association (2012-2015)
- President, Fudan Advanced Materials Laboratory Students Association (2012-2015)
- Volunteer, Shanghai World Expo (2010 Summer)
- Chair, Shanghai Municipal Government World Expo Command Volunteer Team (2010 Spring)

Teaching Interests:

I got the Teaching Certificate Program offered by Stanford in 2019. Aside from my research career, I also have extensive experience teaching at Fudan, Harvard and Stanford undergraduate students and graduate students. My teaching audiences have spanned middle school students to post-doctoral scholars. At Fudan, I have TAed undergraduate-level classes for the pre-medical students in General Chemistry and graduate level classes for chemical engineering student. At MIT and Stanford University, I have guest-lectured graduate-level chemical engineering classes. I have received positive feedbacks from both teaching experiences. I have been taken some Chemical Engineering and biomedical core classes in both undergraduate-level and graduate-level, thus I am willing to study and teach any existing classes, as well as design new classes. Last but not least, I mentored 4 undergraduate students and 4 graduate students in

Chemistry, Chemical Engineering and Biomedical Engineering at Fudan, Harvard and MIT, and they finally got very top university offers, such as Stanford, Princeton, UCSB and Boston University. Coming from a Chemical Engineering and Biomedical Engineering postdoc background at MIT and chemistry PhD background at Fudan, I am interested in teaching most of the major courses for Chemical Engineering, especially Molecular Thermodynamics, Biocatalysis, Transport Mechanics, Heat and Mass Transfer, Fluids Dynamics, and Thermodynamics, Separation Process. I am also interested in other courses including Material Science, Chemistry, Nanoscience and Nanotechnology, Neurotechnology, Biomaterials, Tissue Engineering and Electrochemistry, Soft Materials and Nanomedicine. These experiences have trained my ability to clearly explain difficult concepts in a simple or detailed manner. For six years of teaching and mentoring experience as a teaching fellow, creating a great educational experience for young scientists and researchers is critical for the continued advancement of science and it is with this view that I will approach teaching as a future faculty member.

Distinction (for Teaching and Mentoring)

- Outstanding Teaching Assistant for General Chemistry (Brand Courses Taught in English for Pre-Medical Students Overseas) (2015)

- Mentoring 4 undergraduate students (They succeeded in applying world-class graduate universities and completing several undergraduate research programs, including the Deng-Hui program, the Wang-Dao program, the Hui-Chun Chin & Tsung-Dao Lee Undergraduate Research Endowment and Student's Academic Science and Technology Innovation Action Support Program. Furthermore, they got National Grand Prize winner of "Challenge Cup" national undergraduate academic science and technology event)

571459 3D Printed Bioelectronics for Tissue Engineering and Regenerative Medicine

Research Interests:

Innovating **soft materials** and **fabrication** strategies for 3D **bioelectronic** devices, with a special emphasis on **biomaterial**-based and cell-integrated electronics for **tissue engineering** and regenerative medicine.

Bioelectronic devices serve as an interaction platform between humans and biology. Devices can be used to monitor biological states, such as part of diagnostics, lab-on-chip devices, wearable health tracking, or can be used to influence biology, as assistive devices or therapy in applications such as brain-machine interfaces and rehabilitation neuroprosthetics. Bioelectronic devices are already widely used in hospitals for wearable vital sign monitoring as well as implantable devices (e.g. pacemakers, cochlear implants). New and increased uses of these devices will continue to grow and impact healthcare and biological sciences as advances in materials and devices are made.

As faculty, I will innovate materials and fabrication strategies to build next-generation bioelectronic devices, devices that achieve seamless interfacing by blurring the lines between what is biologic and what is electronic. The Rutz Laboratory will use tissue engineering and biomaterial approaches to design new, soft and hydrated materials for bioelectronic device fabrication. The Rutz Lab will also explore the use of additive manufacturing for achieving unique 3D device formats as well as co-printing cells alongside electronic inks for biological-integration achieved at the fabrication stage.

Teaching Interests:

I have been very active in research mentorship and have advised 17 diverse students to-date, including 9 graduate students and supervising 2 undergraduate theses. I have teaching experience in biomedical engineering and chemistry, including lab and course development, guest lecturing, teaching assistance, and tutoring. As faculty, I am looking forward to teaching introductory and advanced courses on topics of chemistry and biology for engineers, biomaterials/soft materials, and tissue engineering. I am especially interested in developing coursework that explores synthesis, characterization, fabrication and application of materials for medical devices, biomaterials, biotechnology, and bioelectronics.

Research Experience:

Postdoctoral Research: "Liquid-infused elastomers for implantable bioelectronic devices"

Advisor: Prof. George G. Malliaras, Electrical Engineering, University of Cambridge

11/2016-Present

Implantable electronics are a class of medical devices of unrealized potential for monitoring, diagnosis, and treatment of health and disease. In long-term implantations, device efficacy can diminish due to several factors including significant surgical trauma, an aggressive foreign body response, and poor material compatibility with the biological milieu. I have investigated liquid-infused elastomers as a new material for bioelectronics. These materials yield a slippery, hydrophobic surface, and I have explored how these unique material properties could be advantageous for bioelectronic devices. I have shown that these materials are biocompatible, reduce biofouling in vivo and reduce insertion trauma.

Additionally, I have written, advised, and collaborated on several other projects. These include synthesis and application of fluorogels, synthesis of shear-thinning conducting polymer hydrogels, direct write fiber drawing of conducting polymers, and additive manufacturing of stretchable electronics. All of which capture my continued research interests in developing new materials and fabrication methods for bioelectronic devices.

Ph.D. in Biomedical Engineering: "Engineering hydrogel inks for 3D tissue and organ printing"

Advisor: Prof. Ramille N. Shah, Northwestern University, Chicago, IL

09/2011-08/2016

3D printing is a powerful manufacturing method that may herald fabrication of complex tissues and organs for transplantation and tissue models. However, a severe limitation to future growth of 3D tissue printing is the lack of printable biomaterials, "inks". In my doctoral research, I invented a single ink method that can be used to synthesize over 100 natural and synthetic hydrogel inks of customizable material properties and ideal printing dynamics. The method was also extended to biorthogonal reactions to optimize post-printed cell viability, and rheology was used to define ink "printability". I also developed a second ink method and scaffolds for the engineering of an ovary bioprosthesis. Most excitingly, when implanted into mice with surgically removed ovaries, the ovary bioprosthesis restored organ function as observed by the live births of healthy pups, lactation and estrous cycling.

B.S. in Chemistry and Molecular & Cellular Biology, University of Illinois Urbana-Champaign, IL

Thesis: "Hyperbranched polyglycerol for drug delivery"

Advisor: Prof. Steven C. Zimmerman, Chemistry

08/2008-05/2011

Selection of Funding and Publications:

Fellowships:

Marie Skłodowska-Curie Actions Individual Fellowship (€183,000 EUR), 2018-2020

Whitaker International Post-doctoral Scholarship ($100,000 USD), 2016-2018

NSF Graduate Research Fellowship ($150,000 USD), 2013-2016

Grants:

University of Cambridge Engineering for Clinical Practices Grant (£10,000 GBP), 2018

Whitaker International Program's Concluding Initiative Grant ($50,000 USD), 2018

Publications:

E. Zeglio, **A.L. Rutz**, T.E. Winkler, G.G. Malliaras, A. Herland. "Conjugated polymers for assessing and controlling biological functions," ***Advanced Materials*** 2019, 1806712.

J. Pas, **A.L. Rutz**, P.P. Quilichini, A. Slézia, A. Ghestem, A. Kaszas, M. Donahue, V. Curto, R.P. O'Connor, C. Bernard, A. Williamson and G.G. Malliaras. "A bilayered PVA/PLGA-bioresorbable shuttle to improve the implantation of flexible neural probes," ***Journal of Neural Engineering*** 2018 15 (6), 065001.

A.L. Rutz, P.L. Lewis, R.N. Shah. "Toward next-generation bioinks: tuning material properties pre- and post-printing to optimize cell viability," ***MRS Bulletin*** 2017, 42.

A.L. Rutz* and M.M. Laronda*, K.A. Whelan, E.W. Roth, T.K. Woodruff, R.N. Shah. "A bioprosthetic ovary created using 3D printed microporous scaffolds restores ovarian function in sterilized mice," ***Nature Communications*** 2017, 8. ** Article featured on CNN, FOX, CBS, TIME, BBC, NPR and others. Altmetric score = 2241. Times cited = 127

A.L. Rutz, K.E. Hyland, A.E. Jakus, W.R. Burghardt, R. N. Shah. "A multi-material bioink method for 3D printing tunable and cell-compatible hydrogels," ***Advanced Materials,*** 2015, 27 (9) 1607-1614. ***Times cited = 217*

561833 Lipid Nanoparticles for Highly Efficient Non-Viral Gene Editing

Therapeutics based on proteins, RNA and DNA have tremendous therapeutic potential but have been challenging to develop due to delivery problems. Recently, CRISPR/Cas9-based therapeutics show great potentials to revolutionize the treatment of genetic diseases. However, the lack of safe and efficient delivery method is still a big challenge for its clinical translation. We have developed a novel lipid nanoparticles based delivery platform which can deliver Cas9 ribonucleoprotein (RNP) and donor DNA into wide variety of type for gene knockdown, DNA mutation correction and gene knockin. This delivery system shows five times increase of NHEJ based gene knockdown efficiency in HEK 293T cells and three times increase in primary myoblast cells compared with commercially available lipofectamine 2000 based formulation. Also, this delivery system shows efficient knockdown of pcsk9 gene in liver in vivo by intravenous injection, which significantly reduced pcsk9 and cholesterol level in the serum. Meanwhile, by delivering of Cas9 RNP and single strand donor DNA together, this delivery system can efficiently correct the DNA mutation that causes Duchenne muscular dystrophy in mice via local injection, with minimal off-target DNA damage. More importantly, this delivery method can knock in complete cytokine genes into tumor cells which have great potential for cancer immunotherapy.

Research Interests: Nanomedicine, Gene Editing, Protein Delivery

Teaching Interests: Chemical Reaction Engineering, Separation Processes, Transport Phenomena, Chemical Process Computer-Aided Design and Analysis, Process Economics and Analysis

582392 Multiscale Modeling for the Design of Functional Soft Materials

Research Interests:

When contrasted with natural biopolymers, it is clear that our ability to design soft materials with comparable structural, electronic, and ionic functionalities is severely limited. The design of these materials would be greatly facilitated by predictive models that capture multiscale structure-function relationships; however, such efforts require the challenging integration of an ensemble of theoretical and computational techniques to accurately describe the electronic, nuclear, and mesoscopic degrees of freedom that influence material performance. My research focuses on the development of predictive multiscale computational methodologies that accurately characterize structural, electronic, and ionic phenomena in soft materials, with the goal of advancing rational design strategies for functional soft materials.

My research group will operate at the intersection of engineering, chemistry, and physics, enabling us to study fundamental and applied problems at multiple spatial and temporal resolutions. My initial research efforts will comprise three areas:

- I will develop and apply multiscale computational models that treat electronic and reactive phenomena at coarse-grained spatiotemporal resolutions. These methodological developments will dramatically enhance the length and timescales for chemically-predictive coarse-grained modeling of electronic and reactive phenomena. Specific target applications will include mechanically-induced bond-breaking and triboelectricity in polymers, morphology-dependent electronic structure in organic semiconductors, and polymer chemical degradation pathways.
- I will apply multiscale simulation techniques to design the conformational and morphological properties of bioelectronics polymers (conjugated polyelectrolytes), as well as to quantify their combined electronic and ionic transport properties.. The utility of organic polymers for combined electronic and ionic conductivity, and the fundamental mechanisms underlying mixed conductivity in organic materials, are currently unexplored.
- I will apply multiscale computational techniques to understand the mechanisms of contact electrification and design new materials with optimized triboelectric properties. Triboelectricity (static electricity) is one of the least understood and most prevalent physical properties of organic materials. Despite the acute influence of nanoscale structure on contact electrification, the role of bond-breaking, ion transfer, and electron transfer are unexplored at a molecular scale.

Teaching Interests: In addition to my research interests, I am a committed educator. As an undergraduate at Wesleyan University, I co-designed a course on the Physics of Sustainability in addition to serving as a teaching assistant for General Physics I & II. I continued my pedagogical development at Northwestern University by completing the Northwestern Teaching Certificate Program, as well as being a teaching assistant for a variety of graduate and undergraduate chemistry courses. As a post-doc at the University of Chicago, I co-instructed the Introduction to Molecular Modeling course for molecular engineering Ph.D. students.

Background: I received my Ph.D. in 2016 from Northwestern University working as a NSF GRFP fellow and Northwestern Presidential fellow with Profs. Mark Ratner and Lin Chen. My graduate work focused on the development of models characterizing self-assembly, photophysics, and electronic transport in organic semiconductors. I then began a postdoc with Prof. Juan de Pablo in the summer of 2016 and was subsequently awarded the Maria Goeppert Mayer Fellowship in the Materials Sciences Division at Argonne National Laboratory, which allowed me to function as an independent researcher. My postdoctoral work has concerned the development of statistical sampling and coarse-graining algorithms, machine learning-enhanced simulation, and the structure and dynamics of polymeric and glassy interfaces. This postdoctoral work has been recognized by the 2019 MRS Postdoctoral Award and the 2019 ACS PHYS Young Investigator Award.

Successful Proposals: Argonne LDRD Named Fellow Call, NSF Graduate fellowship, UChicago Research Computing Cluster, Argonne Laboratory Computing Resource Center.

Selected Publications:

1. Jackson, N.E.; Bowen, A.S.; Antony, L.A.; Webb, M.A.; Vishwanath, V.; de Pablo, J.J. "Electronic Structure at Coarse-Grained Resolutions from Supervised Machine Learning," *Adv.*, 2019, 5 (3); eaav1190.
2. Yu, J.*; Jackson, N.E.*; Xu, X.; Morgenstern, Y.; Kaufman, Y.; Ruths, M.; de Pablo, J.J.; Tirrell, M. "Multivalent counterions diminish the lubriity of polyelectrolyte brushes," *Science*, 2018, 360 (6396), pp 1434-1438.
3. Jackson, N.E.; Chen, L.X.; Ratner, M.A. "Charge Transport Network Dynamics in Molecular Aggregates," *Natl. Acad. Sci.*, 2016, 113 (31), pp 8595-8600.
4. Jackson, N.E.; Kohlstedt, K.L.; Savoie, B.M.; Olvera de la Cruz, M.; Schatz, G.C.; Chen, L.X.; Ratner, M.A. "Conformational Order in Aggregates of Conjugated Polymers," *J. Am. Chem. Soc.*, 2015, 137 (19), pp 6254-6262.

582014 Fabrication of Highly Sensitive LSPR Biosensor for the C-Reactive Protein (CRP) Detection Based on the Immunocolloidal Gold Nanoparticles

Research Interests: Biological Engineering, Biosensors / Devices

Biosensor technology for the rapid and sensitive detection of disease and pathogenic bacteria generally requires the sample preparation steps such as pre-concentration and desalting, and the process of labeling with fluorescence dyes and chemical reporters. Recently, there has been a growing interest in non-labeled biological detection techniques that can be detected without additional sample preparation steps. Here, we demonstrate a simple method for the fabrication of plasmonic-active chip which are uniformly deposited in a single Au layer. The glass based biosensor chip that can be effectively SPR driven was prepared via the control of the chemical ligand concentration, temperature change, and surface modification. As such, we examined the correlation between the enhanced plasmonic effect and surface structure of the resulting substrate to determine the optimum conditions for highly sensitive LSPR sensor chip. To confirm the possibility of the non-labeled LSPR biosensor, we apply it to the quantitative detection of C-reactive protein associated with cardiovascular and renal risk factors. We report a limit of detection for CRP of 1 ng/mL. Therefore, our LSPR biosensor chip could provide a significant contribution to the ongoing efforts to fabricate non-labeled biosensor chips for fields such as diagnostic, medical, and environmental and food.

Teaching Interests: Biosensors / Devices

580705 Multiscale Biomechanics of Platelet-Driven Blood Clot Contraction and Intracellular Mechanisms of Its Termination

Research Interests and Experience:

My primary research goals are directed toward understanding the basic mechanisms in biological systems using engineering principles, computational modeling, and biophysics. As a faculty research associate and postdoctoral fellow at the University of Pennsylvania, UC Riverside, Argonne National Lab, and University of Notre Dame, I have balanced experiments and modeling approaches to study transport, mechanical and structural properties of biopolymer networks, biomechanics of blood clot retraction and interaction of cancer cells with the extracellular matrix. I have co-authored over twenty-five peer-reviewed publications, including **seventeen first-author papers**, with the most recent ones published in *Nature Communications, Biomaterials, Haematologica, Oncogene, Acta Biomaterialia, Journal of the Royal Society Interface, Matrix Biology, Soft Matter, Scientific Reports, Biomedical Optics Express, Langmuir.*

Successful Proposals: AHA Scientist Development Grant (PI), Walter Cancer Foundation Grant (PI), U01 NIH (Role: Co-I)

Teaching Experience:

As a visiting research assistant professor at the University of Notre Dame in 2012-2014, I had an opportunity to design and lecture my own interdisciplinary course of Topics in Applied Mathematics for graduate students (fall semester 2012). This course introduced graduate students to basic modeling approaches for solving various problems in biomedical engineering such as hemodynamics, tissue optics, drug delivery, characterization of biosensors as well as analysis of experimental data. I have found this an ideal opportunity to merge teaching with my current research interests. I was very pleased to teach the same course in the spring semester 2014.

Furthermore, I successfully taught large undergraduate courses on Applied Linear Algebra (Fall 2014) and Probability and Statistics (Spring 2016) at the Department of Applied and Computational Mathematics and Statistics at Notre Dame. The objective of the first course was to impart the fundamental knowledge of linear algebra and computational linear algebra that is needed to solve matrix algebra problems that arise in various applications. In the Probability course I have taught students concepts in statistics and probability with practical applications to various problems in science.

Future Directions:

Extracellular cues have a profound effect on a wide range of cellular behaviors, including motility, growth, adhesion, differentiation, apoptosis, and signal transduction. The long-term goal of my future research is to develop and use novel interdisciplinary approaches to study the interactions between cells and extracellular matrix in biological systems which importance is evident from its relation to a variety of disorders including but not limited to blood clotting and cancer as well as to the design of novel biomaterials.

One of the short-term goals would be to delineate and quantitatively characterize multiscale biomechanical and structural mechanisms of platelet-induced clot contraction. My projects will study the impact of cellular biomechanical function and metabolic state on the structural and mechanical properties of blood clots and plasma-derived biologically active biomaterials. In another project I will examine the role of biomechanical forces in cancer progression and metastasis.

Selected Publications:

1. **Kim O.V.**, Nevzorova T.A., Mordakhanova E.R., Ponomareva A.A., Andrianova I.A., Le Minh G., Daminova A.G., Peshkova A.D., Alber M.S., Vagin, O., Litvinov R.I., Weisel J.W. (2019). Fatal dysfunction and disintegration of thrombin-stimulated platelets. *Haematologica*, 104(9),1-13.
2. **Kim O.V.**, Litvinov R.I., Alber M.S., Weisel J.W. (2017) Quantitative structural mechanobiology of

platelet-driven blood clot contraction. *Nature Communications*, 8, 1274.

3. Britton, S., **Kim, O.**, Pancaldi, F., Xu, Z., Litvinov, R.I., Weisel, J.W. and Alber, M., (2019). Contribution of nascent cohesive fiber-fiber interactions to the non-linear elasticity of fibrin networks under tensile load. *Acta Biomaterialia*, 94, 514 – 523.
4. Klymenko Y.*, **Kim O.***, Loughran E., Yang J., Lombard R., Alber M., Stack M.S. (2017) Cadherin composition and multicellular aggregate dynamics in organotypic models of epithelial ovarian cancer intraperitoneal metastasis. *Oncogene*, 36, 5840–5851
5. **Kim O.V.**, Litvinov RI, Chen J, Chen DZ, Weisel JW, Alber MS. (2017) Compression-induced structural and mechanical changes of fibrin-collagen composites. *Matrix Biology*, 60, 141-156.
6. **Kim O.V.**, Liang X., Litvinov R.I., Weisel J.W., Alber M.S., Purohit P.K. (2015) *Foam-like compression behavior of fibrin networks. Biomechanics and Modeling in Mechanobiology*, 1-16.

7. **Kim O.V.**, Litvinov, R. Weisel J., and Alber M.S. (2014) Nonlinear structural mechanics of fibrin networks under compression, *Biomaterials*, 35, 6739-6749

582941 Hierarchical Self-Assembly and Biological Interactions of Functional Synthetic and Natural Supramolecular Systems

Research Interests:
Macromolecules in the biological milieu, such as peptides and proteins, play a critical role in the function of cellular systems. For example, cytoskeletal proteins control adhesion, migration, and transcellular material transport. Such functions heavily rely on the structure and assembly dynamics of the self-assembling components of the cytoskeletal networks. Taking inspiration from these natural structures, I am interested in creating organic supramolecular nanomaterials with engineered function towards cellular interactions. My work as a graduate student involved the development of one-dimensional peptidic nanostructures appended with organic electronic units, which exemplify tunable photophysical, mechanical and biochemical properties. This molecular design approach merges the biological relevance of self-assembling peptides and the optoelectronic function of organic semiconducting units—creating a bioelectronic scaffolding platform compatible with electrically-active cells such as neurons and cardiomyocytes. The findings from this work enabled new ways to effectively generate localized electric fields within the internal electronic conduits of peptide-based nanomaterials. Inspired by photosynthetic systems and by understanding the dynamic structure-function relationships for these materials, semiconducting peptide monomers that can be assembled as aligned hydrogels were successfully built according to design principles that allowed for a directed photonic energy transport, a sequential electron transport in a multicomponent system, and the transmission or equilibration of voltage or current when incorporated in a transistor device. These soft scaffolding materials, with rationally tunable properties, offer a unique tissue engineering platform towards locally delivering electronic, topographical and biochemical cues to cells.

Currently, in my postdoctoral work, I am continuing to investigate the multi-scale interactions of nanomaterials with biological systems. We are particularly exploring the nano-bio interface by understanding how engineered nanomaterials interact with biological barrier tissues (such as the endothelium and dermal layers) and how the biochemically entangled supramolecular networks of cytoskeletal proteins in barrier tissues reorganize in response to nanomaterial exposure. Together with these investigations, our team continuously develops new platforms to assess the impacts of these nanomaterials to tissue function, such as by using "organs-on-chips" systems and utilizing label-free microscopy techniques. Moving forward, as an independent researcher, my goal is to build a research group that will pioneer next generation
functional supramolecular assemblies that can be used as "adaptive probes" for monitoring biological investigations. Beyond the capability of soft organic materials to serve as biocompatible substrates for regenerative medicine applications, I am interested to utilize such functional nanomaterials based on synthetic or natural biomacromolecules as tools for investigating cytoskeletal reorganization processes and cellular signaling events mediated by ion channels. From the library of compounds that will be built, we will engineer appropriate organic nanomaterial-based sensors coupled with device platforms that can enable the investigations on various dynamic biological phenomena at multiple spatiotemporal scales. Developing bioinspired nanomaterials for investigating biological mechanisms and processes opens the possibilities for unraveling new concepts governing cellular mechanotransduction and interactions at the biotic-abiotic interface.

Successful Proposals/ Fellowships: Irving S. Sigal Postdoctoral Fellowship (American Chemical Society), International Student Research Fellowship (Howard Hughes Medical Institute), Faculty for the Future Fellowship (Schlumberger Foundation)

Postdoctoral Supervisor: Prof. Kevin Kit Parker, Disease Biophysics Group, John A. Paulson School of Engineering and Applied Sciences, Wyss Institute for Biologically Inspired Engineering, Harvard University, Cambridge, MA

Ph.D. Supervisors: Prof. John D. Tovar, Department of Chemistry; and Prof. Hai-Quan Mao, Department of Materials Science and Engineering, Translational Tissue Engineering Center (TTEC), and Whitaker

Biomedical Engineering Institute, Johns Hopkins University, Baltimore, MD

Teaching Interests:

Throughout my academic career, I have served in multiple classroom teaching and laboratory mentoring posts. I was as a laboratory instructor in University of the Philippines-Diliman and a teaching assistant in Johns Hopkins, both for undergraduate introductory and organic chemistry laboratory and lecture classes for chemistry majors and non-majors. I was also as a guest instructor at Harvard for a cellular engineering course for engineering majors. Aside from classroom-setting teaching, I have an extensive undergraduate and graduate student research mentoring experience in graduate school and as a postdoctoral researcher. Four of these undergraduate students were co-authors in publications that resulted from the research projects that we worked on. I have also participated as a mentor in a multitude of STEM-related outreach activities involving students across different elementary to high school levels. These collective teaching and mentoring experiences equip me to teach graduate- and undergraduate-level courses in organic materials/biomaterials, which is my area of research expertise. I am also interested in teaching courses which are related to fundamentals of materialschemistry to nanotechnology. In addition, I would welcome the opportunity to develop new courses, particularly on topics that involve supramolecular materials, optoelectronics, as well as and their applications in engineering and biomedical sciences. I plan to promote an active project-based learning environment for my potential students in these courses. Importantly, I will ensure a learning-centered teaching/ problem-based learning and an inclusive pedagogy practice that aims to foster the best environment where students can excel.

583369 Genetically Engineered Probiotics Designed at the Interconnect of Synthetic Biology and Metabolic Engineering

My proposed research platform will focus on three areas: 1) developing new genetic tools for cellular communication in prokaryotes and eukaryotes; 2) establishing new hosts for the production of bioactive compounds and 3) unifying my investigations in dynamic regulation and natural product synthesis towards the application of genetically engineered probiotics to treat gastrointestinal diseases.

During my PhD, I established a foundation in synthetic biology, a rapidly advancing discipline that impacts many aspects of society. With my background in mathematical modeling during my MS in chemical engineering, I sought to bring engineering rigor to my studies in cellular communication under the guidance of Dr. William Bentley. I elucidated the mechanisms in intra- and interkingdom communication systems and leveraged this understanding to genetically engineer bacteria for bioengineering and biomedical applications.

As a postdoctoral scholar in Dr. Jay Keasling's laboratory, I have learned to design and manipulate the enzymes and pathways that lead to high-value products. I was awarded the NIH F32 National Research Service Award (NRSA) fellowship to develop an in-depth expertise in polyketide synthases (PKSs), a modular assembly-like enzyme that is responsible for the production of many pharmaceutical drugs. Informed by phylogeny and based in chemoinformatics, my work has advanced a new paradigm in manipulating these enzymatic pathways, and we have applied this knowledge to programmatically produce high-value chemicals in *Streptomyces albus*.

By marrying my graduate school expertise of dynamic regulation with my knowledge in metabolic pathways, I am developing a generalizable platform to produce disease-fighting molecules in the gastrointestinal system. Through a just-in-time release mechanism, genetically engineered probiotics will release therapeutics only at the disease site and only when sufficiently accumulated. I submitted an NIH K99 Transition to Independence application as a therapy to treat colon cancer, which would represent a vertical advance in cancer probiotic therapy.

Teaching Interests

As the product of universities that excel in both education and research, I have a deep appreciation for the role these twin aims in the mission of academia. My experiences together as a student and educator have helped me develop a teaching style that has been efficient and effective in both the classroom and laboratory. My basic educational beliefs can be distilled down to two distinct goals: (1) to inspire students and convey the fundamental and eventually more specialized aspects of science and engineering; and (2) to prepare students for a successful career that fosters independent thinking, ingenuity and effective implementation of the scientific process.

My diverse background in chemical and biomolecular engineering enables me to explain concepts and provide context in a multidisciplinary manner to retain the interest of diverse audiences. I am comfortable teaching undergraduate courses in traditional chemical engineering topics such as chemical kinetics and thermodynamics. and in bioengineering courses such as biofluids and genetic circuits. My foundation in these broad topics will allow me to provide a unique perspective on education in this interdisciplinary field.

Selected Publications (22 total, 12 first-author, 1 corresponding)

1. **Amin Zargar***, S Chang, AK Kothari, AM Snijders, JHua Mao, T Bivona, JD Keasling, "Overcoming the challenges of cancer drug resistance through bacterial-mediated therapy", (Chronic Diseases and Translational Medicine, submitted) ***Corresponding author**
2. **Amin Zargar**, JF Barajas, R Lal, JD Keasling. "Polyketide synthases as a platform for chemical product design", *AICHE Journal*, (2018
3. Jesus Barajas, **A Zargar**, B Pang, V Benites, J Gin, E Baidoo, C Petzhold, N Hillson, JD Keasling,

"Biochemical Characterization of b-Amino Acid Incorporation in Fluvirucin B2 Biosynthesis", *ChemBioChem* (2018).

4. **Amin Zargar**, DN Quan, N Abutaleb, E Choi, WE Bentley, "Biocompatible capsules to control quorum sensing", *Biotechnology & Bioengineering* (2016)
5. **Amin Zargar**, DN Quan, WE Bentley, "Reducing stochasticity in cell populations: rewiring quorum sensing to engineer dynamic gene expression with population-level control", *ACS Synthetic Biology* (2016)
6. **Amin Zargar**, DN Quan, M Emamian, CY Tsao, HC Wu, CR Virgile, WE Bentley, "Rational design of 'controller cells' to manipulate protein and phenotype expression", *Metabolic Engineering* (2015). 30: p. 61-68.
7. **Amin Zargar**, DN Quan, KK Carter, M Guo, HO Sintim, GF Payne, WE Bentley, 'Bacterial secretions of nonpathogenic *E. coli* elicit inflammatory pathways: a closer investigation of interkingdom signaling", *mBio* (2015). 6:2 e00025-15.

583400 Quantitative Single-Cell Analysis of RNA Regulation at the Single Molecule Level

Key words: Interdisciplinary (Systems Biology, Biophysics, Bioengineering, Deep learning, etc.), quantitative, single molecule, single mammalian cells

Research Interests:

I am passionate about **quantitatively** investigate important but previously inaccessible **molecular and cellular biology** questions, *in vitro* and *in vivo*, at the **single molecule level** in **individual mammalian cells**. My **interdisciplinary** background allows me to integrate diverse approaches (from molecular biology to deep learning, from time-lapse microscope to mammalian cell engineering, from photonics to integrated circuits like Arduino, etc.) to **develop more sensitive and powerful single molecule tools** to probe fundamental biological questions.

Selected Graduate Work: Single molecule manipulation --- in David Bensimon lab (ENS de Paris)

Motivation: Base-pairing sequence-structure relationship is essential for gene regulation. In particular, the competition between a hairpin and an adjacent hybrid is involved in many significant biological processes, such as replication arrest, CRISPR recognition, DNA origami, etc. However, due to the technical difficulty to monitor this dynamic competition process, the governing rules remains unclear.

Developed toolbox: I developed a method to characterize the dynamics of adjacent hybrids by monitoring the extension of (hybrid-blocked or closed) hairpin, using magnetic tweezers.

Achievement 1: My work provided a foundation to understand base-pairing competition. More specifically, I systematically quantified the dynamics of adjacent hybrid displacement at various conditions and proposed a stochastic step-by–step model to describe underlying mechanism.

Achievement 2: My work invented a novel 4th generation sequencing method, providing the impetus and foundation for a biotech startup, Depixus@. The first commercial platform will be launched next year.

Selected Postdoc Work: Single cell RNA splicing regulation --- in Michael Elowitz lab (CalTech)

Motivation: In eukaryotes, splicing is ubiquitous. During splicing, nascent pre-mRNAs are processed to remove introns and/or exons. A key function of this splicing process is the ability to increase gene-product diversity by generating multiple distinct isoforms from a single gene, through alternatively splicing particular exons and(or) introns. However, the majority of individual introns (and exons) are constitutively spliced. It has remained unclear what functional roles such constitutive splicing provides.

Developed toolbox 1: I developed a method to measure splicing efficiency at the transcriptional active site in individual cells, using multiple-channel intensity-based single-molecule fluorescence in-situ hybridization (smFISH), by adapting skills from stellar photometry in crowded star fields.

Developed toolbox 2: I developed a label-free automatic segmentation pipeline using deep learning.

Achievement 1: My quantitative single-cell splicing analysis characterized the dependence of splicing efficiency on transcription level in individual cells and revealed that splicing exhibited an unexpected 'economy of scale' behavior. These results indicate constitutive splicing as a non-linear filter that amplifies the expression of genes when they are more strongly transcribed, playing an active functional role in modulating gene expression.

Achievement 2: I am collaborating with two other labs and using the developed methods to quantify splicing efficiency within the proximity of long noncoding RNA Malat1 (with Mitchell Guttman's lab at CalTech), and compare allele-specific transcription level within the proximity of XIST (X-inactive specific transcript) (with Katherine Plath's lab at UCLA), both at the single cell level.

Future directions:

Motivation: Single-cell RNA study has revealed many previously hidden or unappreciated events in population-based research, like the significance of cell-cell heterogeneity during embryonic and immune systems, the fundamental transcriptional noise in most cell types, etc. However, many important RNA species (e.g. circular RNAs, RNA modifications) remains challenging to study at the single cell level. This is because those RNAs have either too short length of specific sequences or too small expression in individual cells, causing technical difficulty for quantitative measurements. In my future lab, I aim to develop a new single-molecule RNA detection toolbox, identify those currently inaccessible RNA species and investigate their regulation at the single cell level.

Why me: My quantitative interdisciplinary background puts me in an optimal position to execute the proposed research program. My PhD training (base-pairing molecular programming *in vitro*) allows me to design reliable DNA probe-set combination, which is the foundation to develop the new RNA detection toolbox. My Postdoc work (quantifying single-cell splicing efficiency by smFISH) equips me with extensive experiences of smFISH technique and knowledge of single cell RNA regulation. More generally, my interdisciplinary training experience (including statistical physics, bioengineering, molecular biology, synthetic biology, machine learning, etc.) allows me to connect modeling and experimental systems in my work and communicate/collaborate with experts from different background.

Teaching Interests:

My interdisciplinary background allows me to teach or co-teach a broad range of topics, including biology-related interdisciplinary courses (e.g. Bioengineering and Biophysics, Computational Biology, Systems Biology), core biology courses (e.g. Molecular Biology, Genetics, Cell Biology), and many other math or physics courses that are useful for studying biological questions (e.g. Statistics, Mathematical Methods in Physics and Engineering, Classical Mechanics, Programming for Beginners [C, Python, MATLAB, Deep Learning], Optics). I can also teach advanced courses that merge modeling and experiments, with topics drawn from modern research, including but not limited to single molecule manipulation, quantitative single cell analysis, and novel bioengineering tools.

Education and Training:

1. California Institute of Technology, Postdoc, in Biology and Bioengineering, Pasadena, CA
2. École Normale Supérieure de Paris, Ph.D. in Biophysics, Paris
3. Massachussetts Institute of Technology, Master internship in Bioengineering, Boston, MA
4. École Normale Supérieure de Paris, Licence3, in Physics, Paris
5. Nanjing University, B.S. in Physics (rank 1st out of 221), Nanjing, China

Invention Disclosures (**Depixus @ has licensed all patents**) :

1. Bensimon D, Croquette V, Allemand JF, Manosas M, **Ding F**, Method of DNA sequencing by polymerization, EP2390350A1, CA2800637A1, WO2011147929A1, EP2576822A1, CN103097551A, US20130171636A1, EP2576822B1
2. Bensimon D, Allemand JF, Manosas M, **Ding F**, Croquette V, Method of DNA sequencing by hybridization, EP2390351A1, CA2800639A1, WO2011147931A1, EP2576818A1, CN103052718A, US20130137098A1, et al.
3. Bensimon D, Allemand JF, **Ding F**, Croquette V, Method of DNA detection and quantification by single-molecule hybridization and manipulation, CA2859913A1, WO2013093005A1, EP2794915A1, US20150031028A1, et al.
4. Bensmion D, Croquette V, Gouet G, Allemand JF, **Ding F**, Process for detection of DNA modifications and protein binding by single molecule manipulation, CA2898151A1, WO2014114687A1

Selected Publications:

1. **Ding F**, Elowitz M, Constitutive splicing and economies of scale in gene expression, ***Nature Structural & Molecular Biology***, (2019), 26, 424–432.
2. **Ding F**, et al., Single cell dynamics and functions of negative autoregulatory splicing, ***elife*** (in process)
3. **Ding F**, et al., Displacement and dissoci-ation of oligonucleotides during DNA hairpin closure under strain, ***Nature Chemistry*** (in process)
4. **Ding F***, Levine JH*, Yi D, Linton J, Elowitz M, Label-free cell segmentation and tracking in various cell lines and imaging conditions by deep learning (in preparation)
5. **Ding F**, et al., Single-molecule mecha-nical identification and sequencing, ***Nature Methods***, (2012) 9(4), 367-372. (**News and Views**: S Linnarsson, ***Nature Methods***)
6. Manosas M, Spiering MM, **Ding F**, Bensimon D, Allemand JF, Benkovic SJ, Croquette V, Mechanism of strand displacement synthesis by DNA replicative polymerases, ***Nucleic Acids Research***, (2012) 40 (13), 6174-6186.
7. Manosas M, Spiering MM, **Ding F**, Croquette V, Benkovic SJ, Collaborative coupling between polymerase and helicase for leading-strand synthesis, ***Nucleic Acids Research***, (2012) 40 (13), 6187-6198.
8. Meglio A, Praly E, **Ding F**, Allemand JF, Bensimon D, Croquette V, Single DNA/protein studies with magnetic traps, ***Opin. Struct. Biol.*** (2009) 19(5):615-22.
9. Ferrer J*, **Ding F***, Tarsa P, Brau R, Lang M, Interlaced Optical Trapping and Fluorescence Excitation improves fluorophore longevity, ***Pharm. Biotechnol.*** (2009) (5): 502-7. (*: authors contributed equally to publication)
10. **Ding F**, et al., Discussion and modification of experiment of measuring the heat conduction of coefficient of gas based on the hot-wire method, ***Physics Experimentation*** (2004) 24(12):39-41

583420 Theory and Modeling of Biopolymers and Bio-Inspired Soft Materials

Research interests:

My research concerns the understanding of biopolymers and the design of bio-inspired soft materials using theoretical and computational tools. Life is largely materialized by biopolymers such as proteins and DNA. The understanding of these biomolecules and their interactions, as well as their responses to environment is not only crucial for human health, but also a great source to inspire the development of novel artificial materials aiming to solve real-world problems such as water purification and clean energy. While synthetic polymers have been widely used in electronics, medicine and energy devices, there is still a huge gap between artificial and biological employments of polymeric materials. The performance of synthetic polymers is often limited by the lack of delicate molecular design and thence precise control of the functional structures. My research endeavor will be devoted to the optimization of soft materials by learning from nature. I will apply my multiscale modeling skills to investigate complex biological processes such as chromatin folding and nucleocytoplasmic transportation, and to extract guiding principles for the rational design of smart artificial materials.

Research experience:

- Modeling of the three-dimensional genomic organization
- Theory of contact scaling for random fractal polymers
- Molecular theory of the nuclear pore complex
- Design of smart artificial nanopores/nanochannels
- Investigation of the curvature effect on the micro-phase-separation of end-tethered polymers
- Enhanced sampling of free energy landscape of patterning polymers
- Non-equilibrium statistical theory of liquid-solid friction and flow boundary
- Simulation of hydration, hydrophobic interaction, and specific ion effects
- Simulation of Brownian motion and diffusion near liquid-solid interfaces
- Simulation of acoustic biosensors

Teaching Interests:

With abroad educational background covering Engineering Physics, Materials Science, and Biomedical Engineering, I will be comfortable with teaching many undergraduate-level courses including Thermodynamics, Statistical mechanics, Fluid mechanics, Quantum mechanics, Polymer physics, Numerical methods and molecular modeling. I am also interested in developing graduate-level courses focusing on specific topics based on my research experience. One such course could be "Bio-inspired materials", in which I will deliver students with recent breakthroughs in this growing field, and the exciting advancements for practical applications.

Selected Publications:

1. Huang*, Y. Li, A. Shim, R. Virk, V. Agrawal, A. Eshein, R. Nap, L. Almassalha, V. Backman* and I. Szleifer*, "Physical and data structure of 3D genome", ***bioRxiv***, doi: https://doi.org/10.1101/596262 (2019), under review of ***Science Advances*** (*co-corresponding author)
2. Huang, M. Tagliazucchi, S.H. Park, Y. Rabin and I. Szleifer, "Molecular model of the nuclear pore complex reveals a thermoreversible FG-network with distinct territories occupied by different FG motifs", ***bioRxiv***, doi: https://doi.org/10.1101/568865 (2019), under review of ***Biophysical Journal***
3. Tagliazucchi, K. Huang and I. Szleifer, "Routes for nanoparticle translocation through polymer-brush-modified nanopores", ***Journal of Physics: Condensed Matter***, 30, 274006 (2018)
4. Huang and I. Szleifer, "Design of multifunctional nanogates in response to multiple external stimuli using amphiphilic copolymer", ***Journal of the American Chemical Society***, 139, 6422-6430 (2017)

5. Huang, S. Gast, C. D. Ma, N. Abbott and I. Szlufarska, "Comparison between the free and immobilized ion effects on hydrophobic interactions: a molecular dynamics study", ***The Journal of Physical Chemistry B***, 119, 13152-13159 (2015)
6. Huang* and I. Szlufarska*, "Effect of interfaces on nearby Brownian motion", ***Nature Communications***, 6:8558 doi: 10.1038/ncomms9558 (2015) (*co-corresponding author)
7. Huang* and I. Szlufarska*, "Green-Kubo relation for friction at liquid-solid interfaces", ***Physical Review E*** 89, 032119 (2014) (*co-corresponding author)
8. Huang and I. Szlufarska, "Friction and slip at solid-liquid interface in vibrational systems", ***Langmuir*** 28, 17302 (2012)

582166 Understanding Physically Crosslinked Polymer Networks to Rationally Design Hydrogel Biomaterials

Research Interests:

Structure-property relationships in physically crosslinked polymer networks are poorly understood. This makes it extremely challenging to rationally design and engineer hydrogels for injectable drug delivery, 3D bio-printing, and self-healing applications. Current design approaches are limited to costly and time consuming trial and error strategies. Using my interdisciplinary training in engineering design and biomaterials, my research will be focused on (1) investigating the fundamental structure-property relationships of physically crosslinked polymer networks and (2) creating new design paradigms for developing hydrogel biomaterials. Physically crosslinked polymer networks have complex structures and relaxation dynamics that endow them with unique rheological and diffusive material properties (functions). Designing the rheological behavior (including solid to liquid transitions –yielding–, dynamic network formation, relaxation times, and viscoelasticity), as well as the diffusion in these materials is currently ineffective. I will measure the effects of composition including physical group modifications, polymer molecular weight, and concentration on the rheological and diffusive materials properties (functions) to understand structure-property relationships. For this purpose, I will employ linear and non-linear rheological techniques to probe viscoelastic and flow behavior of physically crosslinked hydrogels. Diffusion measurements, using fluorescence recovery after photobleaching and microrheology, will be used to explore the diffusive behavior of the polymer networks which impacts the macroscale materials behavior. Initial research thrusts in my group will focus on establishing structure-property relationships for hydrogel biomaterials, creating design paradigms for injectable physically crosslinked systems, and developing novel associative self-healing materials platforms.

Research Experience:

Designing Polymer-nanoparticle hydrogels for Drug Delivery

Department of Materials Science and Engineering, Stanford University (advised by Eric Appel)

My postdoctoral research has focused on investigating the rheological behavior of physically crosslinked hydrogels and on designing them as material solutions for complex challenges. Specifically I investigated polymer-nanoparticle hydrogels (PNPs), crosslinked by dynamic hydrophobic interactions, and their broad applicability in adhesion prevention after myocardial infarction, wildfire prevention, drug-delivery, and cell transplantation. While these applications are seemingly disparate, careful analysis reveals that they have similar design criteria, benefiting from a material capable of dynamic rearrangement (flow) followed by localization and permanence. As yield stress fluids, PNPs are capable of a solid to liquid (yielding) transition when sheared followed by a rapid recovery to a solid and were demonstrated to satisfy the design requirements of each application. I investigated (1) how processes can impose constraints on potential materials solutions and (2) how the hydrophobic modifications affect shear-thinning and dynamic network formation for PNPs. For example, we found that injectable drug delivery strategies impose design constraints on the viscosity, shear thinning, network formation kinetics, and diffusion of the material being injected. Adhesion prevention imposes constraints on the yield stress and sprayability of the material being used. A non-monotonic relationship between clinical efficacy and yield stress was observed. Large amplitude oscillatory shear was used measure the effects of composition and hydrophobic modification on the non-linear viscoelastic response, yielding, and formation kinetics of the PNPs.

Development and Characterization of Poly(phthalaldehyde) for Transient Applications *Department of Mechanical Science and Engineering* (advised by Scott R. White)

My thesis research developed a novel autonomous material platform capable of disintegrating on-demand, enabling the invention of solid state transient (destructible) polymer films and devices. Metastable polymers that depolymerize in response to environmental stimuli to change shape, form, or function have garnered increased research interest with applications in signal amplification, temporary electronics, recycling, and drug delivery. These *transient* applications take advantage of the large transformation in properties that accompany the depolymerization of high molecular weight polymers into their monomeric components. My PhD research consisted of a detailed investigation of the metastable polymer cyclic polyphthalaldehyde (cPPA). Depolymerization of cPPA was controlled by incorporating light and thermal responsive agents that destabilize the cPPA upon exposure to the corresponding stimuli. cPPA was triggered to depolymerize into its monomeric component by exposure to heat or acid. Its degradation in response to these stimuli was chemically characterized with fourier transform infrared spectroscopy, confocal Raman spectroscopy, and NMR. The effects of degradation on the mechanical performance was explored using dynamic mechanical analysis. With precise control over the triggered depolymerization of cPPA, transient electronic devices were created that could be destroyed on-demand. Thermal triggering was shown to volatilize the transient substrates, leaving only 15% of its mass as residue after thermal triggering. Furthermore, we investigated a long standing challenge for the synthesis and stabilization of cPPA by investigating the effects of initiator concentration on film stability, resulting in a repeatable and reliable synthesis of cPPA that could be thermoformed to create bulk three dimensional cPPA plastic components.

Selected Publications:

- L. M. Stapleton, A. N. Steele, H. Wang, **H. Lopez Hernandez** et al. "An Effective Post-Operative Pericardial Adhesion Barrier Utilizing Polymer-Nanoparticle Hydrogels". Nature BME. 2019 *In Press.*
- E. M. Lloyd, **H. Lopez Hernandez**, A. M. Feinberg, M. Yourdkhani, E. K. Zen, E. B. Mejia, N. R. Sottos, J. S. Moore, S. R. White. "Fully Recyclable Metastable Polymers and Composites". Chem. Mater. 2019, 31, 398.
- **H. Lopez Hernandez**, S. K. Takekuma, E. B. Mejia, C. L. Plantz, N. R. Sottos, J. S. Moore, S. R. White. "Processing-dependent mechanical properties of solvent cast cyclic polyphthalaldehyde." Polymer. 2019, 162, 29.
- **H. Lopez Hernandez**, A. K. Grosskopf, L. M. Stapleton, G. Agmon, E. A. Appel. "Non-Newtonian Polymer-Nanoparticle Hydrogels Enhance Cell Viability during Injection". Macromol. Biosci. 2019, 19, 1800275.
- **H. Lopez Hernandez**, O. P. Lee, C. M. Possanza Casey, J. A. Kaitz, C. W. Park, C. L. Plantz, J. S. Moore, S. R. White. "Accelerated Thermal Depolymerization of Cyclic Polyphthalaldehyde with a Polymeric Thermoacid Generator". Macromol. Rapid Commun. 2018, 39, 1800046.
- **H. Lopez Hernandez***, A. M. Feinberg*, C. L. Plantz, E. B. Mejia, N. R. Sottos, S. R. White, J. S. Moore. "Cyclic Poly(phthalaldehyde): Thermoforming a Bulk Transient Material". ACS Macro Lett. 2018, 7, 47.
- O. P. Lee, **H. Lopez Hernandez**, J. S. Moore. "Tunable Thermal Degradation of Poly(vinyl butyl carbonate sulfone)s via Side-Chain Branching". ACS Macro Lett. 2015, 4, 665.
- C. W. Park, S.-K. Kang, **H. Lopez Hernandez**, J. A. Kaitz, D. S. Wie, J. Shin, O. P. Lee, N. R. Sottos, J. S. Moore, J. A. Rogers, S. R. White. "Thermally Triggered Degradation of Transient Electronic Devices". Adv. Mater. 2015, 27, 3783.
- **H. Lopez Hernandez**, S.-K. Kang, O. P. Lee, S.-W. Hwang, J. A. Kaitz, B. Inci, C. W. Park, S. Chung, N. R. Sottos, J. S. Moore, J. A. Rogers, S. R. White. "Triggered Transience of Metastable Poly(phthalaldehyde) for Transient Electronics". Adv. Mater. 2014, 26, 7637.

* Equal Contributions

Selected Awards:

2018 Stanford Polymer Collective Art of Science Award

2017 -- 2018 NSF AGEP Postdoctoral Fellowship (California Alliance)

2016 -2017 Mavis Future Faculty Fellowship

2016 Mechanical Science and Engineering Teaching Fellow

2016 1st Place Poster at SACNAS Conference

2015 -2016 Beckman Institute Graduate Research Fellowship

2015 1st Place Graduate Student Presentation Award at SACNAS Conference \\

2015 Academic Excellence and Good Citizenship Award, University of Illinois Graduate College

Teaching Interests:

I am excited about the opportunity to teach and mentor up and coming classes of engineers. I have been active in my mentorship training, mentoring several undergraduates during my graduate and postdoctoral research positions. I have also pursued formalized teaching training at the University of Illinois and Stanford University, where I am pursuing the teaching certificate. As mechanical engineer, I am excited and qualified to teach heat transfer, thermodynamics, mathematics and fluid mechanics. I am also willing and capable of teaching other core curriculum courses. I am looking forward to developing and teaching courses in polymer physics, rheology, and polymer science and engineering (solid state). In the laboratory, I am looking forward to mentoring my postdoctoral researchers, graduate student researchers, and also any undergraduates that are searching for research experience. I will help prepare them for their career path, providing them with the resources and guidance needed to achieve their goals. My laboratory will take a multidisciplinary approach, using principles from engineering, chemistry, biology, physics, and materials science to tackle our research questions.

Teaching Experience:

- Stanford Teaching Certificate, *in progress*
- Guest Lecturer, *Organic and Biological Materials, Entropic Elasticity and Rheology Lectures.* MATSCI 210. Department of Materials Science and Engineering, Stanford University, Spring 2019.
- Guest Lecturer, *Organic and Biological Materials, Rheology Lectures*, MATSCI 210. Department of Materials Science and Engineering, Stanford University, Winter 2018.
- Guest Lecturer, *Organic and Biological Materials, Rheology Lectures*, MATSCI 210. Department of Materials Science and Engineering, Stanford University, Spring 2018.
- *Teaching and Leadership Course*, ENG 598. University of Illinois at Urbana-Champaign, Spring 2017.
- Teaching Fellow Lecturer, *Experimental Stress Analysis*, Lecture and Laboratory Course, TAM 456. Department of Mechanical Science and Engineering, University of Illinois at Urbana-Champaign, Spring 2016.

582171 Biocompatible Zwitterionic Polymers for Medical Applications

Research Interests:

Polymer biomaterials have been extensively used in various biomedical applications, such as (i) increasing the biocompatibility of implantable tissue-/blood- contacting artificial medical devices, (ii) engineering functional tissues and promoting the healing process in tissue regeneration, and (iii) providing spatial and temporal control over the release of therapeutic agents in drug delivery and nanomedicine. Conventional polymer biomaterials such as poly(2-hydroxyethyl methacrylate) (poly(HEMA)) and poly(ethylene glycol) (PEG) have been investigated for decades and their potential issues have gradually emerged. As novel alternatives, bioinspired zwitterionic polymer materials such as poly(2-methacryloyloxyethyl phosphorylcholine) [poly(MPC)], poly(carboxybetaine methacrylate) [poly(CBMA)], and poly(sulfobetaine methacrylate) [poly(SBMA)] show unique and prominent performance. With 10 years of research experience in the zwitterionic polymer field and a Ph.D. in Materials Engineering from The University of Tokyo (Japan) under the guidance of Prof. Kazuhiko Ishihara and a postdoctoral training in the Department of Chemical Engineering in the University of Washington (USA) under the guidance of Prof. Shaoyi Jiang, I am always seeking to design novel zwitterionic materials, aiming to further develop promising healthcare technologies and deliver cutting-edge biomedical breakthroughs. My main research interests include, but not limited to the following areas:

- *Biocompatible surface zwitterionization of tissue-/blood- contacting medical devices*

I started my research with exploring the nonfouling mechanisms of zwitterionic polymers at the molecular level. The project involved developing, evaluating, and optimizing super-hydrophilic zwitterionic biocompatible polymers to serve as antifouling coatings for various biomedical devices, proving healthcare professionals with innovative tools for addressing medical needs without posing additional risks to patients. Medical implements come into contact with human blood and other fluids, as well as non-human cells, tissues, and proteins on a regular basis. Traditional tools and devices containing hydrophobic surfaces could cause negative biological reactions in the patient being treated, while also degrading the performance and lifespan of medical implements over time. I have developed super-hydrophilic zwitterionic polymers that are especially useful as coatings for medical implements, as they significantly reduce the number of undesirable reactions that traditionally occur between medical instruments, especially implantable devices, and the patient's surrounding biological environment.

- *3D crosslinked zwitterionic polymer hydrogel bioelectronics for regenerative medicine*

Polymer hydrogels infiltrated with high water contents have been extensively studied in tissue engineering due to i) resemblance to biological tissues and ii) tunable physical modules to minimize mechanical mismatch with various biological tissues. Biocompatible zwitterionic polymer hydrogels with remarkable flexibility in the design of their electrical, mechanical, and biological properties render hydrogels a unique bridging material to the biological world. In this project, I have developed two unique hydrogels: (a) a redox-active polymer hydrogel able to reversibly and spontaneously encapsulate living cells while preserving high biocompatibility, providing a mild environment to explore the fate of encapsulated stem cells/cancer cells/bacteria for regenerative medicine/cancer therapy/biofuel cells/biosensor under electrical stimulations; (b) a biocompatible, stretchable, conformable and highly electronic conductive zwitterionic hydrogel with tunable mechanical strength serve as an implanted artificial tissue with no foreign body reaction to alternate injured tissue and/or promote/sensor the healing of injured tissue via electrical signals. The 3D printing technology will be used to rapidly manufacture personalized tissue engineering scaffolds and even directly print functional tissue and organs.

- *Zwitterionic polymer nanoparticles as non-invasive vectors for nanomedicine*

This project includes understanding of fundamental interaction mechanisms between zwitterionic nanomaterials and lipid-bilayer membrane of living cells, and it has involved molecular-level design of a zwitterionic nanoprobe to monitoring the dynamic distribution of biomolecules (e.g., mRNA) in living cells in real time which could aid in the early detection of cell pathogenesis. Standard *in vitro* assays such as DNA microarrays, reverse transcription polymerase chain reaction (RT-PCR), and Northern blotting can quantify changes in gene expression levels of a cell population. However, none of these methods can be applied to living cells. Cells must be lysed to extract specific macromolecules for these procedures and thus can only offer a static view of cellular events at a given moment. Therefore, to overcome the limitations and enable the live imaging of mRNA in cells, I have developed a non-invasive nanoprobe through the conjugation of zwitterionic polymer and molecular beacons with a special designed nucleotide base sequence. This new material can provide: (a) a quick non-invasive cell membrane penetration in minutes which is faster than any other polymer biomaterials found so far; (b) a high degree of sensitivity as well as high spatial and temporal resolution; (c) remarkable flexibility in the design of base sequences to specific binding other molecules *in vivo*.

Teaching Interests:

With my education background and research experience, I am capable of teaching core courses at both undergraduate and graduate levels, including Chemical Engineering Principles, Chemical Engineering Laboratory, Thermodynamics and Kinetics, Transport Processes, Surface and Colloid Science, and Surface Analysis. In addition, I would be delighted in teaching the following courses as well: Chemical Kinetics and Reactor Design, Advanced Surface Analysis, Biomaterials/Nanomaterials, Polymer Chemistry & Physics, Principles of Molecular Engineering, Polymer Processing & Characterization, Polymer Science and Engineering, and Electrochemical Engineering. Furthermore, I would like to develop courses at the intersection of chemical engineering, materials science and engineering, and bioengineering, such as Bioinspired Design of Materials, Advanced Topics in Biomaterials, and Engineering Materials for Biomedical Applications.

I will create an active classroom/lab for my students and encourage students across diverse backgrounds to discuss each other to inspire their learning. I will set aside regular office hours to help my students on their study, scientific research and campus life. In addition, undergraduate students who are interested in scientific research will be encouraged to enter the lab to practice what they have learned in class or start a co-project with seniors under my supervision. For graduate students, I will create a lab environment fulfilled with curiosity to stimulate their research interest. I will also provide them with plenty of opportunities to attend domestic and international conferences. I believe that my passion, endeavor, experience and international education background will inspire me to motivate and mentor the next generation of students in their careers.

581613 Electrochemistry Enabling New Frontiers in Biomaterials Engineering

Electrochemistry enabling new frontiers in biomaterials engineering

Sina Jamali

Mechanical, Materials, Mechatronic and Biomedical Engineering, University of Wollongong, Australia

Research Interests:

Owing to a multi-disciplinary research and education background and completion of two PhDs in UK and Australia, my broad research interests are in areas of polymeric materials, biomedical engineering, electrochemical sensing, and surface engineering. In particular, I am fascinated by the application of electrochemistry in studying how biomaterials interact with a biological system or how cells and organs operate and communicate in the body. Understanding the physiological processes and chemical interactions at the cellular level is the gateway to future medical advancements, such as improving on the functionalities of biomedical implants, designing more effective drugs and drug delivery systems or new frontiers in regenerative medicine.

As a doctoral student and a postdoctoral research fellow, I have implemented advanced electrochemical techniques in the field of biomedical engineering and materials for in vitro detection and interactive mapping of analytes, for characterization of electro-active interfaces with the purpose of neural cells growth stimulation and for degradation monitoring of biomedical implant materials and functional surface coatings. I also pioneered a method for detection and mapping of hydrogen on surfaces through electro-oxidation of hydrogen which has been widely adopted by others and applied to hydrogen sensing in simulated physiological systems as well as energy devices.

The ultimate aim of my research is to advance a versatile platform for highly sensitive electrochemical bio-sensing, from data acquisition to signal processing, that can be adapted to monitor biochemical processes In Vitro and In Vivo in a variety of applications. Electro-active sensors can be used in monitoring electrically signalling disorders and analytes in a metabolic system while applications in the field of targeted drug delivery are endless. With the ever-growing market of biomedical implants, the tremendous projected investment by the government and industry in this sector and the demand for advanced multi-functional medical implants, my research will make a tangible contribution toward such advancements.

A review of my recent works (summarized in the image below) on chemically and electrochemically interactive characterization of biomaterials in vitro will be the focus of my oral presentations at this meeting.

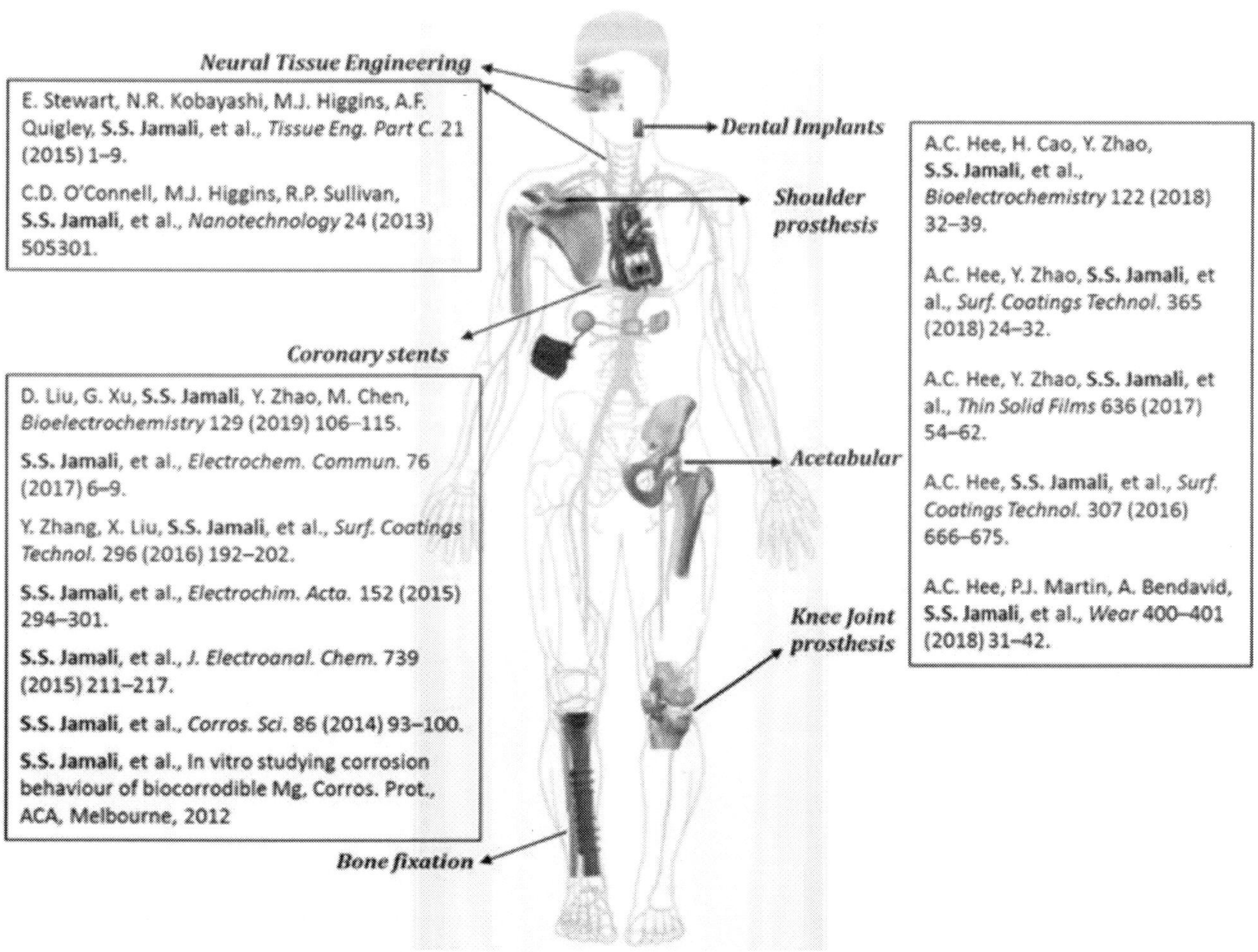

Teaching Interests:

I am passionate about teaching a variety of courses as a faculty member, in particular, General and Organic Chemistry, Surface and Interface Sciences, Electrochemistry, Biomaterials, Wear and Degradation, and Polymer Science and Engineering. I have been teaching Surface Engineering as well as Wear and Degradation subjects at the undergraduate and graduate levels which also involved curriculum development and gained me outstanding teaching evaluation records. As a passionate public speaker, I value practical and engaging communication of science in the classroom. I exploit techniques such as experimental demonstrations (where safe and applicable), interactive online lecture contents, field trips and lab visits, guest lectures, and more, to engage students with the content and deliver an effective subject.

Selected Publications

D. Liu, G. Xu, **S.S. Jamali**, et al., *Bioelectrochemistry* 129 (2019) 106–115.

A.C. Hee, H. Cao, Y. Zhao, **S.S. Jamali**, et al., *Bioelectrochemistry* 122 (2018) 32–39.

S. S. Jamali, T. Suesse, S. Jamali, D. J. Mills, Y. Zhao, *Prog. Org. Coatings* 2017, *108*, 68.

M. H. Khan, **S. S. Jamali**, et al., *Adv. Mater.* 2017, *29*, 1603937.

S. S. Jamali, D. J. Mills, *Prog. Org. Coatings* 2016, *95*, 13.

S. S. Jamali, D. J. Mills, *Prog. Org. Coatings* 2016, *95*, 13.

S. S. Jamali, et al., *Surf. Interface Anal.* 2015, *47*, 1187.

E. Stewart, N.R. Kobayashi, M.J. Higgins, **S.S. Jamali**, et al., *Tissue Eng. Part C.* 21 (2015) 1–9.

S.S. Jamali, et al., *Electrochim. Acta.* 152 (2015) 294–301.

S.S. Jamali, et al., *J. Electroanal. Chem.* 739 (2015) 211–217.

575057 Engineering Microorganisms and Their Environment for Increased Performance in Biorenewable Applications

Research Interests: Microbes are useful tools in a number of relevant engineering applications including biorenewable chemical and fuel production. In my graduate work, I investigated microbial robustness to biomass-derived feedstocks and products for improved performance in biorenewable applications. Key outcomes of my research were an emulsion formulation which enabled *P. putida* KT2440 to grow in pyrolytic bio-oil and a unique disk diffusion assay which can be used to simultaneously assess tolerance and scavenging abilities of lignin utilizing microbes. In my future work, I aspire to continue in metabolic engineering and biotechnology while addressing gaps between academic and industry for instance by seeking collaborations with industry. Although the future of biorenewable fuels and chemicals is precarious, investigating and improving upon the principles of using a tunable and flexible microorganism is well worth the research. Applications of microorganisms span everything from crop productivity, food preservation, and human health.

Teaching Interests: In graduate school, I was a teaching assistant, guest lecturer, and research mentor. I also participated in a unique fellowship called Trinect. In Trinect, I teamed up with an elementary teacher and a student teacher to teach science and engineering to Des Moines public schools. These opportunities inspired my teaching philosophy which centers on effective, student-centered teaching of chemical engineering courses. I am motivated to value all students, to provide students with the tools they need to fulfill their goals, and to help students discover their potential. I aspire to be the type of teacher that I always admired when I was a student-the kind that takes the time to help any and all students including the ones who are struggling. Part of valuing each student is to recognize their goals while also helping them to fulfill the goals of the course. As part of the program at Iowa State called Preparing Future Faculty, I practiced designing courses so that they are aligned with learning goals, and I am confident in applying what I learned in the future. Although I think it is important to address both individual and course goals, I also want to reach beyond and help students discover their full potential. Even in bigger classes, I look forward to using teaching strategies such as team-based learning in order to allow each student to feel like they are getting the feedback they need to succeed. I am grateful for the teachers and mentors in my life who have supported me in my academic and professional career to realize my dreams. I will continue educating and helping students gain the tools they need to realize their own dreams.

581882 Nano-Optical and -Electronic Devices with Machine Learning for Biomarker Discovery and Diagnostics in Personalized Medicine

Research Interests:

For personalized medicine to be fully realized, progress must be made towards developing more rapid and inexpensive point-of-care devices to probe and understand the multi-omics biomarkers that signal disease susceptibility. Such devices will provide clinicians with the data they need to prescribe patients with timely, effective treatments necessary for proper care and survival. This would be a large step forward compared to cumbersome laboratory testing that can take days to weeks or diagnoses from qualitative signs and symptoms alone. New diagnostic devices have employed a range of chemical and biological detection techniques with photonic, electronic, and other advanced materials approaches for increasing sensitivity and selectivity. Despite the progress, significant improvements are still needed. Decreasing sample preparation and pretreatment steps employing complex biological extraction, amplification, and chemical labeling is critical for making devices simpler, faster, and cheaper. Additionally, emerging devices should be engineered for detecting multiple targets and multiplexed samples for broad-spectrum diagnostics.

My PhD research has focused on exploring new techniques for nucleic acid sequencing, tailored towards addressing the issues with current personalized medicine diagnostics. Sequencing not only provides a way to diagnose biomarkers on the genomic, transcriptomic, and epigenomic levels, but also serves as a platform for discovery of new biomarkers from analysis of sequenced data. While current next-generation sequencing-by-synthesis and nanopore methods have paved the way for cheaper and faster sequencing of large datasets, they still suffer from complex sample preparation requiring enzymatic amplification and chemical labeling, as well as being ensemble-based measurements with limited resolution towards single-cell studies. Through my work, I have used experimental and computational tools to develop nano-electronic and -optical spectroscopic techniques with the potential for truly single-molecule, label free nucleic acid sequencing. In two electronic methods, a scanning tunneling microscope was used for performing scanning tunneling spectroscopy and conductance measurements that probe the molecular orbitals of single nucleotides within conformationally-constrained DNA and RNA macromolecules. Robust machine learning algorithms relying on biophysical parameters were designed and implemented for identifying the nucleobases A, G, C, and T/U from their electronic signals, shown to reach accuracy levels >90%. In an optical method, surface-enhanced Raman spectroscopy measurements from nanopatterned surfaces and nanoparticle plasmonic substrates were used to measure nucleobase content in DNA k-mer fragments. A powerful bioinformatics algorithm was then designed for mapping and scoring the k-mer content measurements to a biomarker database for probabilistic identification of genes. This optical sequencing platform was used to correctly identify a β-lactamase antibiotic resistance gene and confirm the *Pseudomonas aeruginosa* pathogen of origin from <12 content measurements (<15% coverage) of the gene. The sequencing techniques developed during my PhD each have potential to facilitate rapid and inexpensive diagnostics. They are all label-free platforms requiring little to no amplification and pretreatment steps that provide broad-spectrum detection, paving the way for implementation into personalized medicine.

Future Research Directions:

As an independent investigator, I will combine my computational and experimental background in machine learning and nanoscale spectroscopic methods to build a multidisciplinary lab with the goal of discovering and detecting new targets for personalized medicine. Specifically, my lab will strive to...

1. Design and apply new machine learning algorithms and bioinformatic tools for identifying novel disease biomarkers from next-generation sequencing data.
2. Engineer new nano-optical and -electronic devices to detect disease biomarkers at higher resolution, pushing the limits of single-molecule sensing.

PhD Dissertation:

"Quantum Techniques for Single-Molecule Nucleic Acid Sequencing" Under the supervision of Prashant Nagpal and Anushree Chatterjee, Department of Chemical and Biological Engineering, University of Colorado Boulder

Successful Proposals:

National Science Foundation Graduate Research Fellowship (NSF GRFP)

Graduate Assistantship in Areas of National Need (GAANN), University of Colorado Boulder – Biomaterials

Education:

Ph.D., Chemical Engineering, University of Colorado Boulder, Expected May 2020

M.S., Chemical Engineering, University of Colorado Boulder, 2017

B.S., Chemical Engineering, University of Nebraska-Lincoln, 2014

Teaching Interests:

During my PhD, I have had the opportunity to TA for a general chemistry course for first- and second-year multidisciplinary engineering students, and a tissue engineering and medical devices elective course for upper-class chemical engineers. As an advanced TA for these classes, I was involved in grading assignments, reports, and exams along with leading recitation sections, office hours, and giving guest lectures. These experiences have shaped my teaching values and motivations. I believe that students can benefit most when guided and taught a general process for problem solving (i.e., approaches applicable to many problems), and how to ask the right questions when approaching a new problem with unknown solution. This idea is very applicable to the fundamental chemical engineering curriculum in reactor design, thermodynamics, kinetics, and fluid mechanics. I think it is detrimental for students to follow specific solution procedures, where they become comfortable with redundantly and blindly applying the same set of steps for routine, elementary solutions. Students should be prepared for the unexpected, and often non-trivial problems encountered in the real world, whether in a laboratory R&D setting or industry.

Future Teaching Directions:

As a faculty member, I will strive to establish new elective courses within the fundamental chemical engineering curriculum. I will use my background in machine learning and nanobiotechnology to create courses where students can become familiar with state-of-the-art technologies and their applications. I believe that more hands-on experience, specifically with applying computer science aspects to the traditional chemical engineering skillset, is vital for students to excel in today's workforce where computational data analysis is becoming a required expertise. I would design a course where students can become familiar applying an array of machine learning algorithms towards analysis of real datasets, such as those from next-generation sequencing. Additionally, I will require students to read and critically analyze reports from scientific journals. As a graduate student, I have already begun developing such courses. I am currently designing a hands-on set of experiments with guided data analysis protocols, aimed at chemical and biological engineering senior design labs. The lab module will incorporate experiments employing the optical sequencing platform developed in my PhD work. The hope is that this will not only highlight the transformative research taking place within the department, but inspire students to pursue research careers or further education.

Service and Outreach:

In addition to teaching, I have taken part in numerous service and outreach efforts in my graduate work. I have helped organize and lead demonstrations for a science and engineering discovery day for middle school students at my university, with activities ranging from exploring Newton's laws of motion to polymer chemistry. As part of a multi-department cohort, I helped design a four-week nanoscience lecture and experimental demonstration series aimed at motivating high school seniors to purse STEM careers. Students were introduced to general ideas in molecular forces, self-assembly chemistry, and the everyday uses of nanotechnology from filtration membranes to surface coatings. I have also volunteered with the Junior Achievement Program, where I led elementary students in activities teaching the importance of economics within the community while providing specific input from an engineer and scientist perspective. As a faculty member, I will establish new outreach opportunities within my research group and department to motivate and excite the next generation of engineers and scientists.

Publications:

[1] **Korshoj, L. E.**; Nagpal, P. Diagnostic Optical Sequencing. *bioRxiv preprint: 10.1101/685792.*

[2] **Korshoj, L. E.**; Nagpal, P. BOCS: DNA k-mer Content and Scoring for Rapid Genetic Biomarker Identification at Low Coverage. *Comput. Biol. Med.* **2019**, *110*, 196–206.

[3] Abel, Jr., G. R.; **Korshoj, L. E.**; Otoupal, P. B.; Khan, S.; Chatterjee, A.; Nagpal, P. Nucleotide and Structural Label Identification in Single RNA Molecules with Quantum Tunneling Spectroscopy. *Chem. Sci.* **2019**, *10* (4), 1052–1063.

[4] Sagar, D. M.; **Korshoj, L. E.**; Hanson, K. B.; Chowdhury, P. P.; Otoupal, P. B.; Chatterjee, A.; Nagpal, P. High-Throughput Block Optical DNA Sequence Identification. *Small* **2018**, *14* (4), 1703165.

[5] **Korshoj, L. E.**; Afsari, S.; Chatterjee, A.; Nagpal, P. Conformational Smear Characterization and Binning of Single-Molecule Conductance Measurements for Enhanced Molecular Recognition. *J. Am. Chem. Soc.* **2017**, *139* (43), 15420–15428.

[6] Afsari, S.; **Korshoj, L. E.**; Abel, G. R.; Khan, S.; Chatterjee, A.; Nagpal, P. Quantum Point Contact Single-Nucleotide Conductance for DNA and RNA Sequence Identification. *ACS Nano* **2017**, *11* (11), 11169–11181.

[7] **Korshoj, L. E.**; Afsari, S.; Khan, S.; Chatterjee, A.; Nagpal, P. Single Nucleobase Identification Using Biophysical Signatures from Nanoelectronic Quantum Tunneling. *Small* **2017**, *13* (11), 1603033.

[8] **Korshoj, L. E.**; Zaitouna, A. J.; Lai, R. Y. Methylene Blue-Mediated Electrocatalytic Detection of Hexavalent Chromium. *Anal. Chem.* **2015**, *87* (5), 2560–2564.

581987 Biosafe, Eco-Friendly Levan Polysaccharide Toward Transient Electronics

Research Interests: Biomedical Engineering, Eco-friendly applications

Teaching Interests: Photolithography, Material Diffusion Phenomenon, Surface reaction

PRESENTATION TYPE: Oral Presentation Preferred

APPLIED SYMPOSIUM: Food, Pharmaceutical & Bioengineering Division (FP & BE)

Abstract

TITLE: Biosafe, Eco-Friendly Levan Polysaccharide toward Transient Electronics

AUTHORS (FIRST NAME, LAST NAME): Kiyoon Kwon[1], Tae-il Kim[1]

INSTITUTIONS (ALL): 1. Sungkyunkwan Univ, Suwon, Korea (the Republic of).

ABSTRACT BODY:

Bio-integrated electronics is of great interest in emerging fields of bio-injectable and bio-medical applications. One of recent technologies, transient electronics, has peculiar characteristics that suggest the different behavior: it completely dissolves/disappear over time in a prescribed and/or triggered manner. Electronic system that is 'transient' in this sense has unique advantages that cannot be addressed with conventional electronics, such as biomedical implants that can be operational for clinically useful time span, then physically degrades/resorbs into the body. One of key properties, the control of entire lifetime of the system still has limitation. Previous researches have been trying to overcome this drawback with regulating pH level, passivating electronic components with other materials, and varying thickness of transient materials. In spite of these efforts to control the lifetime of devices, those strategies are limited to solution-based control. Here, we suggest a new approach that transient devices are successfully compromised by external heat source using near infrared (NIR) light, and all experiments are confirmed by in vivo and in vitro tests, and simple analytical (theoretical and FEM) simulations.

582104 Constitutive Modeling of Complex Biomaterials

Research Interests:

Biofluids and biomaterials typically exhibit rich and complex behavior under deformation which arises from the dynamic adjustments to the internal microstructure. Moreover, these materials are extremely diverse ranging from low viscous fluids like blood to stretchable elastic materials like tendons to compact, solid materials able to support large stresses like bones. The complex behavior and function of each of these materials has been well engineered through evolution for millions of years. However, our understanding of how to characterize these materials and their overall relation to health is typically not well understood. I am interested in progressing this understanding through advanced, yet computationally inexpensive, modeling based off precise and physiologically motivated experiments. This can be achieved through carefully designed constitutive models which link the microstructure to macroscopically observable features. Constitutive modeling is advantageous to microscopic based computational techniques in that constitutive models can easily be coupled with more traditional continuum approaches like computational fluid dynamics (CFD) simulations and can be implemented in complex geometries with unsteady boundary conditions without increasing the computational cost significantly. Through more accurate modeling of these materials, we can improve our understanding of the structure and function of biomaterials, enhance health monitoring and surgical techniques, and begin to move toward synthetic, biocompatible replacement materials.

Teaching Interests:

Teaching and mentorship are largely what motivate me to pursue a career in academia. My primary interests are in fluid mechanics and transport. However, I would also hope to teach advanced courses involving rheology, colloid science, and/or computational methods. As an instructor, I would hope to incorporate more practical problems and projects motivated by current industrial and academic research. In addition to scholastic teaching interests, I am also excited to mentor future researchers. Through mentoring, I hope to challenge my students and motivate them to ask questions and subsequently search for answers via the scientific method.

Research Experience:

I am currently a 5th year PhD candidate working with Dr. Antony Beris and Dr. Norman Wagner at the University of Delaware with a dissertation titled "An experimental and theoretical investigation of blood rheology". Thus far in my research, I have helped to standardize approaches to measuring the rheology of blood and have pioneered experiments that are better representative of *in vivo* flow behavior. Using the experimental results, I have developed a transient constitutive model for the thixotropic behavior of blood. The model features several physiologically relevant parameters and can enable insight into the microstructure of the sample at any point in time and with any previous strain history. Additionally, the model significantly outperforms previous models that have been established for blood rheology. Currently, I have been working to include inhomogeneous effects in the model and incorporate it into CFD simulations that can be compared with microfluidic experiments that we are conducting. In parallel to this, I have been working on measuring blood across species to elucidate the interspecies hemorheological differences and move towards understanding the individualized evolutionary design of blood. Although my primary research focuses on measuring and modeling blood rheology, I have also worked on side projects to model flow of complex materials including viscoelastic simulation of polymer extrusion in 3D printing alongside Dr. Michael Mackay and an industry funded project on flow of dilute polymer solutions through a hyperbolic contraction.

Teaching Experience:

Throughout my undergraduate and graduate studies, I have pursued various teaching opportunities. As an undergraduate at Cornell University, I served as a teaching assistant for several core engineering courses including process control strategies, fluid mechanics, and heat and mass transfer. Additionally, I led an academic excellence workshop as a supplemental course for engineers taking an introductory course on computing with MATLAB. At the University of Delaware, I have also enjoyed many teaching positions. For two consecutive years, I served as a teaching assistant for the undergraduate fluid mechanics course and was selected to receive the Robert L. Pigford Teaching Assistant Award my second year. As a teaching assistant, I led office hours, wrote homework solutions, and helped assign grades. Moreover, I helped to develop a final project that brought in elements indirectly related to my thesis that tested and taught numerous fluid mechanics concepts. After completing my teaching assistant requirements, I also was selected as a Fraser and Shirley Russel Teaching Fellow for undergraduate fluid mechanics for the following year. This fellowship, starting in the Fall of 2019 enables me to be directly involved with the course by giving lectures, designing homework and exam problems, and holding office hours. Outside of the classroom, I have also served as a research mentor for six different undergraduate students and one high school student.

Future Direction:

Moving forward, I want to use the tools I've developed throughout my thesis to model other complex biomaterial systems. Various other biological materials, including synovial fluid, saliva, mucus, and vitreous humor or more elastic materials like skin, tendons, and ligaments, have been shown to demonstrate interesting rheological characteristics. However, the modeling of these fluids is lacking partly due to oversimplified or inaccurate models and partly because of a lack of reliable experimental data. After first characterizing the samples through physiologically motivated shear and extensional rheological experiments, I hope to extend my constitutive modeling framework to apply to this variety of fluids. Subsequently, I want to incorporate these models into continuum simulations to achieve a better understanding of how these biomaterials function. Finally, I hope to compare the simulation results to a range of experimental data involving but not limited to scattering, atomic force microscopy, tribology, microfluidics, and tensiometry whether obtained in my research group or through collaborations with other experimental research groups.

Publications:

1. **JS Horner**, MJ Armstrong, NJ Wagner, and AN Beris, "Investigation of blood rheology under steady and unidirectional large amplitude oscillatory shear," *Rheol.*, 62(2), 577-591 (2018).
2. M Armstrong, **J Horner**, M Clark, M Deegan, T Hill, C Keith, and L Mooradian, "Evaluating rheological models for human blood using steady state, transient, and oscillatory shear predictions," *Acta*, 57(11), 705-728 (2018).
3. **JS Horner**, AN Beris, DS Woulfe, and NJ Wagner, "Effects of ex vivo aging and storage temperature on blood viscosity," *Hemorheol. Microcirc.*, 70(2), 155-172 (2018).
4. **JS Horner**, MJ Armstrong, NJ Wagner, and AN Beris, "Measurements of human blood viscoelasticity and thixotropy under steady and transient shear and constitutive modeling thereof," *Rheol.*, Under Revision.
5. M Armstrong, T Helton, S Rogers, and **J Horner**, "Analysis of transient human blood rheology using sequence of physical processes (SPP)," *Acta*, Under Revision.
6. **JS Horner**, NJ Wagner, and AN Beris, "A review on recent advances in modeling and simulation of blood flow," *Soft Matter* (Invited), In preparation.

Successful Grants/Fellowships:

- Graduate Fellowship – NASA / Delaware Space Grant Consortium (DESGC) (Awarded: 05/2019)

581699 Integrating Molecular Modeling with Experimental Work to Propel a Bioproducts Pipeline

Research Interests: Throughout my studies I have explored different aspects of bioproduct development, from using biotechnology to harvest biologically based materials that are valuable all the way to designing industrial processes. As a researcher I seek to engage a diverse group of peers and students in developing sustainable chemical products that can compete, in terms of quality and economic value, with existing products on the market. A dual focus on molecular modeling and experimental work will accelerate fundamental science related to biomolecule recovery. Being able to extract, separate, and purify valuable biomolecules from plants or biomass is a critical initial step in developing bioproducts like pharmaceuticals or sustainable chemicals. Using biotechnology – including biostatistics, molecular modeling, and advanced instrumentation in the laboratory – is critical to building a bioproducts pipeline that can ultimately allow novel sustainable products to make it to market. To continue to establish myself as a researcher in the bioproducts development space, I would seek to train and engage students in laboratory work and experiments while also teaching those who are interested in how specialized software can allow them to carry out virtual experiments and aid in process design. I envision valorization of underutilized biomaterials by developing them into building blocks for production of new consumer-ready sustainable bioproducts. This includes developing pharmaceuticals, ingredients (food, beverage, cosmetics), and value-added chemicals from sustainable sources through environmentally-friendly processes. A dual focus on modeling and experimentation is useful in carrying out fundamental research and propelling the results toward real-world applications.

Teaching Interests: My goal as an educator is to develop a curriculum and classroom atmosphere that engages students from diverse backgrounds in the learning process through dynamic lessons and activities. Throughout my experiences teaching students at different stages of their academic careers and in settings including small groups, laboratories, and large lecture halls, I have developed my teaching philosophy and continue to look for innovative methods to create an inclusive classroom environment for students to learn. My education and background in chemical and biological process engineering gives me the versatility to teach a variety of courses in CHE or CHBE curricula, including courses on (bio)separations, process design, computational techniques, and more. In order to prepare CHE students for diverse career paths, I will set goals for student learning, use a variety of teaching methods, and use different assessment tools to make sure student learning objectives are met. I aspire to engage students in and out of the classroom and help them reach their full potential.

PhD Dissertation:

Molecular Modeling of High Oleic Soybean Oil as an Extraction Solvent for Removing Hydrogen Sulfide from Natural Gas; advised by Dr. Abigail Engelberth, Purdue University.

Selected Grants & Awards:

2019 Wadsworth Graduate Mentoring Award (Purdue), 2019 Outstanding Service Award from ABE/College of Engineering (Purdue), 2017 Society for Industrial Microbiology Diversity Travel Grant, 2016 International Countercurrent Chromatography Conference Travel Grant

Selected Publications:

Brace, E.C., & Engelberth, A.S. (2017). Enhancing silymarin fractionation using the conductor-like screening model for real solvents. *Journal of Chromatography A, 1487*, 187-193.

Zhang, H., **Brace, E.C.,** & Engelberth, A.S. (2016). Selection of a nonaqueous two-phase solvent system for fractionation of xylo-oligosacchardie prebiotics using a conductor-like screening model for real solvents. *Journal of Liquid Chromatography & Related Technologies, 39(14)*, 666-673.

573566 Advancing Technologies for Prenatal and Women's Health

Research Interests:

My primary research interest is using biomaterials, drug delivery, and microfluidic techniques to develop maternal and prenatal treatments and non-invasive diagnostics.

Ph.D. Thesis: "Investigating molecular interactions of the placenta for prenatal treatment and testing" Advisors: Prof. Anita Shukla and Prof. Anubhav Tripathi, Center for Biomedical Engineering, Brown University

The focus of my Ph.D. research is developing methods to better understand the maternal-fetal interface and enhance non-invasive prenatal testing. Placental trophoblast cells have great potential to provide information about pregnancy complications both fundamentally and clinically during pregnancy. I have developed a placental lipid bilayer model to investigate small molecule and environmental toxin interactions. Using this placental mimetic biomaterial, I aim to screen new therapies and potential hazardous material to inform risk to a developing fetus. Additionally, we have investigated methods for enrichment of placental trophoblast cells in clinical samples to provide the entire fetal genome non-invasively. Our enrichment method has led to a 700% purity increase of placental trophoblast cells within heterogenous cervical samples, which has further led to successful isolation of single fetal trophoblast cells using differential cell settling and automated cell picking.

As a National Science Foundation (NSF) graduate research fellow, I had the opportunity to develop my research project and choose a Ph.D. research area that I am passionate about. Within my Ph.D. and future research, I seek to fulfil an unmet need for chemical and biomedical engineering techniques towards advancing prenatal and women's health. As a future faculty member, I aim to develop biomaterials to investigate trophoblast implantation mechanisms for informing treatment of maternal-fetal health complications, such as preeclampsia and infertility. I seek to identify fetal- originated material from non-invasive clinical samples, such as blood, saliva, and cervical samples, for improved prenatal diagnostics. Finally, I aim to develop an improved nutrient transport system for preterm birth.

Teaching Interests:

As an undergraduate student, I was often focused on the answer, whether it be the answer to my problem set, test question, or lab experiment. What I have realized during my more senior years and graduate studies is that being able to ask well thought questions is arguably more valuable and leads to the development of critical thinking skills. I aim to bring this philosophy to my teaching and mentoring by increasing the focus on having students ask questions. I have had valuable experience within project focused classrooms and believe that this is an excellent way to allow students to question.

I have previous experience teaching middle and high school students during summer programs at my undergraduate university. I reorganized the biomedical summer program to allow for team innovation development. As a graduate student, I have been a teaching assistant for four semesters for core courses including Transport and Biotransport Processes and Biomedical Engineering Design and Innovation. I have also mentored six undergraduate researchers; many have presented at national conferences and received honors such as the NSF graduate research fellowship. I am most proud to see my undergraduate students become independent in the lab and confidently present their work.

With degrees in chemical engineering and biomedical engineering, I am prepared to teach core courses including Chemical Engineering Fundamentals, Transport Phenomena, Fluid Mechanics, and Unit Operations Labs. I am also passionate about developing and teaching elective courses including Nano- and bio- engineering, Biomaterials, and Biomedical imaging and diagnostics for chemical engineers.

Selected Publications:

Bailey-Hytholt, C.M., Sayeed, S., Kraus, M., Joseph, R., Shukla, A., Tripathi, A., A Rapid Method for Label-free Enrichment of Rare Trophoblast Cells from Cervical Samples, in review.

- Bailey, C.M., Tripathi, A., Shukla, A., Effects of Flow and Bulk Vesicle Concentration on Supported Lipid Bilayer Formation, Langmuir, 33(43), 11986-11997, 2017.
- Bailey, C.M., Nagarajan, R., Camesano, T. A., Designing polymer micelles of controlled size, stability and functionality for siRNA delivery, Control of Amphiphile Self-Assembling at the Molecular Level: Supra-Molecular Assemblies with Tuned Physicochemical Properties for Delivery Applications, pp. 35-70, American Chemical Society, 2017. Featured figure on book cover.
- Bailey, C.M., Kamaloo, E., Waterman, K. L., Wang, K. F., Nagarajan, R., Camesano, T. A. Size dependence of gold nanoparticle interactions with a supported lipid bilayer: A QCM-D study. Biophysical chemistry, 203, 51-61, 2015.

580989 Biofilm Prevention By UVC Side Emitting Optical Fibers

I have two main career goals that are related to (i) research and (ii) teaching. The first is to develop chemical free point of use water treatment device to ensure quality drinking water in both the United States and developing countries. I will work at the interface of basic science and industry to develop innovative and green technologies. The second goal is to educate diverse communities in STEM through a focus on water treatment and water resources. I will become a tenure track professor to both educate and develop a research team with a focus on photo-assisted water treatment.

Research

The lack of appropriate disinfection strategies results in an estimated 502,000 diarrheal deaths each year. Germicidal ultraviolet-C (UVC) irradiation does not produce chlorinated DBPs and requires only electrical power for disinfection. UVC Light emitting diodes (LED) have promising characteristics for water treatment (e.g. lack of warm up time, tunable radiation, no degradation at on/off cycles and longer life of use). However, their small surface area of irradiation limits their integration into treatment reactors. To progress development and integration of photon driven water treatment my research interests includes (i) better distributing light from UV lamps, (ii) improving the quantum yields of reactions, and (iii) understanding consumer perception and desires related to water treatment technologies.

First I seek to integrate optical technologies into water treatment to increase uniform light delivery. By using a UV-C glowing optical fiber (Provisional Patent), I illustrated an increase of > 600 times in surface area of irradiation as part of my dissertation. As a tenure track faculty, I will develop point of use devices that use these optical fibers, and other highly reflective surfaces in collaboration with current industrial partners (i.e. AquaSense). **Second, I will improve reaction quantum efficiency by fundamentally understanding wavelength specific photolysis mechanisms.** Proteins, DNA, and organic bonds (pharmaceuticals) are prone to lysis at specific frequencies. By using LEDs, I will explore single and synergistic wavelength effects on both microorganisms and organic contaminants of interest. This understanding will lead to development of highly efficient LED water treatment reactors. **Last, I seek to understand how community perception of water quality effects technology adaption and integration of point of use systems.** I have established strong relationships with the director of health a downriver community of Colombia (Mompox). I have and will continue to work with them to ensure the technologies I develop are truly beneficial to consumers who do not have access to reliable municipally treated water. My research will transpose into (i) advancement of water treatment science and technology and (ii) increased quality of life in developing communities.

Teaching

My main contribution as a professor will be to train the next generation of engineers to be leaders in both industry and academia. A tenure track position will enable me to accomplish this through both mentoring and teaching. By developing a research team, I will **mentor** graduate students to grow as well-rounded researchers and humans. I believe a dissertation or thesis is one piece of the graduate degree pie. I will encourage my students to fill the remaining pieces through opportunities such (but not limited to) (i) national and international internships (ii) leading outreach and service events, (iii) communicating their work with the general public, (iv) and organizing conference sessions. In addition to developing crucial soft skills, these opportunities facilitate knowledge transfer by allowing the student to better understand and communicate the broader impact of their fundamental research.

I hope to **teach** fundamental chemical engineering courses to tie technical concepts to my **application driven teaching philosophy.** I will challenge students to understand ethical and political dilemmas that are directly tied to technical classroom lectures. For example, in unit operations, I will use power plant CO2 scrubbers as examples to drive equation development of gas absorption. I will illustrate how manipulating the equation variables could influence both the efficiency and cost of CO2 scrubbers. I

will end the lecture with a brief discussion how politics has both encouraged and limited their integration in power plants over the years.

I will use the same philosophy in developing courses in nanotechnology, photo-assisted reactions, and advanced water treatment. My global experience with water treatment, strong industrial network, and active participation in the NSF Nanotechnology Enabled Water Treatment (NEWT) allow me to add impactful examples and discussions in these topics. The critical thinking skills developed through a chemical engineering degree is unparallel to any other. Those fundamental concepts rooted in broader impact can empower students to solve global problems.

583459 Force Response Defines Both Subcellular Architectures and Dynamic Protein Binding

Research Interests:

Understanding the dynamics of biological processes is often approached through the lens of biochemical equilibria, wherein a healthy balance of reactants and products is actively maintained within enclosed bioreactors (cells and organelles). However, an equally compelling case can be made that there is also a dynamic and controlled balance of forces (and subsequent motion) within the cell. Altered force response between cytoskeletal elements can result in the stiffest of cellular superstructures or flexible transportation networks that allow for trafficking of organelles. Additionally, physiological processes that alter osmotic pressure or membrane tension can subsequently control phenomena as diverse as subcellular protein-protein phase separation or protein functionalization of organelle surfaces. This balance of forces within the cell is often mediated through intrinsically disordered proteins (IDPs), or proteins that lack a secondary structure and behave as biopolymers.

Through work by myself and others, there has been a growing consensus that IDPs give rise to unique force-responsive structures, combining a template of soft matter principles with sequence-encoded biologically-specific interactions. Exhibiting a dynamic regulation beyond most engineering methods today, these IDPs can be actively reprogrammed via post-translational modifications (much as biochemical equilibria are maintained via enzymatic activity) to alter or instill new functions beyond their naturally encoded state. Additionally, just as an unbalanced biochemical equilibrium is often associated with disease, external insults or internally aberrant perturbations to the homeostatic balance of forces are definitively linked to neuropathologies such as Alzheimer's and Parkinson's disease.

Despite the critical role of IDPs in maintaining a balance of forces within a cell, traditional biological and biochemical methods have had little success in bridging emergent mechanical properties afforded by IDPs and the molecular modifications to IDPs that occur in physiology (and disease). Requiring an innovative and multidisciplinary approach, understanding the conformational landscape of these proteins could not only yield insight into possible therapies but allow us to repurpose those emergent functionalities to design a new class of biologically-programmable nanomaterials.

Postdoctoral Project: *"Force response defines both subcellular mediated architectures and dynamic protein binding."*

Under supervision of Ka Yee C. Lee, James Franck Institute, Chemistry, and Institute for Biophysical Dynamics, University of Chicago

PhD Dissertation: *"Probing forces generated and architectures mediated by Tau on microtubules"*

Under supervision of Cyrus R. Safinya, Physics, Materials, and Molecular, Cellular and Developmental Biology, University of California Santa Barbara

Research Experience:

Having found my passion in defining the molecular parameters of neurodegenerative diseases and pursuing biologically-inspired materials, I have taken my formal training in physics and engineering (at UC Berkeley and UC Santa Barbara) and have sought to describe previously intractable biological phenomena through the framework of colloidal and interfacial science. As a graduate student, not only did I train extensively with molecular biologists to express and purify proteins, I developed synchrotron X-ray scattering techniques to investigate the structure and force response of cell-free reconstitutions of microtubules, dynamically-assembly protein nanotubes, and Tau, an intrinsically disordered protein unequivocally linked to Alzheimer's.

As a Kadanoff-Rice Postdoctoral Fellow at the University of Chicago, I have expanded my biological skillset while developing new experimental platforms and synchrotron-based techniques. In turning my focus to a-synuclein (an intrinsically disordered protein linked to Parkinson's), I mastered newly developed chemical engineering techniques and molecular cloning to more faithfully represent the healthy and diseased states of the protein. As a-synuclein is known to bind to membrane organelles, I pioneered new membrane-nanoparticles that closely mimicked said organelles to precisely measure protein/membrane interactions. Additionally, this platform allowed for the use of X-ray photon correlation spectroscopy, a technique previously limited to soft matter physics but with which I have validated and advanced for biological studies. My intent has always been to become a complete scientist, undergirded by my belief that only a multidisciplinary approach can unveil physiological function and diseased dysfunction of cellular machinery.

Teaching Interests:

As a graduate student, I had the opportunity to teach three undergraduate-level physics courses, personally delivering 2-3 lectures a week to over 200+ students and supervising 3-5 teaching assistants in associated laboratory courses. As these courses in physics were designed for undergraduates majoring in biology, I understand the need to effectively communicate ideas in a manner most familiar to the audience, not the lecturer. Having taken both undergraduate and graduate engineering courses myself, I am more than capable and happy to teach introductory courses in chemical engineering and also teach (and if necessary, add) advanced courses in colloid/interfacial science, statistical mechanics, advanced X-ray scattering methods, protein expression/purification/modification methods, and soft condensed matter physics – all fundamental tools and knowledge requisite for an advanced training in chemical engineering with a focus in the applied life sciences.

Teaching is not just limited to the classroom but also extend to the laboratory and community. Having successfully mentored a total of 15 undergraduate and 4 graduate students during my graduate and postdoctoral work, I have the tools to create a laboratory environment that is both mutually respectful and curiosity driven. I also served as the student coordinator for Research Internships in Science and Engineering from 2013-2015, an undergraduate summer research program that targeted minority-serving institutions. Especially gratifying was the opportunity to work with students who clearly took to science and engineering but had been traditionally marginalized, with over 55% being female and over 40% being underrepresented minority students. Given the interdisciplinary nature of my research and extensive background in outreach programs, I am well equipped to work with students who come from a variety of backgrounds, academic or otherwise.

Proposed Future Research: Leveraging my unique training in soft matter physics and neurobiology, I propose a research program to apply precision force, structural, and dynamics measurements of minimal biological reconstitutions to recover the emergent mechanics conferred by IDPs. In particular, my research will be organized along three thrusts:

- *What emergent mechanical and transport properties arise as a function of IDP sequence and cellular conditions?* While IDPs are understood to give rise to complex cytoskeletal and subcellular architectures, the biological function of these architectures remains nebulous. While some IDP-mediated architectures seem to facilitate intracellular trafficking (Tau-mediated microtubule networks), others alter spatial localization of organelles and other biomacromolecules (synapsin-1 liquid droplets). Using optical, scattering, and force techniques I have developed, I aim to understand the emergent mechanical and transport properties of these architectures and the IDP sequence motifs/cellular conditions that drive architecture formation.
- *How does cellular reprogramming alter IDP functionalization, especially modifications linked to neurodegenerative disease and injury?* External insults (mechanical trauma in traumatic brain injury) or aberrant biochemical modifications (hyperphosphorylation of Tau in Alzheimer's disease) are known to collapse or abrogate function of IDP-mediated architectures. Perturbing the basal state of these architectures both mechanically and biochemically will not only let us understand how diseased states alter architecture function, but examining the biological control of materials properties to affect transport properties may yield insight for other active matter systems that have been of intense, recent interest.

How can we adapt these emergent properties to design a new class of biologically-programmable materials? Indeed, many of the properties conferred by IDPs can be naively modeled as polymers, ranging from grafted polymer brushes to polymer coacervates. However, that ignores the rich behavior that evolution has granted through biological specificity and reprogramming of IDPs, ranging from polymer brushes that attractively interact with each other to coacervates that specifically recruit certain membrane organelles but not others. By utilizing the principles that drive these properties, we can create new functional and programmable materials through cellular expression as opposed to traditional synthetic approaches.

Biology offers a rich landscape of optimized and actively controlled forces in colloids and fluids. By understanding this landscape, we can elucidate how the homeostatic steady state is perturbed into disease (yielding insights into possible therapies) and potentially repurpose this control for materials development, establishing an experimental framework that extends chemical engineering and soft matter physics into cellular biology.

Selected Publications:

Chung, P.J.*; Zhang, Q.*; Hwang, H.L.; Leong, A.; Dufresne, E.M.; Narayan, S.; Adams, E.J.; Lee, K.Y.C. a-synuclein sterically stabilizes spherical nanoparticle-supported lipid bilayers. ***ACS Applied Bio Materials***, 2 (4), 1413-1419, 2019.

Chung, P.J.; Hwang, H.L.; Dasbiswas K.; Leong, A; Lee, K.Y.C. Osmotic shock-triggered assembly of highly-charged, nanoparticle-supported membranes. ***Langmuir***, 34 (43), 13000-13005, 2018.

Chung, P. J.; Song, C.; Deek, J.; Miller, H. P.; Li, Y.; Choi, M. C.; Wilson, L.; Feinstein, S. C.; Safinya, C. R. Tau Mediates Microtubule Bundle Architectures Mimicking Fascicles of Microtubules Found in the Axon Initial Segment. ***Nature Communications***, 7, 12278, 2016.

Chung, P. J.*; Choi, M. C.*; Miller, H. P.; Feinstein, H. E.; Raviv, U.; Li, Y.; Feinstein, S. C.; Safinya, C. R. Direct Force Measurements Reveal Protein Tau Confers Short-Range Attractions and Isoform-Dependent Steric Stabilization to Microtubules. ***Proceedings of the National Academy of Sciences***, 112 (47), E6416–E6425, 2016.

583866 Engineering Biomimetic Materials and Cells for Diagnostic and Therapeutic Applications

Research Interests:

Nature represents the largest reservoir from which people can learn to better understand biological processes and design strategies to interfere with these processes. My career objective is to apply concepts from nature to rationally engineer bioinspired and biomimetic materials and cells to interface with biology and tackle grand challenges in medicine. Especially, my research focuses on making biomimetic materials and engineered cells as effective approaches to both diagnose and treat diseases. My goal is to run a multidisciplinary research group that mainly focuses on two broad research areas. 1) Engineer biomimetic nanomaterials for immunoengineering and immunomodulation. In this area, I envision that engineering nanomaterials to mimic and interfere immunological processes can achieve systemic or local immunomodulation for disease treatment. One particular interest is to design the next-generation biomimetic nanovaccines to modulate the immune response for treating diseases like addiction to abusive drugs and cancers. 2) Engineer circulating blood cells for the diagnosis and treatment of diseases. In this area, one particular interest is to specifically deliver therapeutics, imaging agents, and nanoparticles to target sites by hitchhiking them to circulating blood cells. Especially, erythrocytes, platelet, and monocytes are three main cell types I am especially interested in. Another particular interest is to surface engineer cells to make them perform like another cell type to solve unmet clinical needs.

PhD Dissertation: "Biomimetic nanomaterials for designing the next-generation vaccines against abusive-drugs and enhancing cancer drug delivery"

Advisor: Chenming Zhang, Ph.D., Department of Biological Systems Engineering, Virginia Tech

Postdoctoral Project: "Cell-, ionic liquid-, and nanocarrier-based strategies for drug delivery"

Advisor: Samir Mitragotri, Ph.D., John A. Paulson School of Engineering and Applied Sciences, Wyss Institute of Biologically Inspired Engineering, Harvard University

Research Experience:

My background as a chemical/biomedical engineer, along with my training in biomaterials, immunology and drug delivery, equipped me with skills to contribute substantial progress in the field. Working with Chenming Zhang during my Ph.D. at Virginia Tech, I worked on engineering the next-generation biomimetic nicotine nanovaccines for immunotherapy against nicotine addiction. Inspired from viruses that are highly immunogenic, we designed hybrid nanoparticle-based nicotine nanovaccines that mimic the structure of viruses. This biomimetic nicotine nanovaccine outperformed conventional conjugate nicotine vaccines that have failed in clinical trials. In addition, by conducting systemic mechanistic studies, we elucidated how fine tuning each component of a nanovaccine could modulate its immunological efficacy. Working with Samir Mitragotri during my postdoc training at Harvard University, I learned how to tackle significant biological barriers for effective drug delivery using biomimetic cell- and material-based approaches. One aspect of my postdoc work has been focused on engineering erythrocytes for drug delivery and immunomodulation. In one direction, we engineered an erythrocyte hitchhiking approach to achieve systemic administration enabled local drug delivery. Based on this technology, we developed Erythrocyte Leveraged Chemotherapy (ELeCT) that could enable conventional chemotherapy for effective treatment of lung metastasis and Erythrocyte Anchored Systemic Nano-immunotherapy (EASY) that could restore local tumor immune microenvironments at the metastatic sites. In another direction, we engineered an erythrocyte hitchhiking approach for systemic immunomodulation. The second aspect of my postdoc work has been focused on tackling the skin barrier using ionic liquid-based materials. Specifically, we engineered ionic liquids that could enhance transdermal delivery of a range of immunomodulatory agents for skin immunomodulation.

Teaching Interests:

I believe that rigorous education forms the basis for innovative research and thus teaching is one of the central part of a great research university. My teaching experience includes serving as a Teaching Assistant in 4 courses. In this capacity, my responsibilities entailed developing and delivering review lectures, grading all assignments, leading class discussions, designing and supervising laboratory experiments, and meeting with students individually. In addition, I have mentored and trained > 20 undergraduate, master students, or Ph.D. students. I am prepared to teach both in chemical/biomedical engineering, my area of research expertise, and in a wide range of courses in bioengineering and pharmaceutics, at the introductory, intermediate, and advanced levels. I especially enjoy mentoring students in active project-based learning in medium classes, and I would welcome the opportunity to develop new courses according to departmental needs, especially on advanced drug delivery technologies.

Education:

Ph.D., 2017, Biological Systems Engineering (Biomolecular Engineering concentration), Virginia Tech

B.S., 2013, Biochemical Engineering, East China University of Science and Technology

Selected Publications (see my C.V. for a complete publication record):

Zhao Z., Ukidve A., Gao Y., Kim J., Mitragotri S. Erythrocyte Leveraged Chemotherapy (ELeCt): Nanoparticle assembly on erythrocyte surface to combat lung metastasis. *Science Advances* 2019.

Zhao Z., Ukidve A., Krishnan V., Fehnel A., Pan D., Gao Y., Kim J., Evans M., Guo J., Mitragotri S. Erythrocyte Anchored Systemic Nano-immunotherapy for restoring local tumor immune microenviroment. *Nature Nanotechnology*, in revision.

Zhao Z., Harris B., Hu Y., Harmon T., Pentel P., Ehrich M., Zhang C. Rational incorporation of molecular adjuvants into a hybrid nanoparticle-based nicotine vaccine for immunotherapy against nicotine addiction. *Biomaterials* 2018.

Zhao Z., Powers K., Hu Y., Raleigh M., Pentel P., Zhang C. Engineering of a hybrid nanoparticle-based nicotine nanovaccine as a next-generation immunotherapeutic strategy against nicotine addiction: A focus on hapten density. *Biomaterials* 2017.

Zhao Z., Hu Y., Harmon T., Pentel P., Ehrich M., Zhang C. Rationalization of a nanoparticle-based nicotine nanovaccine as an effective next-generation nicotine vaccine: A focus on hapten localization. *Biomaterials* 2017.

Zhao Z., Hu Y., Hoerle R., Devine M., Raleigh M, Pentel P., Zhang C. A nanoparticle-based nicotine vaccine and the influence of particle size on its immunogenicity and efficacy. *Nanomedicine: Nanotechnology, Biology and Medicine* 2017.

Zhao Z., Ukidve A., Krishnan V., Mitragotri S. Effect of physicochemical and surface properties on in vivo fate of drug nanocarriers. *Advanced Drug Delivery Reviews* 2019.

Zhao Z., Ukidve A., Dasgupta A., Mitragotri S. Transdermal immunomodulation: Principles, advances and perspectives. *Advanced Drug Delivery Reviews* 2018.

584388 Multi-Modal Biofabrication Approaches for Biomaterials Development and Tissue Engineering

Research Interests: Tools and Methodologies in Bioengineering

Three dimensional bioprinting and biofabrication techniques have enormous potential for advancing regenerative medicine, with the ultimate goal of producing personalized artificial organs and tissues on demand, and more immediate goals of providing physiologically relevant models of healthy and diseased tissues. However, current methods and materials do not yet address significant limitations in speed, feature resolution, construct size and biofunctionality necessary to realize clinical relevance of engineered tissues.

My research addresses three main questions:

1. How can the **development of multi-modal biofabrication techniques** address limitations of existing methods?
2. How can these new biofabrication techniques enable **creation of functional biomaterials**?
3. How can these techniques and materials support **tissue engineering applications**?

The first aspect centers around development of platform technologies, integrating established biofabrication modalities (e.g. extrusion-based bioprinting, stereolithography, electrospinning) to create hybrid techniques, and investigating applications of novel techniques, such as acoustic patterning, in biofabrication.

The second explores the interplay of biomaterials development and biofabrication. Multi-modal methodologies will be used to create microstructured biomaterials, particularly hydrogels, with emergent material properties. These will include both single-component and composite systems. I anticipate that hybrid biofabrication techniques can be used to confer additional functionality to the materials via introduction of multi-scale complexity to the internal architecture of the materials.

The third forms the basis of a collaborative program to explore translational routes for these new materials and methods, and to answer questions relevant to tissue engineering. Clinical and industrial partnerships will be emphasized. Initial directions I will investigate include:

- Vascularization of engineered tissues
- Cellular mechanotransduction
- Cell-material interactions

I enjoy exploring the application of engineering paradigms to develop new tools and utilizing existing tools and methodologies in novel ways. I find it rewarding to establish new techniques, rather than rely solely on commercially available equipment, which can be prohibitively expensive for early career researchers, and restrictive and inflexible to research needs. Researchers in my group will be encouraged to "hack" equipment and to develop their own instrumentation.

I am also interested in exploring **robust scale-up of tissue engineering processes,** and intend to be involved with development of standards and regulatory pathways as these methods reach maturity.

My research group will be a safe, positive, and inclusive working environment and I welcome diverse backgrounds. While providing opportunities to work on existing projects, I will encourage students to pursue independent lines of inquiry. I also support undergraduate participation in research and will encourage students to actively contribute to the body of knowledge through "Data in Brief"-type publications.

Research Experience:

I earned my PhD in Engineering in 2017 as part of the National Institutes of Health (NIH) – Cambridge Scholars Program, in which I conducted research at two of the leading research institutions in the world, the NIH (Bethesda, Maryland, USA), and the University of Cambridge. With these experiences, I have maximized my ability to work in an international collaborative environment and have started building an extensive professional network. From the beginning of my research experience as an undergraduate earning my BSc in Chemical Engineering from the University of Kentucky (2011), I have sought opportunities to apply engineering principles to biological problems. I am dedicated to pursuing an academic career in engineering of biological systems— primarily tissue engineering—the natural intersection of these fields.

My PhD research incorporated both biology and materials engineering; in my current postdoctoral research position at the University of Bristol, I have expanded to elements of acoustic and electrical engineering, and rapid prototyping and device fabrication, leading to the invention of a new, in-air acoustic particle manipulation and patterning technique. As a result, I am proficient in working across disciplines, and capable of managing multiple projects simultaneously.

Postdoctoral Research:

- *A Novel Biofabrication Approach Using Ultrasound-Guided Particle Manipulation*

Advisor: Mike Fraser, University of Bristol, Bristol, UK

- *Acoustically-Triggered Hydrogel Cross-Linking for 3D Bioprinting*

Pump-priming grant in collaboration with Avinash Patil, University of Bristol

PhD Dissertation:

Hydrogel Substrate Effects on Protein Kinase A and Osteogenesis

Completed as part of the National Institutes of Health – University of Cambridge Scholars Program.

Advisors: Constantine A. Stratakis, National Institute of Child Health and Human Development, National Institutes of Health, Bethesda, MD, USA; Michelle L. Oyen, University of Cambridge, Cambridge, UK.

Teaching Interests:

Undergraduate Courses:

Foundational Chemical and Materials Engineering Courses (e.g. Principles of Chemical Processes, Thermodynamics, Fluid Mechanics, Introduction to Materials Science, Polymers)

Undergraduate and Graduate Courses:

Biomaterials, Biofabrication, Tissue Engineering, Engineering Approaches in Regenerative Medicine

I approach engineering education by emphasizing the translational nature of the material. Engineering is a practical subject – a large quantity of the material can be described in the form of authentic problems. It is vital that students be able to relate the subject matter to some practical use or application. Not only does this maintain interest, but also initiates the valuable process of applying knowledge learned in the classroom to issues one might later see in the workplace.

Outside of the classroom, I intend to organize "skill share"-type events for students and researchers, in order to support expansion of individual practical skills while fostering a collaborative and collegial environment. This will also a teaching opportunity for PhD students and postdoctoral researchers and a chance to develop their transferable skills.

Teaching Experience and Pedagogy:

University of Cambridge Engineering Department, 2013-2014

Supervisions (small group teaching):

- Undergraduate (second year) engineering materials science
- Undergraduate (second year) engineering biomaterials

Demonstrations (lab practicals):

- Undergraduate (second year) engineering materials science
- Undergraduate (third year) engineering biomaterials

University of Kentucky College of Engineering, 2009-2011

Peer Tutoring:

- Undergraduate introductory physics, calculus, chemistry
- Undergraduate introductory chemical engineering courses
- Associate Fellow of the UK Higher Education Academy via 2013-2014 Teaching Associate Programme, University of Cambridge
- Participant in "Scientists Teaching Science" workshop, May 2013, Bethesda, MD, USA
- Participant in "College Teaching for the 21st Century" workshop, June-July 2013, Bethesda MD, USA

Selected Funding Proposals and Awards:

- December 2018, University of Bristol Faculty of Engineering Pump-Priming Fund (£5000)
- August 2011- March 2016, National Institutes of Health-Oxford/Cambridge Graduate Partnership Program (fully-funded PhD program, covering University course and laboratory fees, travel, and stipend, approximate value of £200,000)
- 2011 National Science Foundation Graduate Research Fellowship (Declined offer)
- 2010 Barry M. Goldwater Scholarship Recipient (full tuition award)

Publications:

1. **Shapiro JM** and Oyen ML. Engineering Approaches for Understanding Osteogenesis: Hydrogels as Synthetic Bone Microenvironments. *Hormone and Metabolic Research*, 48(11) (2016), 726-736. DOI: 10.1055/s-0042-100469
   - **Invited Review**
2. Liu S, **Shapiro JM**, Saloustros E, Stratakis CA. Bone abnormalities in mice with protein kinase A (PKA) defects reveal a role of cyclic AMP signaling in bone stromal cell-dependent tumor development. *Hormone and Metabolic Research,* 48(11) (2016), 714-725. DOI: 10.1055/s-0042-117111
3. Bush BG, **Shapiro JM**, DelRio FW, Cook RF and Oyen ML. Mechanical measurements of heterogeneity and length scale effects in PEG-based hydrogels. *Soft Matter*, 11(36) (2015), 7191-7200. DOI: 10.1039/C5SM01210D
4. **Shapiro JM** and Oyen ML. Viscoelastic analysis of single-component and composite PEG and alginate hydrogels. *Acta Mechanica Sinica*, 30(1) (2014), 7-14. DOI: 10.1007/s10409-014-0025-x
5. **Shapiro JM** and Oyen ML. Hydrogel Composite Materials for Tissue Engineering Scaffolds, *JOM* 65 (2013), 505-516. DOI: 10.1007/s11837-013-0575-6
6. Meenach SA, **Shapiro JM,** Hilt JZ and Anderson KW. Characterization of PEG–iron oxide hydrogel nanocomposites for dual hyperthermia and paclitaxel delivery, *Journal of Biomaterials Science, Polymer Edition*, 24(9) (2012), 1112-1126. DOI: 10.1080/09205063.2012.741321

In Review:

1. Khoo W, Chung SM, Lim SC, Low CY, **Shapiro JM** and Koh CT. Fracture Behavior of Multilayer Fibrous Scaffolds Featuring Microstructural Gradients. In review for *Materials & Design*.

In Preparation:

1. **Shapiro JM**, Drinkwater BW and Fraser M. Sonolithography: An in-air ultrasonic patterning technique. Intended for submission to *Nature Communications*.
2. **Shapiro JM**, Gregson F, Reid JA, Drinkwater BW, Fraser M. Experimental determination of particle size threshold for node/anti-node localisation in acoustic pressure field. Intended for submission to *Lab on a Chip*.
3. **Shapiro JM**, Liu J, Drinkwater BW, Fraser M. Microscale, size based, in-air particle separation via sonolithography. Intended for submission to

Outreach, Engagement and Service:

- February 2019 – Present: Postdoctoral Research Associate Representative for the University of Bristol School of Computer Science, Electrical and Electronics Engineering and Engineering Mathematics
- December 2018 – Present: Health and Safety Manager of Bristol Interaction Group workshop
- November – December 2018: University of Bristol Academic Staff Development "Cultivating Leadership" course
- Operated a research fair stand for the 2018 European Researcher's Night (Bristol FUTURES) to present aspects of my research to schoolchildren and adult members of the community
- Interviewed in 2014 for the National Institute of Child Health and Human Development newsletter to discuss my experience with interdisciplinary research https://science.nichd.nih.gov/confluence/pages/viewpage.action?pageId=86443503
- 2014 SET for Britain Finalist – Selected to present a poster of my PhD research to Members of UK Parliament
- Published an "In Person" editorial on *Science Careers* website, entitled *Can I Get a Ph.D. in Collaboration?* (2013). http://sciencecareers.sciencemag.org/career_magazine/previous_issues/articles/2013_09_11/caredit.a1300196
- 2013 and 2015 National Institutes of Health Graduate Student Retreat Committee
- Poster judge for 2013 and 2015 National Institutes of Health Post-Baccalaureate Poster Day, Bethesda, MD, USA

584984 An Unsteady State Reactor Engineering Model for Spatially Heterogeneous Wine Fermentations

Red wine fermentations are performed in the presence of grape skins and seeds to ensure extraction of color and other phenolics. The presence of these solids results in two distinct phases in the fermentor, as the solids float to the top to form a "cap." Modeling of red wine fermentations is, therefore, complex and must consider spatial heterogeneity to predict fermentation kinetics. We have developed a reactor engineering model for red wine fermentations that includes the fundamentals of fermentation kinetics, heat transfer, diffusion, and compressible fluid flow via Finite Element Analysis in COMSOL. A kinetic model for the extraction of phenolics from grapes as a function of temperature and ethanol concentration was derived, incorporating the release of phenolics from grape cells, the adsorption equilibria of phenolics onto grape cell wall material, and the reaction of phenolics in solution. This extraction model was then combined with the spatial-temporal fermentation reactor engineering model, enabling prediction of phenolic extraction as a function of fermentation progression and winemaking conditions.

Research Interests: Dr. Miller's research interests include bioprocess engineering (both fermentation and downstream processing), food process engineering, process intensification, and separations processes.

Teaching Interests: Dr. Miller's teaching interests center around process engineering, such as separations, reaction kinetics, unit operations laboratories, process design, and biochemical engineering.

585377 Optoelectronic Devices and Smart Biomaterials

Research Interests

My work involves the development of active biomaterials and bioelectronic devices to understand and promote tissue repair and introduce interfaces between biological tissues and electronics. I aim to build a laboratory that presents tools and platforms that can be shared with and be accessible to researchers of diverse specialties. Our work will also overcome the inherent limitations of electrical stimulation for tissue modulation, which has limited the development of closed loop systems by utilizing optical stimulation and smart biomaterials when necessary.

I have established myself in the field of nerve guidance scaffolds (8 papers, 2 patents). I have also developed optoelectronic devices in the last 3 years of my training as a postdoctoral fellow at MIT's bioelectronics laboratory. More specifically, my work led to the development of an easily reproducible device to enable studying optical stimulation in the spinal cord and the peripheral nerves.

I aim to bridge the gap between biomaterials and optical/electronic devices with 3 main research areas:

1. Study the effects of optically stimulating selective neuron types on nerve repair in combination with nerve guidance scaffolds. Pursue a combinatorial approach to spinal cord repair through a collaborative effor
2. Develop a neural interface to connect amputated limbs with active prosthetics. Nerves from the amputed limb will grow into an active nerve guidance scaffold consisting of multiple electrodes to provide dense neural recording and stimulation. The recordings are communicated to and from robotic limbs to enable closed-loop motions.
3. Bioelectronic device design for bladder control after spinal cord injury. Closed-loop actuators with integration of smart biomaterials for closed-loop monitoring and control.

Experiences related to each proposed research area:

1) At MIT, we developed fully implantable LED-embedded neural probes. With the aid of an implanted programmed microcontroller, the LED turns on and off based on pre-determined parameters. This invention enables the transformation of using genetically-precise optical stimulation to the peripheral nerves and spinal cord.

I see a future where we have understood the role of major if not all neuronal types in the spinal circuitry. We will have also built strong collaborations in which different promising approaches currently explored in isolation from one another including cell grafts, pharmacology and closed loop systems are combined to find a combinatorial therapy for spinal cord injury.

2) I have developed nerve guidance scaffolds at unprecedented open volume percentage and introduced a technique to mass-produce scaffolds from a variety of materials that can mimic the complex outer geometry of a targeted nerve without limitation in size. We will develop the next generation of microchannel scaffolds consisting of electrodes that interface with active prosthetics.

3) Bladder control is one of the main obstacles individuals with paralysis live with. Most research has focused on electrical stimulation of innervated nerves to functionalize the organs through a feedback mechanism. However, electrical stimulation leads to pain and eventually becomes ineffective. The delivery of biochemicals to excite and inhibit neurons suffers from shortcomings of no adequate local drug delivery approach and side effects caused by the biochemicals. I will develop smart biomaterials that cause the relaxation and contraction of the involved muscles through a feedback system to enable voiding on-demand.

Selected Publications/Patents:

- **Dena Shahriari**, Gabriel Zi Loke, Ian Tafel, PoHan Chiang, Yoel Fink, Polina Anikeeva. Scalable

Fabrication of Porous Microchannel Nerve Guidance Scaffolds with Complex Geometries. *Advanced Materials*. 1902021. (2019).

- Polina Anikeeva, **Dena Shahriari**, Yoel Fink, Gabriel Loke, Ian Tafel. "Structures with Complex Geometries and Controlled Porosity in Micrometer to Meter Dimensions Produced at Large Scale". *US Provisional patent*. 62/772968. 2018.

- Jeff Sakamoto, **Dena Shahriari**, Mark Tuszynski, Wendy Campana, Jacob Koffler. "High microchannel volume scaffolds for nerve repair". *US patent*. 62/238,506. 2017.
- Kendell Pawelec*, Jacob Koffler*, **Dena Shahriari**, Angelica Galvan, Mark Tuszynski, Jeff Sakamoto. "Microstructure and in vivo Characterization of Multi-channel Nerve Guidance Scaffolds". *Biomedical Materials*. 13 (4), 044104. (2018).
- **Dena Shahriari**, Masataka Shibayama, Daniel Lynam, Kayla Wolf, Go Kubota, Jacob Koffler, Mark Tuszynski, Wendy Campana, Jeff Sakamoto. "Peripheral nerve growth within a hydrogel microchannel scaffold supported by a kink-resistant conduit". *Journal of Biomedical Materials Research Part A*. 105 (12), 3392. (2017).
- **Dena Shahriari**, Jacob Koffler, Wendy Campana, Mark Tuszynski, Jeff Sakamoto. "Hierarchically ordered porous and high volume poly caprolactone (PCL) microchannel scaffolds enhanced axon growth in transected spinal cords". *Tissue Engineering Part A*. 23 (9-10), 415. (2017).
- **Dena Shahriari**, Jacob Koffler, Daniel Lynam, Mark Tuszynski, Jeff Sakamoto. "Multichannel alginate scaffolds for spinal cord repair: An *in vitro* and *in vivo* assessment". *Journal of Biomedical Materials Research Part A*. 104 (3), 611. (2015).
- Daniel Lynam, **Dena Shahriari**, Kayla Wolf, Phil Angart, Jacob Koffler, Mark Tuszynski, Christina Chan, Patrick Walton, Jeff Sakamoto, "Brain Derived Neurotrophic Factor Release from Layer-by-Layer Coated Agarose Nerve Guidance Scaffolds". *Acta Biomaterialia*. 18, 128. (2015).

Proposals:

K99 Pathway to Independence. Not awarded. Role: PI

Massachusetts Department of Public Health grant. Not awarded. Role: co-PI Craig Neilsen Spinal Cord Injury Postdoctoral Fellowship. Awarded. Role: PI Hanna Grey Postdoctoral Fellowship. Not awarded. 2018. Role: PI

Selected Awards:

2018 Craig Neilsen Spinal Cord Injury postdoctoral fellowship

2017 MIT's Materials Day best poster award

2016 Biomaterials Innovation Research Center award for scientific excellence

2015 Univ. of Michigan best poster award in polymer engineering

2014 Univ. of Michigan/Michigan State Univ. chemical engineering blue/green seminar best poster award

2010 NSF graduate student fellowship

Teaching Interests

I would be interested in teaching core chemical engineering courses as well as courses related to materials science and neural engineering both at the graduate and undergraduate level. The courses I would be particularly enthusiastic about teaching are chemical reaction engineering, transport processes, biomaterials and biosurfaces, polymer science, biological processing, soft materials and chemical products. I would also like to develop courses about optical, electrical and magnetic properties of materials and neuroelectronic interfaces.

586987 Molecular Engineering Approaches Towards Platform Immuno-Biomaterials

Research Interests:

My vision is to create an independent research laboratory specializing in molecular engineering approaches towards creating platform immuno-biomaterials for medical application. More specifically, my research group will hybridize my expertise in nanomaterial development, soft matter creation, and immunology to address unmet needs in the biomaterials and drug delivery communities. Toward this end, three of the initial research projects from my group will involve the development of RNA nanomaterials for cancer immunotherapy applications, the engineering of shear-thinning soft material platforms to address autoimmune disorders, and the generation of anti-fibrotic hydrogels to evade the foreign body response.

The vision for my independent research laboratory was shaped by my experience as a NSF GRFP Ph.D. student and postdoctoral fellow in the laboratories of Professors Robert Langer and Daniel Anderson at the *Massachusetts Institute of Technology*. During my Ph.D. studies at the *Massachusetts Institute of Technology*, I gained expertise in nanomaterials science from the standpoints of material development, formulation, as well as *in vitro* and *in vivo* administration. More specifically, I employed molecular engineering approaches throughout my graduate work to create non-viral nanomaterials capable of delivering therapeutic RNAs (including messenger RNAs, short interfering RNAs, and antisense RNAs) to target cells within the body. Therapeutic RNAs can be used for a range of biomedical applications including genomic engineering, cancer immunotherapy, and protein replacement strategies. However, their serum instability, immunogenicity, and limited ability to passively transfect cellular membranes limits their clinical potential. Toward this end, our work explored how fundamental properties of RNA encapsulating nanomaterials (including size, encapsulation efficiency, zeta potential, and particle morphology) influenced the transport and biological mechanism of our nanomaterials to/within target cells in the liver, lung, and spleen. Further highlights of this work involved the development of patented nanotechnologies that are currently being pursued as lead delivery vectors for the treatment of human disease with a pharmaceutical partner.

After completing my Ph.D. studies, I remained in the laboratory of Professor Robert Langer at the *Massachusetts Institute of Technology* to develop expertise on the molecular engineering of soft matter for immunological research. To date, the most significant research contribution to my postdoctoral studies involved the development of an injectable hydrogel platform for the recruitment of specific immune cell populations to the site of injection. Further, my work also involved the molecular engineering and fabrication of a catalyst and initiator free hydrogel platform with tunable rheological properties for use in 3-dimensional cell culture and tissue engineering applications. We are also currently exploiting the scalability and batch consistency of this second hydrogel platform for biological additive manufacturing processes.

In sum, my future research vision is to create a molecular engineering laboratory that specializes in the creation of platform immuno-biomaterials exploiting concepts from nanomaterials development, soft matter creation, and immunology. In doing so, not only will we aim to better understand the fundamental underpinnings governing biological processes in the body, but we will also work towards the generation of translational technologies for the ultimate goal of treating human disease at the immunological level.

Teaching Interests:

1. *Thermodynamics:* This course will cover the fundamental principles of thermodynamics. Discussion surrounding how changes in temperature, the relationship between heat and work, and transformation of energy will be covered. Particular emphasis will be focused on the understanding of the fundamental underpinnings of thermodynamics with application towards comprehending combustion, fluid flow, and power generation. Select examples of chemical engineering approaches to understanding biological systems will also be provided to supplement representative in class

examples exploring thermodynamic processes.

2. *Introduction to Chemical Engineering*: This course will cover the basic underpinnings of engineering through the analysis of physical and chemical processes. Material and energy balances, kinetics of chemical reactions, staged separations, and material balance will be covered. Current research trends including nanotechnology, biotechnology, and energy will also be discussed to highlight specific concepts from chemical engineering.
3. *RNA delivery from concept to clinic:* This course will cover the historical background, transport principles, and cutting-edge technologies associated with the local and systemic delivery of nucleic acid therapeutics. Proposed mechanisms of cellular uptake as well as the formulation and characterization of polymeric and nanoparticle based materials for non-viral RNA delivery will be discussed. Special units will focus on utilizing immune cell populations as antigen presenting cells and their potential role in developing cancer immunotherapy strategies.

<u>Selected Relevant Publications (25 Total Published Manuscripts and 3 US Patents):</u>

1. Fenton, O.S.; Tibbitt, M.W.; Appel, E.A.; Jhunjhunwala, S.; Webber, M.J.; Langer, R.* "Injectable Polymer-Nanoparticle Hydrogels for Local Immune Cell Recruitment", *accepted*, *Biomacromolecules*, DOI: 10.1021/acs.biomac.9b01129.
2. Fenton, O.S.; Andresen, J.L.; Paolini, M.; Langer, R.* "Β-Amino-acrylate Synthetic Hydrogels: Easily Accessible and Operationally Simple Networks for Biomaterials Application", *Angewandte Chemie*, **2018**, *57*, 16026.
3. Fenton, O.S.; Olafson, K.N.; Pillai, P.; Mitchell, M.J.*; Langer, R.* "Advances in Biomaterials for Drug Delivery", *Advanced Materials*, **2018**, *30*, 1705328.
4. Fenton, O.S.; Kauffman, K.J.; McClellan, R.L.; Kaczmarek, J.C.; Zeng, M.D.; Andresen, J.L.; Rhym, L.H.; Heartlein, M.W.; DeRosa, F.; Anderson, D.G.* "Customizable Lipid Nanoparticle Materials for the Delivery of siRNAs and mRNAs", *Angewandte Chemie*, **2018**, *41*, 13582.
5. Fenton, O.S.; Kauffman, K.J.; Kaczmarek, J.C.; McClellan, R.L.; Jhunjhunwala, S.; Tibbitt, M.W.; Zeng, M.D.; Appel, E.A.; Dorkin, J.R.; Mir, F.; Yang, J.H.; Oberli, M.A.; Heartlein, M.W.; DeRosa, F.; Langer, R.; Anderson, D.G.* "Design and Synthesis of Ionizable Lipid Materials for the In Vivo Delivery of Messenger RNA to B Lymphocytes", *Advanced Materials*, **2017**, *29*.
6. Fenton, O.S.; Kauffman, K.J.; McClellan, R.L.; Appel, E.A.; Dorkin, J.R.; Tibbitt, M.W.; Heartlein, M.W.; DeRosa, F.; Langer, R.; Anderson, D.G. "Bioinspired Alkenyl Amino Alcohol Ionizable Lipid Materials for Highly Potent In Vivo mRNA Delivery", *Advanced Materials*, **2016**, *28*, 2939-2943.

For a complete listing please see: https://scholar.google.com/citations?user=ZSSrYrMAAAAJ&hl=en

586990 Defining Capture and Release Mechanisms of Biochemical Ligands in Multifunctional Biomaterials to Control Cell Function

Research Interests:

Soluble and bound ligands cooperatively act on cells in regulating cell signaling, adhesion, and function. I am interested in bridging the interface of material science and biomedical engineering to design and fabricate a library of synthetic bioactive materials that allow to present combinations of such ligands with tunable binding affinities as a platform to better understand how these cues regulate cell signaling, and as a vehicle to control cellular function. My research plan will focus on developing novel biomaterials that feature capabilities for binding growth factors covalently and noncovalently with defined capture and release mechanisms, and establishing in vitro platforms that can be engineered to optimize the synergistic effects of growth factor signaling and matrix cues to direct cellular behavior in multiple physiological settings. The ability to harness cell behavior using these novel systems will be translated for applications in the area of functional vascularization of tissue constructs, diabetic wound healing, and controlled stem cell or organoid differentiation.

Teaching Interests:

My experiences as a teaching assistant during my graduate school had a significant impact on shaping my goals in pursuing a career in an academic environment. I participated in teaching for both undergraduate-level and graduate-level courses in the Department of Materials Science and Engineering, which included lecture-based "Introduction to Materials Science and Engineering" and laboratory-based "Polymer Synthesis and Characterization Laboratory" classes. Such experience for engaging with students at different levels in various class settings gained myself tremendous passion in teaching. My philosophy and key strategy for teaching include encouraging students in classroom discussion and communication, engaging individual contribution, creating open learning environment, improving core skills such as critical thinking, problem solving and oral/written techniques. I will also focus my energy to create teamwork projects that expose students to fundamental principles for multidisciplinary subjects. My future teaching plan aims to retain and continue to evolve my passion for students growth and the formation of strong mentor/mentee relationships.

Selected Publications: (Total: 16, 11 first author/co-author)

1. **Li, L.**; Stiadle, J. M.; Levendoski, E. E.; Lau, H. K.; Thibeault, S. L. and Kiick, K. L. "Biocompatibility of Injectable Resilin-based Hydrogels", **Journal of Biomedical Materials Research: Part A**, 2018, 106A, 2229-2242.
2. **Li, L.**; Eyckmans, J. and Chen, C. S. "Designer Biomaterials for Mechanobiology", **Nature Materials**, 2017, 16 (12), 1164-1168.
3. **Li, L.**; Stiadle, J. M.; Lau, H. K.; Zerdoum, A. B.; Jia, X.; Thibeault, S. L. and Kiick, K. L. "Tissue Engineering-based Therapeutic Strategies for Vocal Fold Repair and Regeneration", **Biomaterials**, 2016, 108, 91-110.
4. **Li, L.**; Mahara, A.; Tong, Z.; Levenson, E.; McGann, C.; Jia, X.; Yamaoka, T. and Kiick, K. L. "Recombinant Resilin-based Bioelastomers for Regenerative Medicine Applications", **Advanced Healthcare Materials**, 2016, 5 (2), 266-275.
5. **Li, L.**; Luo, T. and Kiick, K. L. "Temperature-Triggered Phase Separation of a Hydrophilic Resilin-like Polypeptide", **Macromolecular Rapid Communications**, 2015, 36 (1), 90-95.
6. **Li, L.** and Kiick, K. L. "Transient Dynamic Mechanical Analysis of Resilin-based Elastomeric Hydrogels", **Frontiers in Chemistry**, 2014, 2 (21), 20-32
7. **Li, L.**; Tong, Z.; Jia, X. and Kiick, K. L. "Resilin-like Polypeptide Hydrogels Engineered for Versatile Biological Function", **Soft Matter**, 2013, 9 (3), 665-673.
8. **Li, L.**; Teller, S.; Clifton, R. J.; Jia, X. and Kiick, K. L. "Tunable Mechanical Stability and Deformation Response of a Resilin-based Elastomer", **Biomacromolecules**, 2011, 12 (6), 2302-2310.

Web Page:

https://www.researchgate.net/profile/Linqing_Li2

https://scholar.google.com/citations?user=nrp3WaoAAAAJ&hl=en

581880 Living Biodevices for Precision Medicine:from Morphing Electronics to Medical Nanorobots

Research Interests:

Biology and medicine have heavily relied on engineering advances to understand how the human body works and to revolutionize traditional medical practices. The rising interest in personalized and precision medicine requires innovative materials and devices which can seamlessly interface and precisely interact within the human body. Our human body is an active and dynamic system with inherent transport, motion, and growth across different time scales. However, synthetic materials and devices are normally made with limited adaptability and mobility, which limit their versatility for precise diagnosis and personalized treatment. This mismatch imposes both challenges and opportunities to develop **active bioelectronic system** which mimic the dynamic living organisms towards personalized and precision healthcare. **My research addresses this problem by developing active bioelectronic system, which can transform, grow, or actuate within body towards both new discovery in both basic science as well as customized medical treatment.** By developing such active bioelectronic system with high adaptability, mobility, and intelligence across nano to macroscales, we will enable a new interface that redefine the interactive relationship between human, environment, other living organisms, environment, tools and intangible data.

Keywords: Bioelectronics, Active Matter, Nanorobotics, Neuroengineering, Soft Robotics, Biomaterials, Wearable Devices, Biosensors, Nanomedicine, Mental Health.

Research Experience:

1. Morphing and elastic bioelectronics for neurological implants and neurochemical sensors, *Department of Chemical Engineering, Stanford University* (advised by Zhenan Bao)

a) Morphing electronics as chronic and adaptable neurological implantable devices

Bioelectronics for modulating nervous system, such as the vagus nerve stimulator, demonstrate great promise in treating various neurological diseases. However, their fixed sizes and shapes cannot accommodate the rapid tissue growth and negatively influence normal developmental functions. For infants, children and adolescents, once the implanted devices are 'outgrown', additional surgeries are usually needed for device removal, followed by replacement. These tedious processes inevitably lead to repeated intervention and elevated complication rates. Although stretchable electronics with high elasticity is ideal for repeated motion and volumetric expansion of organs, they still cannot adapt to developmental tissue growth, which should not be restricted by any strain. We address this limitation with **Morphing Electronics (termed MorphE) which are designed and fabricated to suitably adapt to in vivo nerve tissue growth with zero strain.** This is achieved by viscoplastic electronic materials (VEMs) that eliminate the stress and tissue constraint on growing tissue. In addition, water-insensitive self-healing property of the multilayered MorphE consisting of viscoplastic electrodes and a strain sensor allows fora reconfigurable platform for individualized implantation procedures. Animal study demonstrated that the soft and self-adapting MorphE showed no damage to the rat nerve after grown 2.4 times its initial diameter. This stable neural interface allowed chronic electrical stimulation and monitoring without disruption of functional behavior in developing rats. **Our MorphE creates a new avenue for adaptive pediatric electronics medicine.**

b) Elastic neurochemical interface for *in vivo* interrogation of the gut-brain axis

The bidirectional communication between the gastrointestinal (GI) tract and the brain, commonly dubbed the gut–brain axis, involves various afferent and efferent pathways to regulate aspects of homeostasis such as satiety, hunger, and inflammation. Disruption of the gut–brain axis has been shown to be involved in the pathogenesis of a diverse range of diseases, including depression, Parkinson's disease, and irritable bowel syndrome. Meanwhile, the GI tract has substantial neural interactions with the central nervous system, and this provides opportunities for bioelectric neuromodulation therapy. However, understanding of the gut motility, gut-brain sensory transduction and their roles in pathogenesis remain unknown due to the absense of proper neuroengineering tools that can be used *in vivo*. The GI tract consists of a series of soft organs with inherent peristaltic movement, joined in a long and twisting tube from the mouth to the anus. Unlike the rigid and fragile neuroengineering tools used to study the brain, a bioelectrical device used to study the GI tract in awake, freely moving animals, would need to be soft and elastic in order to be compatible with the GI environment. As a result, conventional electrophysiological and neurochemical recording in the GI tract of awake animals has been impossible because no such device exists. **I developed a bioelectronic toolkit, based on tissue-like elastic materials, for *in vivo* interrogation of the gut-brain axis.** The toolkit includes: (1) neurochemical sensors to monitor gut serotonin, which is a critical link of gut–brain communication; (2) strain sensor to monitor the gut motility; (3) microelectrodes for monitoring gut myoelectric activity and delivering electric stimulation. **This multifunctional device is then used to monitor serotonin dynamics, gut myoelectric activity, and peristalsis movement along the GI tract of freely moving wild-type mice, mice with depression-like or disease-like phenotypes.** The new elastic bioelectronic tool will provide novel insights about the dynamics of gut serotonin and lay the foundation for the development of transformative diagnosis and treatment for patients with psychiatric disorders or GI dysmotility disorders through local intervention in the GI tract.

2. Medical micro/nanorobots for drug delivery. *Department of NanoEngineering, University of California San Diego* (advised by Joseph Wang and Liangfang Zhang)

My PhD research has been focusing on creating a new class of robots for an adventurous journey: micro/nanoscale robots, which mimic the intelligence of microorganism and work as autonomous medicine to fight diseases. **I pioneered the first use of nanorobot to effectively treat disease in a live animal's body, a milestone to realize the dream of "Fantastic Voyage".**

Designing miniaturized and versatile robots of a few micrometers and less would allow access throughout the whole human body, leading to new procedures down to the cellular level and offering localized diagnosis and treatment with greater precision and efficiency. A significant challenge of micro/nanorobots is the exploration energy sources for biocompatible in vivo operation, and then the translation of these inventions into clinical use. I developed a set of microscale machines, with dimensions similar to that of cancer cell or bacteria, which could smartly propel and target disease sites under their own power via reacting with body fluid such as stomach acid and intestinal fluid. Upon arrival inside of the body, these self-powered machines act in many ways—from directly mediating the physiological factor of local medical conditions, to releasing a medical payload at the target site, or penetrating the tissue so that the loaded medicine last longer in the body. After the mission, the microrobots would degrade themselves, leaving nothing toxic behind.

Specifically, my recent work has proven that microrobot-enabled delivery approach is a promising new method for easy and effective treatment of stomach and gastrointestinal tract diseases (published on Nature Communications 2017, 8, 272; Angewandte Chemie International Edition 2017, 56, 2156; Science Robotics 2017, 2, eaam6431). Drugs used to treat stomach diseases are normally taken with additional substances, called proton pump inhibitors, to suppress gastric acid production. But proton pump inhibitors can cause adverse side effects including headaches, diarrhea, or depression. The micromotors developed have a built-in mechanism to neutralize gastric acid and effectively deliver their drug payloads in the stomach, without the use of proton pump inhibitors, and further enhance the drug retention in the body through the tissue penetration. In the disease model of *helicobacter pylori* infection, my results show that treatment with microrobots kill the bacteria 6 times more efficiently than with conventional treatment, and obviates the use of proton pump inhibitor, the most widely sold drugs in the world. **This research on the in vivo operation of micromotors represents the first example of using microrobots to treat disease in live animals. In this emerging research direction, a plethora of continuously successful tests will need to be run before the micro/nanorobots can be used in human trial. These initiative steps of translating new micro/nanorobot from test tubes to practical treatments in live animals, are vital in what could be a commonplace, precise, targeted, and safe alternative to traditional high-dose medications.**

Selected Awards:

2019/06 MIT Technology Review 35 Innovators Under 35 (TR35, Global List)

2018/08 Young Investigator Award, Division of Inorganic Chemistry, American Chemical Society

2018/07 EUDRAGIT Award, North American Region, Evonik Corporation

2017/08 Young Investigator Innovation Award, 1st International Conference on Micro/Nanomachines

2016/09 Siebel Scholar, bioengineering category, Thomas and Stacey Siebel Foundation

2016/05 Dan David Prize Scholarship in Nanoscience, The Dan David Foundation, Israel

2016/03 UCSD Interdisciplinary Research Award

2015/11 A.T. Kearney Scholarship, Falling Wall Foundation, Germany

2015/10 Brain Prize for Breakthrough Ideas in Neurotechnology, San Diego Brain Consortium

2015/04 Rudee Best Poster Award and NanoEngineering Best Poster Award, UCSD Research Expo

2015/04 Materials Research Society Graduate Student Award

Successful Proposals:

1. Stanford Bio-X Interdisciplinary Initiatives Program Seed Grant: "Closed-loop neurochemical sensing and modulation system for treating psychiatric disorders". 2018/07.
2. Stanford Merck Initiative: "A Soft Haptic Interface Based on Liquid Crystal Elastomer". 2018/04.
3. Kavli Institute for Brain and Mind Innovative Research Grant Award (Student PI): "Development of a nanorobotic toolkit for ultrahigh precision neuron targeting and manipulation. 2015/06.

Selected Publications (* denotes equal contribution) :

1. **J. Li** *, Y. Liu *, B. Chen, E. Spear, L. Yuan, L. Beker, J. B.-H. Tok, Y. Cui, A. Habtezion, X. Chen, Z. Bao. "NeuString: an Elastic Neurochemical Interface for Brain and Gut". In preparation.
2. Y. Liu *, **J. Li** *, S. Song *, J. Kang, Y. Tsao, S. Chen, W. Xu, Y.-Q. Zheng, J. B.-H. Tok, P. M. George, Z. Bao. "Morphing Electronics for Growing Tissue". Nature 2019, under review.
3. C. Fang *, **J. Li** *, Y. Zhang, J.Z. Lee, Y. Yang, F. Yang, J. Alvarado, M.A. Schroeder, L. Yang, M. Cai, J. Gu, K. Xu, X. Wang, Y.S. Meng. "Quantifying Inactive Lithium in Lithium Metal Batteries". ***Nature*** 2019. In press.
4. **J. Li** *, P. Angsantikul *, W. Liu, B. Esteban, X. Chang, E. Sandraz, Y. Liang, S. Zhu, Y. Zhang, C. Chen, W. Gao, L. Zhang, J. Wang. "Biomimetic Platelet-Camouflaged Nanorobots for Binding and Isolation of Biological Threats". ***Adv. Mater.*** 2018, 1704800.
5. **J. Li**, B. Esteban, W. Gao, L. Zhang, J. Wang. "Micro/Nanorobots for Biomedicine: Delivery, Surgery, Sensing, and Detoxification". ***Science Robotics*** 2017, 2, eaam6431. Highlighted in *IEEE Spectrum*.
6. **J. Li** *, P. Angsantikul *, W. Liu *, B. Esteban *, S. Thamphiwatana, M. Xu, E. Sandraz, X. Wang, J. Delezuk, W. Gao, L. Zhang, J. Wang. "Micromotors Spontaneously Neutralize Gastric Acid for pH-Responsive Payload Release". ***Angew. Chem. Int. Ed.*** 2017, 56, 2156–2161. Highlighted in *New Scientist, Science Daily*.
7. Esteban *, P. Angsantikul *, **J. Li** *, M. Lopez, D.E. Ramírez, S. Thamphiwatana, C. Chen, J. Delezuk, R. Samakapiruk, V. Ramez, L. Zhang, J. Wang. "Micromotor-Enabled Active Drug Delivery for *In Vivo* Treatment of Stomach Infection". ***Nat. Commun.*** 2017, 8, 272. Highlighted in *NPR, The Walt Street Journal*.
8. **J. Li**, P. Angsantikul, W. Liu, B. Esteban, X. Chang, E. Sandraz, Y. Liang, S. Zhu, Y. Zhang, C. Chen, W. Gao, L. Zhang, J. Wang. "Biomimetic Platelet-Camouflaged Nanorobots for Binding and

Isolation of Biological Threats". ***Adv. Mater.*** 2017, 1704800.

9. **J. Li** *, X. Yu *, M. Xu, W. Liu, E. Sandraz, H. Lan, J. Wang, S. Cohen. "Metal-Organic Frameworks as Micromotors with Tunable Engines and Brakes", ***J. Am. Chem. Soc.*** 2017, 139, 611–614. Highlighted in *Nature Review Chemistry*.

10. **J. Li**, W. Liu, T. Li, I. Rozen, J. Zhao, B. Bahari, B. Kante, J. Wang. "Swimming Micro-Robot Optical Nanoscopy", ***Nano Lett.*** 2016, 16, 6604–6609. Highlighted in *Chemistry World*.

11. **J. Li** *, S. Thamphiwatana *, W. Liu *, B. Esteban, P. Angsantikul, E. Sandraz, J. Wang, T. Xu, F. Soto, V. Ramez, X. Wang, W. Gao, L. Zhang, J. Wang. "Enteric Micromotor Can Selectively Position and Spontaneously Propel in the Gastrointestinal Tract", ***ACS Nano*** 2016, 10, 9536–9542. Highlighted in *Chemical & Engineering News*.

12. **J. Li**, I. Rozen, J. Wang. "Rocket Science on the Nanoscale", ***ACS Nano*** 2016, 10, 5619–5634.
13. **J. Li***, O. Shklyaev *, T. Li *, W. Liu, H. Shum, I. Rozen, A. C. Balazs, J. Wang. "Self-Propelled Nanomotors Autonomously Seek and Repair Cracks", ***Nano Lett.*** 2015, 15, 7077–7085. Highlighted in *PNAS, Chemical Engineering News*.

14. **J. Li***, T. Li *, T. Xu *, M. Kiristi, W. Liu, Z. Wu, J. Wang. "Magneto-Acoustic Hybrid Nanomotor", ***Nano Lett.*** 2015, 15, 4814–4821.
15. Zhu *, **J. Li***, Y. J. Leong, I. Rozen, X. Qu, R. Dong, Z. Wu, W. Gao, P. H. Chung, J. Wang, S. Chen. "3D-Printed Artificial Microfish", ***Adv. Mater.*** 2015, 27, 4411–4417. Highlighted in *Wired, Discovery News, Forbes, Guardian, The Washington Post,* etc.

16. **J. Li***, W. Gao *, R. Dong, A. Pei, S Sattayasamitsathit, J. Wang. "Nanomotor Lithography", ***Nature Commun.*** 2014, 5, 5026.

Teaching and Service

2017/3-2017/6 Teaching assistant of CHEM 240 "Electrochemistry", UCSD

2016/3-2016/6 Teaching assistant of NANO 244 "Nanomachines and Nanorobots", UCSD

2011/9-2012/1 Teaching assistant of laboratory class "Nanofabrication", FDU

2017/9-now Mentor of "Second-Year Mentoring" of Stanford chemical engineering PhD program

2012/9-now Mentored 16 undergrads and 2 high school students, co-authored 30 publications

2016/8-2017/6 Academic planning committee members of UCSD YKAVLI

2016/9 Panelist of UCSD STEM PhD student orientation

2016/8 Outreach demo for high school students of Johns Hopkins Centre for Talented Youth

2016/7 Guest speaker for 5th graders of Ocean Discovery Institute STEM Professional

2016/8 Volunteer of UCSD Moores Cancer Center Luau and Legends of Surfing Invitational

2016/8 NanoEngineering Session Host of 2016 UCSD Summer Research Conference

2012/5 Curator of Fudan NanoArt Exhibition

Teaching Interests:

I am interested in teaching most of the major courses for Chemical Engineering, especially Heat and Mass Transfer, Fluids Dynamics, and Thermodynamics, Separation Process. I am also interested in other courses including Electrochemistry, Soft Materials, Nanotechnology and Nanofabrication, Nanorobotics, Nanomedicine.

581917 Soft, Flexible Tissue-Integrated Chemical Sensors: From Wearable to Implantable Neural Systems

Research Interests:

Postdoctoral Experience.

PI: John Rogers, Northwestern University.

Project: Wireless, battery-free, bio-integrated chemical sensors for wearable and implantable applications.

The rich composition of solutes and metabolites in sweat and the relative ease with which it can be noninvasively collected make this class of biofluid highly attractive for diverse applications in ambulatory health diagnostics and sports performance. Advances in biotechnology, nanoscience, materials science and electrical engineering lay the foundations for recent demonstrations of skin-interfaced devices that continuously monitor biochemicals in human perspiration. However, majority of the reported systems require complex electronics and batteries resulting in bulky, large footprint devices that are cumbersome to the users. I address this issue by developing first examples of wireless, battery-free, skin-integrated sweat analyses and capturing systems that are nearly 20 times lighter and 4 times smaller than the most sophisticated sweat sensors reported to date. Such devices result from an unusual combination of enzymatic biofuel cell-based electrochemical sensors, colorimetric assays, advanced microfluidics with passive capillary burst valves, and near field communication (NFC) protocols.

A widespread route to understanding neural circuitry relies on optical/electrical stimulation at precise regions of the brain, followed by subsequent neurochemical measurements. Unfortunately, at present, stimulation and sensing functions each involve completely distinct platforms and components. The need for inserting multiple separate, wired components, and inability to precisely control the spatial orientation of these components during insertion significantly compromises animal behavior and impedes correct interpretation of the data. I am addressing this important issue by developing a new class of integrated wireless, battery-free neural implants for simultaneous optical stimulation and neurochemical sensing. The microneedle comprises of a μLED for precise optical stimulation of dopaminergic neurons and an electrochemical microsensor for dopamine detection. An unconventional amalgamation of disparate technologies, such as, lithography, laser patterning, electrochemical nano-structuring, polymer science and wireless electronics results in seamless fabrication of the dopamine sensor adjacent to the optical stimulation components. Separately, developed mathematical models help predict the device's sensing characteristics and obtain 3D maps of dopamine release from the neurons. Studies with freely moving mice are presently underway.

PhD Experience.

PI: Joseph Wang, UC San Diego

Project: Novel ink formulations for all-printed stretchable and self-healing electrochemical devices for wearable sensing and energy applications.

Conventional electrochemical systems are rigid and hence cannot be easily conformed to the soft, curvilinear structures of the human skin. I addressed this major issue by pioneering the development of novel printable inks for realizing first examples of highly stretchable and self-healing electrochemical devices. The all-printed carbon nanotube and conducting polymer based stretchable devices developed by me exhibit electrochemical properties similar to conventional, rigid electrodes. However, unlike the traditional electrodes, my devices can easily accommodate strains upto 500% with negligible impact on their electrochemical properties. I leveraged such systems to obtain stretchable glucose and ammonia sensors for real-time sweat sensing. Similarly, I also developed a high power density lactate-based biofuel cell for scavenging energy from the human sweat to power wearable devices. The biofuel generated ~1mW/cm^2 which represents nearly 10 times higher power density compared to previous examples and enabled me to power a Bluetooth radio using the wearable biofuel cell. Separately, I developed a new class of printable inks that result in devices with self-healing properties similar to the human skin. Such smart systems are attractive for wearable applications since they can recover their function autonomously if damaged during the ambulatory functions performed by the body. The self-healing agent encapsulated microcapsules and magnetic microparticles-based systems self-restored their electrochemical properties within seconds of complete mechanical damage. I demonstrated the practical viability of such intelligent systems by developing self-healing sodium sensor and zinc-silver oxide battery for wearable applications.

Future Directions:

As a faculty member I would like to establish a highly interdisciplinary research group that builds upon my expertise in biochemical sensing, biofuel cells, batteries, and printed electronics to address some of the key challenges in wearables, neuroscience, printed electronics and pollution sequestration. (i) Wearable chemical sensors: My group will develop next generation of wearable systems for metabolomic, proteomic and peptidomic analysis of sweat for deeper understanding of the sweat chemistry and its relation to the human physiology. (ii) Neural implants: We will aspire to develop new classes of brain implantable systems for simultaneous stimulation, sensing and targeted drug delivery at specific regions of the brain that will provide us deeper understanding of how the brain functions and help us test new drugs and develop new strategies to address scourge of neurodegenerative diseases such as Alzheimer's and Parkinson's. (iii) Self-assembly enhanced printing of high-performance devices: My group will leverage my expertise in developing nano/micro-particle ink formulations to realize new classes of inks and printing processes that, unlike present inks and printing techniques, will lead to precise orientation and self-assembly of particles within the printed devices for various energy storage and sensing applications. Such printed devices will exhibit superior performance to their counterparts developed using conventional printing methods since the highly ordered nano/micro-particles within the printed systems will play a key role in enhancing their energy storage and sensing characteristics. (iv) Pollution sequestration for fabricating printed electronics: We will develop new strategies that leverage my knowledge of printable inks, fuel cells, batteries, and sensors to sequester particulate matter pollutants from the environment and transform them into printable inks for various printed electronics applications.

Selected Publications

Total papers: 46

First/co-first author papers: 24

H-index: 29

Total citations > 3700

1. "Soft, skin-interfaced microfluidic systems with passive galvanic stopwatches for precise chronometric sampling of sweat", **A. J. Bandodkar** et al, ***Adv. Mater.*** 2019 (DOI: https://doi.org/10.1002/adma.201902109)
2. "Battery-free, skin-interfaced microfluidic/electronic system for simultaneous electrochemical, colorimetric & volumetric sweat analysis", **A. J. Bandodkar** et al, ***Sci. Adv.*** 2019, **5**, eaav3294.
3. "Soft, skin-integrated multi-functional microfluidic systems for accurate colorimetric analysis of sweat biomarkers and temperature", J. Choi, **A. J. Bandodkar** et al, ***ACS Sensors*** 2019, **4**, 379.
4. "Soft, stretchable, high power density electronic skin-based biofuel cells for scavenging energy from human sweat" **A. J. Bandodkar** et al, ***Energy Environ. Sci.***, 2017,**10**, 1581.
5. "All-printed magnetically self-healing electrochemical devices", **A. J. Bandodkar** et al, ***Sci. Adv.*** 2016, **2**, e1601465.
6. "A wearable chemical–electrophysiological hybrid biosensing system for real-time health and fitness monitoring", S. Imani, **A. J. Bandodkar** et al, ***Nat. Commun.*** 2016, **7**, 11650.
7. "Highly Stretchable Fully-Printed CNT-based Electrochemical Sensors and Biofuel Cells: Combining Intrinsic and Design-induced Stretchability", **A. J. Bandodkar** et al, ***Nano Lett.*** 2016, **16**, 721.
8. "All-Printed Stretchable Electrochemical Devices", **A. J. Bandodkar** et al, ***Adv. Mat.*** 2015, **27**,

3060.

Teaching Interests:

Teachers are important pillars in one's success and I owe my accomplishments to the teachers in my past and present who instilled scientific curiosity in me, taught me scientific skills, helped me overcome my weaknesses, and guided me in tackling challenging tasks. A major motivation for me to become a faculty member is to partake in the noble occupation of teaching and guide young minds towards a successful, fruitful and content life. During my PhD and now as a postdoc, I have mentored 23 undergraduate students of which 8 are currently PhD students in top tier schools; 5 work in industry and 10 are pursing undergraduate degrees. I have also been a Lecturer/TA for three graduate level courses (~50 students in each course). Apart from helping students understand the concepts during office hours, on occasions, I have also given lectures on behalf of the concerned professor. Considering my ability to explain complex scientific concepts in a simple manner to people without STEM background and my passion to motivate people towards research, I was invited as a guest lecturer for an undergraduate course (~150 students) on "Designing Information" at the UC San Diego's Fine Arts Department. In this lecture, I illustrated various ways in which wearable devices offer an exciting avenue for achieving personalized Internet of Things and how these devices are having an impact in today's day and age. I was surprised and overjoyed by the overwhelming interest shown by these Fine Art students, whom till then had given little thought to this technology. In fact, one of the students even collaborated with me on a project that combined chemical sensors with fine arts – a truly unique learning experience for both of us. In addition to these, I have also been invited as a speaker at the University's undergraduate outreach events to motivate and encourage young minds to take up careers in the STEM fields. As a faculty member I would be thrilled to teach courses on mass and heat transfer, thermodynamics, electrochemistry, chemical sensors, nanobiotechnology and surface science. Additionally, I would be excited to get involved in committees focused on increasing student participation in STEM fields.

582114 Developing Tools for Emerging Liquid Biopsy Applications

Research Interests:

My Ph.D. work at Texas A&M University focused on developing methods to optically probe surface interactions of microspheres and biomimetic systems such as polymer vesicles. The combination of theoretical, modeling, computational, and experimental approaches produced a significant breakthrough using an interference-based technique to enable fast and nanometer-scale reconstruction of the geometry of curved microscopic objects near surfaces. During my post-docs at the University of California, Los Angeles, and Texas A&M, I have grown as a researcher in the development of next-generation imaging and sensing platforms based on lens-free computational microscopy, label-free quantitative imaging tools, smartphones, and microfluidics, applied to emerging liquid biopsy analysis in my most recent work.

To pursue research as a faculty, I am uniquely equipped to tackle interdisciplinary problems at the interface between engineering, optics, biology, and medicine. In particular, personalized medicine enabled by liquid biopsy of bodily fluids for early disease diagnosis, treatment monitoring, and patient prognosis, faces important technological challenges depending on the targeted biomarker (e.g., bulk fluid, cells, exosomes, and cell-free nucleic acids). For instance, antibody-based approaches for the detection of circulating tumor cells (CTCs), cancer cells in circulation that could be responsible for metastasis at distant sites, typically involve complex workflows that potentially miss metastatic cells not targeted by the antibodies used. Alternatively, label-free interference-based quantitative imaging can be performed at the single-cell level for identification of highly metastatic cells using label-free fingerprints such as adhesion and filopodia dynamics. This approach overcomes limitations of conventional analysis methods while offering great potential for label-free phenotyping of cells in a broader context. Similarly, liquid biopsy is not limited to cancer, with advancements in exosome-based analysis enabling minimally invasive access to a wide range of pathologies, including diabetes, inflammation, acute kidney injury, concussion, cardiovascular disease, alcoholic liver disease, and central nervous system diseases.

The development of point-of-care (POC) tools for analysis of bodily fluids such as blood is thus fundamental in the routine establishment of liquid biopsy in the medical field. Microfluidic platforms that process small sample volumes (microliters) in portable easy-to-operate and low-cost set-ups are called to play a central role in such applications. As an example, blood viscosity analysis at the POC can be enabled by self-driven capillary flow in microchannels that are large enough to be monitored and recorded using a smartphone. Such capillary filling dynamics span an important range of shear rates that can be modulated to accurately probe the rheological behavior of complex fluids. This will enable probing blood viscosity as a potential biomarker in diabetes and cardiovascular diseases, which has been hindered by the complexity associated with conventional methods to measure blood rheology.

Therefore, building on my expertise and skills, I aim to integrate novel and powerful imaging and sensing tools and approaches into portable and low-cost liquid biopsy analysis platforms to facilitate personalized medicine. This will be achieved through fundamental research, a combination of theoretical, modeling, computational, and experimental approaches, and emphasis on the development of platforms for high-throughput analysis of bio-colloidal systems. Recent advances in lens-free computational microscopy and smartphone-based microscopy and sensing aided with artificial intelligence techniques represent extremely valuable tools for POC applications and are expected to play a key role in my future research.

Teaching Interests:

Teaching in chemical engineering offers a unique opportunity for motivating students, especially at the undergraduate level, by providing a variety of real-life scenarios where new concepts can be applied, even to understand natural phenomena around us. This is fundamental in my teaching philosophy, where inquisitive questions (e.g., how do tall trees transport water from their roots to the top?) are presented to the students before a new topic (e.g., capillary phenomena) is introduced. In addition, interactive in-classroom experiences that help measure/visualize/illustrate content being learned are integral part of my approach. For this purpose, smartphones can be readily exploited as powerful scientific/educational tools, capturing images, videos, and incorporating virtual and augmented reality technologies. For instance, I have worked with Honors students in the Fluid Operations course to develop a portable smartphone-enabled capillary-based viscometer. It is also important to recognize that knowledge is not complete, and concepts learned in classrooms could lead to more questions and technological developments. Thus, I finish course materials by making students wonder, highlighting recent progress and challenges ahead in the ever-evolving chemical engineering field.

I feel confident in my ability to teach most of the traditional subjects in the chemical engineering curriculum at the undergraduate level, and my preferences are Transport Phenomena and Process Dynamics and Control. I am also interested in developing an elective course on colloidal science and quantitative imaging methods that can be taken by both graduate and undergraduate students.

582117 Interfacing Cell/Tissue Engineering with Gene Editing Tools and Sequencing Technologies for Regenerative Medicine

Research Interests:

Understanding genetic mechanisms of brain diseases and developing regenerative therapeutics need genetically tractable in vitro models. Such models should recapitulate physiological properties, cellular and architectural complexity, and genetic makeup of the human brain as well as be compatible with genome editing, transcriptome profiling technologies, and gene delivery tools. My research interfaces biomaterials, stem cell biology, gene editing and delivery, single cell sequencing technologies, and computational biology to develop engineered biomaterials templated stem-cell based biomimetic in vitro tissues to model brain diseases/injuries and discover translational therapeutic targets and strategies.

In my PhD research, I focused on developing microscale technologies for spatial and geometrical control of multicellular organizations. I developed dynamic microstructures by utilizing a thermoresponsive polymer. These microscale templates enabled to spatially arrange same or different cell types in different compartments of microgels or directly in microwells without an encapsulating material. This controlled cell organization can lead to complex cell-cell and cell-matrix interactions for better recapitulating native tissues.

For my postdoctoral studies, I developed engineered neural tissues composed of brain-related cells derived from human stem cells and adapted encapsulating hydrogel materials to replicate complex brain-related associations in culture. I conceived a method of differentiating human glial cells, an abundant cell-type in the brain, directly from stem cells in a rapid manner and then engineered material properties enveloping the neurons and glia in a 3D structure. Based on the transcriptome profile of these neural tissues, I investigated how tailoring hydrogel material properties changed the RNA-signature and maturity of neural tissues through comparisons to human brain transcriptome data. Using single-cell sequencing, I demonstrated 3D tissues I developed reflect transcriptional patterns of cell types in the human brain. I also exploited CRISPR-Cas gene editing technologies via viral gene delivery tools to interrogate neurological disease-associated genes in neurons within a 3D matrix.

As I establish my own research program combining biomaterials science, stem cell biology, high-throughput sequencing technologies, CRISPR genome editing and gene delivery tools, and computational biology, I will build on the full spectrum of my training experience to develop leading technologies to replicate brain diseases in vitro and search for therapeutic targets and methodologies that restore the health of the brain.

Education and Training:

- Postdoc, 2013-2018, Broad Institute of Massachusetts Institute of Technology and Harvard, Biomedical Sciences and Genomics
- PhD, 2013, Massachusetts Institute of Technology, Electrical Engineering and Computer Science, (minor in Medical Engineering)
- MS, 2007, Istanbul Technical University, Faculty of Electrical and Electronics Engineering, Control Engineering
- BS, 2005, Istanbul Technical University, Faculty of Electrical and Electronics Engineering, Control Engineering

Research Experience:

- Broad Institute of Massachusetts Institute of Technology and Harvard, October-2013 to July-2018, Postdoctoral Associate, Advisor: Feng Zhang, PhD
- Massachusetts Institute of Technology, January-2013 to July 2013, Postdoctoral Associate, Advisor: Robert S. Langer, ScD
- Massachusetts Institute of Technology, July 2007 to December 2012, Graduate Research Assistant,

Advisors: Robert S. Langer, ScD and Ali Khademhosseini, PhD

Selected Publications:

* Denotes corresponding author(s).

H. Tekin*, S. Simmons, B. Cummings, L. Gao, X. Adiconis, C. C. Hession, A. Ghoshal, D. Dionne, S. R. Choudhury, V. Yesilyurt, N. E. Sanjana, X. Shi, C. Lu, M. Heidenreich, J. Q. Pan, J. Z. Levin, F. Zhang. "Effects of 3D culturing conditions on the transcriptomic profile of stem-cell derived neurons". *Nature Biomedical Engineering*. 2018.

S.R. Shin, H. Bae, J.M. Cha, J.Y. Mun, Y.C. Chen, **H. Tekin**, H. Shin, S. Farshchi, M. Dokmeci, S. Tang, A. Khademhosseini*. "Carbon nanotube reinforced hybrid microgels as scaffold materials for cell encapsulation". *ACS Nano*, 2012, 6 (1): 362-372.

H. Tekin, J.G. Sanchez, C. Landeros, K. Dubbin, R. Langer*, A. Khademhosseini*. "Controlling spatial organization of multiple cell types in defined 3D geometries". *Advanced Materials*, 2012, 24 (41): 5543-5547. (Frontispiece Cover).

H. Tekin, T. Tsinman, J.G. Sanchez, B.J. Jones, G. Camci-Unal, J.W. Nichol, R. Langer*, A. Khademhosseini*. "Responsive micromolds for sequential patterning of hydrogel microstructures". *Journal of the American Chemical Society*, 2011. 133 (33): 12944–12947.

H. Tekin, J.G. Sanchez, T. Tsinman, R. Langer, A. Khademhosseini. "Thermoresponsive platforms for tissue engineering and regenerative medicine". *AIChE Journal*, 2011, 57 (12): 3249–3258. (Cover Article).

H. Tekin, G.O. Ince, T. Tsinman, K.K. Gleason, R. Langer*, A. Khademhosseini*, M.C. Demirel*. "Responsive microgrooves for the formation of harvestable tissue constructs". *Langmuir*, 2011, 27 (9): 5671–5679.

H. Tekin, M. Anaya, M. Brigham, C. Nauman, R. Langer*, A. Khademhosseini*. "Stimuli-responsive microwells for formation and retrieval of cell aggregates" *Lab on a Chip*, 2010, 10 (18): 2411-2418.

Selected Awards/Honors:

2013, JALA Ten Award (The Top 10 Technological Breakthroughs Selected by The Journal of Laboratory Automation)

2012, MIT's Outstanding Undergraduate Research Mentor Award (Graduate Mentor of the Year)

2011, Materials Research Society Best Student Oral Presentation Award

2011, Bioanalysis Highly Commended Young Investigator Award

2005, Werner von Siemens Excellence Award

Teaching Interests:

I am always excited about teaching and mentoring next-generations for building future by teaching them scientific foundations and creating an environment to motivate them. My interest and commitment in mentoring students were recognized by **MIT's prestigious Outstanding Undergraduate Research Mentor Award**. I will leverage all mentoring and teaching skills I built up to educate next-generation of scientists and engineers in my faculty career. I am excited about teaching both basic and applied courses at the undergraduate and graduate level in the scope of engineering cells and biomaterials, quantitative biology, and interdisciplinary biosciences.

582182 Engineering Microsystems for Regulating Cellular Behavior: From Implantable Drug Factories to Novel Platforms for Single-Cell Genomics

Research Interests:

We are built from trillions of cells, but there is a gap in our understanding of how these cells work together. The ability to precisely probe individual cells in the context of their complex environments and control their behavior could open new opportunities for developing next-generation targeted therapies. *My lab will harness microfluidics to create novel platform technologies which can accurately measure biomolecules (DNA, RNA, and proteins) within single cells.* We are particularly interested in applying these technologies for studying *cellular diversity* and *cell-cell interactions* in the context of *cancer and immune diseases*. Finding from these will help us develop new therapies and guide the development of smart biomaterials. My interdisciplinary research training in microsystem engineering, biomaterials development, and single-cell analysis uniquely positions me to integrate approaches in the proposed areas of research

Teaching Interests:

My teaching philosophy has been shaped by my learning experience during my undergrad at IIT and later as a graduate student at MIT. My fundamental philosophy of education is to foster my students' ability to gain a clear grasp over the fundamentals, think critically and gain problem-solving skills. During my graduate school, I have been teaching assistant for two core undergraduate curriculum - **Thermal and Fluid Engineering I & II** (class size of more than 120 students) for three semesters. These courses teach fluid mechanics, heat transfer and thermodynamics, and adopt an integrated approach to designing thermal systems. I had a transformative experience while teaching these courses which helped me develop into a better educator. More recently**, I have been involved in designing and teaching Undergraduate Chemical Design (Course 10.494)** along with my postdoc adviser Prof. Daniel Anderson. I was responsible for coming up with a suitable design project for the students, and give lectures to on the fundamental concepts required for the design project. My lecture was very well received (rated 6.9/7) which gave me immense satisfaction, and confidence for developing new courses in the future. *My strong background in core engineering courses along I took during my undergraduate and graduate work has prepared me well for teaching several core classes in Transport and Fluid Mechanics within the Chemical Engineering program.* I am also interested in developing a new course on - 'Convergent approaches to solving biological problem' where the students will be exposed to current interdisciplinary approaches to solving pressing current problems in biology and medicine.

Research experience:

My graduate and postdoctoral training has allowed me to gain expertise in a broad range of fields ranging from microfluidics, numerical modeling, surface chemistry, biomaterials, pre-clinical animal studies, and next-generation genomics. The breadth of my experience has allowed me to draw original ideas and novel techniques from across disciplines and work effortlessly with chemists, biologist, and clinicians as a team. This ability makes me a unique candidate to execute research at the interface of science and engineering.

I was trained in quantitative mathematical modeling during my undergraduate and master's research, where I used numerical approaches to study thermodynamic system (IJHMT 2008, IJR 2007) and biological transport processes (Biophy. J 2010). In my graduate research, I developed novel bio-inspired microfluidic cell-sorting technologies. These devices exploited the physiological processes of cell homing to isolate specific immune cells (Sci rep. 2014), and harnessed DNA nanotechnology to effectively capture rare tumor cells from blood (PNAS 2013). While working on these technologies, I developed expertise in microfabrication, bio-conjugation chemistry, surface chemistry, and nanotechnology, and learned how to incorporate these approaches to build integrated devices that can process biological samples.

The first couple of years of my postdoctoral training focused on gaining expertise in biomaterials, drug delivery systems, and translational research. Funded by a JDRF Postdoctoral Fellowship, I have developed a biocompatible cell-encapsulation macrodevice that can effectively transplant immunogenic therapeutic cells in animals (in final revision). I discovered that membrane porosity could modulate infiltration of immune cells in vivo, and developed an effective surface coating that can mitigate device fibrosis. This work required innovation in device engineering and material chemistry, along with extensive in vivo studies in several animal models. Currently, enabled by a K99/R00 Award and in collaboration with Prof. Phillip Sharp, I have developed a microfluidic platform to enable high throughput sequencing of microRNA from single cells. The droplet-based microfluidic device performs co-immunoprecipitation to isolate miRNA from individual cells and process them for sequencing using DNA barcoding technology. I am using the platform to study miRNA regulation during differentiation of ES cells, and study heterogeneity of leukemia cells

Selected Awards and Honors:

- Koch Image Award (2019)
- NIH/NIBIB K99 Pathway to Independence Award, MIT (2018-23)
- Future of Science Award, Keystone Single-cell omics, Sweden (2017)
- Top 40 under 40 Healthcare Innovators, MedTech Boston (2016)
- JDRF Postdoctoral Fellowship, MIT (2015-18)

Selected Publications (full list at google scholar):

- **Bose,** N. Xue, A. Navia, E. Yoon, V. Chauhan, A. Shalek, P. Sharp, S. Garg, D. Anderson, Highly parallel single cell profiling of microRNAs for cellular identification in leukemia. (in preparation).
- **Bose**, L. Volpatti, D. Thiono, V. Yesilyurt, C. McGladian, O. Veish, A. Wang, G. Weir, R. Langer, DG Anderson, A biocompatible macrodevice enables long-term survival of therapeutic xenogeneic cells in vivo. (In final revision)
- **Bose**, R. Singh, M-H Hollatz, C. Shen, C-H Lee, D. M. Dorfman, J. M. Karp, R. Karnik: Affinity flow fractionation of cells via transient interactions with asymmetric molecular patterns. Scientific Reports 3, 2013
- Zhao*, C. Cui*, **S. Bose***, D. Guo, C. Shen, W. P. Wong, K. Halvorsen, O. Farokhzad, G. Teo, J. Phillips, D. Dorfman, R. Karnik, J. M. Karp: A bioinspired multivalent DNA network for capture and release of cells. Proceedings of National Academy of Sciences USA, 2012. *equal contributions
- C-H Lee*, S. Bose*, K. Van Villet, J. M. Karp, R. Karnik: Examining the lateral displacement of HL60 cells rolling on asymmetric P-Selectin patterns, Langmuir, 27(1), 2011.
- Bose, S. Das, J. M. Karp, R. Karnik: A semianalytical model to study the effect of cortical tension on cell rolling, Biophysical Journal, 99(12), 2010.

581742 Biologically Inspired, Electrically Active Membranes for Sustainability and Medicine

Research Interests:

Like the metal- and semiconductor-based circuitry that has dominated the past century's technological progress, biological systems also perform complex computation and fast electrical signaling over relatively long distances, but use the transport of ions in aqueous media instead of electrons to mediate signals. Bioelectricity is most impressively employed in specialized "excitable tissues" such as neurons, receptors, muscles, and the electric organs of fish such as the electric eel. The active component of each of these systems is a self-assembled nanoscale membrane containing exquisitely precise molecular machines that allow patches of membrane to toggle between functioning as resistors, capacitors, cables, or batteries in response to intracellular or extracellular stimuli. Excitable tissues are thus able to produce emergent electrical phenomena such as action potentials, defensive discharges, and signal integration while remaining soft, transparent, environmentally benign, and operating within biological constraints, unlike traditional electronics. **In my program of research, I propose to use the design playbook of excitable tissues and the toolkit of responsive polymer chemistry to develop novel active material schemes that transduce and integrate signals and energy from environmental gradients and stimuli. My aim is to develop platform technologies that can be deployed in contexts that leverage the autonomously stimuli-responsive and biologically benign character of the materials involved, such as sustainable building materials, distributed sensors, and medical devices.** My initial efforts will include three main projects:

(1) Active polymer membranes for novel ionotronic devices. Using soft materials such as hydrogels and dielectric elastomers, my group will engineer synthetic schemes inspired by various neuronal architectures that generate and integrate electrical signals from environmental stimuli. Building on my ongoing postdoctoral work on electrical responses from switchably permselective cross-linked polymer membranes as a sensing modality, I intend to develop ionotronic networks that incorporate signal multiplexing and feedback loops in order to engineer complex electrochemical responses (*e.g.* logic gating, oscillations) to environmental stimuli including light, chemical cues, and others. Possible applications range from implantable devices to consumer product packaging materials capable of autonomous homeostatic interventions.

(2) Engineering and electrophysiological study of lipid and block copolymer membranes. In addition to developing devices based around polymer membranes as described above, I intend to apply recent advances in engineering stable lipid bilayers and similar self-assembled membranes for use both in devices and in the fundamental study of the biophysics of membrane-bound molecular machines such as ion channels and transporters. In particular, I am interested in using electroconformational coupling to drive biochemical reactions and transport at the molecular level in the membrane environment in order to elucidate the mechanism of transporter-mediated disease states and to enable new functionality within biomolecular assemblies for use in medicine.

(3) Energy harvesting from unrecognized natural gradients. Between all manner of naturally-occurring aqueous solutions (*e.g.* rivers, oceans, xylem, phloem, blood, sweat, tears) there exist large concentration differences of sodium chloride, other electrolytes, and neutral osmotic drivers. While a body of literature has appeared around extracting usable electrical energy on the kilowatt scale by combining fresh and salt water through schemes such as reverse electrodialysis and pressure-retarded osmosis, less effort has been spent on seeking out untapped biological and/or environmental streams to exploit for smaller-scale power generation. My lab will seek to use existing gradients maintained by biological homeostatic mechanisms or natural fluid flows to implement power generation schemes for low-power electronics such as medical implants or distributed environmental sensors.

Research Experience:

The "biophysics-inspired materials engineering" research agenda that I intend to pursue is the culmination of a highly interdisciplinary set of research experiences and a curiosity and reverence for the way evolved systems accomplish tasks related to their survival.[1] After an undergraduate background in organic chemistry,[2] I joined Prof. Michael Mayer's biophysics group at the University of Michigan while pursuing my Ph.D. in chemical engineering. My work there (and at the Adolphe Merkle Institute in Switzerland, where the group later moved) convinced me of the central importance of membranes as mediators of the flow of energy and matter in biological systems – and, accordingly, of their promise in engineered systems.

During my Ph.D., I first became interested in excitable tissues when my colleagues and I engineered an entirely hydrogel-based electrical power source capable of generating high voltages (>100 V) from salt gradients by replicating the functions of the main components of an electric eel's specialized organs with a series of hydrated polymer analogues.[3] We also designed a pressure generator for soft robotic actuation that relied on osmotic gradients across polymer membranes in a manner similar to the cells responsible for fast motion in plants.[4] Additionally, we performed fundamental[5] and applied[6] studies of self-assembled membranes composed of synthetic bolaamphiphilic lipids of a type found in single-celled hyperthermoacidophiles, as well as electrophysiological studies of the voltage-driven assembly of engineered ion channels in lipid membranes.[7]

My postdoctoral research in Prof. Joanna Aizenberg's bioinspired materials group at Harvard University has involved incorporating various stimuli-responsive chemistries into polymer membranes to induce switchable permeability and lateral signal transmission.

Teaching Interests:

I would be happy to teach thermodynamics and transport at an introductory or advanced level, as well as more specialized classes on polymer chemistry, self-assembly, bioconjugation techniques, biophysics, and nanomaterials. I would also be interested in giving classes or seminars on responsible research conduct and general research practices (*e.g.* keeping a notebook, using tools to review the literature, selecting a reference manager, and writing a journal article). Further, I am excited to teach and mentor students from diverse backgrounds.

I would like to create a classroom environment that leverages real-world examples and demonstrations whenever possible in order to teach the principles of science and engineering while providing real-world context. In particular, biology is teeming with compelling applications of physical principles that could make interesting case studies. Finally, I have found that some of my most interesting moments in the classroom have come from narrative explanations of how we humans have historically figured out the universe's truths. For example, only after enabling developments in electrical engineering (the voltage clamp) and marine biology (the discovery of the squid giant axon) were Hodgkin and Huxley able to gather the data required to hypothesize that voltage-gated ion conductances were the mechanism behind the action potential, for which they won a Nobel prize. Following the luck and logic of our precursors helps us to demonstrate model-building and inference while motivating interdisciplinary work. In the classroom, I'll make sure my students don't lose sight of the giants on whose shoulders we stand.

Selected Publications:

(1) **Schroeder, T. B. H.**; Houghtaling, J.; Wilts, B. D.; Mayer, M. It's Not a Bug, It's a Feature: Functional Materials in Insects. *Adv. Mater.* **2018**, *30*, 1705322. https://doi.org/10.1002/adma.201705322.

(2) Dugal-Tessier, J.; O'Bryan, E. A.; **Schroeder, T. B. H.**; Cohen, D. T.; Scheidt, K. A. An N-Heterocyclic Carbene/Lewis Acid Strategy for the Stereoselective Synthesis of Spirooxindole Lactones. *Angew. Chem. Int. Ed.* **2012**, *51* (20), 4963–4967. https://doi.org/10.1002/anie.201201643.

(3) **Schroeder, T. B. H.**; Guha, A.; Lamoureux, A.; VanRenterghem, G.; Sept, D.; Shtein, M.; Yang, J.; Mayer, M. An Electric-Eel-Inspired Soft Power Source from Stacked Hydrogels. *Nature* **2017**, *552* (7684), 214–218. https://doi.org/10.1038/nature24670.

(4) Bruhn, B. R.; **Schroeder, T. B. H.**; Li, S.; Billeh, Y. N.; Wang, K. W.; Mayer, M. Osmosis-Based Pressure Generation: Dynamics and Application. *PLoS ONE* **2014**, *9* (3), e91350. https://doi.org/10.1371/journal.pone.0091350.

(5) **Schroeder, T. B. H.**; Leriche, G.; Koyanagi, T.; Johnson, M. A.; Haengel, K. N.; Eggenberger, O. M.; Wang, C. L.; Kim, Y. H.; Diraviyam, K.; Sept, D.; et al. Effects of Lipid Tethering in Extremophile-Inspired Membranes on H+/OH− Flux at Room Temperature. *Biophys. J.* **2016**, *110* (11), 2430–2440. https://doi.org/10.1016/j.bpj.2016.04.044.

(6) Eggenberger, O. M.; Leriche, G.; Koyanagi, T.; Ying, C.; Houghtaling, J.; **Schroeder, T. B. H.**; Yang, J.; Li, J.; Hall, A.; Mayer, M. Fluid Surface Coatings for Solid-State Nanopores: Comparison of Phospholipid Bilayers and Archaea-Inspired Lipid Monolayers. *Nanotechnology* **2019**, *30* (32), 325504. https://doi.org/10.1088/1361-6528/ab19e6.

(7) Fennouri, A.; Mayer, S. F.; **Schroeder, T. B. H.**; Mayer, M. Single Channel Planar Lipid Bilayer Recordings of the Melittin Variant MelP5. *Biochim. Biophys. Acta BBA - Biomembr.* **2017**, *1859* (10), 2051–2057. https://doi.org/10.1016/j.bbamem.2017.07.005.

581779 Computational Modeling in Cancer Systems Biology: Stochasticity, Complexity, and Multiscale Dynamics in Disease Progression and Drug Response

Research Interests:

Biological systems differ from human engineered systems in three primary respects: 1) they are highly stochastic, 2) they are extraordinarily complex, and 3) they exhibit dynamics over a wide range of temporal and spatial scales. These factors are especially prevalent in cancer, which is a complex and adaptive disease characterized by genetic mutations, phenotypic plasticity, and interactions among tumor cells and with the tumor microenvironment. Tumors are highly heterogeneous and exhibit complex responses to drugs and other therapeutic interventions. Indeed, patient responses to therapy can vary widely and it is not uncommon for an initial positive response to be followed by tumor recurrence and metastasis, which is often fatal.

It is becoming increasingly apparent that to unravel the complexities of cancer a systems engineering approach is necessary. Cancer can be thought of as a reverse engineering problem: it is fundamentally the result of a breakdown of the cell's natural tumor suppressing machinery. Thus, therapeutic approaches aimed at reestablishing or rewiring this apparatus are of great interest. However, the molecular mechanisms that underlie cancer cell function remain poorly understood. Computational models are powerful tools for addressing this challenge. They are a means by which current understanding of a biological system can be formalized in mathematical terms. Moreover, through an iterative process of model building, prediction, and experimentation, models can be systematically expanded over time to capture increasingly complex behaviors. This is the basis of the cancer systems biology approach.

In this poster, I present results of combined computational and experimental investigations of numerous cancer types treated with therapeutic drugs, including small cell and non-small cell lung cancers and melanoma. My results indicate that tumors employ a "bet hedging" strategy whereby cells diversify across multiple phenotypes in the absence of drug in order to increase the chances of population survival when faced with drug. I hypothesize that surviving subpopulations can persist in a patient for extended periods of time before eventually acquiring genetic resistance mutations that drive tumor recurrence. Therefore, developing a comprehensive understanding of the molecular networks underlying tumor heterogeneity may lead to novel therapies that can reduce or even eliminate these surviving cells before resistance mutations arise, improving patient outcomes. This work is the basis for a recent **K22 grant application (Impact Score: 25)** aimed at building a detailed, mechanistic model of the biochemical pathways underlying cell fate decisions in cancer cells.

Teaching Interests:

My expertise lies in chemical kinetics, thermodynamics, and numerical methods. I also have experience in physical chemistry, biochemistry, statistics, theoretical computer science, and software engineering. During my years as a postdoctoral researcher I have had the opportunity to mentor numerous undergraduate and graduate students and have given numerous lectures on systems biology. I enjoy teaching and make a concerted effort to engage students by providing analogies that convey complex concepts in an easy-to-understand manner. My philosophy of teaching is that nothing is ever as complicated as it seems and the best way to learn is through a hands-on approach. Thus, I aim to guide students in the right direction but ultimately leave it to them to find the solution to a problem. There is nothing more satisfying than when something "clicks" for a student and they have that moment of true understanding. I am prepared and enthusiastic about teaching full semester chemical engineering courses at both the undergraduate and graduate levels and look forward to mentoring graduate students and postdocs in their research endeavors. **As a Hispanic-American**, I am also strongly committed to minority outreach and recruiting underrepresented students into STEM programs.

Select Publications:

B.B. Paudel, **L.A. Harris**, K.N. Hardeman, A.A. Abugable, C.E. Hayford, D.R. Tyson and V. Quaranta, "A non-quiescent 'idling' population state in drug-treated, *BRAF*-mutated melanoma," ***Biophysical Journal*** 114, 1499–1511 (2018).

Z.W. Jones, R. Leander, V. Quaranta, **L.A. Harris*** and D.R. Tyson*, "A drift-diffusion checkpoint model predicts a highly variable and growth-factor-sensitive portion of the cell cycle G1 phase," ***PLoS One*** 13, e0192087 (2018). (**equal authors*)

L.A. Harris*, M.S. Nobile*, J.C. Pino*, A.L.R. Lubbock, D. Besozzi, G. Mauri, P. Cazzaniga and C.F. Lopez, "GPU-powered model analysis with PySB/cupSODA," ***Bioinformatics*** 33, 3492–3494 (2017). (**equal authors*)

L.A. Harris*, P.L. Frick*, S.P. Garbett, K.N. Hardeman, B.B. Paudel, C.F. Lopez, V. Quaranta and D.R. Tyson, "An unbiased metric of antiproliferative drug effect *in vitro*," ***Nature Methods*** 13, 497–500 (2016). (**equal authors*)

L.A. Harris, J.S. Hogg, J.J. Tapia, J.A.P. Sekar, S. Gupta, I. Korsunsky, A. Arora, D. Barua, R.P. Sheehan and J.R. Faeder, "BioNetGen 2.2: Advances in rule-based modeling," ***Bioinformatics*** 32, 3366–3368 (2016).

J.S. Hogg*, **L.A. Harris***, L.J. Stover, N.S. Nair and J.R. Faeder, "Exact hybrid particle/population simulation of rule-based models of biochemical systems," ***PLoS Computational Biology*** 10, e1003544 (2014). (**equal authors*)

581438 Multidimensional Single Cell Analysis: Devices and Technology for Cancer Biology

Research Interests:

My research lab will develop new methods to study single cells in multiple dimensions, to better understand cell heterogeneity and cancer biology. We will focus on combining transcriptomics with phenotypic and phenomenological measurements to generate insights and hypotheses beyond the central dogma. In the field of cancer biology, we will target questions such as cell invasiveness, metabolism, and adhesion, that are largely invisible to the primary –omics methods (genomics, transcriptomics, proteomics), but have great clinical significance.

Single cell analysis has revolutionized biology, and it is perched at the nexus of biology, chemistry, and materials science. Recent advances in microfluidics, DNA preparation, and high-throughput processing have made single cell transcriptomics a benchtop tool. Yet, single cell transcriptomics has provided more exciting questions than satisfying answers to this point, largely because cell function is not clearly derived from knowledge of the transcriptome. To push our understanding deeper, my lab will combine high throughput single cell transcriptomics with the practical tools and workhorses of biology.

Tools like fluorescence microscopy and pulse-chase labeling can effectively describe the phenomenology of biology. When combined with the transcriptome, the functional phenomenology of the cell, i.e. subcellular localization of fluorescent proteins, can be correlated to gene regulatory networks and the control systems of the cell. In order to make multiple orthogonal measurements in a single cell (fluorescence microscopy + transcriptomics for instance), we will develop new devices and materials, chemistries, and bioengineering strategies. My lab will use systems biology and computational approaches to process these novel data sets, and study the heterogeneity of cancer phenomena.

Research Experience:

Postdoctoral: Multi-omic single cell measurements (Jim Heath, Caltech and the Institute for Systems Biology)

During my postdoc I developed new strategies to measure proteins and transcripts from the same single cells. Contrary to current technologies that convert protein to RNA signals, I created DNA mappings between transcriptomes and protein measurements by encoding the location of cells as DNA. This forms the basis for bridging transcriptomics with other measurement tools. The strategy of encoding physical parameters as DNA (location, time) was combined with microfluidic devices to make unorthodox multi-omic measurements in the same single cells, in this case bead-based transcriptomics and fluorescence sandwich immunoassays. This leads to new opportunities in systems biology to study cross-biomolecular species interactions. I also worked on projects in single cell analysis for immunotherapy and cell transitions over time.

Doctoral: Intracellular delivery strategies (Nick Melosh, Stanford University)

During graduate school, I created a new platform for intracellular delivery called nanostraws. Using microfabrication and deposition techniques from materials science, I created nanoscale conduits to cells to deliver different molecular species ranging from ions to small molecules to DNA plasmids. The nanostraws had unique characteristics like temporal control of delivery, and were adapted to platforms for oral ingestion and immune cell studies. This project required a unique perspective on the mechanics and mechanobiology of cells, and the nanobiointerfaces needed to control single cells precisely.

Other: DNA nanotechnology (Irene Chen, Harvard); Virus biotemplating (Angela Belcher, MIT); Cancer phosphoproteomics (Forest White, MIT); Atomic layer deposition of transparent conducting oxides (Jeff Elam, Argonne National Lab)

Awards: F32 Postdoctoral Fellowship, NSF Graduate Fellowship, NDSEG Graduate Fellowship

Teaching Interests:

My teaching interests include nanotechnology, nano-interfaces in biology, biomaterials, surface science, and systems biology. My teaching and learning experience has been highly interdisciplinary, with undergraduate degrees in Materials Science and Math, graduate studies in Materials Science, and a postdoc in a Chemistry and Chemical Engineering department and a Systems Biology institute. This has prepared me to teach courses in thermodynamics, polymers, and biochemistry as topics relevant to Chemical Engineering students. I am most excited to teach a course on nanobiotechnology, where the difference between biotic and abiotic systems becomes blurred. I TA'd a similar course as a graduate student for two quarters. Many nanobiotechnology courses tend to focus on the field of drug delivery, but I would focus on modern biological technologies enabled by nanoscience, to prepare students not just for academia but also for relevant jobs in modern biotech. Primary topics include high-throughput biology, next generation sequencing and DNA nanotechnology, neuroscience, and chemical biology. I am also excited to blend my new experience at the Institute for Systems Biology into my teaching by focusing on the computational aspects of high-throughput chemistry and biology, and how biological insights can be extracted from big data.

My teaching style will be drawn from my interdisciplinary background. In terms of training chemical engineers for their future careers, I can appreciate and build on the strong analytical skills that most chemical engineers possess while exposing them to new topics in the many research topics I have been involved in. Design projects are an effective way to allow students to branch out from coursework, think across disciplines, and be creative, and my courses will encourage that diversity. My own analytical background is strong, and I have placed a heavy emphasis on mentorship, with one of my students publishing a first author manuscript.

Selected Publications:

6 first authors, h-index=14

1. **AM Xu**, Qianhe Liu, Kaitlyn Takata, Sarah Jeoung, Yapeng Su, Igor Antoshechkin, Sisi Chen, Matthew Thomson, James R Heath. Integrated Measurement of Cytosolic Proteins and Transcripts in Single Cells. *Lab on a Chip*, 21, 3251-3262. (2018)

- Front cover

2. **Alexander M Xu**, Derek S Wang, Peyton Shieh, Yuhong Cao, Nicholas A Melosh. Direct Intracellular Delivery of Cell-Impermeable Probes of Protein Glycosylation by Using Nanostraws. *ChemBioChem*, 18, 623-628. (2017)
3. **Alexander M Xu**, Amin Aalipour, Sergio Leal-Ortiz, Armen H Mekhdjian, Xi Xie, Alexander R Dunn, Craig C Garner, Nicholas A Melosh. Quantification of Nanowire Penetration into Living Cells. *Nature Communications*, **5**, 3613. (2014)
4. **Alexander M Xu***, Jules J VanDersarl*, Nicholas A Melosh. Nanostraws for Direct Fluidic Intracellular Access. *Nano Letters,* **12**, 3881-3886. (2011)

- Front cover

5. **Alexander M Xu** and Paul H Huang. Receptor Tyrosine Kinase Coactivation Networks in Cancer. *Cancer Research*, **70**, 3587-3590. (2010)

Co-authored papers appearing in: *Immunity*, *Science Signaling*, *Cancer Research*, *ACS Nano*, *Molecular Biosystems*, *Nano Letters*

579374 The Role of Topographical Cues in Cancer Cell Migration and Metastasis

Research Interests:

My research program will focus on: (1) studying how and why physical forces, especially those related to the architecture or topography of the microenvironment in distinct tissues, influence the spread and formation of metastatic cancers; and (2) developing therapeutics that target interactions between disseminated tumor cells and the physical microenvironment to inhibit the progression of metastatic cancers.

Metastasis describes the process whereby cancer cells move from a primary tumor to establish lesions in distinct secondary organs. Experiments in animal models suggest that thousands to millions of tumor cells per gram of tumor can be shed into the circulatory system per day, but that very few survive to form metastases. If this large quantity of cells enters the circulatory system, perhaps even before the detection of a primary tumor, why do some cancer cells survive and grow into metastases? Furthermore, if dissemination of cells occurs early in the lifespan of a tumor, how can we design and execute treatment methods to check the growth of disseminated tumor cells? It is likely that many rate-limiting steps in metastasis occur early in colonization, before secondary tumors are detectable in patients or in murine models. These patterns of early dissemination are particularly common in breast, esophageal, and renal cell cancers, while microenvironmental preconditioning from the primary tumor may be more important in the metastasis of prostrate, colorectal, and pancreatic cancers.

My research program will initially focus on how topographical cues that disseminating tumor cells receive from the architecture of the microenvironment in distinct organs modulate cell arrest and extravasation. Specifically, I will study the mechanisms by which disseminating cancer cells arrest, survive, and extravasate in biomimetic microfluidic devices and in blood vessels of varying topographical organization in the zebrafish hematopoietic niche and brain. Microfluidic devices will be engineered to recapitulate key physical features of vascular topography found in vivo. Key findings will be confirmed in intravital experiments in the larval *Danio rerio* (zebrafish), an optically transparent and relatively inexpensive model organism. Cytokines associated with tumor progression as observed in murine and human studies are conserved in the zebrafish, and ~70% of human genes have at least one zebrafish orthologue. My postdoctoral work elucidated mechanisms leading to organotropism (the clinical observation that different types of primary cancers preferentially metastasize to different organs) of human and murine brain- and bone marrow-seeking breast cancer cells in a larval zebrafish xenograft model. In this model, brain-targeting cells colonized the brain at greater rates than bone marrow-targeting cells. Bone marrow-targeting cells were more likely to extravasate in the caudal vein plexus (CVP), a hematopoietic niche in the larval zebrafish tail that is analogous to mammalian bone marrow. Functional and mechanistic experiments in microfluidic devices and zebrafish will identify druggable targets that can be exploited during cancer cell dissemination, the exploration of which will form the basis for future grant applications.

My long-term career goals are to leverage an improved understanding of the mechanisms of early cancer cell dissemination, extravasation, and colonization to develop treatments to prevent the growth of metastatic cancer or force metastatic cells into dormancy. In analogy to the role of topography in cancer cell migration, I hypothesize that immune cell access to disseminated cancer cells could depend on the physical environment (ECM organization, outward fluid flow from blood vessels, etc.) around disseminated cells, and that reduced access of immune cells to dormant tumor cells slows their clearance by immune cells. Understanding these forces may help develop drugs and/or immunotherapy strategies that prime the immune system to recognize disseminated cells before they can grow into large metastases. Therefore, future research projects will also assess how microenvironmental topographical cues influence the phenotype of immune cells in various organs.

Teaching Interests:

My teaching goals are to prepare students for a wide range of chemical engineering and biotechnology careers by supporting critical thinking, integration of new information into existing frameworks, and use of available tools to solve real-life problems. I am interested in teaching undergraduate and graduate courses in fluid mechanics, transport, separations, biomechanics, and imaging theory and methods, as well as seminars or graduate courses in microfluidic techniques and technologies. I am also willing to teach online courses. I have had teaching experience as a graduate assistant in Chemical and Biomolecular Engineering Transport I (undergraduate course) and a NanoBioTechnology lab course (graduate course). I have also completed a course on pedagogy sponsored by the National Institutes of Health (NIH) that is designed to prepare scientists for instruction in STEM fields, and I have served as a course instructor and mentor for summer students at the NIH.

577372 Reverse Perfluorocarbon Emulsions for Pulmonary Drug Delivery

Research Interests:

Pulmonary drug delivery using water-in-perfluorocarbon reverse emulsions has the potential to improve traditional treatment methods of lung diseases. Liquid perfluorocarbons (PFC) function like air, providing a medium for gas exchange; and as a delivery vehicle, enabling deeper lung penetration of delivered drugs. Since most drugs are sparingly soluble in PFC liquids, aqueous drug solutions must be dispersed as a water/PFC emulsion. Several studies have confirmed drug-loaded PFC emulsions maintain therapeutic effects; however, few studies have engineered controlled release profiles that cater to specific therapeutic windows or disease states.

Teaching Interests:

Drug transport, Colloids and Surfaces, Physiology, Systems Modeling and Analysis for Biomedical Engineering

578221 Engineering DNA-Polymer Assemblies

PhD dissertation: "Design, Control, and Implementation DNA Origami Mechanisms"

--Advisor: Carlos Castro, Mechanical and Aerospace Engineering, Ohio State University

Postdoctoral research: "Polyelectrolyte Complexation with Biomolecules"

--Supervisor: Matt Tirrell, Pritzker School of Molecular Engineering, University of Chicago

Research Interests:

Chemical and biological systems exist in a mechanical environment. Across multiple scales tissues, cells, proteins, nucleic acids, and other biomaterials generate and are subjected to mechanical signals such as force, stress, and flow. The way these signals affect function and movement is still largely unknown. Modern polymer and DNA nanotechnology provides materials to build custom nanoscale devices with unprecedented control over physical and chemical properties. I aim to engineer devices at the scale necessary to investigate the mechanical and chemical interactions between various biomolecules and their environment. Employing the specificity of DNA assembly and the tunability and robustness of polymer assemblies can help achieve these goals.

My graduate research expanded DNA origami technology by producing dynamic devices with programmable motion, an underpinning for my future research program. I developed nanoscale kinematic joints including hinges and sliders demonstrating rotational and linear motion. Borrowing concepts from macro-scale machine design, I used these fundamental components to build higher order mechanisms with specific 2D and 3D motion. These joints and mechanisms were then used to demonstrate multiple actuation methods used to control the configuration of these nanostructures. A novel actuation method I developed allows for real-time control of DNA-based devices using a network of low affinity connections. This groundwork will enable the construction of complex engineering devices in the future.

My postdoctoral research uses polyelectrolytes and biomaterials as robust building blocks for self-assembled nanoparticles and customizable materials. This work uses charged homo- and block copolymers to complex therapeutic nucleic acids and proteins to develop particles and materials with programmable functionality. A thorough investigation of chemical structure - physical property relationships and their effect on stability of these particles and materials enables efficient design and construction of delivery devices for therapeutic payloads.

Future Research:

As an independent investigator my laboratory will focus on the design, assembly, and implementation of nanodevices made from biomaterials. Building on my experience in DNA nanotechnology and polymer science, I plan to explore the following directions:

1. **Programmable DNA-Polymer Assemblies**. Employing the vast geometric, mechanical, and chemical design space of DNA nanotechnology with robust functional polymer assemblies, I plan to engineer materials and nanoparticles with programmable features useful in measuring, sensing, and delivery. To achieve this goal the extensive design parameters will be utilized for global and local responsiveness to ion conditions, pH, temperature, or the presence of biomolecules. Additionally, static assemblies with structural DNA networks will permit specific placement and patterning of molecules like polymers, peptides, or antibodies to study specific interactions.
2. **Functional Polyelectrolyte Complex Micelles (PCMs) for Delivery of Biomolecules**. Self-assembling hydrophilic neutral-charged block copolymers with charged therapeutically relevant nucleic acids and proteins is a growing, promising area of research. Following extensive development and characterization of simple polymer-based systems, I plan to build tailored delivery devices with their application in mind. Focusing on how delivery vehicles behave and release cargo

in serum, cells, and eventually animals, is the crucial next stage for therapeutic PCMs. This includes design for peptide-driven and pH-triggered endosomal escape, programmed disassembly and cargo release, and the addition of targeting peptides or ligands on the micelle corona.

3. **Multiscale Assembly and Actuation of DNA Mechanisms**. Structural DNA nanotechnology allows engineers to design devices with unprecedented geometric precision. However, large assemblies of multiple DNA structures have poor reproducibility and little control over size and shape. I aim to optimize the macromolecular assembly of DNA nanostructures into micron-scale filaments and networks for materials and biomolecular applications. Further goals for this project include covalent and non-covalent addition of polymers to microscale DNA assemblies for stability purposes and programmable actuation of robotic microscale devices using non-traditional methods including motor proteins.

Teaching Interests:

I am interested in teaching Polymer Mechanics, Biomolecular Design, Systems/Controls, Dynamics, Thermodynamics, and Fluid Mechanics. I have two years of experience as a graduate teaching assistant in both large lecture and laboratory settings and I have taught numerous guest lectures.

Successful Proposals:

- NSF Nanomanufacturing (contributor)
- Presidential Fellowship, Ohio State University (lead)
- Molecular Foundry at Lawrence Berkeley National Lab (2) (lead)
- Advanced Photon Source at Argonne National Laboratory (lead)

Selected Publications (Chronological, of 14 total journal articles, Google Scholar h-index: 10)**:**

- Marras, A.E., Vieregg, J.R., Ting, J.M., Rubien, J.D., Tirrell, M.V. "Polyelectrolyte complexation of oligonucleotides by charged hydrophobic – neutral hydrophilic block copolymers" *Polymers*. 11:83 (2019)
- Marras, A.E., Shi, Z., Lindell, M., Patton, R.A., Huang, C.M., Zhou, L., Su, H-J., Arya, G., Castro, C.E. "Cation-activated avidity for rapid reconfiguration of DNA nanodevices" *ACS Nano. 12:9484-9494 (2018)*
- Lauback, S., Mattioli, K., Marras, A.E., Armstrong, M., Rudibaugh, T., Sooryakumar, R., Castro, C.E. "Real-time magnetic actuation of DNA nanodevices via modular integration with stiff micro-levers" *Nature Communications. 9:1446(2018)*
- Lei, D., Marras, A.E., Liu, J., Huang, C.M., Zhou, L., Castro, C.E., Su, H.J., Ren, G. "Three-dimensional structural dynamics of DNA origami Bennett linkages using individual-particle electron tomography" *Nature Communications. 9:592(2018)*
- Marras, A.E., Zhou, L., Kolliopoulos, V., Su, H.J., Castro, C.E. "Directing folding pathways for multi-component DNA origami nanostructures with complex topology." *New Journal of Physics*. 18:055005 (2016)
- Marras, A.E., Zhou, L., Su, H.J., Castro, C.E. "Programmable motion of DNA origami mechanisms." *Proceedings of the National Academy of Sciences*. 112:713-8 (2015)
    - Highlighted in *Nature Materials*, *Nature Nanotechnology*, *PNAS*
- Zhou, L., Marras, A.E., Su, H.J., Castro, C.E. "Direct design of an energy landscape with bistable DNA origami mechanisms." *Nano Letters*. 15:1815-21 (2015)
- Zhou, L., Marras, A.E., Su, H.J., Castro, C.E. "DNA origami compliant nanostructures with tunable mechanical properties." *ACS Nano*. 8:27-34 (2014)

569124 Synergizing Engineering, Chemical, and Immunological Concepts to Design the Next Generation of Therapeutics for Unmet Clinical Needs

Research Interests:

The overall focus of my research is to develop solutions for unmet clinical needs using engineering, biological, immunological, and chemical techniques. The overall vision is to combine identification of targeting ligands, nanoparticles, and immunologic stimulating payloads to generate therapeutics that allow for controlled activation of an immunologic response against a pathologic region of tissue. I have published manuscripts relating to all three components of this process and have experience with all phases of from design to preclinical validation. Additionally, combining synthetic chemical handles with proteins, separating accumulation and activation phases of pro-drugs, and tissue penetration are also major foci. Synergizing these interests into a single project allows for generation of novel solutions for current clinical needs that isn't apparent by focusing on a single piece along the process of developing next generation therapeutics.

Research Experience:

I have experience identifying targeting ligands using multiple platforms including phage and yeast surface display, site-specific combination of biological and synthetic chemistry products, development of novel immunotherapies for cancer, and reducing adverse events associated with treatments. I have primarily focused on malignant disease including non-small cell lung cancer and glioblastoma multiforme brain cancer including a recent publication that exploited blood brain barrier (BBB) disruption, rather than disease-associated cellular receptors, to deliver therapeutics to neuropathologies. I also have experience with immunological assays, vaccines, peptide synthesis, liposome production, protein production, in vitro tissue culture models, and in vivo rodent models.

Teaching Interests:

Primarily interested in developing and equipping engineering students for the challenges they will encounter the dynamic field of engineering. While basic competencies are required to have a productive conversation, my interest is teaching abstract concepts such as problem solving, rationale decision making, optimization of resources, and interpersonal skills that are needed to succeed in almost any future position. Given the rapidly changing dynamic and the career mobility of current generations these ideas are essential to the success of future engineers.

Teaching Experience:

Mentored high school, undergraduate, and medical science training program (MSTP) students in the laboratory from diverse educational and cultural backgrounds. Lead a team of research technicians and taught group sessions for the first-year graduate students core course.

Future Directions:

My short-term goals are to combine engineering principles with physiologic targeting mechanisms to improve accumulation and penetration of targeting ligands in neuropathologies that exhibit blood brain barrier disruption as well as develop methods to separate the accumulation and activation phases of prodrug therapies. My long-term goals are to engineer delivery systems for precision delivery of therapeutics to biological tissues and develop scalable solutions to produce these platforms.

Selected Publications:

1. **Umlauf BJ**, Clark PA, Lajoie JM, Georgieva JV, Bremner S, Herrin BR, Kuo JS, Shusta EV. Identification of Variable Lymphocyte Receptors that can Target Therapeutics to Pathologically

Exposed Brain Extracellular Matrix. (In Press, *Science Advances*)
2. **Umlauf BJ**, Mix KA, Grosskopf VA, Raines RT, Shusta EV. Site-Specific Antibody Functionalization Using Tetrazine-Styrene Cycloaddition. *Bioconjug Chem.* 2018 May 3.
3. Zorniak M, Clark PA, **Umlauf BJ**, Cho Y, Shusta EV, Kuo JS. Yeast display biopanning identified human antibodies targeting glioblastoma stem-like cells. *Scientific Rep.* 2017 Nov 20;7(1):15840
4. **Umlauf BJ**, Chung CY, Brown KC. Modular Three-Component Delivery System Facilitates HLA Class I Antigen Presentation and CD8+ T cell Activation Against Tumors. *Mol Ther.* 2015 Jun;23(6):1092-1102
5. **Umlauf BJ**, McGuire MJ, Brown KC. Introduction of plasmid encoding for rare tRNAs reduces amplification bias in phage display biopanning. *Biotechniques.* 2015 Feb 1;58(2):81-4.

569707 Engineering Materials to Recapitulate the Stem Cell Microenvironment

Research Interests:

Despite the tremendous potential of stem cells to repair and replace damaged or diseased tissue, little progress has been made in translating stem cell-based therapies from the laboratory to the clinic. Several critical engineering challenges remain that have limited the clinical use of stem cells, including difficulties producing the large number of stem cells needed for therapeutic applications, the efficient and reproducible differentiation of stem cells into mature cell types of interest, delivery of cells to target tissues during transplantation, and maintenance of the viability and function of the cells post-transplantation. Materials science approaches stand to overcome these significant hurdles. By designing synthetic systems to mimic the native stem cell microenvironment, we can exert greater control over cell fate decisions, such as self-renewal and differentiation. I will present work from my PhD studies and postdoctoral training that employs hydrogels to identify crucial material parameters that regulate the phenotype of adult stem cells. My PhD work leveraged the modularity of protein-engineered hydrogels to develop mimics of the extracellular matrix with simultaneous tuning of stiffness, proteolytic degradability, and adhesive ligand concentration without varying other key matrix properties, such as microstructure, swelling, and nutrient transport. This system enabled me to identify matrix remodeling as a previously unrecognized requirement for maintenance of neural progenitor cell (NPC) stemness and regenerative capacity in 3D hydrogels. Whereas stiffness and adhesive ligand concentration did not alter NPC stemness, only NPCs cultured in high degradability gels maintained the capacity to self-renew as stem cells. Furthermore, matrix remodeling was necessary for efficient differentiation and maturation of NPCs into neurons and astrocytes. The precise control of material properties afforded by the ELP hydrogels permitted detailed mechanistic studies into the biochemical basis for NPC stemness maintenance regulation by matrix remodeling, revealing a cell-cell contact mediated pathway that implicated ß-catenin and Yes-associated protein (YAP) signaling. Importantly, these results were generalizable to other engineered materials systems that were proteolytically or physically remodelable, suggesting that culture within remodelable 3D hydrogels is a promising strategy for expansion and differentiation of NPCs for therapeutic applications. While matrix remodeling represents a single time-dependent interaction between cells and their microenvironment, the native ECM is highly dynamic, varying its biochemical and biophysical properties during aging and disease. In my postdoc, I have developed hydrogel platforms that enable temporal control over parameters such as matrix stiffness and composition to model pathogenic processes *in vitro*. By mimicking changes in the muscle stem cell (MuSC) niche, I aim to uncover cell-extrinsic factors that lead to MuSC dysfunction and muscle wasting with aging and disease. Identifying target biochemical pathways that are dysregulated in the pathogenic state may enable pharmacologic interventions to restore muscle function. Overall, hydrogel-based approaches represent a significant advance in realizing the therapeutic potential of stem cells.

Teaching Interests:

Polymer Science; Thermodynamics; Mechanics of Materials; Reaction Kinetics; Biomaterials; Stem Cell Engineering; Tissue Engineering

560307 Ionic Liquids to Overcome Obstacles in Nanoparticle Drug Delivery

Research Interests:

The physical chemistry of *ionic liquids*, which consist of a bulky, asymmetric cation and anion that is liquid at room temperature,[1] can be exploited to accomplish a broad range of tasks, which historically have included synthesis,[2,3] catalysis,[4] electrochemical,[5] and energy applications.[6] In addition, recently ionic liquids have shown great promise as drug delivery agents, increasing cell permeability and navigating epithelial junctions in both the skin and intestines.[7–11] A conservative estimate suggests that 10^8 possible ionic liquid combinations can be fabricated,[12] and within this exists a breadth of manipulatable physicochemical properties leading to shifts in lipophilicity, aromaticity, and intra-ionic interactions,[13] which can be tuned to control interactions between ionic liquid moieties and biological materials.[14–16]

The use of nanoparticles as drug delivery systems has grown dramatically due to their ability to deliver drugs in a more effective, safe, and specific way compared to traditional therapeutics,[17–19] particularly in the context of administering chemotherapy to treat cancer.[20,21] However, the vast majority of nanoparticle technologies do not progress clinically as they face a number of currently insurmountable challenges (some of which are enumerated below), which result in <5 % arriving to their intended destination. My research seeks to revolutionize the field of nanoparticle drug delivery as a whole by solving the big-picture challenges that remain by using ionic liquids. Namely, providing non-invasive and localized delivery avenues (**Project 1**), methods to avoid the formation of a protein corona and subsequent macrophage uptake in the bloodstream (**Project 2**),[22] and improved cellular entry and controlled release once the nanoparticle reaches its destination (**Project 3**). The technology proposed herein could be applied broadly to many existent nanoparticle delivery systems and thus is of high potential impact.

This proposed work unites my previous research interests in ionic liquids, nanoparticles, and drug delivery. My D.Phil. pioneered the electrochemical quantification of polymer-stabilized nanoparticles suspended in ionic liquids. My current postdoctoral work tackles understanding the efficacious transdermal delivery of drugs and proteins using ionic liquids. I have published these results in 34 peer-reviewed journal papers, 16 as first author, including a first-author publication in Angewandte Chemie. I have two more papers that are currently under review.

Teaching Interests:

With respect to mentorship and teaching, I have thus far formally advised 17 students at the University of Oxford and Harvard University. The educational levels of these students have varied from first-year undergraduates, through to second-year Ph.D. candidates. In all cases, I was actively involved in designing the students' projects, ensuring they had the skills and information they needed to succeed, and co-authoring original research papers that were published in well-ranked international journals (13 in total). By mentoring a range of different students at different stages of their academic careers, I have extensive practice of keeping projects on track and ensuring that students feel supported, whilst also developing future researchers with the confidence to approach research problems independently. I have previously taught content at a variety of levels in a variety of contexts, including as a Lecturer, Demonstrator, and Tutor in the UK, Australia, and the USA. I have consistently received very positive feedback from my past students and colleagues on my ability to communicate difficult concepts, retain high levels of student engagement, and spark interest in research.

Whether in a laboratory, lecture theatre, or tutorial room, my pedagogical goal is to create an atmosphere that is both academically challenging and inclusive. In my experience, creating a conversation where the students are active participants is crucial in retaining their interest and attention, and enhancing their understanding of the material. As a tutor at St John's College, Oxford, I did this by primarily working on the whiteboard and asking students to share their working for each question from a problem set with their colleagues. When questions were raised, I guided the discussions, but gave students the opportunity to reason their way to the answer as a group before sharing the solution. My accessibility as an instructor meant that within a few weeks all members of the class felt comfortable engaging with each other and with me.

Retaining typically underrepresented students is another one of my interests as an instructor. In particular, I embrace the opportunity to act as a visible role model for young women in the physical sciences and in engineering, and this is particularly true when I am teaching classes where female students are a minority. I am aware of the existence of my own unconscious biases, and I actively work to combat these and redress systematic inequalities that may prevent students from reaching their full potential as engineers. I am committed to ensuring that students feel a sense of belonging and inclusion in my classes.

Finally, I enjoy sharing with students the scope and impact of chemistry and engineering. In my experience, tying in topics with curated examples from contemporary research excites students, as they begin to see where they might begin to fit in themselves. Whilst teaching Electrochemical Methods at the University of Oxford, I shared examples of electrochemical detection devices involving nanotechnology, including original images from my own research. This stimulated discussion in the lecture and gave the students a feel for the ways in which the theoretical material I was teaching them intersected with current research priorities.

563614 Understanding Nucleation and Crystal Growth of Organic Molecular Materials with Application to Pharmaceutical Manufacturing

Crystallization plays a major role in the production of commodity chemicals, fine chemicals, and pharmaceuticals. In the latter, crystallization is heavily used as a purification method and as a means to isolate Active Pharmaceutical Ingredients (APIs) with the desired structure, habit, and particle size distribution. Despite the widespread use of crystallization processes and the increased research interest within the pharmaceutical sector, the lack of fundamental understanding of nucleation and crystal growth leads to the use of tedious and hardly generalizable process design methodologies.

Pharmaceutical crystallization processes frequently encounter a large number of development challenges. Some of the challenges are system-related, like the presence of structurally similar or toxic impurities, the appearance of new polymorphs, or the inhibition of crystallization by foreign species. Other challenges are related to the industrial setting, either as process requirements or as a short availability of development time and raw materials. In a process where small variations in solution composition will impact both thermodynamics and kinetics, and where the incorporation of impurities follows a wide variety of mechanisms, it is almost mandatory to take process development shortcuts like assuming a negligible impact of solvents and impurities on crystallization kinetics, or predicting impurity incorporation based on system-specific phenomenological models.

Research Interests:

My research interests are in understanding the mechanisms of nucleation and crystal growth, with initial focus on their interplay with foreign species and solvents. This research will lead to the development of novel mechanistic models for crystallization kinetics and impurity incorporation, ultimately assisting the automation of crystallization process development through systematic methods. A better understanding of the mechanisms of nucleation and growth will also contribute to the development of novel crystallization strategies for the control of crystal quality attributes, including the use of additives to control kinetics, impurity incorporation, and polymorphism.

Teaching Interests:

Most of my teaching experience is in undergraduate courses on Unit Operations at the pilot scale (crystallization, drum filtration, ultrafiltration, heat transfer, and reverse osmosis), and graduate courses on Good Manufacturing Practice (GMP) and quality in food, pharmaceutical, and biotech industries. As a teacher, I will develop a graduate course on industrial crystallization and a graduate course on solid state analysis of pharmaceuticals. Both courses include skills that are highly sought after in pharmaceutical research and manufacturing, but rarely offered in chemical engineering departments. My teaching methods involve a fair combination of traditional lectures and active learning, as I believe that student involvement, practical applications and spaced learning facilitate the understanding and long term retention of the taught concepts. I also strive to provide an environment where students are motivated to learn and have access to sufficient resources that meet their learning needs.

579196 A Polymeric Reactor for the Synthesis of Superparamagnetic-Thermal Treatment of Breast Cancer

Research interest: Engineered superparamagnetic iron oxide nanoparticles (SPIONs)

Teaching interest: Nanomedicine

Engineered superparamagnetic iron oxide nanoparticles (SPIONs) have been studied extensively for their localized homogeneous heat generation in breast cancer therapy. However, challenges such as aggregation and inability to produce sub- 10nm SPIONs limit their potential in magneto-thermal ablation. We report a facile, efficient, and roboust *in situ* method for the synthesis of superparamagnetic iron oxide nanoparticles (SPIONs) within polyethylene glycol (PEG) reactor adsorbed onto reduced graphene oxide nanosheets (rGO) via microwave hydrothermal route. This promising modality yields crystalline, stable, biocompatible, and superparamagnetic PEGylated SPION-rGO nanocomposites (NCs) with uniform dispersibility. Our findings show that rGO acts as a breeding ground for the spatially distributed nano-sites around which the ferrihydrite seeds accumulate to ultimately transform into immobilized SPIONs. PEG, in parallel, acts as a critical confining agent physically trapping the accumulated seeds to break down their aggregation and create multiple domains on rGO for the synthesis of quantum-sized SPIONs (9±1 nm in diameter). This dual functionality (rGO & PEG) exhibits a pronounced effect on reducing both the aggregation and the sizes of fabricated SPIONs as confirmed by the STEM images. Toxicity studies indicate that PEGylated SPION-rGO NCs are more biocompatible than PEGylated SPIONs showing no significant induction of cell apoptosis, mitochondrial membrane injury, or oxidative stress. Significantly less lactate dehydrogenase (LDH) release, and hence less necrosis, is observed after 48 hr exposure to high doses of PEGylated SPION-rGO NCs compared with PEGylated SPIONs. NCs induce local heat generation reaching up to 43°C causing significant MCF-7 breast tumor cell ablation of about 68% upon applying an alternating magnetic field.

586545 Developing Methods to Accelerate the Design, Development, and Implementation of Therapeutics, Materials, And Equipment for Diseases with Urgent Unmet Clinical Needs

Research Interests

My long-term research goal is to develop experimental and computational methods to accelerate the design, development, and implementation of therapeutics, materials, and equipment for diseases with urgent unmet clinical needs. A primary limitation to translate these materials into the clinical setting is the cost associated with every step of the biotechnology development process. I plan to approach this challenge via three angles: (1) development of high throughput transformative methods for synthesis of therapeutics; (2) computational evaluation of material performance using systems, agent, finite element, molecular dynamics, and machine learning techniques; and (3) 3D tissue-engineered constructs and microfluidic systems to mimic disease states.

Processing: Many proteins have been discovered with significant therapeutic potential. Unfortunately, the bench-scale techniques used to produce these materials are often difficult to translate to preclinical and pilot-scale studies. Furthermore, traditional protein purification techniques, such as affinity chromatography, can be cost-prohibitive and difficult to scale. Because of these concerns, I have evaluated how methods such as tangential flow filtration and selective sedimentation can be modified to handle complex separation systems. I plan to deploy these techniques to increase the scale and decrease the manufacturing costs of biologically derived materials.

Simulation: Within the tissue microenvironment, complex physical and chemical systems can significantly alter the performance of protein and nanotechnology-based therapeutics. I plan to use modern computational methods to assess how biochemical, intracellular, and physical interactions can influence the therapeutic potency of different therapeutic agents. These models can couple finite element modeling with agent-based cell behavior.

Tissue Engineering: While computational models are a useful tool to assess complex biophysical systems, they are limited by the assumptions and scales of the simulation system. To better evaluate the potential of biologically derived materials as a potential treatment, I will develop 3D tissue-engineered constructs of various disease states. With these systems, I plan to use a microenvironment that contains multiple cell types in physically accurate systems. This will allow me to model both the biochemical and biophysical stresses that may alter the therapeutic behavior of these materials in the disease state.

Research Experience: My previous research focused on the development of hemoglobin-based oxygen carriers for use in cancer therapy. This work spanned the entire product development lifecycle, including bench-scale production, computational evaluation, small animal *in vivo* performance assessment, and pilot-scale process development. Because of this, I have developed a broad skill set from chemical synthesis to high-performance computer modeling. Additionally, I have worked on various products derived from blood products, including solutions intended for use in transfusion medicine, tissue engineering, and organ preservation. This work has often required the implementation of internal quality management and process safety evaluation systems.

Teaching Interests

I believe that the goal of chemical engineering education should be to prepare students for careers by providing them the technical, critical thinking, and communication skills. *To accomplish this goal, I plan to develop and refine coursework that integrates, non-intuitive problem-solving methods, project-based evaluation of student performance, integration of computing techniques, and reinforcement of technical writing and communication.* I am looking forward to teaching courses focused on transport phenomena or process safety. I am also excited to develop a course that integrates computational skills with solving problems in chemical and biological engineering.

Teaching Experience: At Ohio State University, I taught for two years as a teaching assistant in the unit operations laboratory. In this course, I developed lectures, quizzes, exams, and in-class exercises to help students better understand how real systems compare to the theoretical methods taught in the other courses. In my free time, I have also developed a course on high-performance computing for chemical engineering graduate students. This course was designed to integrate object-oriented design as a way to solve complex chemical engineering problems. When possible, I integrated modern open-source computing methods to solve problems in physical and biological systems. Students were instructed on how to apply the fundamentals of analytical solutions to verify and validate that the solutions are representative of real-world systems. Students were also provided with additional modern software tools such as LaTeX and Git for better translation into industrial work environments. Additionally, I have mentored four graduate students and 15 undergraduate students in my research group.

587486 High Performance Materials for Wearable Electronics and Sensors

Previous Research Experience:

I am currently a 6th year PhD candidate in the Department of Chemical and Biomolecular Engineering at Rice University under Matteo Pasquali. Individual carbon nanotubes (CNTs) have outstanding mechanical strength, electrical conductivity, and thermal conductivity. However, obtaining these excellent single molecular properties on the macroscale has proven difficult. My PhD work focuses on the dissolution of CNTs into chlorosulfonic acid and the processing of these solutions into high-performance CNT fibers. I have studied how the rheology and phase behavior of the solutions affect the spinning process and the electrical and mechanical properties of the resulting fibers. I demonstrated that lowering the concentration of CNTs in the spinning solution results in fibers with improved mechanical properties (higher tensile strength, Young's modulus, and toughness) but similar electrical properties. Through wide-angle X-ray scattering (WAXS), polarized Raman spectroscopy, and nano-computed tomography, I determined that there was no significant change in alignment or porosity of the fibers. However, mechanical testing showed that the stronger fibers also had a larger elongation at break. Therefore, I concluded the lower concentration solutions are able to assemble into larger CNT bundles within the fiber structure. These bundles then slide past each other and are the driving mechanism for the tensile strength. The larger bundle size does not affect conductivity as the resistance in the fiber is dominated by the tube-tube junctions and material impurities. Additionally, I have been studying the mechanical behavior of CNT fibers under various strain rates and temperatures. Consistent with other polymeric materials, the tensile strength of CNT fibers decreases with increasing temperature and increases with increasing strain rate. I am currently studying the effects of CNT fiber purity and aspect ratio on these trends to better understand the frictional forces between the bundles of CNTs within the fiber. Finally, I have also studied the application of CNT fibers for use as wearable electrocardiogram (EKG) sensors for continuous heart monitoring. I plied CNT fibers together to make a CNT thread that can be sewn with a standard sewing machine. The material is soft and can be washed in washing machine without any signal degradation making it an ideal wearable sensor. Additionally, because it is carbon based, it can easily interface with skin which allows it to obtain similar quality EKG signals to that of commercial wet electrodes.

Research Interests:

My postdoctoral research interests include developing wearable sensors and flexible electronics for wearable devices. Wearable sensors have gained significant interest in recent years because they can seamlessly monitor and relay information about an individual's wellbeing; this is demonstrated through the popularity of the Apple Watch and the Fitbit. Although these devices are sufficient for recreational use, they are limited in the sensors available (typically heart rate and step count) and they require an additional rigid device. There are applications for individuals such as military personal and first responders that could vastly benefit from lightweight sensors for physiological stressors, environmental hazards, and physical location. Furthermore, wearable sensors would make tracking training regiments easy for athletes and would make monitoring for individuals with chronic illness such as heart conditions or diabetes easier and less invasive. I feel that my background with colloidal processing and carbon nanotubes could help develop devices that are made using scalable techniques and are robust. I have background in this area due to my work on sewable EKG sensors, but I am interested in pursuing this line of research further and expanding my knowledge of the field.

Teaching Interests:

I am the most interested in teaching the Introduction to Chemical Engineering course. As students enter the chemical engineering curriculum for the first time, many students can feel anxious or hindered in their studies due to stereotype threat. I aim to develop and promote an inclusive curriculum with a growth mindset that guides a larger proportion of students towards success in the field. I plan to accomplish this by providing a highly structured course. The course would be composed of quizzes and small projects leading up to the exam so students can continuously gauge their understanding of the material. Furthermore, I would promote transparency by elucidating course expectations and learning goals at the start of the course. I would also like incorporate continuous feedback throughout the course by using anonymous surveys. These short surveys would ask students about the pace of the course and their current understanding of the material. Finally, I feel that it is crucial to also act as mentor for more than just the coursework. The adjustment to college life can be difficult and having a support system can make this transition much easier by helping students overcome financial and personal barriers.

587488 Bionanotechnology: DNA Nanotechnology-based Molecular Devices, Bio-sensors, and Hybrid Materials forBiological and Nanophotonic Applications

Research Interests: Bionanotechnology has emerged as a new paradigm to study biological processes, many of which occur at the nanoscale, and to design next-generation sensors and diagnostics. My laboratory will advance the field of bionanotechnology by developing DNA-based molecular bio-sensors, devices, and hybrid materials that will be used to address biological questions regarding cooperativity amongst cell receptor-binding ligands (Project 1), to solve problems in biological detection, e.g., the detection of low abundant circulating tumor DNAs (ctDNAs) (Project 2), and to further our understanding in nanophotonics by designing programmable nanophotonic materials for application in biology and energy (Project 3). Our ultimate goal is to understand biological processes in a more profound sense, by utilizing multidisciplinary approaches at the intersection of biology, materials, and biophysics and to apply such knowledge to design next-generation therapies and diagnostics.

Project 1: It is known that ligand cooperativity regulates cellular behavior. For example, the cooperativity between RGD (Arginylglycylaspartic acid) and BMP (Bone morphogenetic protein) enhances transcriptional activity that promotes osteogenic differentiation in hMSCs (Human Mesenchymal Stem cells), and RGD and VEGF (Vascular endothelial growth factor) regulates vascularization of endothelial cells. These interactions occur at nanoscale distances. Due to the lack of technology to spatially control the distance, density, and ratio between and amongst ligands, such cooperative phenomena have not been explored and investigated. Using DNA origami, I propose to engineer molecular rulers to investigate these cooperativities and apply the developed tools to study other molecular interactions. **Project 2:** Quantitative detection of circulating tumor DNAs (ctDNAs) in cancer patients is essential for early cancer diagnosis and precision medicine, but they are very difficult to quantify because of the low abundance in the blood. Based on the highly specific binding properties of DNA, I envision to develop DNA sensors as a novel and practical approach to solve this problem. **Project 3:** We are in great need of novel technologies for energy and nanophotonic applications. To address this, my group will develop hybrid-nanomaterials composed of natural photosynthetic systems and/or photonic materials/nanoparticles using self-assembled and highly programmable 3D DNA nanostructures to spatially arrange proteins, fluorophores, nanoparticles, and other molecules and moieties in the nanoscale. Beyond the projects, as mentioned earlier, my group will be interested in exploring other innovative ideas in a broad range of areas, including drug delivery. To advance the research in my group, I will engage in active collaborations with other faculty members, especially with computational groups and biology groups within the department and the university.

Research Accomplishments: During my doctoral research in Prof. Hao Yan's and Prof. Yan Liu's laboratory at Arizona State University, I invented DNA nanostructure-based multi-chromophoric artificial light-harvesting antenna systems and investigated the photonic properties using various spectroscopic methods, including fluorescence spectroscopy, time-correlated single-photon counting, Streak camera, ultrafast pump-probe spectroscopy and other optical spectroscopic methods (JACS 2011, JACS 2014, JACS 2014). These studies furthered our fundamental understanding of multi-chromophoric systems and paved the path for designing and developing more complex and sophisticated artificial light-harvesting antenna systems for energy and bio-switch applications. After finishing my Ph.D., I joined Prof. Yonggang Ke's laboratory in the Department of BME at Georgia Institute of Technology and Emory University, as a postdoctoral fellow, to design and test complex DNA origami structures for nanophotonic (JACS 2016, ACS Nano 2017) and mechanobiological applications (Nano Letters 2018). To gain more biological insights in mechanobiology, I pursued a second postdoc in Prof. Christopher Chen's laboratory at Boston University and Harvard Medical School. After joining, I developed a-actinin tension sensors to investigate the mechanobiology of sarcomerogenesis in iPSC-derived (induced pluripotent stem cell) cardiomyocytes, and I employed DNA origami to spatially control ligand molecules to study cell adhesion and signaling. While working in Prof. Chen's laboratory, I also learned various micro-patterning and microfluidic device design techniques. In conclusion, during my research training, I have gained vast knowledge and acquired a unique skill set in DNA nanotechnology, nanomaterials, biophysics, analytical science, surface/conjugation chemistry, and cell biology, which will enable me to develop an exciting and transformative research program in bionanotechnology.

Funding: I have prior experience in writing NIH-R21 proposal with my previous advisor, which was granted. This experience will help me to write two NIH-R21 proposals for Project-1 and -2, and a NSF proposal for Project 3, immediately after my joining as a faculty. Also, the initial results from these three projects will help me to write proposals for RO1 grants. I will also seek grants from DoD, ARO, and DoE and funding opportunities from pharmaceutical companies and private foundations.

Teaching Interests: Aside from my research career, I also have enjoyed teaching/mentoring. At Arizona State University, I was an Organic Chemistry Teaching Assistant for two semesters. I also have mentored several high school, undergraduate, and graduate students during Ph.D. and Postdoctoral studies. I believe that as a mentor, my role is to motivate the students and stimulate their thinking and problem-solving capabilities. I am fully committed to facilitating these goals by setting up an engaging teaching environment. As a faculty, I would like to teach courses at both undergraduate and graduate levels, covering fundamentals of fluorescence, biophysics, analytical chemistry, and physical chemistry. I would also like to introduce graduate-level courses in bionanotechnology and its applications in various research areas, and instrumentations (spectroscopy, optical microscopy, electron microscopy).

581856 Process Intensification for Sustainable Fuels and Energy Production

Research Interests:

Biomass, coal, natural/shale gas, or carbonaceous wastes can be converted to liquid fuels through X-to-liquids (XTL) processes, some (coal and gas) of which are applied commercially in a number of plants around the world. One of the biggest challenges associated with XTL processes is the high capital costs of these plants. Small-scale XTL have the advantages of being less capital intensive, less risk, and have the potential of utilizing renewable resources such as biomass, sustainable sources such as waste, and gas reserves which previously could not be exploited, i.e. associated and stranded gas reserves. However, the cost per barrel of these processes still remains high, if conventional technology is employed. Novel reactive distillation processes are paving the way for a more sustainable chemical process industry that is profitable, safer and less polluting [1–3].

Research in process intensification (PI) has recently gained considerable attention due to challenges related to energy and the environment, alongside risks in capital investment decisions. These challenges necessitate the development of optimization-based computational tools for process synthesis and design, which enable the integration of multiple phenomena that occur at different scales in an intensified unit. Current efforts in the field are yielding promising results that indicate the transformation of the petrochemical industry. We have recently demonstrated that PI can play a crucial role in the production of fuels, chemicals and electricity using reactive distillation with a Fischer–Tropsch (FT) based process [4,5]. We developed an equation-oriented framework for optimal synthesis of integrated reactive distillation (RD) systems for FT processes in GAMS [6], which has recently been implemented in PYOMO [4].

To produce synthesis gas at the required H_2/CO ratio for the RD process, we are developing a membrane steam-methane reforming (SMR) model. This optimization-based model provides flexibility in both process design and operation. The excess hydrogen from the reformer is used to optimize gasoline and diesel production in the product refining sections. The models are developed in Python and the optimization of the integrated flowsheet is implemented in PYOMO. The use of the SMR with internal hydrogen separation greatly improves conversion and lowers operating conditions. This type of modular design and integration of processes, which is suitable for small-scale plants, also has the potential to ease the scalability of the process. This recent model development will be presented at this conference [P1].

The deployment of large-scale equation-oriented models in industrial Oil & Gas facilities for real-time optimization has made the use of reduced-order models (ROMs) more attractive. I am working on a project with Petrobras to develop accurate and efficient ROMs for their RECAP unit in Mauá, Brazil, and will be presenting our result at this conference [P2].

Research Experience

My PhD Thesis was on the 'Interaction between Reaction and Phase Equilibria in the Fischer–Tropsch Reaction' advised by Prof Diane Hildebrandt and Prof David Glasser at the University of the Witwatersrand in Johannesburg, South Africa. The aim was to develop a Fischer–Tropsch synthesis (FTS) model which formed part of the models used in the design, commissioning and start-up of a coal-to-liquids Pilot Plant in Baoji, China.

I spent a year at the University of Kentucky working on catalyst development, characterization, and testing in slurry-phase FT reactors with Prof Burtron H. Davis. This was part of an integrated process synthesis approach since I wanted to expand my experience from working with fixed beds, CSTRs, and Batch Reactors to working with three-phase slurry reactors. The insight and experienced gained from working with slurry reactors could be applied to modeling biomass conversion or other petrochemical process reactors.

As a faculty member at the University of South Africa, I decided to expand my research area to consider renewable energy sources. I advised a PhD student (amongst others) on the conversion of glycerol and water through solar photocatalysis for the production of hydrogen. This project was part of an integrated approach where the waste product of one process (Biodiesel Production) is used to produce something useful (hydrogen) using a renewable energy source (solar energy). This led to a development of a high activity catalyst using titanium hollow spheres as support.

I spent a year at Texas A&M Energy Institute under the Mentorship of Prof Christodoulos A. Floudas working on hybrid energy systems. This gave me exposure to applying mixed-integer nonlinear programming techniques to optimize processes and to developing mathematical models that make use of advanced computational tools to solve complex engineering problems.

Teaching Interests:

I was a Senior Lecturer (Assistant Professor) in Chemical Engineering at the University of South Africa from 2012–2015, and was an Associate Professor from 2016–2018. I advised Masters and PhD students and taught:

PCT402C: Process Control IV,

RTE4701: Reactor Technology IV.

I am committed to excellence in teaching. With my extensive teaching experience and my strong background in chemical and systems engineering, I am well-suited for the instruction of a wide range of undergraduate and graduate subjects as well as develop my own courses focusing on Process Intensification with its application to Sustainable Energy Processes.

Masters and Doctoral Students Supervision

I have advised one PhD and two Masters students to completion at the University of South Africa; and have co-advised three Masters students to completion at Carnegie Mellon University.

Research Grants as a Principal Investigator

National Research Foundation of South Africa, Thuthuka Grant Holder: 2018–2020

Grant no: 113652; Value per annum: R432,000.00 (approximately $30,857).

National Research Foundation, Knowledge Interchange and Collaboration: 2017

Grant no: 110240, Total value: R23,000.00 (approximately $1,643).

Vision Keepers Support Program: 2016–2018

From: University of South Africa, Value per annum: R177,534.72 (approximately $12,681)

Emerging Researcher Support Program: 2016–2018

From: University of South Africa, Value per annum: R845,804.40 (approximately $60,415)

National Research Foundation of South Africa, Thuthuka Grant Holder: 2015–2017

Grant no: 94090; Value per annum: R494,800.00 (approximately $35,343).

Future Direction

As faculty, I aim to lead a research group that uses process systems engineering concepts and catalysis fundamentals for applications in energy processes and the environment. Our work will rely heavily on the use core chemical engineering principles and advanced computational algorithms designed to enhance decision making in complex engineering systems.

I believe in the great value of collaborations and have demonstrated it in numerous projects with colleagues from both academia and industry. I am especially interested in exploring potential collaborations in catalysis, reaction engineering, thermodynamic modeling, systems design and optimization of energy- and sustainability-related projects.

Service

I have been extensively involved in community engagement activities and have mentored and advised students from previously disadvantaged backgrounds and thus should be able to contribute positively to faculty diversity and to working with minorities. I am an approachable, hardworking, enthusiastic, motivated and focused individual with effective interpersonal and communication skills. I can communicate complex and conceptual ideas to students as well as to peers using high level skills and a range of media.

Presentations at the current AIChE Annual Meeting

[P1] Modular gas-to-liquids process with membrane steam-methane reformer, Fischer–Tropsch reactive distillation and integrated product refining. Topical Conference: *Advances in Fossil Energy R&D: Design and Optimization of Environmentally Sustainable Advanced Fossil Energy Systems.*

[P2] Reduced-order model for a large-scale real-time optimization of a residue fluidized catalytic cracker. Session: *Industrial Applications in Design and Operations*.

Selected Recent Publications

[1] **C.M. Masuku**, L.T. Biegler, Recent advances in gas-to-liquids process intensification with emphasis on reactive distillation, *Curr. Opin. Chem. Eng. 22* (2019) https://doi.org/10.1016/j.coche.2018.12.009.

[2] W.D. Shafer, M.K. Gnanamani, U. Graham, J. Yang, **C.M. Masuku**, G. Jacobs, B.H. Davis, Fischer–Tropsch: product distribution – The fingerprint of synthetic fuels, *Catalysts 9* (2019) 259.

[3] T.W.P. Seadira, G. Sadanandam, T. Ntho, **C.M. Masuku**, M.S. Scurrell, Preparation and characterization of metals supported on nanostructured TiO_2 hollow spheres for production of hydrogen via photocatalytic reforming of glycerol, *Appl. Catal. B Environ. 222* (2018) 133 – 145.

[4] N. He, Y. Hu, **C.M. Masuku**, L.T. Biegler, 110th Anniversary: Fischer–Tropsch synthesis for multiphase product recovery through reactive distillation, *Ind. Eng. Res. Chem. 58* (2019) https://doi.org/10.1021/acs.iecr.9b02352.

[5] Y. Zhang, **C.M. Masuku**, L.T. Biegler, An MPCC reactive distillation optimization model for multi-objective Fischer–Tropsch synthesis, *Comput. Aided Chem. Eng. 46* (2019) 451–456.

[6] Y. Zhang, **C.M. Masuku**, L.T. Biegler, Equation-oriented framework for optimal synthesis of integrated reactive distillation systems for the Fischer–Tropsch processes, *Energy Fuels 32* (2018) 7199–7209.

Keywords: *Fischer–Tropsch Synthesis, Catalysis and Reaction Engineering, Process Synthesis and Optimization, Process Intensification, Sustainable and Renewable Process Design.*

581887 Molecular Modeling and Machine Learning for Catalysis and Separations

Molecular Modeling and Machine Learning for Catalysis and Separations

Research Interests:

As earth's population grows and the standard of living in developing nations improves, demand for energy and water will increase, furthering the strain on global resources. As production increases to meet demand, technologies that save energy and reduce pollution are essential to minimize environmental impact. Design of high-performance materials for catalysis and separations can reduce waste, conserve energy, and enable new green chemistries. Nanoporous adsorbents and catalysts are critical technologies with the prospect of controlling molecular-scale interactions through material and process design. Consequently, progress toward optimal materials and chemical processes can be accelerated with molecular modeling. Simulations of molecular-scale phenomena provide valuable insight into mechanisms of adsorption, transport, and reaction in complex environments. Computer-aided screening across the expansive space of materials and processes can focus experimentalists to synthesize the most promising materials and test them at optimal conditions. As vast libraries of materials are created in lab and in silico, extracting actionable information from these big data can be difficult. Techniques from machine learning and data science can be harnessed to identify critical trends and descriptors, as well as build predictive models for discovery.

Machine learning is experiencing a Renaissance in the chemical sciences, being used to make sense of the vast and diverse spaces of all possible molecules, materials, and reactions. Less attention is focused on characterizing *process space* – the space of industrially-relevant temperatures, pressures, and compositions – in which complex phase behavior, and non-ideal solvent/solute/surface interactions influence process performance. My research group will combine expertise in electronic structure calculations, Monte Carlo and molecular dynamics simulations, force field fitting, thermodynamics, and machine learning to investigate challenging problems in these complex chemical systems. Leveraging these techniques in collaboration with experimentalists will accelerate progress toward a more sustainable future.

Research Experience

During my Ph.D. with Prof. Dion Vlachos, I applied molecular dynamics and electronic structure calculations to investigate fundamental interactions between solvents, solutes, and catalysts in biomass catalysis. In my first project, I discovered how co-solvent interactions protect biomass-derived furans from side reactions in fructose dehydration [1] [2]. I developed structure-activity relationships for homogeneous Lewis acids to probe mechanistic details in glucose isomerization [3] [4]. I also exhaustively screened site geometries for Sn-Beta, identifying the most stable active geometry and proposing a source for its Brønsted acidity [5]. In my final project, which I continued during my post doc, I developed the reaction mechanism for fructose etherification on the hierarchical Sn-SPP zeolite and integrated it with multicomponent liquid-phase adsorption [7].

In my current post doc with Prof. Ilja Siepmann, I am developing and applying Monte Carlo and molecular dynamics simulations in five projects: 1) predicting solution-phase adsorption into hierarchical zeolites under reaction conditions [10], 2) investigating adsorption of organics at the air-water surface [6], 3) calculating the viscosities of novel lubricants [8], 4) predicting thermophysical properties of hydrocarbons [9], and 5) developing molecular modeling and machine learning techniques to predict adsorption isotherms [10].

Teaching Interests:

Excellence in teaching has the potential to multiply progress made toward a more sustainable future through the engineers who will come after us.

I have served as a teaching assistant in both undergraduate and graduate thermodynamics courses, and I have been a guest lecturer in kinetics and reaction engineering at 3 institutions. In addition, I organized a computational chemistry workshop for the graduate reaction kinetics course, to guide students through exercises using the same software I use in my research.

I am particularly excited about introducing machine learning and data science into undergraduate and graduate Chemical Engineering education. By broaching these topics in core courses and going deeper in electives, students will become acquainted with cutting-edge data analysis tools and equipped for careers in data science and its intersection with traditional chemical engineering careers.

Courses of expertise: thermodynamics and statistical mechanics, kinetics and reaction engineering, numerical methods and linear algebra, computational chemistry, and engineering statistics.

Postdoctoral Research Associate

Postdoctoral Project: *Monte Carlo Simulations for Novel Materials and Process Design*

Advisor: Prof. J. Ilja Siepmann, Department of Chemistry and Department of Chemical Engineering and Materials Science, University of Minnesota

PhD Dissertation: *Leveraging First-Principles Simulations for Deeper Insights in Biomass Processing*

Advisors: Prof. Dionisios Vlachos and Dr. Stavros Caratzoulas, Department of Chemical and Biomolecular Engineering, University of Delaware

AIChE Conference Talks:

[10] **Josephson, T. R.**, Sun, Y., and Siepmann, J. I. "Molecular simulations and machine learning for multicomponent adsorption." Session: Faculty Candidates in CoMSEF/Area 1a

[9] **Josephson, T. R.**, Singh, R., Minkara, M. S. , and Siepmann, J. I., **2019**, "Partial molar properties from molecular simulation using multiple linear regression." Session: Recent Advances in Molecular Simulation Methods II

Selected Publications

[9] **Josephson, T. R.**, Singh, R., Minkara, M. S. , and Siepmann, J. I., **2019**, "Partial molar properties from molecular simulation using multiple linear regression." *In Review*

[8] Liu, S., **Josephson, T. R.**, Athaley, A., Chen, Q. P., Norton, A., Ierapetritou, M., Siepmann, J. I., Saha, B., Vlachos, D. G., **2019**. "Renewable lubricants with tailored molecular architecture." Science Advances. doi: 10.1126/sciadv.aav5487

[7] **Josephson, T. R.**, DeJaco, R., Pahari, S., Ren, L., Guo, Q., Tsapatsis, M., Siepmann, J.I., Vlachos, D., Caratzoulas, S., **2018**. "Cooperative Catalysis by Surface Lewis Acid/Silanol for Selective Fructose Etherification on Sn-SPP Zeolite." ACS Catalysis. doi: 10.1021/acscatal.8b01615

[6] Minkara, M. S., **Josephson, T. R.**, Venteicher, C. L., Chen, J. L., Stein, D. J., Peters, C. J., Siepmann, J. I., **2018**. "Monte Carlo simulations probing the liquid/vapour interface of water/hexane mixtures: adsorption thermodynamics, hydrophobic effect, and structural analysis." Molecular Physics. doi: https://doi.org/10.1080/00268976.2018.1471233

[5] **Josephson, T. R.**, Jenness, G. R., Caratzoulas, S., Vlachos, D. G., **2017**. Distribution of open sites in the Sn-Beta zeolite." Microporous and Mesoporous Materials. doi: 10.1016/j.micromeso.2017.02.065

[4] **Josephson, T. R.**, Brand, S. K., Caratzoulas, S., Vlachos, D. G., **2016.** 1,2-H versus 1,2-C-shift on Sn-silsesquioxane." ACS Catalysis, 7. doi: 10.1021/acscatal.6b03128

[3] Brand, S. K., **Josephson, T. R.**, Labinger, J. A., Caratzoulas, S., Vlachos, D. G., Davis, M. E., **2016.** Methyl-ligated tin silsesquioxane catalyzed reactions of glucose." Journal of Catalysis. doi:

10.1016/j.jcat.2016.06.013

[2] **Josephson, T. R.**, Tsilomelekis, G., Bagia, C., Nikolakis, V., Vlachos, D. G., and Caratzoulas, S., **2014**. Solvent-induced frequency shifts of 5-hydroxymethylfurfural deduced via infrared spectroscopy and ab initio calculations." Journal of Physical Chemistry A. doi: 10.1021/jp508340p

[1] Tsilomelekis, G., **Josephson, T. R.**, Nikolakis, V., and Caratzoulas, S., **2013**. Origin of 5-hydroxymethyl furfural stability in water/dimethyl sulfoxide mixtures." ChemSusChem. **Cover article**. doi: 10.1002/cssc.201300786

581953 Scaling up First Principle Simulation in Realistic Environment: Solvent Effects and Excited State Properties in Computational Catalysis

Research Interests:

First principle simulations have been playing an increasingly important role in the study of molecular properties and reactions, especially in computational catalysis. First principle simulations have multiple aspects for researchers to explore: the level of electronic structure method, the system complexity, and the time scale of simulation. Due to computation bottlenecks, improvement of one aspect of the simulation is often achieved at the cost of compromising the others. My research has been focused on improving different aspects of first principle simulations simultaneously in order to understand reactions in the realistic complexes experimental environment both accurately and efficiently. Specifically, I have focused on (1) molecular properties and reactions in solution phase; (2) improving the accuracy of density functional theory (DFT) in predicting transition metal chemistry; (3) accurate and efficient description of excited state properties; and (4) Automated control and analysis of first principle simulations.

During my Ph.D. study, I developed the graphical processing units (GPUs) algorithms to accelerate the widely used implicit solvent model, the conductor-like polarizable continuum model (C-PCM), in first principle calculations to enable the evaluation of solution phase molecular properties and reactivity in both ground state and excited state. I also exploited the GPUs in developing the new ensemble DFT methods that are capable of describing the electronic structure of molecules in photochemistry reactions.

During my postdoctoral studies, I investigated the density delocalization errors in approximate DFT in the studies of transition metal chemistry. Transition metal containing molecules play essential roles in catalysis. However, they are challenging to study with conventional computational modeling using DFT. My works on benchmarking density the delocalization errors and spin splitting energy evaluations provide broad recommendations for computational modeling of open-shell transition metal containing systems.

Currently, I focus on building workflows to improve the automated control and analysis of transition metal chemistry simulations. A crucial step to obtain accurate results for transition metal containing systems is to choose between single-reference and multi-reference based methods. I develop an open-source module, MultirefPredict, to automate the calculation of widely used multi-reference diagnostics, which will then be embedded into the inorganic chemistry toolkit, molSimplify, to enable automated recommendation of first principle methods and computing resources.

Teaching Interests:

Based on my background in physical chemistry, I am pleased to teach kinetic, thermodynamics, numerical analysis, physical chemistry, and courses specialized in atomistic simulations, including electronic structure methods (e.g. DFT), molecular dynamics, etc.

581962 Elucidating Atomic Dances through Reaction Landscapes

Research Interests:

The vision of my research group will be the discovery and elucidation of relationships between catalysts and the reactions they catalyze. How do atoms come together and why do they come together the way they do? Transformation of molecules is at the core of energy and environmental chemistry. An understanding of how they transform at the atomic level accelerates progress in catalyst development. Research projects will be chosen due to environmental impact, industrial importance, and potential for the ideas and theories of the chemistry to be generalized and applicable across disciplines. My research group will collaborate with other research groups whose expertise deviates from our own - for instance, with computational groups and synchrotron facilities at national laboratories. Themes my research group will focus on are:

- Discovery of active sites and Elucidation of catalytic mechanisms at the atomic level
- Design and Validation of *in-situ* and *operando* reactors

PhD Research: (Purdue University, School of Chemical Engineering, Advisor: Fabio H. Ribeiro)

"Synthesis and Characterization of Copper-Exchanged Zeolite Catalysts and Kinetic Studies on NO_x Selective Catalytic Reduction (SCR) with Ammonia"

My PhD research involved unraveling atomic details of the mechanism for model Cu-SSZ-13 zeolite solids to reduce NO_x from diesel engine exhaust with NH_3 and O_2 to environmentally benign H_2O and N_2 at 200°C. This research involved collaboration with computational groups (Schneider Group from the University of Notre Dame and Greeley Group at the Purdue University), industrial partners (Cummins Incorporated), Argonne National Laboratory, and international partners (Universidad de Antioquia, Colombia). Both experimental and computational groups paved the road to discovery. We first determined from kinetic and spectroscopic evidence that precursors to the active sites were Cu^{2+} ions, and that even though the material is a solid catalyzing a gas phase-reaction, the Cu^{2+} behaves as if it is on a leash in a liquid solution. Additional quantitative spectroscopy, titration experiments, and density functional theory (DFT) computation presence of two types of active Cu^{2+}, namely bare Cu^{2+} and $(CuOH)^{1+}$. Computational insight in addition to *operando*, *in-situ*, and *ex-situ* x-ray absorption spectroscopy (XAS) experiments at the Advanced Photon Source at Argonne National Laboratory gave oxidation-state and coordination information, which were used to piece together how these active Cu^{2+} catalyzed the reaction.

Postdoctoral Research: (Leiden University, Leiden Institute of Chemistry, Advisor: Marc T.M. Koper)

"Understanding Atomistic Details of the Hydrogen Evolution Reaction (HER) on Pt(111) and Graphene/Pt(111) in H_2SO_4"

My postdoctoral research involves unraveling atomic details for the mechanism of the hydrogen evolution reaction (HER) in aqueous H_2SO_4 on Pt(111) and Graphene/Pt(111) electrodes under ambient temperatures and pressures. A monolayer of graphene was grown on a Pt(111) surface using chemical vapor deposition (CVD) and characterized using scanning electron microscopy (SEM) and Raman spectroscopy. Cyclic Voltammogram (CV) sweeps indicated that the graphene monolayer is selectively permeable to H^+ ions in the electrolyte, allowing only H^+ ions into the confined layer between Pt(111) and graphene. The impact of this confinement on the HER mechanism was probed using reaction kinetics (rates, Tafel slopes, reaction orders).

During my postdoc, I refined my skillset in collecting kinetics in both gaseous and aqueous systems and contemplated differences in the theories between thermal-activated and electron-activated catalysis. In both my PhD and postdoctoral research, probing and understanding how flow of electrons affected the local active site environment and vice versa under catalytic conditions was essential in rationally piecing together mechanistic details.

Teaching Interests:

My teaching experience includes being a teaching fellow twice for the undergraduate fluid mechanics course at Purdue University, a structured study group leader for the undergraduate general chemistry course at the University of Michigan and a tutor for the undergraduate material and energy balances course, also at Michigan. During my studies at Michigan, I developed educational materials with Professor H. Scott Fogler for his books "Elements of Chemical Reaction Engineering, 5th Edition", and "Strategies for Creative Problem Solving". I also took a course on educational methods in engineering with Professors Phillip Wankat and Matthew Ohland at Purdue and attended a workshop hosted by Professors Richard Felder and Rebecca Brent. I am competent to teach any core chemical engineering course. In terms of course development, I would like to develop a 1-credit "Design and Construction of Tubing and Piping Systems" course targeted for senior undergraduates and first-year graduate students if facilities and funds are available. To tie my research into teaching, I am enthusiastically willing to teach or develop a course in materials characterization and reaction kinetics in catalysis targeted for first and second-year graduate students. I believe that a display of enthusiasm and joy of learning will engage and motivate students throughout courses. Giving students more responsibility will increase my teaching efficiency without coping out while ensuring that the students are challenged, simultaneously allowing me to retain time required to maintain my research group and scientific output.

Diversity Viewpoint:

Diversity supports resilience and facilitates work where several perspectives combine to create something strong and robust. Though diversity about gender and race dominate headlines, other forms of diversity such as skillset, nationality, generation, and much more are also important. In an increasingly globalized world, the need to be proactive in fixing the leaky pipeline for all those who wish to explore the world through science is ever so strong. During my PhD, I mentored two students from the Universidad Nacional de Colombia (one of whom is earning her PhD in Paris) in addition to four high school teachers (three from Indiana and one from a Native American reservation in New Mexico). From these experiences, I observed first hand that different viewpoints from a diverse group of scientists not only challenge outdated assumptions, but also ask questions that would otherwise been overlooked. I have hope, but the one thing we need more than hope is action. Portions of the pipeline are improving, but certain portions are still abysmal. Policies are a good start, but what really changes things are consistent actions.

581976 From 1- and 2-Dimensional Materials to Architectural Properties in Catalysis: Rationalizing, Predicting and Designing through First-Principles Methods

Research Interests:

Electronic-structure methods, such as density functional theory (DFT) calculations, are powerful tools for the rationalization of reaction mechanisms and for guiding the synthesis of catalysts with tailored properties. The most part of computational investigations in the catalysis field usually deals with reaction path analysis. However, a rigorous and creative use of simulations makes possible to move away from more common reaction mechanism studies, and to focus on less treated aspects of catalysis, such as *architectural properties*.

In my research, starting from DFT calculations, I extrapolate the principles that guide the stability and catalytic activity of materials over a wide range of size-domains. Among these:

- Supported single atom catalysts (SACs) for catalytic and electrocatalytic applications;
- Sub-nanometer metal clusters for the upgrade of biomass-derived molecules;
- Stable, free-standing 2D Bismuthene monolayers for CO_2 reduction reaction;
- Surface reconstructions and dynamics under electrocatalytic operating conditions;
- Non-conventional Moiré catalytic architectures.

Postdoctoral Project

"Computational studies of 1- and 2-D materials for catalytic and electrocatalytic applications" Under the supervision of Prof. M. Mavrikakis, University of Wisconsin-Madison (USA).

PhD dissertation

"Theoretical insights into heterogeneous metallic catalysts for biomass-based hydrogen production" Under the supervision of Prof. D. Duca, University of Palermo (Italy).

Teaching Interests:

During my post-doctoral and doctoral experience I served as mentor for young PhD, Master, and undergraduate students. I have served as Teaching Assistant and lecturer for catalysis, kinetics, and inorganic chemistry courses. My chemistry (PhD) and chemical engineering (Post-doc) background is suitable to teach courses as: i. Reaction Engineering; ii. Intro to Chemical Engineering; iii. Thermodynamics; iv. Catalysis and kinetics; v. Computational and Theoretical Chemistry; vi. Inorganic Chemistry; vii. Organic Chemistry; viii. Physical Chemistry; ix. Surface Science; x. Materials Science

Funding:

Helped with the preparation of successful grant applications and renewals:

- Atomic-Scale Design of Metal and Alloy Catalysts: A Combined Theoretical and Experimental Approach (DOE-Funded project; PI: Prof. Manos Mavrikakis)
- NSF Travel Support: 2019 Gordon Research Seminar (GRS) and Gordon Research Conference (GRC) on Chemical Reaction at Surfaces; Ventura, California (PI: Prof. Manos Mavrikakis)

Selected Publications:

- Cai H. †, <u>Schimmenti R.</u>†, Nie H., Mavrikakis M., Chin Y.-H., The Unique Role of Proton-Hydride pair in Heteroarene Catalytic Hydrogenation, *submitted*;

† † †

Yang F. , Elnabawy A. , Schimmenti R. , Yao S., Deng R., Song S., Lin Y., Xu W., Mavrikakis M., Bismuthene for Highly Efficient Carbon Dioxide Electroreduction Reaction, *submitted*;

- Cortese R., Schimmenti R., Godina L., Prestianni A., Ferrante F., Duca D., Murzin D. Yu., A Combined Theoretical and Experimental Approach for Platinum Catalyzed 1,2-Propanediol Aqueous Phase Reforming, *Phys. Chem. C*, 121, 14636, 2017;
- Schimmenti R., Cortese R., Duca D., Mavrikakis M., Boron Nitride-supported Sub-Nanometer Pd_6 Clusters for Formic Acid Decomposition: A DFT Study, *ChemCatChem*, 9, 1610, 2017;
- Cortese R., Schimmenti R., Ferrante F., Prestianni A., Decarolis D., Duca D., Graph-Based Analysis of Ethylene Glycol Decomposition on a Pd Cluster, *Phy. Chem. C*, 121, 13606, 2017;
- Ferrante F., Prestianni A., Cortese R., Schimmenti R., Duca D., Density Functional Theory Investigation on the Nucleation of Homo-and Heteronuclear Metal Clusters on Defective Graphene, *Phys. Chem. C*, 120, 12022, 2016;
- Schimmenti R., Cortese R., Ferrante F., Prestianni A., Duca D., *Chem. Chem. Phys.*, 18, 1750, 2016.

[†]These authors contributed equally.

581981 Dynamic Heterogeneous Catalysis to Enhance Turnover Frequency Via Surface Resonance

Research Interests:

(i) Dynamic surface resonance for heterogeneous catalysis via microkinetic modeling and experiments. The greatest rate limitation in catalysis, the Sabatier maximum, is overcome by imposing a dynamic variation of catalyst surface electronic characteristics, leading to surface coverages and surface reaction rates that exceed the existing classes of known metal catalysts. Catalytic Resonance Theory (CRT) also permits unique capability for tuning individual reactions and selectively controlling reactions for small-scale, high-throughput reactors for energy, water and fuel applications.

(ii) Experiments on automated Lewis acid and redox site counting and characterization (i.e. redox potential, Lewis acid strength). Lewis acidic and redox active catalysts play key roles in industrial chemistries such as selective catalytic reduction of NOx, alkene oxidation, and aldol condensation. Fundamental comparisons between different catalyst formulations are difficult because there are few to no methods that can quantitatively assess Lewis acidity or redox activity. In-situ treatment with acidic oxygenate molecules under FTIR, UV-vis, or gravimetric observation can selectively titrate Lewis acid centers, allowing for site counting and strength assessment.

(iii) Molecular modeling and experiments on synthesis of chemoselective solid Lewis base and Lewis acid catalysts using layer by layer methods such as atomic layer deposition (ALD), chemical vapor deposition (CVD), and chemical liquid deposition (CLD). Lewis basic catalysts are highly efficient hydrolysis, isomerization, and Knoevenagel condensation catalysts. Their properties, structures, and behavior are largely unknown as solid acid catalysts are frequently more popular in the academic community. Using a combination of periodic DFT (VASP) and experiment: Lewis basic single sites, oligomers, and clusters are studied.

Postdoctoral Project:

"Principles of Dynamic Heterogeneous Catalysis: Enhanced Turnover Frequency and Conversion via Surface Resonance"

Under the supervision of Paul J. Dauenhauer, Department of Chemical Engineering and Materials Science, University of Minnesota

PhD Dissertation:

"Depositing SiO_2 on Metal Oxides to Synthesize Tunable Solid Acid Catalysts"

Under the supervision of Justin M. Notestein, Department of Chemical and Biological Engineering, Northwestern University

Undergraduate Research:

"Reaction Engineering to Optimize Furfural Yields from Real and Ideal Xylose Feedstock"

Under the supervision of James A. Dumesic and David Martin Alonso, Department of Chemical and Biological Engineering, University of Wisconsin - Madison

Research Experience:

My academic career has concentrated on the field of heterogeneous catalysis with complementary DFT, microkinetic modeling, and experimental projects. Most of my formal training is in controlled synthesis of metal and metal oxide catalysts, reaction engineering of liquid phase biomass reactions, and using vapor phase reactions/spectroscopy techniques to characterize solid acid catalysts. I have collaborated often with computational groups including a project with the Broadbelt group at Northwestern, focusing on the homogeneous and heterogeneous epoxidation of limonene with environmentally friendly H_2O_2 or industry standard organic peroxides. As a result of my collaborations and experiences, I have versatile experience in materials science, reactor design, sol-gel processing, infrared spectroscopy, NMR, adsorption/chemisorption, and DFT.

Teaching Interests:

My teaching experience consists of four teaching assistant quarters with three being for the senior Process Design course and one being for the senior Chemical Engineering Practice Laboratory. I won the Chemical and Biological Engineering department TA award in 2015 for my work as a Process Design TA because in Fall 2014; I redesigned/delivered 12 lectures on pumps, distillation, and reactors as the primary professor was on leave for that quarter. I am interested in teaching/designing the curriculum for Process Design, the Chemical Engineering Practice Laboratory, and perhaps a specially designed course on Heterogeneous Catalysis and/or Reaction Engineering.

Future Directions:

As a faculty member I want to explore and expand on dynamic heterogeneous catalysis and 'Catalytic Resonance Theory' as it has the potential to transform the entire field of heterogeneous catalysis. In particular, I want to leverage chemistry and synthetic knowledge from my PhD to investigate the use of chemical modifiers to change binding energies on the catalyst surface. Careful control of the solvent environment, partial pressure, and catalyst structure is needed to achieve optimal performance. These projects will necessitate the combined application of many techniques with DFT catalyst/adsorption screening, microkinetic modeling, and experiments working together to explain complex adsorption/reaction phenomena.

In parallel, I want to develop new classes of catalytic materials leveraging my PhD experience with catalysts synthesis using layer-by-layer techniques. Innovation is needed for supported metal catalysts, solid bases, and bifunctional materials where catalytically inactive sites assist with adsorption. Automated and precise characterization techniques will be developed in tandem to compare new formulations to standard catalysts. My overall philosophy will be to use DFT catalyst screening in combination with precise catalysts synthesis/materials science to select catalytic systems with high adsorption affinity for the reactants of interest (i.e. small molecules, methane, methanol, ethanol) and high reactivity/productivity.

Selected Publications:

M. A. Ardagh, O. A. Abdelrahman, and P. J. Dauenhauer, 2019. "Principles of Dynamic Heterogeneous Catalysis: Surface Resonance and Turnover Frequency Response". ACS Catalysis. doi: https://doi.org/10.1021/acscatal.9b01606 [online ahead of print]

M. A. Ardagh, Z. Bo, S. L. Nauert, and J. M. Notestein, 2016. "Depositing SiO_2 on Al_2O_3: a Route to Tunable Brønsted Acid Catalysts". ACS Catalysis, 6 (9), 6156-6164.

Z. Bo, S. Ahn, M. A. Ardagh, N. M. Schweitzer, C. P. Canlas, O. K. Farha, and J. M. Notestein, 2018. "Synthesis and Stabilization of Small Pt Nanoparticles on TiO_2 Partially Masked by SiO_2". Applied Catalysis A: General, 551, 122-128.

582004 Sustainable Fuel and Chemical Synthesis Via Catalytic Valorization of Abundant and Renewable Resources

Research Interests:

Sustainability is a fundamental challenge of our society and will remain one, since we rarely limit our consumption until the impacts of diminishing resources become both clear and present. A sustainable, independent economy operates more effectively when taking advantage of abundant feedstocks including both grown and waste resources such as sugars, lignocellulose, plant/algal oils, industrial off-gases, and commercial wastes. My interest is in understanding the fundamental catalysis behind the conversion of such feedstocks to fuels and chemicals to rationally develop new materials and processes, practically influenced by current and projected demands and constraints imposed by society. My research has thus far targeted production of heavier fuels such as jet and diesel, demand for which is projected to increase while that for lighter fuels such as gasoline is projected to decrease. Commercial technologies for lignocellulose-derived jet and diesel fuels thus far have not been developed, creating a large need for research into this area.

I synthesized a low-cost base metal catalyst for the conversion of sugars to mono-oxygenates which can be used to produce renewable diesel and jet fuels. Previously only precious metal catalysts were used for this reaction. The cobalt catalyst used in this study is over 50 times cheaper than previously used precious metal catalysts and was used to achieve higher mono-oxygenate yields than previously reported. I also found evidence of a possible oligomer-mediated pathway for production of these species.

I have additionally developed a novel technology for the conversion of ethanol to distillate-range fuel molecules. I developed an extensive kinetic model to elucidate the reaction chemistry that occurs in a key stage of this process using experimental data I collected for ethanol condensation over hydroxyapatite. Alcohol structure (i.e. branching and chain length) and undesired olefin production are related to both fundamental catalytic parameters and natural consequences of condensation (e.g. increasing water pressure, decreasing alcohol pressure). This model shows the significant role dehydration of C_{4+} alcohols plays in limiting alcohol yields at increasing conversions. We have used this model to aid in the development of more selective catalysts and to better understand how this step affects the overall ethanol-to-distillate technology.

Teaching Interests:

As a professor, the greatest impact I can have is in educating and inspiring future generations of scientists and engineers. My passion is fueled by the reward of seeing understanding develop in a student's eyes, knowing that I was the difference. I am a classically trained Chemical Engineer and am excited to share my passion for chemical engineering with future generations. I am most interested in teaching the following classes: *Reaction Kinetics and Reactor Design*, *Thermodynamics I*, *Thermodynamics II*, and *Chemical Engineering Fundamentals*. In the classroom I have served as a teaching assistant twice for *CBE 250 Process Synthesis* and once for *CBE 311 Thermodynamics of Mixtures*, receiving two department awards for my teaching. I used my discussion sections to reflect upon practices proven to be effective in literature to find myself as a teacher. Here I have been guided by three main philosophies: (1) teaching as a relatable, collaborative facilitator, (2) providing a learning environment that addresses different learning styles and backgrounds while emphasizing collaborative efforts and real-world applications, and (3) rationally designing curriculum based on backward design with clearly-established learning outcomes, feedback-directed adaptation, and holistic examination design. As an instructor I will apply these philosophies in the classroom and continuously reflect upon and refine my teaching to inspire my students to become the next generation of innovators, armed with mindsets and life-long skills that will serve them far beyond the classroom. Additionally I have been a research mentor to four undergraduate students and a high school student. Each of these students learned differently and possessed a unique perspective on research, which has given me an ability to adapt my mentoring style to best benefit my students.

582053 Discovery and Optimization of Processes and Catalysts for the Organic Electrosynthesis of Carbon-Neutral Fuels and Chemicals

Past and Current Research:

The electrochemical reduction of carbon dioxide (CO_2) offers a potential means for synthesizing carbon-neutral fuels and chemicals using intermittent renewable electricity. Unfortunately, the state-of-the-art electrocatalysts for CO_2 reduction are neither active nor selective enough to make the process commercially viable. Additionally, research efforts aimed at discovering novel electrocatalysts with superior activity or selectivity compared to pure metallic Cu have failed to identify promising candidates. My research activities to date have sought to alleviate this issue.

I investigated many different aspects of the electrochemical CO_2 reduction reaction during my doctoral studies under the guidance of Prof. Alexis. T. Bell at the University of California at Berkeley. These research efforts were focused on the development of analytical instruments, the elucidation of reaction mechanisms, the establishment of structure-activity relationships, and the discovery of novel electrocatalysts. I developed an analytical instrument capable of quantifying the composition of the local reaction environment in the immediate vicinity of the electrocatalyst by sputter depositing it directly onto a pervaporation membrane that was coupled to a mass spectrometer. Using this approach, I observed a surprisingly high abundance of multi-carbon aldehydes relative to alcohols within the local reaction environment while conducting CO_2 reduction over Cu. This observation led to the hypothesis that primary aldehydes are intermediates to the corresponding primary alcohols during CO_2 reduction over Cu. Furthermore, I discovered that the reduction of these aldehyde intermediates could be suppressed by introducing small amounts of Ag into the Cu surface by Galvanic exchange. The incorporation of relatively large Ag atoms into the Cu surface produces compressive surface strain that inhibits the ability of Cu to reduce these aldehyde intermediates further by reducing its oxophilicity and the steady-state coverage of adsorbed H.

I have begun to develop electrocatalysts for hydrocarbon and alcohol evolution beyond Cu since beginning my postdoctoral research under the guidance of Prof. Ib Chorkendorff at the Technical University of Denmark. Thus far, I have developed a rational protocol for designing Cu-free intermetallic electrocatalysts that utilizes the d-band valence electronic structure of the electrocatalyst as the descriptor of electrocatalytic activity. The d-band valence electronic structure of a transition metal determines its binding energies for reactive chemical species. Unsurprisingly, the d-band valence electronic structure of Cu is highly unique. I have demonstrated that the d-band valence electronic structure of a transition metal can be systematically modified to resemble Cu via the formation of strong intermetallic bonds with electronically dissimilar metals. Furthermore, I have demonstrated that composition gradients form in the near-surface region of such intermetallic alloys upon air exposure using both x-ray photoelectron and ion scattering spectroscopies. These previously unidentified near-surface composition gradients cast doubts on the conclusions drawn from prior studies of CO_2 reduction over similar intermetallic electrocatalysts and potentially explains the lack of promising alternatives to Cu. Finally, I have developed a novel experimental methodology that prevents the formation of these near-surface composition gradients.

Future Research Interests:

My future research interest is to discover and optimize processes and catalysts for the organic electrosynthesis of carbon-neutral fuels and chemicals. Beyond the reduction of CO and CO_2, I am interested in investigating the partial oxidation of chemicals derived from natural gas (such as methane and ethane), the selective reduction of biomass-derived platform molecules (such as furfural and levulinic acid), and the upgrading of organic molecules via electrochemical carbonylation. I am interested in establishing the structure-activity relationships for these reactions and using them to conduct hypothesis-driven electrocatalyst discovery. Concurrently, I will utilize a combination of mass spectrometry and infrared spectroscopy to elucidate reaction mechanisms. Finally, I will also design and validate working devices to bridge the gap between fundamental science and commercial implementation.

Teaching Interests:

I have a strong interest in sharing my passion for science with others, both in the classroom and in the laboratory. I was the founding president of the Renewable Energy and Energy Efficiency Club (RE^3) during my undergraduate studies. This club combined stimulating lectures from local academic and industry experts with hands on projects designed to cultivate student interest in renewable energy and energy efficiency technologies. During my doctoral studies I served as a graduate student instructor for courses on transport and separations processes and chemical process design. Furthermore, I volunteered with Students for Environmental Energy Development, an organization that aims to attract high school students into STEM careers through self directed research projects. I have also mentored 4 undergraduates and 2 masters students through academic research projects. Several of these students become co-authors of published manuscripts for their intellectual contributions and have since gone on to pursue doctoral degrees on their own. Finally, I have recently co-authored a textbook chapter on the fundamentals of heterogeneously catalyzed electrochemical CO_2 reduction in order to increase the breadth of my impact further. As a classroom educator, I am interested in teaching students the fundamentals of thermodynamics, kinetics, and transport phenomenon and how they can be employed to tackle challenging real world problems. Furthermore, I am interested in developing new courses that cover material more closely aligned with my research interests, such as electrochemistry, catalysis, and emerging energy technologies.

582079 Enabling New Chemistries through Catalyst Design

Enabling New Chemistries through Catalyst Design

Marcella Lusardi

Research Interests:

Economies revolve around carbon-based chemicals and their derivatives. The concentrated distribution of traditional carbon resources (e.g., petroleum) renders the consumer goods, food and agriculture, and transportation industries, among others, vulnerable to fluctuations in a global carbon economy and volatile geopolitical relations. Capitalizing on unused, ubiquitous carbon feedstocks—like biomass, which most often finds its way into waste streams—addresses these economic challenges, while also adding a foundational brick to the construction of a closed carbon cycle. Unfortunately, existing processes for producing valuable chemical intermediates are only suited to handle the specific physicochemical properties of traditional feedstocks, which are starkly different (e.g., molecular weight, oxygen content) from those of alternative feeds. Thus, practically leveraging these diverse feeds requires extensive development of fundamentally new chemical processes. My work addresses this by designing the equally diverse set of catalysts that are critical to such conversion technologies.

My research approach is to understand the fundamental origin of activity within catalytic materials through rigorous materials characterization, and then use this insight to design catalysts with physicochemical properties tuned for specific chemical routes. *This characterization-synthesis-reactivity cycle systemizes the discovery of structure-activity relationships across a matrix of structure parameters. A key component to my research efforts is the strong focus on in-depth, tailored synthesis techniques that enable the creation of materials with desired catalytic properties.*

Catalytic material development requires a hand-in-hand partnership between synthesis and characterization. To this end, my research has focused on using intensive syntheses of organic, inorganic, and hybrid materials that carefully control the interaction among solvents, structure-directing agents/ligands/surfactants, and molecular precursors to tailor properties like porosity and active site distribution. These techniques, in conjunction with rich *in-situ* and *ex-situ* characterization methods (e.g., IR, NMR), alter and elucidate catalytically-dependent electronic and spatial environments of physicochemically-diverse materials. Examples of applications of this methodology from my prior research, along with the corresponding impact, include the following:

- **Production of xylene analogs from an ethanol derivative [1]:** Next generation catalysts can use insight regarding the specific roles of acid and base centers to optimize site coordination for selective production of the *para-* isomer, which could be a direct drop-in intermediate in the synthesis of terephthalic acid (40 MT/yr demand for PET plastics and polyester fabrics).
- **Application of SSZ-13 as a dimethyl ether carbonylation catalyst [2]:** The turnover rate for methyl acetate production over this commercially-available material (aluminosilicate with the chabazite framework) rivals that measured over the state-of-the-art zeolite carbonylation catalyst, mordenite.
- **Quantitative mapping of solvent-surfactant-precursor recipes for porous carbon nitrides [3]:** The facile, one-pot synthesis method developed in this work yielded the highest reported surface area for a soft-templated carbon nitride (C/N ~ 0.75). These textural properties facilitate enhanced molecular and electronic transport in photocatalytic applications.

Physicochemical properties that impact catalytic reactivity expand beyond surface chemistry and porosity. The design of hybrid hydrophilic-hydrophobic materials, for example, could provide an avenue for executing conversion reactions that are often thwarted by the presence of water in nonconventional carbon feedstocks. Such hybrid materials could be realized through the synthesis of acid centers connected to nonpolar organic linkers, encapsulated in an inorganic scaffold of hydrophobic domains. Research in my group moving forward will continue to approach practical challenges in catalyst development for new chemical routes through optimized synthesis protocols and extensive structure elucidation.

Teaching Interests:

Teaching both undergraduate and graduate students is one of the main reasons I am interested in pursuing a career in academia. I am qualified to teach introductory chemical engineering courses, materials and energy balances, and upper-level undergraduate and graduate courses in thermodynamics. Additionally, I am interested in building a new course centered on materials characterization. Often, techniques like XRD and SEM are not encountered unless a student's research directly requires it. Exposure in a classroom environment (in an upper-level undergraduate/graduate course) to such methods could excite and attract students who might otherwise not get exposed to a major component of materials education and research. The course would be a combination of theoretical background (e.g., the mathematics of crystal periodicity), instrument design (e.g., incident/diffracted x-ray beam, goniometer), and examples for practical application (e.g., crystallography in pharmaceuticals, catalysis, semiconductors). This course would benefit students interested or currently conducting research in a variety of chemical engineering subdisciplines.

References:

[1] Lusardi, M., Struble, T., Teixeira, A., and Jensen, K. F. "Identifying the roles of acid-base sites in formation pathways of tolualdehydes from an ethanol derivative over MgO-based catalysts." *In submission.*

[2] Lusardi, M., Kale, M., Kang, J. H., Chen, T., Neurock, M., and Davis, M. E. "The origin of catalytic reactivity in chabazite-type zeolites in the carbonylation of dimethyl ether to methyl acetate." *In preparation.*

[3] Peer, M.,† Lusardi, M.,† and Jensen, K. F. "Facile soft-templated synthesis of high-surface area and highly porous carbon nitrides." *Chem. Mater.* 2017, 29, 1496–1506. †*indicates co-first authorship.*

582095 Precision Synthesis to Control Catalyst Surface Structures for Improved Reactivity and Performance

Research Interests:

Catalytic reactions play a vital role in the production of fuels and chemicals for our growing world. The development of new catalysts for chemical transformations is hindered, however, by a reliance on empirical knowledge due to the complexities of surface-adsorbate interactions. Without detailed structure-activity relationships, an experimental trial and error method is often used to identify appropriate catalyst formulations for a variety of reaction systems. By elucidating the controlling factors of catalyst performance, we can design the best catalyst structure for a particular transformation. The ability to rationally design catalysts for any given reaction would facilitate important improvements to reactions for energy, chemical, and environmental applications.

Well-defined catalytic materials with a uniform distribution of active site structures offer one approach to the development of structure-activity relationships. My Ph.D. research under Professor James Dumesic has focused on the synthesis of well-controlled bimetallic surface structures to tune catalytic performance. Conventional catalyst synthesis techniques, such as impregnation, can result in catalysts with monometallic species within the bimetallic system. I demonstrated that uniform, bimetallic active sites can be formed by selectively depositing a promoting metal onto a parent catalyst. By using controlled surface reactions to synthesize bimetallic catalysts, I have elucidated structure-activity relationships for Pd-based bimetallic catalysts for selective hydrogenation, hydrodechlorination, and amination reactions. Using FTIR, STEM-EDS, and XAS techniques, I concluded that the resulting Pd structure varies with catalyst composition, and that highly active, isolated Pd species can be formed at low Pd loadings. Additionally, using both in-situ and ex-situ characterization, we identified important changes to the catalyst surface in the presence of adsorbates, highlighting the dynamic nature of the catalyst.

By combining synthesis of well-defined bimetallic active sites and advanced characterization techniques with fundamental reaction kinetics studies, we can develop structure-activity relationships which inform further catalyst improvements. My future research program will use these approaches to understand the active site and develop novel catalytic materials for a variety of chemical transformations.

Teaching Interests:

Throughout my graduate studies, I have had a variety of experiences through which I have developed my interest in teaching and gained a variety of tools to be an effective instructor and mentor. I have served as a teaching assistant for undergraduate *Chemical Kinetics and Reactor Design* and *Transport Lab* courses. Additionally, I was a teaching assistant for graduate level *Kinetics and Catalysis* where I developed a range of course assignments and assessments. I have also taken courses in the Delta Program for Teaching and Learning and taught *Entering Research,* a seminar course for students conducting research for the first time. I have come to see teaching as an extension of my own interest in solving global problems through engineering and science. My goal in teaching is to motivate and excite students so that they have authority over their own learning and become independent critical thinkers and problem solvers. I aim for my students to develop the skills to solve technical problems through application of fundamental principles in team-based environments. Based on these goals and my teaching experiences, my approach to teaching rests on three main principles: a foundation of student engagement, a student-centered classroom, and an inclusive learning environment that values all perspectives.

I believe that incorporating active learning approaches is key to ensuring effective learning. The next generation of chemical engineers will face a wide range of challenges in changing feedstocks, the need for improved efficiency, and mitigating effects of climate change, and I aim to guide student development of problem-solving skills through hands-on laboratory experiences and project work. I am deeply passionate about teaching and plan to continue to apply best practices from literature on teaching and learning to my own classroom. I am interested in teaching material balances and unit operations courses, as these serve as the foundation for a comprehensive chemical engineering education. I am also interested in teaching kinetics and catalysis courses at both the undergraduate and graduate level and would be interested in developing elective courses on microkinetic modeling or material science for chemical engineers. By incorporating topics from current research into my classroom, I aim to inspire students to build upon their chemical engineering education and develop the skills needed to be the next generation of problem solvers.

582106 Reshaping the Carbon Cycle with Catalysis: Selective Activation of Chemical Bonds for Producing Carbon Neutral Fuel and Chemicals

Research Interests:

A grand challenge for catalysis this century is to accelerate the carbon cycle–wherein fossil fuels derived from biomass formed over 300 million years ago are extracted from the ground and then used to produce energy, fine chemicals, and environmentally damaging greenhouse gases–to the speed of catalytic chemistry. While renewable energy sources like solar or wind power are promising, they do not meet the energy density requirements for, specifically, automobile propulsion. Additionally, many chemicals crucial to the production of pharmaceuticals, plastics, and adhesives also use fossil fuel sources. In order to sustainably address these energy and chemical needs, new fundamental chemistries and technologies are required that will enable creation of new "oil" and chemical feedstocks from carbon neutral sources like biomass.

There is a *critical need* for fundamental insights into the chemical identity of catalytic active sites and the interplay between such sites and the reaction environment. By integrating quantum mechanical simulations and state-of-the-art microscopic, spectroscopic, and kinetic experiments, I will interrogate reaction events at the molecular level and establish the catalytic rules that lead to selective activation of chemical bonds on materials ranging from extended surfaces to nanoparticle-support interfaces under realistic reaction environments (e.g. high pressures and solvents). This research program contrasts with the data science approach. As purely computational methods often fail to replicate experimental data in even the simplest systems, there is no reason to believe scaling up of computational data via data science techniques (where errors can potentially be magnified) will provide any further fundamental insights into catalysis. My specific approach will be to use kinetic and spectroscopic experiments to reduce the parameter space of highly complex catalytic systems of promise wherein congruence to experimental data provides the means to navigate any computed data that are within the known errors and limitations of computational methods. This methodology will allow me to probe the nature of catalytic active sites, transitions states, and reaction steps in a truly atom resolved manner without relying solely on often error prone or ambiguous computational data. Furthermore, my background in merging computational chemistry and experimental spectroscopic and kinetic techniques prepares me to tackle highly challenging catalytic systems covering homogeneous and heterogeneous catalysis in both vapor and condensed phases. Overall, my goal is to identify molecular level properties at chemically active interfaces which can then be leveraged to enhance catalytic performance of reactions used for the sustainable production of carbon neutral fuels and fine chemicals.

Successful Proposals:

Environmental Molecular Sciences Laboratory General Cycle Proposal, 2019

Proposal Title: Uncovering the Atomistic-Scale Catalytic Roles of Hydrogen Transfer at Liquid-Solid Interfaces for the Targeted Removal of Oxygen from Bio-Oil Compounds

U.S. Department of Energy Office of Science Graduate Student Research Award, 2015

Proposal Title: Efficient Development of Earth Abundant Catalysts: Elucidating the Aqueous Phase Deoxygenation Mechanism of Phenolics for Upgrading Bio-oils to Usable Biofuels

Center for Nanoscale Materials at Argonne National Lab High-Impact Nanoscience and Nanotechnology User Proposals, 2013-2016

Selected Proposal Title: Identifying Trends in the Nanoscale Interactions between Noble Metal Dopants and Fe Oxide Surfaces which Predict Dopant Hydrodeoxygenation Activity

Postdoctoral Projects:

"Hydrodeoxygenation of Phenolics at Solvent-Metal Interfaces: Enabling New Catalytic Pathways by Modifying the Reactive Hydrogen Species"

Under supervision of Ya-Huei (Cathy) Chin, Department of Chemical Engineering and Applied Chemistry, University of Toronto (UofT)

"Low Temperature CO Oxidation with a Single Site Pt Catalyst Supported on Thin Film Cu Oxide: Elucidating the Chemical Nature of the Pt Active Sites"

Under supervision of Jean-Sabin McEwen, The Gene and Linda Voiland School of Chemical Engineering and Bioengineering, Washington State University (WSU)

PhD Dissertation:

"Atomistic Elucidation of the Synergy within Noble Metal Promoted-Fe Catalysts for the Hydrodeoxygenation of Phenolic Compounds"

Under supervision of Jean-Sabin McEwen, The Gene and Linda Voiland School of Chemical Engineering and Bioengineering, Washington State University (WSU)

Research Experience:

My research focuses on the atomistic level chemical properties of solid surfaces, specifically, how the local environment—i.e. surface structure and chemical composition as well as solvent-solid interfaces—influences the macroscopic catalytic performance for energy and environmental applications. Atomistic insight is critical for discovering new materials and chemistry and for tuning the reactivity and selectivity of homogeneous and heterogeneous catalysis in both vapor and condensed phases. To this end, I specialize in the merging of computational catalysis techniques (e.g. density functional theory, transition state theory, coverage dependent model construction, and molecular dynamics) with spectroscopy (e.g. x-ray photoelectron, infrared, Raman, and nuclear magnetic resonance), scanning tunneling microscopy, and kinetics (e.g. rate dependencies and isotopic effects). An example of the power of such a combined computational and experimental approach can be seen in my work developing Pt single atom supported catalysts for lower temperature CO oxidation, a reaction key to reducing automotive emissions during the cold start-up period. I was able to unambiguously show that single Pt atoms supported on thin film Cu oxides perform CO oxidation much more efficiently than more traditional catalysts and at a much lower temperature range. Furthermore, I identified key chemical properties of the single atoms and support material that lead to their enhanced catalytic performance, providing a clear path forward for the development of highly economic single atom catalysts. This research approach allows me to precisely decipher the nature of catalytically active sites, determine the effect of solvent-surface interfaces, and identify reactive intermediates through detailed quantum chemical analyses. Thus far, I have published 26 peer reviewed papers with over 700 citations.

Selected Publications:

Therrien, A.J.; **Hensley, Alyssa J.R.**; Marcinkowski, M.D.; Zhang, R.; Lucci, F.R.; Coughlin, B.; Schilling, A.C.; McEwen, J.-S.; Sykes, E.C.H., *An Atomic-Scale View of Single-Site Pt Catalysis for Low-Temperature CO Oxidation*. Nature Catalysis 2018, 1, 192-198.

Hensley, Alyssa J.R.; Wang, Y.; Mei, D.; McEwen, J.-S., *Mechanistic Effects of Water on the Fe-Catalyzed Hydrodeoxygenation of Phenol - The Role of Brønsted Acid Sites*. ACS Catalysis 2018, 8, 2200-2208.

Hensley, Alyssa J.R.; Wöckel, C.; Gleichweit, C.; Gotterbarm, K.; Papp, C.; Steinrück, H.-P.; Wang, Y.; Denecke, R.; McEwen, J.-S., *Identifying the Thermal Decomposition Mechanism of Guaiacol on Pt (111): An Integrated X-ray Photoelectron Spectroscopy and Density Functional Theory Study*.

Journal of Physical Chemistry C 2018, 122, 4261-4273.

Hensley, Alyssa J.R.; Wang, Y.; McEwen, J.-S., *Phenol Deoxygenation Mechanisms on Fe (110) and Pd (111)*. ACS Catalysis 2015, 5, 523-536.

Hensley, Alyssa J.R.; Hong, Y.; Zhang, R.; Zhang, H.; Sun, J.; Wang, Y.; McEwen, J.-S., *Enhanced Fe_2O_3 Reducibility via Surface Modification with Pd: Characterizing the Synergy within Pd/Fe Catalysts for Hydrodeoxygenation Reactions*. ACS Catalysis 2014, 4, 3381-3392.

Teaching Interests:

My teaching interests lie in exposing chemical engineering graduate students to the experimental and theoretical principles and tools that allow us to probe and control atomistic level chemical phenomena. Using my time as a postdoc at WSU as an example, I noticed that the majority of entering chemical engineering graduate students never had a full course on quantum mechanics and other advanced topics. Knowledge of such topics is crucial to correctly interpreting computational, spectroscopic, and kinetic results. To address this need, I developed and taught a "Fundamentals in Atomistic Computational Methods" course covering the foundational principles of quantum mechanics, atomistic modeling theory, statistical mechanics, and solid state physics. This course was taught at WSU in the Spring 2018 semester and combined theoretical principles with practical applications in the form of molecular model construction and simulation for catalytic reactions. Students were evaluated based on four written exams and a final oral exam. The development of this course gave me much needed hands-on experience in the preparation and implementation of graduate level classes.

One of the most challenging and enjoyable aspects of my time in academia has been mentoring junior scientists on original research projects. Watching a new student grow beyond the rote memorization of a subject to a deep understanding is immensely gratifying. During my time at WSU, I directly mentored six graduate students and four undergraduates from four different countries. These students have authored or co-authored nine publications and given 16 presentations on their original research. As a woman in STEM myself, the support and female representation from my undergraduate advisor and UofT postdoctoral supervisor was critical to my success and I found it very rewarding to in turn do the same for two female graduate students and three female undergraduate students.

Service Experience:

It is a fundamental belief of mine that anyone can become a scientist or engineer if given sufficient support, tools, and opportunities, and I have worked to improve representation in STEM fields through my outreach and volunteerism. I was a project judge for WSU's Imagine Tomorrow for 3 years and an event organizer and judge for Science Olympiad for 8 years. These one-day events are regional and national STEM design- and research-competitions that together reach 16,000 middle and high school students annually across the United States. Through these activities, hundreds of middle and high school aged women have directly been exposed to my leadership and presence during these activities as they explored scientific thought, inquiry, and research. In this way, they have been able to see that a career in STEM is eminently achievable for themselves.

582151 Development of Thin Film Deposition & Etching Processes for Challenging Materials

Research Interests:

The ability to efficiently scale up production of novel material systems for future applications remains a challenging bottleneck for the adoption of a variety of new technologies. Pressing examples include: a) Topological insulators for low power switching, non-volatile memory, and energy harvesting applications, b) Complex metal oxides for photo- and electro-catalytic applications, c) Chromium free anti-galvanic barrier coatings for durability in extreme environments, among many others. A common theme among many of these materials is that precise control over stoichiometry, microstructure, and crystallinity (frequently in the form of epitaxial thin films) has only been achieved at high temperatures and using deposition techniques not well suited to large areas.

Magnetron Sputter Epitaxy (MSE) presents itself as a viable alternative to pulsed laser deposition and molecular beam epitaxy for its ability to retain precise control over stoichiometry over larger areas. MSE may be further improved with the adoption of pulsed deposition practices such as High Power Impulse Magnetron Sputtering (HiPIMS). HiPIMS offers many advantages in that process developers are capable of controlling and tuning ion fractions and ion energy during deposition. This control over ion energy ensures control over film stress, crystallinity, and many other metric of interests at room temperature.

Using HiPIMS as an enabling technology, I plan to devote a portion of my early career to furthering HiPIMS research as a reactive sputtering technique, in particular for material systems that require epitaxial heterostructures. My belief is that a significant number of these material systems may be deposited epitaxially at room temperature in a manner that will offer flexibility on substrate choice as well as enable rapid commercialization given the prevalence of conventional magnetron sputtering in high volume manufacturing. Concurrently, I envision participating in the development of HiPIMS as a highly energy tailored etching technique. The ability to control the ion energy of reactive species is a crucial component of atomic layer etching (ALE) process development for future of the semiconductor processing industry.

Research Experience

My research experience has spanned from basic science and phenomena discovery in electronic materials to applied research in coordination with both established industrial partners and start-up technology companies for products close to market release for high volume manufacturing. My doctoral work with Dr. Edmund Seebauer in the Department of Chemical and Biomolecular Engineering at the University of Illinois focused on the formation of medium range order in amorphous TiO_2 and the transport of photogenerated charge carriers across TiO_2/perovskite interfaces for heterojunction photocatalysis applications. The work began a deep dive into materials characterization that has followed me beyond graduate school. My postdoctoral research moved me from thermal Chemical Vapor Deposition (CVD) to plasma processes in Dr. David Ruzic's laboratory. Topics of research there centered around the implementation of secondary plasma sources for improving existing processes including radical sources for reactive sputtering, laser texturing of metal surfaces, and surface wave plasma antennas to assist reactive ion etching processes. Additionally, I gained exposure to a variety of plasma diagnostics including Langmuir probes, radical probes and retarding field energy analyzers.

During Dr. Ruzic's recent sabbatical, I extended my time at the Center for Plasma-Material Interactions (CPMI) as a Visiting Research Scientist. My duties were expanded to include research advising for eight graduate students and nearly 30 undergraduate students, proposal writing in response to government agency solicitations, as well as initiating new projects with industrial sponsors. In addition to authoring four grants to date at three different agencies, I also independently attracted a sponsored research agreement with DuPont as a principal investigator.

Teaching Interests:

One of the greatest sources of satisfaction for myself when teaching an engineering curriculum comes from introducing students to totally foreign concepts in a way that leaves them stimulated and eager to continue their journey as scholars and future engineers. During graduate school, I served as a teaching assistant (TA) for three semesters covering thermodynamics, kinetics, and senior design/capstone class. Desiring to continue student interactions in such a role, I led discussion sections and taught substitute lectures for my advisor in both thermodynamics as well as an elective class of his development focused on the principles of semiconductor processing. I have discovered that in teaching both core and elective classes, the ability to make the material relatable through connections with industrial processes, current research questions, and personal experiences is invaluable. Such method allows for a certain story telling element that breaks up the potential monotony of a lecture, while offering students perspective and improving their ability to apply concepts outside of their coursework.

In addition to teaching, I look forward to the opportunity to train, advise, and mentor both graduate and undergraduate researchers in my research group much as I have as a research scientist. I plan to continue participating in demonstrations for primary and high school student groups as I have since working as an undergraduate researcher. Given the opportunity as a professor, I would seek to develop and offer plasma engineering and/or semiconductor processing as elective classes.

582154 Unifying Principles in Thermally and Electrochemically Driven Catalytic Reactions

Research Interests:

Overview of past and current Research:

At the atomic level, catalytic reactions involve the making and breaking of chemical bonds between reacting molecules and active sites. This has led to an understanding of heterogeneous catalysis in terms of active site-adsorbate interactions. However, as our ability to characterize the complex nature of active sites advances, it has become apparent that the reaction environment can also have important influences. This includes influencing reactivity through environment-induced restructuring of the catalyst surface, as well as direct participation of the solvating medium in the reaction. Appreciation of this added complexity is necessary for a complete description of catalytic reactions, and once understood, can yield additional tools for designing more efficient catalysts. Optimizing the structure of both the binding site and the non-reactive components in its vicinity can give us exquisite control over chemical reactions, as elegantly demonstrated by nature's catalysts, enzymes. Throughout my Ph.D. and postdoc, I have sought to understand the impact of these secondary effects on both electrochemical and thermally activated catalytic reactions.

Ph.D. Research- Electrolyte effects on electrochemical CO_2 reduction

(University of California, Berkeley; Advisor: Alexis Bell)

The composition of the electrolyte in which electrocatalysts are tested has been observed to influence reactivity; however, detailed understanding of these effects is still lacking. Through a combination of experimental and theoretical studies, we elucidated the effects of electrolyte cation and anion identity on the activity and selectivity of metal catalysts for this reaction. We showed that cations in solution act as a built-in promoter with a function analogous to alkali metal promoters for gas phase reactions, interacting electrostatically with adsorbates to influence reaction energetics. Anions can also affect CO_2 reduction rates, acting both as a buffer and a proton source, increasing rates of hydrogenation reactions.

Postdoc Research- Developing atomic-scale structure property relations for isolated metal catalysts

(University of California, Santa Barbara; Advisor: Phil Christopher)

Atomically dispersed metals on oxide supports have recently gained significant interest as catalytic materials, but relationships between their structure and properties remain unknown. Through a combination of atomic level and sample averaged *in-situ* characterization, we demonstrated that the local coordination of the metal to the support determines the ability of the surface site to form additional bonds with adsorbates. We have demonstrated the consequences of these coordination effects for reactions important for producing sustainable fuels.

Future research:

Significant environmental issues associated with the use of fossil fuels, along with advances in technologies to generate renewable electricity have made it likely that future energetic inputs to drive chemical transformations will be electrical in addition to thermal. However, the use of electrocatalytic processes for sustainable fuel and chemical production requires scientific breakthroughs in the design of catalysts and reaction systems to drive these reactions. Understanding the fundamental concepts that govern reactivity in these systems can enable their rational design and facilitate these breakthroughs. In my research, I plan to establish parallels between well studied thermal catalytic reactions and new electrocatalytic processes to provide design rules for catalyst design and highlight what unique opportunities the electrochemical system provides. I plan to understand the diversity of chemical transformations possible by electrochemical processes, beyond the conversion of small molecules such as water and carbon dioxide. To design catalysts for these transformations, I plan to optimize the dynamic structure of both the binding site and the non-reactive components in its vicinity to maximize selectivity and activity.

Teaching Interests:

Teaching, both in the classroom and laboratory, is a major driver in my decision to pursue an academic career. My goal in teaching is to develop a student's deep understanding of a few core Chemical Engineering concepts and use them as a guide to approach any problem they are faced with. Both as an undergraduate and graduate student, I served as a student instructor for core chemical engineering courses. I was awarded the departmental Outstanding GSI award for the graduate Kinetics Course at Berkeley. Based on these enjoyable and instructive experiences, as a graduate student I independently developed and taught an elective course *Fundamentals of Electrocatalysis*. As a faculty member, I am excited to teach any of the core courses of Chemical Engineering, particularly courses in Kinetics and Reaction Engineering. Given the opportunity, I am also eager to develop new courses such as *Electrochemical Engineering*, or an introductory course on current and emerging energy transformations. These courses would aim to expose students to interesting applications of the foundational principles of thermodynamics, kinetics, and transport phenomena. In the laboratory, I have served as a mentor to undergraduate and junior graduate students. Serving as a mentor and aiding in the development of students in research will be an important focus throughout my career as a faculty member.

Selected references:

1. **J. Resasco**, N.P. Dasgupta, J.R. Rosell, J. Guo, P. Yang, *J. Am. Chem. Soc.* **136**, 10521-10526, (2014).
2. **J. Resasco**, H. Zhang, N. Kornienko, N. Becknell, H. Lee, J. Guo, A. L. Briseno, P. Yang. *ACS Cent. Sci.* **2,** 80-88, (2016). Cover Article
3. **J. Resasco**, L. Chan, E. Clark, C. Tsai, C. J. Hahn, K. Chan, T. F. Jaramillo, A. T. Bell, *J. Am. Chem. Soc.* **32**, 11277–11287, (2017).
4. **J. Resasco**, Y. Lum, E. L. Clark, J. Z. Zeledon, A. T. Bell. *ChemElectroChem* **5**, 1064-1072, (2018).
5. E. L. Clark*, **J. Resasco***, A. Landers, J. Lin, L. Chung, A. Walton, C. Hahn. T. F. Jaramillo, A. T. Bell. *ACS Catal.* **8,** 6560-6570, (2018).
6. **J. Resasco**, S. Dai, G. Graham, X. Pan, P. Christopher. *J. Phys. Chem. C* **122,** 25143-25157 (2018). Cover Article
7. L. DeRita*, **J. Resasco***, S. Dai*, A. Boubnov, H. V. Thang, A. S. Hoffman, I. Ro, G. W. Graham, S. R. Bare, G. Pacchioni, X. Pan, P. Christopher. *Nat. Mater.* **18**, 746–751 (2019).

*Denotes equal contribution

581653 Fixed Feed Temperature Program Modulation (FFTPM) - a Versatile Method for Extracting Adsorption Thermodynamics for Weakly Adsorbing Systems over a Range of Adsorbates

Research Interests:

Energy Conversion, Chemical Energy Storage, Heterogenous Catalysis, Ads/Desorption measurements for atypical adsorbates, Energy Efficiency and Cycle Improvements, Adsorption Heat Pumps, Dehumidification sorbents and cycles, Combined Heat/Power, Electrochemical Separation/Compression, Catalytic Membrane Reactors, Waste to energy, Waste Valorization (in particular: CO2/CH4)

Research Approach:

Material & Energy Balances on well-characterized system(s), typically packed bed reactors with controlled feed(s), pressure(s), and temperature(s). Quantitative analysis of reactor feed/effluent via MS and/or GC.

Ideal collaboration is with a materials synthesis group and/or a modeling/simulation group.

Teaching Interests:

Material and Energy Balances

Thermodynamics I,II

Kinetics

Catalysis Elective

Teaching Approach and Philosophy:

Active learning/small group settings

Strong foundation of mass conservation and First/Second Laws

Well developed logic for problem setup (exact solution is less important than the critical thought necessary to set up the problem)

581614 Mechanistic Study and Engineering of Vapor/Solid Interface *in-Operando*: Applications in Heterogeneous Catalysis for Energy Upgrading Reactions

I obtained my PhD in physical chemistry at the University of Notre Dame in 2016, followed by postdoctoral research in the Department of Chemistry at University of Illinois Urbana-Champaign and Department of Chemical Engineering at UC Berkeley and Lawrence Berkeley National Laboratory. I have expertise in heterogeneous catalysis, surface science, and reaction kinetics under *in-operando* conditions. Below my research and teaching interests are outlined. It is appreciated that AIChE could offer me a platform showing my research and teaching plans while meeting recruiters/searching committee members in person.

Research Interests: My lab aims to advance the fundamental understanding and modulation of general vapor/solid interface in terms of their chemical and electronic properties at molecular levels. In particular, reactant (vapor)/catalyst (solid) interfaces in thermal and photo-assisted heterogeneous catalysis will be investigated *in-operando*, i.e., characterization while a catalytic event is occurring. Information regarding adsorption energetics, elementary steps that mediate catalytic sequences, identity and evolution of kinetic-relevant surface species, and electronic characteristics (band bending, working function, valence band, etc.) will be obtained to unravel the reaction mechanism for design and engineering of better catalysts for energy upgrading-related chemical processes.

Mechanistic study of heterogeneous catalysts *in-operando*, assisted by vigorous kinetic analysis, offers a robust venue to unambiguously capture, elucidate and manipulate the physiochemical processes at the vapor/solid interface. Information regarding elementary steps that mediate catalytic sequences, identity, adsorption/desorption energetics and evolution of kinetic-relevant surface species, and electronic characteristics including band structure at the vapor/solid heterojunction will be obtained to elucidate the reaction mechanism and develop better catalysts in energy upgrading-related chemical processes.

Teaching Interests: Teaching perpetuates knowledge and inspires learning. Being able to teach drives me to apply for a professorship over a research-only position. Here, teaching as a professor has a two-fold format, passing knowledge to college students in classrooms and training of future chemists in research-based environments.

For classroom teaching, I believe in customized teaching tailored to individuals or individual groups within an audience. Unambiguous descriptions of ideas maximize the engagement of audience and allow one to intellectually challenge and inspire learners. A variety of approaches including schematics, videos, and texts with a combination of textbook and examples from publications or even my own research will be employed, as long as they serve to teach efficiently and enjoyably. In a research team, the passion and learning efficiency of students are largely driven by the communicating style of the principle investigator, where teaching customized down to individuals becomes indispensable. Teaching becomes a fulfilling process if the teacher is able to (and willing to) apply the most inspiring approach towards an individual and maximize his/her potential.

Given my expertise in physical chemistry, surface science, and heterogeneous catalysis, I am confident and excited to teach general chemistry, physical chemistry and related courses concerning topics on heterogeneous catalysis, surface science, vacuum science and scientific instrumentation. I am also interested in developing advanced courses in surface chemistry, solid state physics and heterogeneous catalysis (reaction kinetics and thermodynamics), bridging together three interconnected areas and introducing students to the world of surface and catalysis. For undergraduate courses I plan to follow departmental established syllabi and ensure the coverage of required concepts and knowledge. For advanced level courses I will use a highly research-oriented approach with a combination of introductions to general concepts and intellectually deep understandings of a particular research area.

581473 Understanding and Controlling Multielectron Transfer Chemistry for Sustainable Energy Technologies

Research Interests:

The synthesis of fuels and chemicals from sustainable feedstocks and energy sources is one of the crucial challenges we face today. The coupling of renewable electricity sources, such as solar or wind, to the generation and interconversion of economically and environmentally relevant products (e.g. hydrogen, ammonia, methanol) is an intriguing prospect but requires the careful coupling of electrical energy to chemical reaction.

Electrochemical methods offer a powerful method for promoting reactivity. A change in the driving force of a reaction that would require the application of hundreds of atmospheres of pressure can be achieved with less than a volt of applied potential. However, control over the driving force of a reaction does not necessarily come concurrently with control over the selectivity of the electron transfer. A multitude of products, as well as catalyst and conductive support corrosion, are often observed during electrochemical reactions, limiting the efficiency of the desired process. With this in mind, *the overarching goal of my research be to understand and promote efficient and selective photo- and electrochemically-driven multielectron transfer reactions with an emphasis on understanding how electrocatalyst surface chemistry and electrolyte conditions influence the efficiency and selectivity of the reaction.*

My specific aims will be toward improved understanding and enhanced efficiency of electrochemical transformations involving nitrogen-based species (e.g., NO_3^-, NO_x, N_2), including the reduction of nitrates as well as both the reduction and oxidation of N_2. The nitrogen cycle plays a pivotal role in maintaining the health of our planet and society, and the efficient, sustainable interconversion of these species would have exciting applications in a range of fields, including commodity chemical synthesis, fuel generation, and environmental remediation. We will rationally develop novel catalytic materials and electrolytes for these reactions, and the relationship of catalyst surface structure to performance will be characterized by careful *ex situ* and *in operando*study. These new electrocatalysts will then be integrated into device structures to evaluate them under practical conditions, and the performance of these devices, both with respect to catalyst activity and stability, will inform the development of the next generation of new catalysts. As our understanding of the fundamentals of the reduction/oxidation of nitrogen species evolves over time, we will couple these reactions to photoactive materials to develop solar-powered devices that can be readily deployed without external power sources. This cyclic strategy of catalyst design, characterization, and device integration should drive us to progress both fundamentally and practically toward real-world solutions to challenging problems in renewable energy, environmental chemistry, and other fields.

Teaching Interests:

My teaching experience includes the mentorship of multiple graduate and undergraduate students in the laboratory. Additionally, I have acted as the head instructor for advanced graduate courses in photoelectrochemistry during my PhD, and have served in teaching assistant roles. My training has prepared me to ably serve a Chemical Engineering department in the instruction of undergraduate and graduate students both in the laboratory and in the classroom. I feel immediately qualified to teach a range of undergraduate-level chemical engineering courses, including kinetics, thermodynamics, mass/energy balances, and other physical/analytical courses. In the future, I hope to have the opportunity to develop (1) a course directed toward graduate students and senior undergraduates aimed at understanding electrochemistry and its practical application to electrocatalysis of energy-rich molecules, as well as (2) a course designed around semiconductor electrochemistry, photoelectrochemistry, and applications to emerging renewable energy technologies.

Research Experience:

Postdoctoral Scholar 2016-Present

Department of Chemical Engineering, Stanford University

Advisor: Thomas F. Jaramillo

Project: Synthesis and Characterization of N_2Reduction Electrocatalysts

- Investigated the dependence of electrocatalytic NH_3production rate on electrolyte composition and catalyst identity
- Developed and applied analytical techniques and best practices for identifying electrochemically generated NH_3in non-aqueous electrolytes

Graduate Research Assistant 2010-2016

Division of Chemistry and Chemical Engineering, California Institute of Technology

Advisor: Nathan S. Lewis

Ph.D. Thesis: *Chemical and Photoelectrochemical Behavior of Graphene-Covered Silicon Photoanodes*

- Investigated the chemical and photoelectrochemical stability of Si and GaAs photoelectrodes covered by atomically thin graphene layers in aqueous electrolytes
- Explored the effect of graphene interfacial layers on the electronic properties of silicon/electrolyte junctions related to solar cell device performance
- Developed methods and classification scheme for the characterization of the efficiency of photoelectrochemical and photovoltaic solar-to-fuels devices

Undergraduate/Graduate Research Assistant 2005-2010

Department of Chemistry, University of Virginia

Advisor: W. Dean Harman

M.S. Thesis: *Laying the Groundwork for a Tantalum Dearomatization Agent*

B.S. Thesis: *Synthesis and Application of an η^2N-acetylpyridinium Complex*

- Synthesized tungsten- and tantalum-based organometallic reagents used to modify and enhance the reactivity of stable aromatic arenes and pyridines
- Investigated the relationship between reactivity enhancement in aromatic ligands and electrochemical properties of the associated organometallic complex

2019 Calendar Year Publications *(As of June 21, 2019)*

1. **Nielander, A.C.**; McEnaney, J.;Schwalbe, J.; Baker, J.; Blair, S.; Wang, L.; Pelton, J.; Bent, S.; Cargnello, M.; Chorkendorff, I.; Jaramillo, T. A Versatile Method for Ammonia Detection in a Range of Relevant Electrolytes Via Direct NMR Techniques. *ACS Catalysis* **2019**,9, 5797-5802.

2. Andersen, S.; Colic, V.; Yang, S.; Schwalbe, J.; **Nielander, A. C.**; McEnaney, J.; Enemark-Rasmussen, K.; Baker, J.; Singh, A.; Rohr, B.; Statt, M.; Blair, S.; Mezzavilla, S.; Kibsgaard, J.; Vesborg, P.; Cargnello, M.; Bent, S.; Jaramillo, T.; Stephens, I.; Norskov, J.; Chorkendorff, I. Assessing the Current State of Catalyst Development for the Electrochemical Reduction of $N_2$to NH_3. *Nature* **2019**,DOI: 10.1038/s41586-019-1260-x.

3. Hellstern, T.*; **Nielander, A.C.***; Chakthranont, P.*; King, L.A.; Willis, J.; Xu, S.; MacIsaac, C.; Hahn, C.; Bent, S.; Prinz, F.; Jaramillo, T.F. Nanostructuring Strategies to Increase the Photoelectrochemical Water Splitting Activity of Silicon Photocathodes.ACS Applied Nano Materials **2019**,2(1) 6-11.

4. Wang, L.; Nitopi, S.; Wong, A.; Snider, J.; **Nielander, A. C.**; Morales-Guio, C.; Orazov, M.; Higgins, D.; Hahn, C.; Jaramillo, T. Directing Selectivity in Catalysis with Surface Area: Electrochemically Converting Carbon Monoxide to Liquid Fuels. *Nature Catalysis* **2019**, DOI*:* 10.1038/s41929-019-0301-z.

5. Boyd, M.; Latimer, A.; Dickens, C.; **Nielander, A.C.**; Hahn, C.; Higgins, D.; Jaramillo, T. Electro-Oxidation of Methane on Platinum Under Ambient Conditions. *ACS Catalysis*, *Accepted*, Jun 19 **2019**.

580856 Computational-Accelerated Materials Design for Negative Emission Technologies

Research Interests:

Aiding limit the increase of the global average temperature, the 2015 Paris Agreement proposed the reduction of carbon dioxide (CO_2) emissions throughout this century; however, recent studies suggest that even a rapid atmospheric CO_2 removal of energy-related sources is insufficient to keep global heating below 2 °C. Most of climate stabilization scenarios suggest that negative emission technologies (NETs) are required to keep global warming well below 2 °C. NETs are cost-effective mitigation techniques to remove and sequester CO_2 from the atmosphere **[1, 2]**. Herein, we propose to develop NETs that sorb CO_2 from the atmosphere and chemically convert it into more useful compounds.

Regarding the CO_2 removal, adsorption-based technologies may be considered a cost-efficient option compared with the current prototypical technologies, which are based on aqueous KOH (as capture media) and $Ca(OH)_2$ (as recovery media), which are causticand thus cost and complex to regenerate capture media in addition to its maintain **[3, 4]**. Porous materials such as metal-organic frameworks (MOFs) have been demonstrated as promising CO_2 sorbents **[5]**. MOFs are particularly attractive for applications such as post-combustion cleanup of coal-fired power plants, where the CO_2 concentration is large (ranging from 12 to 70 %); however, the low relative concentration for CO_2 in the atmosphere (less than 0.04%) makes designing sorbents for CO_2 removal from the atmosphere significantly more challenging. To capture atmospheric CO_2, sorbent material must be designed to exhibit high selectivity for the CO_2. Coordination pillared-layers (CPLs), which are also porous materials, are excellent candidates for this task and have been shown to demonstrate high selectivity for CO_2 over N_2, O_2, and even CH_4 **[6]**. Moreover, their manifold design allows their textural properties (void volume, surface area, and pore volume) to be systematically tuned, which is critical for optimizing gas adsorption performance. However, property-function relationships that describe how the textural properties of CPLs correlate to their CO_2 adsorption performance is lacking. Herein, we propose to develop such relationships in silico. Specifically, we propose to identify CPL-like porous materials, where small amounts of these materials are capable of preferentially sorbing low concentrations of CO_2. Furthermore, since the CPL pore-network can act as a host or encapsulating medium for ultra-small metal nanoparticles (< 5nm), such nanoparticles when combined with CPLs can act as catalysts to convert captured CO_2 into valuable chemicals. Further it has been demonstrated that the pore-networks of CPLs can prevent agglomeration of the metal nanoparticles, which is critical for the catalyst performance and also promotes shape selectivity. Designing CPL-encapsulated catalysts will require systematic evaluation of hundreds to thousands of materials, including discovery of new materials with unique properties. In the last three decades, the computational materials research community has devised solid rational design/engineering strategies enabling atomic-level control of the structure and composition materials, and molecular simulations have played a pivotal role in accelerating the material properties screening. Herein, we propose to develop high throughput screening strategies based on established methods in rational sorbent and catalyst design, and computational tools—such as machine learning—in order to design materials that possess both gas adsorption and heterogeneous catalytic properties. We will leverage our research expertise in molecular simulations, including Quantum Mechanics, Molecular Dynamics, Monte Carlo to design NETs and other materials that can play pivotal roles in solving energy-related challenges.

Postdoctoral Project:

- Molecular dynamics studies of liquid water/solid catalyst surfaces
- Development of a Force Field to model CH_3OH and H_2O dynamics over a Pt (111) catalyst
- Quantum chemical study of H_2O adsorption on NpO_2Surfaces

Under supervision of Rachel B. Getman, Chemical and Biomolecular Engineering, College of Engineering Computing and Applied Sciences, Clemson University.

PhD Dissertation:

- Quantum mechanics calculations of the CO_2 interaction on adsorbents based on porous coordination pillared-layers (CPLs)
- Monte Carlo analysis of the physical adsorption of CO_2 on adsorbents based on porous coordination polymers (CPLs)

Other projects:

- Determination of Stable Intermediates on CO_2 Adsorption Onto Metal(Salen) Complexes: Quantum mechanics calculations

Under supervision of Maria C. Curet-Arana, Chemical Engineering Department, University pf Puerto Rico – Mayaguez Campus.

Research experience:

In my academic career path, I started, as undergrad, doing experimental research studying coal pyrolysis and coal pyrolysis waste reutilization. I earned expertise in managing laboratory equipment for synthesize materials such as zeolites, for water hardness removal, and coal-pyrolysis furnace, to improve coal calorific power. And characterization techniques such as, x-ray diffraction, thermogravimetric analysis, and scanning electron microscopic. However, starting my Ph.D. I did a significant turn to computational chemistry simulations. Since then, I have gained literal expertise in simulation techniques applicable to multiple length and time scales, e.g., Density Functional Theory (femto), Møller–Plesset perturbation theory (MP) (femto), grand canonical Monte Carlo (nano), and Molecular Dynamics (nano). I acquired proficiency modelling the CO_2 interaction and adsorption with porous material. During my post-doc, I have earned expertness in modelling surface catalytic intermediates under aqueous phases conditions. My vision for my independent research career is to develop new paradigms in CO_2 conversion to chemical commodities in catalyst materials based on encapsulating ultra-small metal nanoparticles in CPL-like porous frameworks. All projects I have worked on have been collaborations with experimentalists and other theoretical groups. Moreover, as result of intense training in various supercomputer center (high performance computing facility at university of Puerto rico–Nanobio, Boqueron and kimkelen–the National Energy Research Scientific Computing Center–NERSC–and in clemson university– Palmetto cluster) I have developed proficient skills performing computer modeling.

Teaching Interests:

Along with my research career, I also have teaching experience. I worked for three terms (one semester plus one summer) at the Portland Community College teaching general chemistry courses. During grad school and post-doc, I TAed and guest-lectured undergrad and grad lecture and laboratory course in Chemical Engineering. I have mentored 10+ undergrad and grad students, conducting training/troubleshooting in programs like Gaussian, VASP, RASPA, and LAMMPS. Moreover, as grad student and post-doc I actively participated in CREST and $WIPR_2EM$ social educational activities, as well as, REU programs. I am able to teach all of the core classes in the undergraduate chemical engineering curriculum and would excel at teaching courses in the line of reaction kinetics and catalysis, adsorption fundaments in nanostructures, thermodynamics, numerical methods, data visualization, or computer coding, as well as courses oriented to teaching computational molecular simulation methods for study catalysis and adsorption phenomena.

Publications:

Dasetty, Siva; ***Meza-Morales Paul J***; Getman, Rachel B; Sarupria, Sapna; Simulations of interfacial processes: recent advances in force field development; Current Opinion in Chemical Engineering; 2019, 23,

138-145.

Meza-Morales, Paul J; Gomez-Gualdron, Diego A; Arrieta-Perez, Rodinson R; Hernandez-Maldonado, Arturo J, Snurr, Randall Q. Curet-Arana, Maria C, CO_2 adsorption-induced structural changes in coordination polymer ligands elucidated via molecular simulations and experiments; Dalton Transactions; 2016, 45, 17168–17178.

Meza-Morales, Paul; Alberto Santana-Vargas, Curet-Arana, Maria C; DFT Analysis of Coordination Polymer Ligands: Unraveling the Electrostatic Properties and their Effect on CO_2 Interaction; Adsorption; 2015, 21, 533-546.

García-Ricard, Omar J; ***Meza-Morales, Paul***; Silva-Martínez, Juan C; Curet-Arana, Maria C; Hogan, John A; Hernández-Maldonado, Arturo J; Carbon dioxide storage and sustained delivery by $Cu_2(pzdc)_2L$ [L= dipyridyl-based ligand] pillared-layer porous coordination networks; Microporous and Mesoporous Materials, 2013, 177, 54-58.

Curet-Arana, Maria C; ***Meza-Morales, Paul***; Irizarry, Radames; Soler, Rafael; Quantum Chemical Determination of Stable Intermediates on CO_2 Adsorption Onto Metal (Salen) Complexes; Topics in Catalysis, 2012, 55, 260-266.

References:

[1]Eisaman, M. D., Rivest, J. L. B., Karnitz, S. D., de Lannoy, C. F., Jose, A., DeVaul, R. W., & Hannun, K. Indirect ocean capture of atmospheric CO_2: Part II. Understanding the cost of negative emissions. *International Journal of Greenhouse Gas Control, 70*, 254–261, 2018.

[2]de Lannoy, C. F., Eisaman, M. D., Jose, A., Karnitz, S. D., DeVaul, R. W., Hannun, K., & Rivest, J. L. B. Indirect ocean capture of atmospheric CO2: Part I. Prototype of a negative emissions technology. *International Journal of Greenhouse Gas Control*, 70, 243–253, 2018.

[3]Koytsoumpa, E. I., Bergins, C., & Kakaras, E., The CO_2 economy: Review of CO_2 capture and reuse technologies. *Journal of Supercritical Fluids*, 132, 3–16, 2018.

[4]Keith, D. W., Holmes, G., St. Angelo, D., & Heidel, K. A., Process for Capturing CO_2 from the Atmosphere. *Joule*, 2, 1573–1594, 2018.

[5]Trickett, C. A., Helal, A., Al-Maythalony, B. A., Yamani, Z. H., Cordova, K. E., & Yaghi, O. M., The chemistry of metal-organic frameworks for CO_2capture, regeneration and conversion. *Nature Reviews Materials*, 2, 17045, 2017.

[6]GarcíaRicard, Omar J., Silva-Martínez, Juan C., Hernandez-Maldonado, Arturo J., Systematic evaluation of textural properties, activation temperature and gas uptake of $Cu_2(pzdc)_2L$ [L = dipyridyl-based ligands] porous coordination pillared-layer networks. *Dalton Trans*., 41, 8922-8930, 2012.

580922 Catalyst Design through Simulations and Machine Learning

Research Interests:

A design of efficient catalysts comprised of earth-abundant materials is crucial for the development of new devices for the sustainable energy future. Unfortunately, no catalyst is available that satisfies activity, stability and cost criteria at the same time, sparking an immediate need for innovative catalyst design strategies. My research group will focus on a computational catalyst design: applying quantum-chemistry simulations and machine learning techniques to develop novel catalysts through understanding dynamics at heterogeneous interfaces and performing a high-throughput screening. Two research topics that my group would initially pursue are:

1. Integrated computational framework for an accurate prediction of catalytic properties
2. High-throughput computational screening of catalysts using automated density functional theory (DFT) calculations and machine-learning.

Beyond catalyst developments, I am also interested in applying simulations and machine learning to other energy conversion/storage technologies such as solar cells and secondary batteries.

Teaching Interests:

My teaching philosophy derives from past experiences in my graduate years. In graduate school classes, my interest in research was sparked by professors asking how scientific concepts related to real-world chemistry and encouraging us to tackle current challenges. My curiosity, originating from intriguing classes, became a strong force driving me to look up recent research articles and chat with students from different departments. These fruitful experiences have helped me to research enthusiastically and stay in academia, and as a faculty member I wish to spread similarly inspiring experiences to motivate students and undertake research with them.

With a strong background in core Chemical Engineering subjects and computational experiences, I will motive students to utilize computers to solve Chemical Engineering problems. I will also develop my own courses, particularly focusing on the introduction of atomic simulations in Chemical Engineering and their correlation to experimental observations.

Research Experience

1. **Graduate Student (KAIST)**

Graduate school of EEWS (Energy, Environment, Water and Sustainability), Korea Advanced Institute of Science and Technology (Sep. 2013 – Aug. 2017)

Supervisor: Yousung Jung

- Studied a selectivity determining factor in CO_2 electrochemical reduction reaction.
- Systematically studied effects of Au nanoparticle/nanowire size on CO_2 reduction reaction.
- Investigated single atom catalysts for applications in CO_2 reduction reaction.

2. **Postdoctoral Researcher (Stanford University)**

SUNCAT Center for Interface Science and Catalysis, Department of Chemical Engineering, Stanford University (Sep. 2017 – Aug. 2018)

Supervisor: Jens K. Nørskov

- Studied oxide-metal interfaces for applications in O_2 reduction reaction

Studied single metal atom catalysts for O_2 reduction reaction

- Collaborated with an experimental group (Prof. Xiaolin Zheng (Stanford Univ.)) to develop ZnO catalyst for H_2O oxidation to H_2O_2.
- Collaborated with an experimental group (Prof. Haotian Wang (Rice Univ.)) to study a local environment effect of single atom catalysts on O_2 reduction reaction.

3. **Postdoctoral Researcher (Carnegie Mellon University)**

Department of Chemical Engineering, Carnegie Mellon University (Sep.2018 – present)

Supervisor: Zachary W. Ulissi

- Developed a method to automate density functional theory (DFT) calculations of metal oxides for high-throughput catalyst screening applications.
- Developed a machine learning method to predict binding energies on various catalyst surfaces.

Selected Publications (Total Publications = 30, Total First Author Publications = 15, Total Citations = 795, h-index=14)

1. Back, K. Tran and Z. Ulissi, "Towards a Design of Active Oxygen Evolution Catalysts: Insights from Automated Density Functional Theory Calculations and Machine Learning", *ACS Catal. (submitted),*
2. Back, J. Yoon, N. Tian, W. Zhong, K. Tran, Z. Ulissi, "Convolutional Neural Network of Atomic Surface Structures to Predict Binding Energies for High-Throughput Screening of Catalysts", *J. Phys. Chem. Lett. (submitted)*
3. Back, M. Yeom and Y. Jung, "Active Sites of Au and Ag Nanoparticle Catalysts for CO_2 Electroreduction to CO", *ACS Catal., 5, 9, 5089 (2015)*
4. Back, J. Lim, N. Kim, Y. Kim, Y. Jung, "Single-atom Catalysts for CO_2 Electroreduction with Significant Activity and Selectivity Improvements", *Chem. Sci.*, 8, 2, 1090 (2017)

Honors

1. \$ 80,000, Global Ph.D. Fellowship, National Research Foundation of Korea (NRF), 2014 – 2017
2. \$ 40,000, Postdoctoral Fellowship, National Research Foundation of Korea (NRF), 2017 – 2018

580966 Accelerating Net-Zero Carbon Emissions By Electrochemical Catalysis: Understanding and Controlling the Reactions at Interfaces

Research Interests:

Overview. Electrochemical conversions of CO_2 to $C_xH_yO_z$, H_2O to H_2, and N_2 to NH_3 /NO^{3-} are promising technologies for future energy conversion and storage processes. These electrochemical catalytic reactions all involve the transfer of multi-electrons and protons to or away from the absorbed substrates. Unfortunately, high activation energies as well as the possibility of undesired side products are caused by the multi-step of electrons/protons transfers. My research interest lies in elucidating the operative mechanisms of these important electrochemical reactions with an emphasis on understanding the electrocatalyst surface chemistry including the reaction pathways, adsorbate/adsorbate interactions, field effects, ligand effects and many others which influence the efficiency and selectivity of these reactions. Electrocatalyst materials capable of efficiently driving these important reactions can thus be designed and synthesized based on the understanding of the mechanistic details. The insights gained should further be useful in any field that deals with the careful control of electron/proton transfers, for instance electrochemical biomass conversion and organic synthesis.

Overall Goal: Develop effective electrocatalysts and processes to produce fuels and chemicals by renewable electricity without carbon emissions. Specific aims are set to achieve these goals: **(a)** understand kinetics and reaction mechanisms on the electrode surface, **(b)** develop new catalysts based on the kinetic modelling and activity descriptor, **(c)** develop processes/reactors to explore economically feasible paths to sustainably produce chemicals and fuels.

Scientific Goal: Understand the reaction mechanism at the interfaces of solid-liquid and/or gas-solid-liquid, and further control the reaction environment of the active sites to achieve high performance.

The first three projects I intend to work on will be: **(i)** electrochemical CO_2 or cascade CO_2/CO reduction to beyond 2e$^-$ products; **(ii)** Electrochemical synthesis of NH_3 and NO_3^- from N_2 and H_2O, and exploring the reaction intermediates on the surfaces; **(iii)** Develop new electrocatalysts for the selective conversion of organic substrates to value-added products under mild conditions. Over time, these will evolve to include additional reactors and catalytic systems design that provide a set of sustainable methods to produce commodity chemicals and fuels.

Successful Proposals:

The Swedish Knut and Alice Wallenberg Foundation Postdoctoral Scholarship Program at Stanford, 2016.

PhD Research. My PhD thesis work was conducted under the supervising of Prof. Licheng Sun at KTH Royal Institute of Technology, Sweden. It introduced innovative concepts for the rational design of molecular catalysts based on comprehensive kinetic and mechanistic studies. Of particular relevance, I demonstrated that radical coupling of the metal oxide intermediates is the rate limiting step for water oxidation on ruthenium complexes with axial imidazole/DMSO ligands. Based on this finding, more efficient catalysts were obtained through the rational design of the axial ligands which facilitate the radical coupling pathway.[1-7] The mechanistic insights gained while working on homogeneous catalysis inspired me to develop a new heterogeneous nickel-hydroxide catalyst with remarkable activity for the oxygen evolution reaction.[8] Later, I also introduced a polymer dots as an effective photocatalyst for hydrogen evolution, opening the door for the use of organic polymer dots as photocatalysts for the green hydrogen production from water.[9]

Postdoctoral Research: During my postdoctorate in Prof. Thomas. Jaramillo's group at Stanford, I extended my primary goal to develop electrocatalysts that can selectively reduce CO_2 into liquid fuels. I started with CO reduction under alkaline conditions, important selectivity trends on the function of electrode surface area and electrode potentials were established based on comprehensive kinetics analysis, outlining key principles for designing CO and CO_2 electrolyzers that are able to produce valuable products with high energy efficiency.[10-11]

Teaching Interests: I look forward to the exciting opportunity to mentor students of all backgrounds in order to help them develop their potentials to the fullest. I am capable of teaching all the core chemical engineering courses that offered by the department based on my education background, besides my research background ensures me the ability to teach all levels of kinetics, reaction engineering, physical chemistry, and data analysis courses. My TAing experience also gives me confidence in teaching chemistry and chemical engineering laboratory courses. Moreover, I would be very interested in developing courses on Electrochemical Energy Conversion Systems that introduce principles, current research efforts, existing challenges and opportunities. Students would learn in this course how to design experiments and solve problems in fuel cells, batteries, (photo)electrochemical cells using equivalent circuits of physical significance.

Selected Publications:

1. "Towards Controlling Water Oxidation Catalysis: Tunable Activity of Ruthenium Complexes with Axial Imidazole/DMSO Ligands"
   L. Wang, L. Duan, B. Stewart, M. Pu, J. Liu, T. Privalov, L. Sun. ***J. Am. Chem. Soc.*** 2012, 134, 18868-18880.
2. "Visible light-driven water oxidation catalyzed by mononuclear ruthenium complexes"
   L. Wang, L. Duan, L. Tong, L. Sun. ***J. Catalysis***. 2013, 306, 129-132.
3. "Highly Efficient and Robust Molecular Water Oxidation Catalysts Based on Ruthenium Complexes"
   L.Wang, L.Duan, Y. Wang, M. Ahlquist, L.Sun. ***Chem. Commun***., 2014, 50, 12947.
4. "Sensitizer-Catalyst Assemblies for Water Oxidation"
   L. Wang, M. Mirmohades, A. Brown, L, Duan, F. Li, Q. Daniel, R.Lomoth, L. Sun, L. Hammarstöm,. ***Inorg. Chem.***, 2015, 54 (6), pp 2742–2751
5. "Electrochemical Driven Water Oxidation by Molecular Catalysts *in situ* Polymerized on the Surface of Graphite Carbon Electrode"
   L. Wang, K. Fan, Q. Daniel, L.Duan, F. Li, B. Philippe, H. Rensmo, L. Sun. ***Chem. Commun.,*** 2015, 51, 7883.
6. "Nickel (II) Complex as an Electrocatalyst for Water Oxidation"
   L. Wang, L. Duan, R. Ambre, Q. Daniel, H. Chen, J. Sun, B. Das, A. Thapper, J. Uhlig, P. Diner, L. Sun. ***J. Catalysis***. 2016, 335, 72–78.
7. "Towards Efficient and Robust Anodes of Water splitting Device: Functionalizing Electrodes with Ru catalyst by *in situ* polymerization"
   L. Wang, K. Fan, H. Chen, J. Sun, L. Sun. ***Catalysis Today,*** 2017, 290, 73.
8. "Promoting the Water Oxidation Catalysis by Synergistic Interactions between Ni(OH)2 and Carbon Nanotube"
   L. Wang, H. Chen, L. Duan, Q. Daniel, B. Philippe, H. Rensmo, L. Sun. ***Advance Energy Materials,*** 2016, 201600516.
9. "Organic Polymer Dots as Photocatalyst for Visible Light-Driven Hydrogen Generation"
   L. Wang, R. Fernández-Terán, L. Zhang, D. Fernandes, L. Tian, H. Chen and H. Tian. ***Angew. Chem. Int. Ed.*** 2016, 55(40):12306-10.
10. "Electrochemical CO Reduction on Poly-Cu: Effects of Potential and pH on Selectivity toward

Multicarbon and
Oxygenated Products"
L. Wang, S. A. Nitopi, E. Bertheussen, M. Orazov, C. G. Morales-Guio, C. Hahn and T. F. Jaramillo. ***ACS Catalysis.,*** 2018, 8, 7445–7454.

11. "Electrochemically converting carbon monoxide to liquid fuels by directing selectivity with electrode surface area"
L. Wang, S. A. Nitopi, E. Bertheussen, M. Orazov, A. Wong, D.C. Higgins, C. G. Morales-Guio, C. Hahn and T. F. Jaramillo. ***Nature Catalysis,*** 2019, *DOI: 10.1038/s41929-019-0301-z.*

582605 Design of Dynamic and Solvated Reaction Environments for Enhanced Conversion and Selectivity with Heterogenous Catalysts

Research Interests:

My research interests combine experimental and theoretical approaches to understand heterogeneous catalytic phenomena in complex solvation environments and reaction environments subjected to oscillatory perturbations of temperature and pressure setpoints or feed composition. Through fundamental investigations of these reaction environments, my aim is to identify additional routes for improving chemical synthesis reactions including: 1) tuning of the solvent-catalyst interface and 2) optimizing each step in the catalytic cycle via cyclic variation of reaction conditions.

Solvent selection and heterogeneous catalyst design are often optimized independently, however, the observed reactivity of a given catalyst may vary significantly in the presence of different solvents. In addition to solvating reactants and products, solvents can enhance or inhibit reactions at heterogeneous catalysts surfaces by interacting with adsorbates and transition states and by changing the electronic structure of the catalyst surface. Addition of a third phase, such as a gas phase or a secondary liquid phase, could form unique multi-phase interfaces at the catalyst surface, creating asymmetric reaction sites that enhance stereoselectivity or enantioselectivity. I am interested in characterizing reactions at these solid-liquid and multi-phase interfaces to generate design criteria for catalysts and condensed-phase reaction environments to be used for chemical synthesis.

Variable-volume reactors and valve configurations that allow alternating pulse injections of reactants could provide routes for overcoming conversion-selectivity trade-offs that arise during steady-state operation of heterogeneously catalyzed reactions. The temporal variation of pressure, temperature, and feed concentration could be designed such that each step of the catalytic cycle for the desired reaction is optimized while the conditions that lead to byproduct formation are minimized. I am interested in characterizing adsorption, surface reaction, and desorption phenomena for these reactor systems and designing catalytic materials that are optimized for operating under dynamic reaction conditions.

Teaching Interests:

I am very eager to develop and teach courses at both the graduate and undergraduate level and provide mentoring to students and post-docs in a research group. Using examples from my own teachers and mentors and my experiences as a graduate student instructor, I look forward to applying active and cooperative learning techniques to help students master technical content. I am comfortable teaching all the undergraduate chemical engineering courses, and I have experience teaching Separation Processes, Reaction Engineering, and undergraduate Chemical Engineering Lab. In addition to helping students master technical skills in chemical engineering, my goal in both the classroom and research group is for students to appreciate the societal impact of engineering careers and develop soft skills such as clear and respectful communication, creative thinking, leadership, and teamwork, which will aid them in their future careers and relationships.

Postdoctoral Projects:

"*Methane activation and upgrading to value-added chemicals*"

"*Characterization of surface species during benzene hydroxylation over a NiO on Ceria-Zirconia catalyst*"

"*Kinetic modeling of cyclopentadiene dimerization in the presence of C_5 alkenes and alkadienes*"

2018-present, Georgia Institute of Technology, Atlanta, GA

Postdoctoral Advisors: Carsten Sievers and Charles Liotta

Doctoral Dissertation:

"*Heterogeneous Catalyst Stability in Hydrothermal Media*"

2013-2018, University of Michigan, Ann Arbor, MI

Doctoral Advisors: Phillip E. Savage and Levi T. Thompson

583218 Sustainability towards the Future:bridging the gap between catalysis science and reaction engineering

Research Interests:

Catalysis science is the central component and primary driver for sustainable chemical and energy production in the future. The ongoing transition from fossil-based resources to alternative sources, such as biomass and shale gas, urgently requires the development of next-generation catalysts and catalytic processes. To move forward, it is imperative to bridge the relatively wide gap between fundamental research of catalysis science and the ability to transfer knowledge to practical innovation, which is essentially based on the discipline of chemical reaction engineering. My research program will address these forefront challenges by developing highly selective and coke-resistant catalysts for methane conversion to C_2 oxygenates and biomass transformation to premium quality aviation fuels.

The modern-day catalysis science emphasizes the precise synthesis of catalytic materials (e.g. mesoporous and zeolite supports), dynamic characterization under working conditions (e.g. *in situ/operado* and ambient pressure approaches), and identification of active site requirements. Chemical reaction engineering (CRE), on the other hand, aims at investigating and optimizing catalytic reactions to reach the most optimal scaled-up reactor design, which typically includes the interactions between intrinsic kinetics with flow phenomena, mass and heat transfer. The key factor to bridging the gap between catalysis science and CRE is to investigate structure-activity relationships of heterogeneous catalysts and determine turnover frequency normalized by well-defined active sites under dynamic working conditions, followed by modeling of mirco- and macro-kinetics of heterogeneous catalytic reactions. This experimental contribution will be the primary focus in my future research group. Since a rich meta-data library exists for catalytic materials and catalyst/reactor performance, data science tools, like machine learning, reinforcement learning and artificial intelligence, may discover hidden trends and periodicities within catalysis data and reactor performance, and thus shorten the development time for next-generation of catalysts. These targets require the combination of controlled materials synthesis, in situ/operado characterization, precise measurements of isotopic kinetics, and rigorous mathematical modeling. The above research efforts in my research group will be fostered by collaboration with professionals of nanotechnology, first-principles and data science, which will leverage multi-disciplinary research to modern catalysis science and reaction engineering research.

My research group will focus on developing transitional-ready catalytic processes with excellent activity of reactants, selectivity towards target products, and turnover frequency over well-defined active sites and stability with long catalyst lifetime for a sustainable future by bridging the gap between catalysis and reaction engineering.

Future Research Directions:

- Conversion of Methane to C_2 Oxygenates (e.g. ethanol, ethylene oxide) over Highly Selective and Coke-Resistant Catalytic Materials
- Conversion of Algae to Premium Quality Bio-Jet Fuels over Robust Bifunctional Catalysts
- Data Science for Reactor Safety and Modeling of Complex Catalytic Kinetics

Successful Proposals:

1. "Direct Catalytic Oxidation of Methane to Liquid Oxygenates," (PI: Dr. Arvind Varma) $353,950, 10/2017 - 9/2022; NSF; part of Engineering Research Center titled "Center for Innovative and Strategic Transformation of Alkane Resources (CISTAR)".
2. "Catalyst Design and Reactor Runaway: Selective Oxidation of Alcohols," (PI: Dr. Arvind Varma) $50,000, 7/2019 - 6/2021 Purdue Process Safety & Assurance Center (P2SAC)

Selected Projects:

Catalytic Nonoxidative Coupling of Methane (NOCM)

- Designed, prepared (wet impregnation), characterized (ICP, XRD, XPS, EDX, etc.) and tested Pt-Bi bimetallic catalysts
- Achieved over 90% selectivity to C_2 species from methane coupling over the new catalyst
- Proposed the explanation of the coke-resistant role of Bi promoter by conducting mechanistic investigations of the above catalyst, including DFT calculations, TPO, TPR and TPSR

Parametric Sensitivity in Chemical Reactors for Selective Oxidation of Alcohols

- Measured differential kinetics for methanol selective oxidation in a fixed-bed reactor
- Gave instructions for the control of reactor stability by developing a new kinetic and reaction model including heat transfer and parametric sensitivity

Selected Patents: - *out of 20 patents/patent applications*

1. Yang Xiao and Arvind Varma, Non-oxidative Production of Hydrocarbon from Methane, U.S. Patent (officially granted), US20190084904A1, September 19, **2018**.
2. Yang Xiao and Arvind Varma, Method of Producing Formaldehyde from Methanol, U.S. Patent (officially granted), US20190194106A1, December 19, **2018**.
3. Yang Xiao and Arvind Varma, Catalytic Deoxygenation of Bio-Oils Using Methane, U.S. Patent, No. US10023809 B2, July 17, **2018**.

Selected Publications: - *out of 22 peer-reviewed journal papers*

1. Yang Xiao, Yuan Wang and Arvind Varma, Low-Temperature Selective Oxidation of Methanol over Pt-Bi Bimetallic Catalysts, *Journal of Catalysis*, **2018**, 363, 144-153.
2. Yang Xiao and Arvind Varma, Highly Selective Nonoxidative Coupling of Methane over Pt-Bi Bimetallic Catalysts, *ACS Catalysis*, **2018**, 8 (4), 2735-2740.
3. Yang Xiao, Jeffrey Greeley, Arvind Varma, Zhi-Jian Zhao and Guomin Xiao, An Experimental and Theoretical Study of Glycerol Oxidation to 1,3-Dihydroxyacetone Over Bimetallic Pt-Bi Catalysts, *AIChE Journal*, **2017**, 63 (2), 705-715. **(as Most Read Article)**
4. Yang Xiao and Arvind Varma, Conversion of Glycerol to Hydrocarbon Fuels via Bifunctional Catalysts, *ACS Energy Letters*, **2016** 1 (5), 963-968.

Teaching Interests:

Because of maintaining passion for teaching, I actively sought to build my repertoire of teaching skills to prepare for pursuing my career goal as a professor. In Spring 2019, I taught four lectures at one of our core chemical engineering courses for graduate students - Chemical Reaction Engineering (ChE660) in the Davidson School of Chemical Engineering of Purdue University. Those lectures include *Experimental methods in catalytic kinetics*, *Stability & multiplicity of reactors*, and *Heterogeneous models for packed-bed reactors*. Furthermore, I published a teaching-related journal article "Applications of Chemical Simulation Softwares in the Course of Transport Phenomena" (*Chemical Industry Times*, **2012**, 26(10), 53-59), which describes how we can use powerful computer tools to facilitate teaching. In addition to being a co-instructor or teaching assistant in the classroom, I have also mentored more than 10 undergraduate students on research since 2010 when I started my graduate school. Most of them have published co-author peer-reviewed papers, and continued their own research career in prestigious graduate schools in the U.S. As a senior research associate in Dr. Varma's lab, I also co-mentored three graduate students in the past 3 years. I am glad to see that now they can work relatively independently, making great progress on research and preparing manuscripts for peer-reviewed journals.

My education, research and teaching experiences have prepared me to teach the core courses, including ***Thermodynamics***, ***Chemical Reaction Engineering/Reactor Design*** and ***Mathematical Methods in Chemical Engineering***, in the Department of Chemical Engineering. Additionally, since data science is becoming increasingly critical in chemical engineering, I am interested in developing new undergraduate elective or graduate-level courses such as ***Data Science for Chemical Engineering*** and ***Methods in***

Catalysis Research to incorporate emerging data science topics into chemical engineering. I believe my great passion for teaching, various past teaching experiences, strong academic background and communication skills will make me an excellent teacher.

581474 Rational Catalysts and Process Development for Light Hydrocarbons Upgrading from Natural Gas/Shale Gas

Research Interests:

Recent advances in natural gas/shale gas technology, especially as practiced in the United States, has significantly improved the economics for producing ethylene and has revolutionized manufacturing approaches to basic chemicals, polymers and materials. Ethane is second to methane as the major hydrocarbon component of natural gas/shale gas, serving as the precursor to ethylene. Ethylene is one of the most important building blocks in the chemical industry and provides the starting point for many essential consumer products such as performance polymers, plastics, and functional chemical intermediates. The greatest challenge for any natural gas/shale gas application is direct production of high value chemicals with limited separation requirements.

The selective oxidative dehydrogenation (ODH) process provides an alternative to ethylene production via naphtha or ethane steam cracking. Steam cracking is the most energy-intensive process in the chemical industry, using approximately 8% of all energy consumed by the chemical industry. The exothermic ODH reaction can significantly reduce the energy consumption compared to steam cracking with a much lower reaction temperature (< 400°C for ODH and > 800°C for steam cracking). ODH of ethane to ethylene also provides higher selectivity to ethylene compared to steam cracking and is not equilibrium-limited.

My previous study has shown that with novel composition and synthesis method, mixed metal oxide catalyst can offer high selectivity and conversion for ethane ODH reaction. The orthorhombic M1 catalyst ($MoVTeNbO_x$) with optimized composition offers > 90% selectivity with > 70% conversion. However, it is still difficult for the ODH process to compete with steam cracking of ethane due to all currently existing steam cracking plants and the economy of scale for steam crackers, which are typically 2 – 4 Blbs/yr in capacity. Price differences of natural gas and ethylene also vary with location and season to make the ODH process less competitive. Further, ODH of ethane from natural/shale gas often requires separation of methane, unreacted O_2, and other hydrocarbon impurities before the ethylene purity is high enough for downstream applications such as polyethylene synthesis.

My future work and research aims to lead the development of novel catalysts by unique composition and structure for catalytic conversion of light hydrocarbons to high value chemicals specially for ethylene oxide and ethylene glycol from natural/shale gas. Novel catalysts synthesis and development will be combined with a rational process for direct production of ethylene oxide (EO) from natural/shale gas using the combination of oxidative dehydrogenation (ODH) of ethane to ethylene and epoxidation of ethylene to EO. Ethylene oxide-derived products are especially relevant to feed the ever-growing market for EO which is at ~4%/year; approximately 10 - 15 Blbs/year of new EO capacity are scheduled to come on-line by 2020 on the Texas-Louisiana gulf coast.

The core components of this study are rational catalysts with high selectivity and long lifetime for both ODH and EO processes, as well as two successive reactor systems for direct production of high value chemicals (ethylene oxide and ethylene glycol) from natural gas. With the price difference between EO/EG and natural gas-derived ethylene, there is a significant profit potential for direct production of EO from natural gas/shale gas, particularly if separations and gas recycle can be avoided. This will make the new process more competitive compared to the existing cracking plants. Another important approach is to limit separation requirements with the addition of the EO process after ODH. Using the full composition of the gas stream from the ODH reactor provides a great process advantage, since in principle no separation should be required between the ODH and EO reactors.

The success of this selective oxidation research can fundamentally change the utilization of remote natural gas fields in the United States, not only with respect to EO as a desired chemical product, but possibly for other products as well, such as hydration of ethylene to fuel grade ethanol. With limited separation requirements, direct production of high value chemicals and use of modular, portable reactors, this potentially low cost technology can be used for most natural gas fields in the United States. The ultimate result of the research will be significant, impacting the value chain and U.S. competitiveness for natural gas-derived feedstocks. Additionally, the scientific knowledge acquired will advance other manufacturing technologies, such as selective oxidation of methane and other hydrocarbons.

Funded projects:

(Total amount of projects as PI and Co-PI is more than $750,000. Specific amount for Dr. Diao is more than $290,000)

Principal Investigator:

1. U.S. National Science Foundation (NSF), Industry–University Cooperative Research Centers Program (IUCRC) of Center for Rational Catalyst Synthesis (CeRCaS), "Boron nitride (BN) supported metal and metal oxide catalysts for selective oxidation reactions", NSF 1464630-PJ15, 06/2018-12/2019
2. U.S. Department of Energy (DOE), Laboratory Directed Research and Development (LDRD), "Catalysts development for water splitting reactions using solid oxide electrolysis cells (SOECs)", DE-AC07-051D14517-216845, 03/2019-09/2019

Co-Principal Investigator:

1. U.S. Department of Energy (DOE), Laboratory Directed Research and Development (LDRD), "Kinetic-based Scale-Up Science for an Energy Efficient Route to Ethylene", 16P6-002FP, 06/2016-09/2017
2. University of South Carolina, Advanced Support for Innovative Research Excellence (ASPIRE) III, "*In situ* high temperature and variable gas environment X-ray Diffractometer for study of effects of temperature and gas environment on stability of catalysts and other novel materials", 15510-18-48251, 07/2018-09/2019
3. U.S. National Science Foundation (NSF), Industry–University Cooperative Research Centers Program (IUCRC) of Center for Rational Catalyst Synthesis, "Computational and experimental analysis of Ag catalysts with GNP or bimetallic materials on direct propylene oxide synthesis", NSF 1464630-PJ22, 06/2019-09/2020

Teaching Interests:

I passionately believe that teaching is a key and enjoyable part of being a new faculty member. I view teaching not only providing knowledge and information, but more about inspiring students who are future engineers, scientists and professors. I am excited about the opportunity to teach both undergraduate and graduate level classes as well as mentoring graduate and undergraduate students laboratory experiments. As a post-doc at Idaho National Laboratory and a research assistant professor at University of South Carolina, I have supervised graduate and undergraduate students in the catalysis and energy related projects with more than five publications.

Also, I have served as lecturer and teaching assistant for undergraduate and graduate courses during my PhD, post-doc and research assistant professor studies since 2011. I heavily participated in giving lectures, designing course contents and preparing grading exam questions and homework problems for courses including Chemical Process Principles, Chemical Engineering Thermodynamics, Heterogeneous Catalysis, and Chemical Engineering Undergraduate Laboratory. I am excited to share the fundamental science in chemical engineering and its significant impact on latest technology in industry and our daily life with undergraduate students to help cultivate their interests for science and engineering, and contribute to their education and professional career development.

Specific courses I would particularly enjoy teaching include Chemical Engineering Fundamentals, Chemical Process Principles, Chemical Reaction Engineering, Materials Characterization, and Chemical Engineering Thermodynamics at both undergraduate and graduate levels. In addition, with background of heterogeneous catalysis, I also plan to develop an elective course introducing the fundamentals of catalysis, including catalyst preparation and evaluation, reactor design, kinetic studies, mass/heat transfer limitations, and mechanism study using advanced characterization.

581805 Rational Design of Alloy Catalysts By Building Atomic-Scale Structure-Property-Activity Relationships

Research Interests: One of main challenges in the rational design of alloy catalysts is the difficulty of fully characterizing the atomic arrangements using experimental methods. It is also prohibitively expensive to assess all possible atomic arrangements using theoretical calculations due to the combinatorial explosion in the number of possible structures in alloys that allow substitutional disorder. To address this challenge, I present new developments in the use of the cluster expansion method, trained by of *ab-initio* calculations, to rationally design alloy catalysts by building atomic-scale structure-property-activity relationships. I use Pt–Ni alloy catalysts for the oxygen reduction reaction (ORR) as examples to illustrate my methods, which can be generally applied to other alloys for other chemical reactions. I will demonstrate how I develop a lattice-parameter-dependent slab cluster expansion to computationally map the structure and catalytic activity of a Pt–Ni(111) alloy surface at every point in the bulk phase diagram. These maps make it possible to identify the synthesis conditions likely to result in highly active catalysts. I also will introduce a method to calculate the activity for alloy nanoparticles with experimentally relevant sizes (5 nm – 10 nm) by building a quaternary Pt–Ni–OH-Vacancy cluster expansion to explicitly predict *OH binding energies. Using this model, I evaluate how different parameters of Pt–Ni nanoparticles affect ORR activity: size (2 nm – 10 nm), Pt composition (50% – 100%), and shape. This approach provides theoretical insights into how to tune the structures of alloy nanoparticles to optimize catalytic activity.

Teaching Interests: During my Ph.D. period at Johns Hopkins University, I worked as a teaching assistant for two courses, General Physics and General Physics Lab for undergraduate with class sizes of 24. As a TA, I accumulated rich experience in teaching by preparing and giving weekly presentations, and holding weekly office hours. For my teaching interests, provided I have B.S and Ph.D. degrees in Physics (major in Condensed Matter Physics), I believe I am qualified for teaching courses related to condensed matter physics, such as Solid State Physics, Electromagnetics, Quantum Mechanics, and Statistical Mechanics, at both undergraduate and graduate levels. I also plan to develop a graduate-level lab-oriented course to introduce commonly used computational methods closely related to my research on nanomaterials for energy storage and conversion.

581978 Combining First-Principles Modeling and Nanomaterial Synthesis to Understand and Improve Environmental Catalysis

Research Interests:

My research combines first-principles-based simulations under realistic conditions and synthesis of nanomaterials with controlled structures to understand and develop heterogeneous catalysis for emission control and aqueous contaminants removal. My postdoc work under the supervision of Dr. William Schneider is an example of a system where I integrate density functional theory calculations with my own microkinetic code to develop quantitative models suitable for direct comparison with experiments. These models are aware of surface coverage of reaction intermediates and water solvent to provide improved predictions. My PhD is a self-driven effort focusing on synthesis and characterization of size and composition-controlled nanometals for catalytic reduction of model contaminants in water. I elucidated multiple structure-activity relationships for catalytic water treatment.

My expertise is in both computational simulation and experimental catalysis and I plan to couple my skills and expand my research to the following fields:

1) Development of molecular understanding of heterogeneous catalysis in aqueous phase by building sufficient solvent-aware models and validating the predictions with in-situ surface vibrational spectroscopy at solid-liquid catalytic interfaces.

2) Catalyst screening by modeling & simulation and preparation of robust metal–metalloid alloys for environmental catalysis in complicated matrices.

3) Development of plasma-electron-driven processes coupled to heterogeneous catalysis for aqueous reactions relevant to recalcitrant contaminant destruction and chemical synthesis.

Teaching Interests:

My goal is to educate future generations of engineers and my teaching philosophy is to stimulate self-learning and prepare students for current environmental and energy challenges. I look forward to teaching core undergraduate courses in Chemical Engineering, including thermodynamics, reaction engineering and chemical engineering labs, and graduate courses such as computational chemistry and math methods for engineering. While a graduate student, I have served as a teaching assistant for five semesters covering topics on thermodynamics, transport and kinetics. I am particularly interested in training undergraduate students on engineering programming with emphasis on data science and teaching graduate students with essentials of molecular modeling. I have also enjoyed mentoring several graduate and visiting students and I have realized that teaching is my passion throughout my academic career.

References:

1. H. Ma, S. Li, H. Wang, and W. F. Schneider, "Water-mediated reduction of aqueous N-nitrosodimethylamine on Pd," Environmental Science & Technology, 2019, in press.
2. H. Ma and W. F. Schneider, "Structure- and Temperature-Dependence of Pt-Catalyzed Ammonia Oxidation Rates and Selectivities," ACS Catalysis 2019, 9, 2407-2414.
3. H. Ma, H. Wang, P. Burns, B. McNamara, E. Buck, and C. Na, "Synthesis and preservation of graphene-supported uranium dioxide nanocrystals," Journal of Nuclear Materials, 2016, 475, 113-122.
4. H. Ma, H. Wang, T. Wu, and C. Na, "Highly active layered double hydroxide-derived cobalt nano-catalysts for p-nitrophenol reduction," Applied Catalysis B: Environmental, 2016, 180, 471-479.
5. H. Ma, and C. Na, "Isokinetic temperature and size-controlled activation of ruthenium-catalyzed ammonia borane hydrolysis," ACS Catalysis, 2015, 5, 1726-1735.
6. H. Ma, H. Wang, T. Wu, and C. Na, "Microwave-assisted optimization of platinum-nickel nanoalloys for catalytic water treatment," Applied Catalysis B: Environmental, 2015, 163, 198-204.

7. H. Wang,* H. Ma,* W. Zhen, D. An, and C. Na, "Multifunctional and recollectable carbon nanotube ponytails for water purification," ACS Applied Materials & Interfaces, 2014, 6, 9426-9434. * These authors contributed equally.

569391 Process-Informed Design of Electrocatalysts for Sustainable Chemical Transformations

The widespread adoption of increasingly affordable renewable energy sources, in tandem with the urgent need to mitigate CO_2 emissions, provides exciting perspectives for the electrification of processes in the chemical industry. In this context, my research is focused on the discovery, understanding and optimization of electrocatalysts for reactions of practical relevance, ranging from the synthesis of bulk chemicals and fuels from sustainable feedstocks (e.g., electrochemical CO_2 reduction) to investigating the use of electrocatalysis in the production of fine chemicals and pharmaceuticals.

Research Interests:

My main research interests comprise (a) the discovery of new materials and modification strategies in electrocatalysis, (b) the rationalization of catalytic behavior by unraveling reaction mechanisms and structure-performance relationships, and (c) the integration of catalysts into electrochemical devices that achieve high reaction rates. To this end, I aim to carry out a highly collaborative research program that benefits from a close synergy with computational and advanced characterization studies, and from the interaction with implementation-oriented researchers to understand constraints and requirements from a process point of view.

Teaching Interests:

I have developed my teacher skills as a TA in laboratory courses, and I am currently a co-lecturer in a graduate-level course at DTU. My aim as a teacher is to prioritize student engagement by adopting a hands-on, project-driven approach. I am particularly interested in using electrochemical devices as a platform to convey practical concepts related to heat and mass transfer, reaction kinetics, CAD, and manufacturing processes. I also aim to promote interest in chemical engineering and electrocatalysis among minority students and, as a group leader, I will strive to be an agent of change to make academic research a more open and welcoming environment for scientists from underrepresented groups.

Successful funding proposals:

- European Union Marie Skłodowska-Curie Actions (MSCA-IF) postdoctoral fellowship

- ETH Zurich Research Grant graduate fellowship

Education:

- Ph.D., Chemical Engineering, ETH Zurich, Switzerland

- M.Sc., Process Engineering, ETH Zurich, Switzerland

- B.Sc., Chemical Engineering, Universidad Simón Bolívar, Venezuela

Selected publications:

- **G.O. Larrazábal**, T. Shinagawa, A.J. Martín, J. Pérez-Ramírez. "Microfabricated electrodes unravel the role of interfaces in multicomponent copper-based CO_2 reduction catalysts". *Nature Communications*, **2018**, 9, 1477

- **G.O. Larrazábal**, A.J. Martín, J. Pérez-Ramírez. "Building Blocks for High Performance in Electrocatalytic CO_2 Reduction: Materials, Optimization Strategies, and Device Engineering". *Journal of Physical Chemistry Letters*, **2017**, 8, 3933-3944

- **G.O. Larrazábal**, A.J. Martín, S. Mitchell, R. Hauert, J. Pérez-Ramírez. "Enhanced Reduction of CO_2

over Cu-In Electrocatalysts: Catalyst Evolution is the Key". *ACS Catalysis,* **2016**, 6, 6265-6274

- A.P Amrute, **G.O. Larrazábal**, C. Mondelli, J. Pérez-Ramírez. "$CuCrO_2$ Delafossite: a Stable Copper Catalyst for Chlorine Production". *Angewandte Chemie International Edition,* **2013**, 52, 9772-9775

- T. Pielhop, **G.O. Larrazábal**, P. Rudolf von Rohr. "Treament of lignocellulosic biomass with aromatic chemicals as scavengers". European Patent EP 3158128 (B1), Issued March 06, 2019

581315 Development of n-Doped Oxides As an Electron Rich Support: Bridging Homogeneous and Heterogeneous Catalysis

Research Interests:

As the need for sustainable chemical processes is rapidly increasing, a renewed focus has been placed on increasing the energy efficiency of chemical processes. By far the most energy intensive process in any homogeneously catalyzed process is the recovery of the catalyst, usually a precious noble metal, from the product mixture. Typical processes which utilize homogeneous catalyst, such the creation of active pharmaceutical ingredients (APIs) and olefin hydroformylation to make block co-polymers, yield poor energy efficiencies due to their intense separation processes, and often yield detrimental effects to the product if residual catalyst remains. One of the key factors to active homogeneous catalysts is their rich electron density, often achieved via the use of intricate ligands that allow the clearly defined active site to freely undergo the required redox cycles to form the desired product. Due to the innate variability of active sites on heterogeneous catalysts, their use in processes that require highly pure products has been limited; however, if the supporting material can be engineered to have energetically preferential grafting sites for metal precursors, not only can the heterogeneity of the surface be mitigated but the functionality of the support can dramatically be changed.

To develop next generation supporting oxides for heterogeneous catalysis the use of n-doped oxides will be explored. Doping semiconducting materials, such as silicon, is a mature field; however, these novel applications which alter the electronic properties of the support have not been carried into heterogeneous catalysis. By incorporating n-doped species, such as phosphorus, we can not only provide preferential adsorption sites for metal precursors during synthesis, but also provide electron rich ligands to promote metal support interactions, thus facilitating the intricate redox cycles required for homogeneous catalysts without the need for extensive energy use needed for catalysts recovery. In this proposal, the use of rhodium catalysts for ethylene hydroformylation will be explored as a model reaction for the use of phosphorus doped silica and compared against typical homogeneous catalyst, therefore providing a bridge between heterogeneous and homogeneous catalysts. This work will lay the foundation to attempt to address one of the key challenges laid out by the Catalysis Science focus area of the Chemical Science, Geoscience, and Biosciences Division of the Department of Energy, which is the convergence of heterogeneous and homogeneous catalysts. The initial work, focusing on simpler model systems such as phosphorus doping into silica over Co/Rh catalysts for light olefin hydroformylation, will then be carried out to more complex systems such as industrially relevant reactions like the enantioselective hydroformylation of styrene to vinyl acetate over several electronically modified supports such as *n/p*-doped silica, titania, or zirconia.

During my PhD at the University of South Carolina under the supervision of Prof. Jochen Lauterbach, I have developed a broad skillset to develop precise catalytic materials with well controlled active sites. Furthermore, by employing various characterization techniques, such as *in situ* X-Ray Absorption, infrared, and Raman spectroscopy; I was able to probe the catalysts under working conditions. The intricate use of spectroscopic techniques allowed me to probe the CO_2 hydrogenation reaction mechanism over facet controlled cobalt nanorods, discern the local structure of single site cobalt and ruthenium catalysts under relevant conditions, and gain insight into the structure of mixed metal oxide catalysts for ethane partial oxidation. Additionally, under the supervision of Prof. Kiyotaka Asakura at the Institute of Catalysis at Hokkaido University, I was able to further refine my knowledge of XAFS by carrying out *operando* High Energy Resolution Fluorescence Detection (HERFD) XANES and simultaneously developed a corresponding model of the catalyst under working conditions via FEFF8, helping verify our experimental findings. This study was carried out on a platinum-gold bimetallic electrocatalyst where HERFD XAFS was coupled with *ab initio* FEFF8 calculations to discern the surface species present on the Pt surface. Working with diverse chemical systems and catalyst compositions at these distinct research institutions has given me perspective on how to independently develop novel projects, where I have learned how to leverage my skillset to explore a diverse set of catalytic systems.

Relevant Publications († denotes equal contribution first author):

1. **Jimenez, J.D**.; Jung, S.; Biddinger, E. J., Ionicity Analysis of Silylamine-Type Reversible Ionic Liquids as a Model Switchable Electrolyte. *Journal of the Electrochemical Society* **2015,** *162* (7), H460-H465.
2. **Jimenez, J.**; Bird, A.; Santos Santiago, M.C.; Wen, C.; Lauterbach, J., Supported Cobalt Nanorod Catalysts for CO_2 Hydrogenation. *Energy Technology* **2017**, *5*, 884-891
3. Wu, J.; Wen, C.; Zou, X.; **Jimenez, J.**; Sun, J.; Xia, Y.; Fonseca Rodrigues, M.-T.; Vinod, S.; Zhong, J.; Chopra, N.; Odeh, I. N.; Ding, G.; Lauterbach, J.; Ajayan, P. M., Carbon Dioxide Hydrogenation over a Metal-Free Carbon-Based Catalyst. *ACS Catalysis* **2017**, 4497-4503
4. **Jimenez, J.D.**†; Mingle, K.†; Bureerug, T.; Wen, C.; Lauterbach, J. Statistically Guided Synthesis of MoV- Based Mixed Oxide Catalysts for Ethane Partial Oxidation. *Catalysts* **2018**, 8, 370-386. **Selected as a Featured Paper**
5. **Jimenez, J.D.**; Wen, C.; Lauterbach, J.; Design of Highly Active Cobalt Catalysts for CO_2 Hydrogenation via the Tailoring of Surface Orientation of Nanostructures. *Catalysis Science & Technology*, **2019**, 8, 1970-1979
6. Fan, M.†; **Jimenez, J.D.**†; Shirodkar, S.N.; Wu, J.; Chen, S.; Royko, M.; Song, L.; Zhang, J.; Guo, H.; Cui, J.; Zuo, K.; Wang, W.; Zhang, C.; Vajtai, R.; Qian, J.; Yang, J.; Yakobson, Y.I.; Tour, J.M.; Lauterbach, J.; Sun, D.; Ajayan, P.M. Ru Atoms Immobilized on Porous h-BN for Highly Active and Selective CO_2 Methanation. *ACS Catalysis, In Revision*
7. **Jimenez, J.D**; Wen, C.; Royko, M.; Kropf, A.J.; Segre, C.; Lauterbach, J. Influence of coordination environment around anchored single-site cobalt catalyst for CO_2 hydrogenation. *In Preparation*.

Teaching Interests:

As teaching at both the undergraduate and graduate level is an essential part of being a faculty member, it is something I have always taken interest in. Beginning in my undergraduate at the City College of New York holding introductory chemistry workshops, I learned early on the importance of effectively communicating various topics. My interest in teaching was further expanded during my PhD at the University of South Carolina, where I served as a Teaching Assistant for both undergraduate and graduate courses, often holding both recitations and full classes for the graduate level Mass Transfer course at USC. One of the key things I have learned in the classroom, through both my experience teaching and being a student, specifically a first generation Hispanic student, is the need for individualized teaching and reaching out to students to ensure the content is not only understood; but that students are properly mentored on their prospective career paths. To this end, I have made it a priority to mentor students all throughout my graduate studies, both undergraduates and junior graduate students, not only showing them the required theories and techniques to succeed, but to instill in them a strong sense of scientific independence. Furthermore, as part my National Science Foundation Integrated Graduate Research Traineeship (NSF-IGERT) I have developed a strong set of soft skills, that include consideration of the patent process and its various levels, collaborating with local politicians to push public policy reform, and presenting highly technical work to a diverse non-technical audience. With this in mind, I would incorporate public policy outreach into my classes, such as including public policy discussion into senior level undergraduate design courses. Additionally, I can host a new class that is taught at both the undergraduate and graduate level that addresses technoeconomic analysis and would include a considerable amount of outreach to local government so that students can see how policy shapes what can and can't be implemented commercially.

581947 Fundamental Design of Sustainable Catalytic Reactions at the Water-Energy-Food Nexus

Research Interests:

Many of the grand challenges facing society are linked by energy and resource tradeoffs. Production of fertilizers and clean water consumes vast amounts of energy. Synthesis of fuels and chemicals depends on fossil-derived compounds and clean water sources, yet reliance on fossil resources to meet energy needs generates deleterious atmospheric emissions. Combined technological approaches can be used to overcome classical challenges for problems that are at the intersection of multiple fields. Engineering the sustainable synthesis of chemicals and fuels via reactions facilitated by catalyst surfaces comprises a key vehicle for approaching these problems. Fundamental mechanistic insights are required to drive rational catalyst design and materials discovery.

Renewable reagents may be converted to energy-dense fluids in order to serve as the bridge between cheap, intermittent renewable energy sources and reliable, transportable energy delivery. For instance, CO_2 may be selectively converted to useful products using tailored catalysts. In situ surface characterization enables development of non-precious metal catalysts for efficient conversion of CO_2 to CO, a chemical building block for methanol and other chemicals. During my PhD work under Prof. Jingguang Chen, Ni/CeO_2 catalysts were found to exhibit extensive oxygen exchange between the CeO_2 support and CO_2. Effects of Ni particle size and oxidation state on reaction selectivity were determined, and non-precious NiFe bimetallic catalysts were developed in order to tune reaction selectivity to CO at high activity. In order to eliminate reliance upon fossil-derived H_2 for methanol synthesis, a one-step process to convert an underutilized natural gas co-product with CO_2 to methanol was sought. Since no thermal route is thermodynamically allowable, non-equilibrium plasma activation was employed to achieve direct alcohols synthesis. As an NSF fellow, I was able to design and build an in situ FTIR plasma-catalysis reactor in order to combine observation of surface intermediates with kinetic and isotope experiments. Coupled with kinetic modeling, the results revealed the role of the catalyst surface and the reaction mechanism. Similar techniques were employed to overcome classical limitations to ammonia synthesis. Ammonia – the major component in fertilizer and a carbon-free alternative fuel – faces century-old challenges to efficient and sustainable synthesis due to thermodynamic constraints, catalyst scaling relations, and dependence upon fossil-derived H_2. As I will discuss in my oral presentation at this year's meeting, in situ characterization provided the first direct observation of surface reactions in the presence of plasma activation for ammonia synthesis as well as evidence of reaction intermediates not observed in the thermal catalytic process.

Future plans for my independent research group will use hybrid technologies to innovate sustainable chemical synthesis processes, while pursuing fundamental understanding of mechanisms and surfaces in order to guide materials discovery. Projects may couple catalytic surfaces with thermal, electrochemical, and plasma activation, in addition to linking with membrane separation processes for water purification and possible interfacing with biological approaches. Studies will employ in situ X-ray and infrared characterizations as well as isotopic and kinetic experiments in order to gain mechanistic and materials insights. These insights can enable development of descriptors for advancing and guiding research, especially in conjunction with computational modeling. I seek an interdisciplinary environment that encourages collaboration across areas of expertise.

Teaching Interests:

I am especially interested in teaching courses in kinetics, thermodynamics, and introductory chemical engineering mass and energy balances. I would be excited to develop a course on surface reactions and kinetics with applications to experimental surface characterization techniques. I would also like to develop a course on engineering sustainability at the water-energy-food nexus, with an emphasis on large-scale energy and mass balances, approaches to engineering problem-solving, and exposure to frontiers in water-energy-food research.

My teaching experience includes training undergraduates, PhD students, and visiting scholars and professors in the laboratory; serving as a teaching assistant; and mentoring more than 30 recently immigrated students on independent research projects at a resource-strapped high school through a program I established and for which I successfully secured funding and organized a poster presentation event at Columbia University.

My goal is to guide students toward having a stake in their learning, which facilitates internalization of content and critical thinking. I employ independent projects that drive students to ask their own questions, group work to encourage engagement in shared efforts, and active learning through in-class problem solving and use of the Socratic method of teaching by systematic questioning. Given the importance of communication in scientific advancement, I also incorporate exercises to develop students' written and oral communication skills.

582064 Electrosynthesis for Sustainable Chemical Production

Research Interests:

Achieving sustainable production pathways for fuels and chemicals is one the great challenges of the 21st century. Propelled by the rapid deployment of renewable electricity generation, electrochemical technologies such as water splitting and CO_2 electrolysis have great promise for green chemical synthesis. However, significant hurdles in theoretical understanding, electrocatalyst design, and process engineering must be overcome for wide-scale commercialization of these technologies. My research interests involve addressing these challenges through a combined approach, where fundamental insights facilitate the discovery of new materials and processes, which are then integrated into pilot-scale electrochemical devices to identify engineering challenges and analyze process feasibility.

During my Ph.D. under Professor Feng Jiao, I have applied this integrated approach to the development of CO_2 electrolysis systems for the electrosynthesis of multi-carbon (C_{2+}) chemical feedstocks such as ethylene and alcohols. Early on, we completed a techno-economic analysis to identify key performance metrics required for commercialization.[1] We then developed a novel highly-porous copper electrocatalyst which enabled CO_2 conversion to C_{2+} products at industrially relevant rates.[2] After device-level testing identified key challenges in operating CO_2 electrolysis in alkaline electrolyte, we decoupled the process through the carbon monoxide (CO) intermediate and demonstrated CO electrolysis to C_{2+} products with high selectivity.[3] Additionally, we identified the role of electrolyte and catalyst for acetate formation,[3,4] and utilized these mechanistic insights to develop a new process for the electrosynthesis of acetamides.[5,6] We are currently developing novel electrolyzer architectures to produce concentrated product streams through CO electrolysis and scaling them up to stack level systems to integrate with a CO_2 electrolyzer.[7,8] During the poster session, I will expand upon the foundational experience I have gained working on these projects and present my future research plans for developing new electrosynthesis technologies.

Teaching Interests:

At Delaware, I served as teaching assistant for an engineering mathematics course where I was responsible for providing lectures to students with a wide range of experience on using Matlab to solve engineering problems. Additionally, I gave guest lectures in the graduate elective courses "Electrochemical Engineering" and "Special Topics in Energy". Given my research background, I am particularly passionate about these topics, and hope to develop courses which inspire the next generation of students to work on these critical challenges. Furthermore, I have had the pleasure of mentoring several undergraduate and visiting graduate students in electrochemical systems research.

Selected Publications & Patents:

1. **M. Jouny,** W. Luc, & F. Jiao. General techno-economic analysis of CO_2 electrolysis systems, *Industrial & Engineering Chemistry Research* 57, 2165-2177 (2018). [WoS Highly Cited Paper]
2. J. Lv,* **M. Jouny,*** W. Luc, W. Zhu, J.J. Zhu, & F. Jiao. Highly porous copper electrocatalyst for carbon dioxide reduction. *Advanced Materials* 30, 1803111 (2018)
3. **M. Jouny,** W. Luc, & F. Jiao. High-rate electroreduction of carbon monoxide to multi-carbon products, *Nature Catalysis* 1, 748-755 (2018). [Selected as Cover Story]
4. W. Luc,* X. Fu,* J. Shi, J. Lv, **M. Jouny**, B. Ko, Y.B. Xu, Q. Tu, X.B. Hu, J.S. Wu, Q. Yue, Y.Y. Liu, F. Jiao, & Y.J. Kang. Two-Dimensional Copper Nanosheets for Electrochemical Reduction of Carbon Monoxide to Acetate. *Nature Catalysis* **2**, *423-430 (2019).*
5. **M. Jouny,*** J. Lv,* T. Cheng,*, B. Ko, J. Zhu, W. Goddard, & F. Jiao. Formation of carbon-nitrogen bonds in carbon monoxide electrolysis. *Nature Chemistry* (under review)
6. F. Jiao, **M. Jouny**, & J. Lv. Electrochemical generation of valuable chemicals from carbon dioxide

and carbon monoxide. *U.S. Patent Application No. 62/757,785.* 2018
7. W. Luc, **M. Jouny**, J. Rosen, & F. Jiao. Carbon dioxide splitting using an electro-thermochemical hybrid looping strategy. *Energy & Environmental Science* 11, 2928-2934 (2018).
8. **M. Jouny,** G. Hutchings, & F. Jiao. Carbon Monoxide Electroreduction: An Emerging Platform for Carbon Utilization. *Submitted.*

*equal contribution

582098 Study of Catalytic Reactions for the Production of Fuels and Chemicals from Renewable Feedstocks

Research Interests:

My research focuses on the ***development of environmentally-friendly processes for the valorization of renewable feedstocks.*** More specifically, I am interested in fundamental studies involving the use of catalysts for the conversion of renewable feedstocks into high-value fuels and chemicals, therefore, creating new alternatives to replace conventional petrochemical routes. Specific topics of my research include the ethylene oligomerization over solid catalysts and the catalytic hydropyrolysis of biomass feedstocks. Contributions in both fields are focused on the development of reaction pathways, kinetic modeling, and the study of reaction mechanisms, in addition to design, development, and screening of new catalysts. Additionally, my contribution to the field of catalysis includes the application of supercritical fluid conditions to increase the conversion of reactants into products and, at the same time, use supercritical ethylene as an environmentally-friendly solvent to remove carbon deposits from the catalyst surface. As an Assistant Professor, I expect to continue focusing on answering fundamental questions associated with catalytic systems. There is a great interest in the conversion of non-conventional feedstock, such as lignin and food waste, to high-value fuels and chemicals, where the development of novel catalysts can contribute to making the process feasible. I am also interested in the valorization of low-value products, such as syngas and biogas, which can be a potential feedstock for the production of high-value products via catalytic conversion pathways. Finally, the knowledge that I obtained developing reaction pathways, mechanism, and kinetics for the ethylene oligomerization and pyrolysis processes can be applied not only to other chemical processes but also to a wide spectrum of areas, such as electrochemistry and biology, therefore, creating opportunities for collaboration in research projects with other faculty.

Teaching Interests:

I am interested in teaching courses related to Chemical Engineering fundamentals, such as Transport Phenomena and Chemical Reaction Engineering. As a graduate student, I worked as a Teaching Assistant (TA) for one quarter, teaching a Transport Phenomenon course for undergraduate juniors enrolled in Bioresource Science and Engineering at the University of Washington. I was responsible for teaching one recitation class every week, in which I reviewed the material covered during the previous week, and solved practice problems with the students. I also taught a full lecture, which enriched my experience as a graduate student. The experience as a TA helped to define my interest in teaching, and seek a position as an Assistant Professor after the conclusion of my Ph.D. degree.

Additionally, I had the opportunity to work with students outside the classroom, mentoring four undergraduate students, who worked under my supervision in two research projects. I believe this experience will be crucial to work with graduate students in research projects.

Successful proposal: I contributed to the writing of the proposal entitled: "The role of supercritical ethylene on oligomerization reactions with nickel-based heterogeneous catalysts," which was submitted to the American Chemical Society - Petroleum Research Fund. *Amount awarded: $ 110,000.00.*

Additional grant writing experience: I contributed to the writing of proposals submitted to NSF CAREER, DOE, and Washington Sea Grant.

Ph.D. Dissertation: "Kinetics and Mechanisms of the Ethylene Oligomerization under Supercritical and Subcritical Fluid Conditions"

Under the supervision of Dr. Fernando L. P. Resende, Bioresource Science and Engineering, University of Washington.

M.S. Dissertation: "Study of the Catalytic Reactions of Ethylene Oligomerization in Subcritical and Supercritical Media over a NiBEA Catalyst"

Under the supervision of Dr. Fernando L. P. Resende, Bioresource Science and Engineering, University of Washington.

Selected Publications:

Seufitelli, G. V. S.; Resende, F. L. P."Effect of Coke Solubility on Product Yield on the Ethylene Oligomerization under Supercritical Conditions." *Manuscript under preparation*

Seufitelli, G. V. S.; Resende, F. L. P."Kinetics of the Ethylene Oligomerization under Supercritical Fluid Conditions." *Manuscript under preparation*

Seufitelli, G. V. S.; Park, J.J.W.; Tran, P.N.; Dichiara, A.; Resende, F. L. P."Kinetics of ethylene oligomerization over Ni-H-Beta catalyst." *Under submission at ACS Catalysis.*

Chandler, D.S.; **Seufitelli, G. V. S.**; Resende, F. L. P. "A novel catalytic route for production of alkanes from hydropyrolysis of biomass." *Manuscript under preparation*

Li, Z.; Zhonga, Z.; Zhanga, B.; Wanga, W.; **Seufitelli, G. V. S.**; Resende, F. L. P. "Effect of alkali-treated HZSM-5 zeolite on the production of aromatic hydrocarbons from microwave assisted catalytic fast pyrolysis (MACFP) of rice husk". *Under submission at ACS Sustainable Chemistry & Engineering.*

Seufitelli, G. V. S.; Resende, F. L. P."Study of the catalytic reactions of ethylene oligomerization in subcritical and supercritical media over a NiBEA catalyst." *Appl. Catal. A Gen.* 2019, vol. 576, pp. 96–107.

Jan, O.; Marchand, R.; Anjos, L. C. A.; **Seufitelli, G. V. S.**; Nikolla, E.; Resende, F. L. P. "Hydropyrolysis of lignin using Pd/ HZSM-5". *Energy Fuels* 2015, 29, 1793−1800.

582107 Direct and Non-Oxidative Conversion of Methane to Value Added Products

Research Interests:

Global energy demand has been continuously increasing at a rapid rate and tremendous research is being done to develop new sources of energy supply. The low cost of natural gas (predominantly contains methane) has created an interest in its conversion to high value chemicals and fuels. This methane can be converted into valuable chemicals by both direct and indirect routes. The indirect conversion of methane by Fisher-Tropsch (FT) to methanol followed by methanol-to-gasoline (MTG) is currently a more commercially viable process than the direct routes. Direct routes have barriers to commercialization including limited conversions, poor selectivity to desired products and catalyst deactivation.

My research has provided me an excellent learning opportunity to understand and work in one of the most interesting and highly debated problems in the field of heterogenous catalysis - catalytic conversion of methane under non-oxidative conditions. My research is focused on addressing the challenges faced in two different processes - i) **Methane dehydroaromatization (MDA),** a direct route to produce aromatics, primarily benzene by non-oxidative conversion of methane and ii) **Catalytic decomposition of methane (CDM),** a technique that involves production of high purity (CO_x free) hydrogen directly from methane.

The mechanism of MDA is still in debate and this work is focused on understanding the fundamental nature of catalysts used in this reaction through decoupling of the Mo/HZSM-5 catalysts to better understand the contribution of each component. The reaction mechanism was therefore studied by activating methane over Mo/SiO_2 and Mo/HZSM-5. The fresh and the spent catalysts were analyzed using BET, SEM, XRD and TGA to understand surface properties before and after the MDA reaction. The effect of H_2 pretreatment on the nature of active site formation and on induction time for benzene formation was investigated in detail. Results showed that H_2 reduction at 750 °C helped in the formation of metallic Mo sites which turned into active Mo_2C sites during methane activation and formed ethylene and aromatics products without any induction time. The significance of zeolites (porous catalyst materials) in MDA is revealed by decoding the mechanism of MDA. Further, our research involved in improving the performance of Mo/HZSM-5 catalyst by addition of promoter metals. A second metal (Fe or K or Rh) was added to Mo/HZSM-5 and its influence on methane conversion and benzene selectivity was studied.

Another part of this research includes the effect of different catalyst materials and their lifespan on hydrogen (% yield) produced from the methane decomposition reaction. A rare mixed metal oxide supported Fe catalyst ($Fe/CeZrO_2$) was subjected to methane activation and its activity for hydrogen production was investigated. Influence of a second metal (Co & Mo) and the effect of aging of catalyst after several cycles of regeneration on hydrogen yield was also studied. Addition of Co or Mo to the Fe catalyst has increased the H_2 yield by ~35% compared to the unpromoted Fe catalyst.

Future research direction:

- Perform molecular level studies (density functional theory) to support the experimental claims on Mo and Fe based catalysts used for non-oxidative methane conversion.
- Synthesis of zeolite-based catalysts to study their catalytic properties to design novel catalysts for different reaction pathways.

Teaching Interests:

Right from the early days of my bachelor's program, teaching has always been my passion. During my Undergraduate studies, I have been a math tutor for high school students and fellow undergraduate students. This tutoring experience made me realize that teaching could be my profession. As a teacher, I believe it is my responsibility to train students to develop their critical thinking and problem-solving ability. Explaining the fundamental concepts in a simple way and having lectures in an interactive way help students to develop interest in the subject. During my PhD, I have mentored 3 undergraduate students who worked with me on my research project. In addition to providing training in technical aspects, I also guided them in developing their soft skills including creative problem solving and presentation skills.

I am interested in teaching Chemical Engineering core courses including reaction kinetics, heat and mass transfer, process calculations, fluid mechanics, crude oil refining and industrial catalysis.

Awards and achievements:

- Peer reviewed journals: 2 published (first author)
- Conference presentations: 8
- Champan scholarship award, The University of Tulsa (2017)
- Honorary mention, TU Research colloquium, The University of Tulsa (2017, 2019)
- Bellwether Fellowship award, The University of Tulsa (2018)
- Graduate student Research grant, The University of Tulsa (2018)

Peer reviewed publications:

- **Ramasubramanian, V**., Ramsurn, , & Price, G. L. (2018). Methane dehydroaromatization–A study on hydrogen use for catalyst reduction, role of molybdenum, the nature of catalyst support and significance of Bronsted acid sites. *Journal of Energy Chemistry*. (https://doi.org/10.1016/j.jechem.2018.09.018)
- **Ramasubramanian, V**., Lienhard, D.J., Ramsurn, H., & Price, G. L. (2019). Effect of addition of K, Rh and Fe over Mo/HZSM-5 on methane dehydroaromatization under non-oxidative conditions. *Catalysis letters*. (https://doi.org/10.1007/s10562-019-02697-8)

Selected presentations:

- The Role of Molybdenum during the Conversion of Methane to Liquids Under Non- Oxidative Conditions, **Vaidheeshwar Ramasubramanian.**, et al., Annual AIChE meeting, Minneapolis MN, 2017
- Effect of Addition of Promoters to the Mo/HZSM-5 during the Non-Oxidative Conversion of Methane to Aromatics, **Vaidheeshwar Ramasubramanian.**, et al., Annual AIChE meeting, Pittsburgh, PA, 2018

582174 One-Step Synthesis of Oxide Catalysts Using Aerosol Reactors for Environmental Applications: Theoretical and Experimental Study

Research Interests: I am interested in the active phases of catalysts and catalytic performance on those active phases. During my PhD, I have been working on catalytic O_2 removal from oxy-coal combustion system by using metal oxide catalysts and noble metal doped metal oxide catalysts. Here is the detail of my PhD research:

CO_2 emission from fossil fuel combustion systems is one of the major contributors to global warming. Oxy-coal combustion has been developed as a promising technique to effectively capture CO_2 in power plants. Because pure O_2 is used to combust coal, a concentrated CO_2 stream (~63%) can be achieved in an exhaust. The concentrated CO_2 stream can be further utilized for enhanced oil recovery (EOR), however, the O_2 level in the stream (~ 3%) should be reduced to satisfy the requirement for EOR (~ 100 ppmv) [1]. To resolve this problem, catalytic O_2 removal with CH_4 has been recently applied with noble metal doped metal oxide catalysts and high O_2 conversions were reported [2, 3]. Although high O_2 conversions were obtained, there is still debate on the active phase of metal oxide catalysts and kinetics during O_2 removal reaction. For enhancing the catalytic performance, a fundamental understanding of the active phase of catalysts and kinetics during O_2 removal reaction is required.

Metal oxide catalysts and noble metal doped metal oxide catalysts are synthesized using aerosol reactors in which a continuous one-step process can be applied and so additional processes such as drying and calcination are not necessary. Active phases of the synthesized catalysts are investigated based on both analytical (catalytic properties) and experimental results (kinetic characteristics). A correlation between the active phase of catalyst and reaction rate constant is obtained. The catalytic properties are optimized by combining theoretical study and experiments. An autocatalytic effect of CO_2 (balance and product) on O_2 removal is also found, which is supported by density functional theory calculations and experimental results. These findings help me to fully understand both active sites of catalysts and kinetics during O_2 removal.

References:

[1] DOE/NETL, Quality guidelines for energy system studies: CO_2 impurity design parameters, (2013).

[2] Q. Zheng, S. Zhou, M. Lail, K. Amato, Oxygen removal from oxy-combustion flue gas for CO_2 purification via catalytic methane oxidation, Ind Eng Chem Res, 57 (2018) 1954-1960.

[3] A. N. Kuhn, Z. Chen, Y. Lu, H. Yang, Sequential oxygen reduction and adsorption for carbon dioxide purification for flue gas applications, Energy Technol-Ger, (2018).

Teaching Interests: I am interested in courses related with chemical kinetics and catalysis.

581506 Engineering the Catalytic Environment: Synthetic, Mechanistic, and Spectroscopic Approaches for Developing Design Principles

Research Interests:

A grand challenge that society faces is satisfying the demand for energy and chemicals using efficient and sustainable carbon resources. Catalysis is inevitable in the solution of this problem; however, viability hinges on the intimate relationship between activity and selectivity. My research will focus on understanding the reaction mechanism for desired and undesired pathways and identifying and manipulating the catalytic active site and reaction environment. My group will combine the controlled synthesis of porous materials (e.g., zeolites and zeotypes), modern *in situ* and *in operando* spectroscopic techniques (e.g., Raman, IR, XAS, neutron), and *intrinsic* kinetic measurements to understand how complex interactions at the catalytic interface can affect the rates and selectivities for a given reaction pathway. To minimize our carbon footprint and contribute to lasting and sustainable solutions, my group will specialize in the conversion of greenhouse gases and the upgrading of emerging (e.g., shale gas) and renewable (e.g., bio-derived) resources. Initial research areas will include: 1) understanding the effects of confinement on thermal and electrochemical CO2 reduction ; 2) the upgrading of bio-derived platform chemicals through thermal- and electrochemical pathways; 3) the development of catalytic systems for low temperature ammonia synthesis; and 4) the discovery of tandem catalytic systems for light alkane upgrading.

Research Experience:

Methods for Understanding Inner- and Outer-Sphere Interactions at the Liquid-Solid Interface: Quantitative Structure-Activity Relationships for Oxidation Reactions; *Department of Chemical and Biomolecular Engineering, University of Illinois at Urbana-Champaign* (advised by David W. Flaherty)

Catalytic reactions that proceed at the liquid-solid interface formed between solvent mixtures and the surface of a heterogeneous catalysts possesses all of the complexities of catalysis at the gas-solid interface but are further complicated by interactions with solvent molecules. My Ph.D. research has focused on the development of quantitative relationships that describe the complex interactions between solvent structures, confining surfaces, and reactive intermediates during alkene epoxidation. Through the combination of controlled zeolite synthesis, *in situ* spectroscopic techniques, and rigorous kinetic measurements, we made several significant contributions to the field of liquid-phase oxidations:

- The functional Lewis acid strength of early transition metal-substituted zeolites is intimately correlated with the electrophilicity of reactive species formed upon activation of hydrogen peroxide. The difference in Lewis acid strength leads to a 100,000-fold difference in rates of alkene epoxidation (and sulfide oxidation) and 100-fold difference in selectivity among groups 4 and 5-substituted zeolites.
- The presence of a confining void (i.e., within a zeolite) selectively stabilizes styrene epoxidation transition states and leads to a 20-fold increase in rates and selectivities. The pore does not, however, influence the electronic structure of the active site or reactive intermediate.
- Polar functional groups within a micropore can spontaneously nucleate small clusters of H_2O that interact with surface intermediates (e.g., transition states, adsorbates), which leads to a 100-fold change in the rates and selectivities of *n*-alkene epoxidation.

Collectively, our findings have shown experimental evidence for the free-energy relationships that describe these complex interactions and have shown that orthogonal modifications of the solvent, surface, and active site can all be used to control catalysis at the liquid-solid interface.

My postdoctoral research will be performed with an expert in electrochemistry and electrocatalysis. This is by design, so that my independent career research can couple modern thermal- and electro-catalytic methods to study and why reactions proceed in response to chemical- and electrical stimuli.

Selected Publications:

- **Bregante, D.T.;** Johnson, A.M.; Patel, A.Y.; Ayla, E.Z.; Cordon, M.J.; Bukowski, B.C.; Greeley, J.; Gounder, R.; Flaherty, D.W.; *Journal of the American Chemical Society* **2019**, *141*, 7302-7319.
- **Bregante, D.T.;** Thornburg, N.E.; Notestein, J.M.; Flaherty, D.W.; *ACS Catalysis* **2018**, *8*, 2995-3010.
- **Bregante, D.T.;** Patel, A.Y.; Johnson, A.M.; Flaherty, D.W.; *Journal of Catalysis* **2018**, *364*, 415-425.
- **Bregante, D.T.;** Flaherty, D.W.; *Journal of the American Chemical Society* **2017**, *139*, 6888-6898.
- **Bregante, D.T.;** Priyadarshini, P.; Flaherty, D.W.; *Journal of Catalysis,* **2017**, *348*, 75-89.
- **Bregante, D.T.;**[‡] Wilson, N.M.;[‡] Priyadarshini, P.; Flaherty, D.W.; *Catalysis*, **2017**, *29*, 122-212.

Selected Awards:

- University of Illinois Dissertation Completion Fellowship (2019 – 2020)
- National Defense, Science, and Engineering Graduate Fellowship (2016 – 2019)
- ACS Graduate Student Award in Environmental Chemistry (2018)
- School of Chemical Sciences Graduate Teaching Award (2018)
- Frederic and Edith Mavis Future Faculty Fellowship (2017 – 2018)
- AIChE Catalysis and Reaction Engineering Travel Award (2018)
- Richard J. Kokes Travel Award (2017)
- Samuel Parr Graduate Fellowship (2015 – 2016)

Teaching Interests:

Teaching, both in the classroom and as a mentor in the lab, is a major driving force for my desire to become a professor. As a chemical engineer by training, I am excited and qualified to teach core undergraduate coursework in chemical kinetics and reactor design, thermodynamics, fluid mechanics, heat and mass transfer, and separations, in addition to graduate-level applied mathematics, transport phenomena, and chemical kinetics. I also plan to leverage my research experience and personal interests to develop and teach a senior-undergraduate/graduate elective on the various facets of catalysis (i.e., heterogeneous, homogeneous, electro, and enzyme), which will complement a graduate-level chemical kinetics course.

My teaching experience extends beyond the classroom and into the laboratory, as students and postdoctoral researchers within my research group will utilize an interdisciplinary approach to problem solving that leverages techniques and thought processes from chemistry, materials science, and physics.

Teaching Experience:

- Kinetics and Reactor Design (Olivet Nazarene University; Spring 2019) – *Ad hoc* Instructor
- Chemical Kinetics and Catalysis (Illinois; Fall 2018) – Teaching Assistant and Guest Lecturer
- Mass Transfer and Operations (Illinois; Fall 2017) – Teaching Assistant and Guest Lecturer
- Chemical Structure and Reactivity (UC Berkeley; Summer 2014) – Teaching Assistant
- Organic Chemistry I (UC Berkeley; Fall 2013) – Teaching Assistant

Service:

I have served as an organizer, abstract reviewer, and session chair for the Catalysis Club of Chicago's 2019 symposium and the 2019 AIChE National Conference (Section 20a). Aside from this service, I am passionate about developing and encouraging young scientists to pursue careers in STEM to solve some of the world's most-important problems. A glaring problem within most STEM fields, is the large disparity in the representation of different groups (e.g., gender and racial), which suggests that, as a society, we are not developing or utilizing the population to its full potential. Addressing this issue requires that STEM be made tractable and *fun* at an early age so that these traditionally-underrepresented groups are encouraged to pursue it. Throughout my undergraduate and graduate career, I have participated in numerous student groups (Berkeley Engineers and Mentors; Girls Adventures in Mathematics, Engineering, and Science) whose goal is to promote science and engineering to young students. In my independent career, I will also serve as a faculty mentor to similar groups (or encourage my students to start one), so that we may increase the pool of young scientists. I also routinely serve on student panels for undergraduate students who are interested in pursuing graduate studies.

At Illinois, I have mentored 4 undergraduate students through a NSF-funded REU that aims to increase female participation in research. Additionally, I have served as a research mentor in the summer pre-doctoral institute here at Illinois, whose mission is to immerse incoming graduate students from underrepresented minorities in research with a senior graduate student to aid in navigating a Ph.D. program. As a professor, I will make it a priority to host or participate in similar programs to develop researchers from all walks of life. Overall, I hope to continue my engagement-with and development-of the scientific community.

For additional information, please see Breganteresearch.org

575330 First-Principles Approaches for Catalyst Design: Novel Descriptors and Strategies for Materials Discovery

Background:

Catalytically driven chemical processes are ubiquitous in a variety of chemical industries and renewable energy technologies. The rational design of catalysts to optimize for activity, selectivity, stability, and cost has the potential to dramatically improve many of these processes. In the past two decades, the ability to understand and design catalyst materials *in silico* using computational electronic structure theory has revolutionized the field of catalysis. Descriptor based approaches to catalyst design have successfully been used to both understand experimental trends, and to design and discover improved catalysts. In one example, the appropriate descriptor for electrochemical reaction energetics is identified to be not the work function, but the total surface charge, solving a standing problem in the field and allowing for investigation of alkaline conditions for the first time. In a second example, a machine learning approach was used to identify important descriptors for dissociation reactions of diatomic gas phase molecules on metal surfaces.

Research Interests:

A critical challenge in computer guided understanding and design of catalysts is the identification of correct descriptors for activity. My research focuses on understanding catalytic processes through the use of first-principles (*ab initio*) electronic structure theory in combination with microkinetic modeling. From these insights, I will identify new descriptors for surface reactivity, particularly for electrochemistry where fascinating new behaviors have been observed that cannot be explained by the traditional d-band model.

Due to the ever increasing availability and power of supercomputing resources, over the past decade there has been an explosion of data in the space of computational materials science. By leveraging the rapid growth of materials databases, I will use a machine learning approach not only to identify new materials, but to bring new physical insights to surface reactivity descriptors. *These descriptors for activity can be leveraged to screen for catalysts and design interfaces in a rational bottom-up approach.*

My future work will therefore have three primary directions:

1. leveraging understanding of electronic structure theory to identify new descriptors for catalyst activity, especially for electrocatalytic processes
2. machine learning approaches for identification of not only new materials, but new physics
3. application of identified descriptors to screen catalysts for activity to a variety of reactions relevant for sustainable energy storage/conversion and chemical synthesis.

Teaching Interests:

I am excited to teach and develop courses at both the undergraduate and graduate level. I have found my experiences teaching to be a gratifying experience on many levels. With my background and teaching experience, I would be confident teaching any fundamental chemical engineering course. However I have a particular interest in teaching courses that are synergistic with my research field and expertise. In a course such as thermodynamics, kinetics, or mass and energy balances, I would easily be able to relate class concepts to real world applications.

I have a background in curriculum development as a graduate student, and I would especially be interested in leveraging this experience to develop a course titled Fundamentals of Heterogeneous Catalysis. This course would be designed to be offered to upper level undergraduate students and first or second year graduate students, but could be easily modified to be an upper division graduate level course. Ultimately, my goal and commitment when teaching and mentoring is to bring students to a point where they have confidence in their understanding and abilities that will serve them throughout their life.

583538 Novel Approaches to Biomass Upgrading for a More Sustainable Chemical Industry

Research Interests:

Biomass, particularly forest products, constitute one of the most plentiful sources of renewable carbon. Despite the high interest, many challenges remain for biomass to replace petroleum as the dominant source of fuels and chemicals. One of these challenges is the inherent complexity and diversity of the naturally occurring feedstock, which will require additional processing necessary to produce a uniform product. As such, advances in reaction engineering, process intensification, and novel applications and materials are necessary to make the fractionation, depolymerization, and valorization of lignocellulosic biomass efficient and economically viable. From this, I have identified three primary research interests: (1) the design of novel reactor/analytical methods for probing the fundamental reactions occurring during catalytic reactions with biopolymers and biomass derived chemicals, (2) the engineering of reactor systems that allow for process intensification for more efficient upgrading, and (3) the design and production of new base chemicals and materials that take advantage of the inherent functionality of biomass derived compounds.

My current research experience focuses broadly around the valorization of lignin (one of three main biopolymers of woody biomass, along with cellulose and hemicellulose). This includes a study on the impacts that industrially relevant isolation methods have on the structure of the polymer. More core to my Ph.D. thesis, I have studied the fundamentals and applications of mechanocatalysis for lignin depolymerization. These reactor systems use mechanical energy (e.g. collisions in a ball mill) to drive catalytic reactions. The appeal of these systems is their ability to perform reactions without solvents, operate under ambient conditions, and be implemented industrially with minimal infrastructure. Fundamental studies have focused on modeling reactive conditions in a ball mill reactor as well as understanding the transformation of the catalyst during ammonia synthesis. Using mechanocatalysis in a more applied setting, I have focused on studying various reaction systems for solvent free lignin depolymerization and product upgrading.

Teaching Interests:

Over the course of my research as a graduate student, I have mentored an undergraduate student and a master's students. While this number is on the low side, working so closely with them, somewhat paradoxically, allowed them to work much more independently. Each had a well-defined project which they largely directed themselves, and when they did need my input and guidance, I was able to provide the highest levels of my attention. The projects these two have worked on have have resulted in paper being submitted and a communication in the works. Both students are on their way to being Ph.D. candidates and conducting their independent research. This experience and the success of these two students have highlighted the importance creating a framework that lets students teach themselves. I plan to take this revelation with me, as a professor, when I will be designing my own courses and instructing Ph.D. students in my lab.

584454 Catalyst Design with Atomic Precision for Fuel Gas Processing and Pollution Purification Reactions

Research Interests:

Heterogeneous catalysis is a key enabler in identifying and navigating sustainable energy flow through our fragile ecosystem. For almost a century, the exploration of cost-effective supported metal catalysts is at the heart of both fundamental and industrial catalysis research. In order to allow such a large family of materials to match the elegant and promising chemistry of their corresponding homogeneous catalytic prototypes, perhaps the ultimate design goal of supported metal catalysts is to simultaneously maximize the dispersion of the supported metals and to display desired intrinsic chemistry per supported metal atom. Working towards this design goal presents both daunting tasks and new opportunities.

The emerging topical research area of single-atom catalysis is a subset of these efforts, and I have been very fortunate to participate in the journey of the birth, debate, expansion, and evolution of the single-atom catalysis research under the guidance of Prof. Maria Flytzani-Stephanopoulos at Tufts as a Ph.D candidate and at GM R&D as an independent researcher. Single-atom catalysis dramatically reduces the usage requirements of expensive and rare metals by stabilizing the supported metal atoms in a fully dispersed state as isolated bonded species that serve as active sites on various carrier (support) materials. These single-atom catalytic materials can also serve as appropriate platforms for many fundamental studies, where the detailed electronic structure, coordination condition, adsorption/desorption behavior, and broader scopes of geometry and ligand effects to the supported metal species can now be much more explicitly probed than in many previous heterogenous catalysis systems. My research work has focused on synthesizing, characterizing, and optimizing various supported single-atom catalysts for industrially important reactions, such as the water-gas shift reaction, self-coupling reactions of methanol, alkane and alcohol dehydrogenation, purification of airborne pollutants, and oxygen-reduction reaction in fuel cells. The main contributions are:

1. Unified the once elusive fact that similarly structured single-atom Pt_1- and Au_1-$O(OH)_x$species are the true catalytic centers for the low-temperature water-gas shift reaction on various support oxides, and thus broadened the material choices of making single-atom catalysts;
2. Invented an environmentally benign and facile synthesis route for single-atom-gold oxo-clusters in either supported or unsupported forms to selectively catalyze hydrogen upgrading and methanol self-coupling reactions, and thus bridged the gap between homogeneous and heterogeneous catalysis of the gold-oxo species;
3. Tackled the limitation of lacking neighboring metal atoms and their associated activity in the classic single-atom catalyst system by creating paired Pt-O-Pt ensembles while retaining 100 % metal dispersion. This new group of catalysts finds an alternative pathway for oxygen activation, which leads up to 1000 times higher activity than the Pt_1 single atoms for several emission control reactions.
4. Developed, patented, and licensed technologies of making sinter-resistant precious metal catalysts. The implementation of the commercialized technology grown from fundamental studies leads to about $0.2 billion savings of precious metal per year for General Motors.

The above research output extends across the clean energy generation sector as well as the energy-consumption processes that target zero emissions. The successful experience in these areas creates a foundation to construct my future research framework in designing various supported metal catalysts that aims to unlock new activity and selectivity for the utilization of several key chemical commodities, such as hydrogen, nitrogen, methanol, light olefins, oxygen and water. The new catalysts to be developed will inherit the design philosophy of atomic precision in my previous work. More importantly, I will utilize various single atoms, not necessarily as direct catalytic centers, but as building blocks to subsequently assemble catalytic centers comprising purposefully paired atoms. Understanding how the various support properties respond to these atomic-level changes of the catalytic centers in given reactions will be another important new research domain. Indeed, well-defined multi-atom catalytic species have been discovered on ideal crystal surfaces and under ultrahigh vacuum conditions. With the fast development of single-atom catalysis for practical catalytic reactions, now is the time to bridge these once isolated research areas and to bring the atomically precise multi-atom catalytic species into real-world engineering operations. The above research interests are also well aligned with the following technology focus areas of the nation:

1. Develop new catalytic systems to reduce hazardous molecules and greenhouse gas emissions that are hard to be regulated today;
2. Search for durable and affordable fuel cell catalysts to boost the competitiveness of the H_2-powered zero-emission vehicles;
3. Seek breakthrough catalytic technologies to stimulate the shale gas economy by making value-added products from saturated light alkanes;
4. Develop sustainable catalytic pathways of making renewable fuels by designing inorganic-organic hybrid catalysts that utilize environmentally benign energy input such as heat, photon, electricity, plasma and their combinations.

My independent grant application experience includes the *International Precious Metal Institute Student Research Award* in 2014 and the *Internal Start-Up Research Award by General Motors* in 2016. The interdisciplinary catalysis work has also stimulated my collaborations with synchrotron, microscopic and computational modeling experts from national laboratories and universities. I have greatly benefited from these inclusive and collaborative research interactions, and I will continue practicing such a spirit in my research collaborations with future colleagues and friends.

Teaching Interests:

Teaching has been an integral part of my career since 2010. I believe an important mission of education should provide the guidance to help our students become academically competent individuals as well as caring and inclusive human beings.

Besides working as a teaching assistant at Tufts, I have taught as an adjunct faculty in chemistry at South University in Michigan, and have mentored two visiting Ph.D students, three Master students, and three summer-intern students to carry out their research work in my laboratory at General Motors. I help the students, including those from underrepresented groups, cultivate their interest to learn, desire to work hard, and ambition to succeed, which in turn ignites my passion and commitment for higher education. As an extension from the in-class activities, my research work in catalysis serves as a versatile vessel for students to understand and practice their learning in several core courses of chemistry and chemical engineering, such as General Chemistry, Analytical Chemistry, Physical Chemistry, Materials Science, Reaction Engineering, and Transport Phenomena. With the university's support in the future, I am very interested in sharing my education and research resources in the form of online academic videos and 3D virtual laboratory operation experience. This will also increase the engagement of the STEM field within the community it is serving, and sparkle the interest among the perspective students.

My education background in both chemistry and chemical engineering and work experience in both academia and industry have prepared me as a future educator who embraces diversity and equity. Conveying these values to our students is another critical mission of higher education. As the serving president of the Michigan Catalysis Society, where many college and graduate students gather at the monthly technical seminars, I have been taking initiatives to advocate emerging research domains and achievements from early-career professors, women scientists, and industrial researchers. These events, along with the associated social media channels, have become an important venue for new generations of students to find career role models in their field of research, as well as to celebrate and appreciate research findings from diverse fields.

Selected Publications:

Corresponding-Author Research Reports

1. "Surpassing the single-atom catalytic activity limit through paired Pt-O-Pt ensemble built from isolated Pt_1 atoms", *Nature Communications* **2019**, 10: 3808.
2. "Single-site Pt/La-Al_2O_3stabilized by barium as an active and stable catalyst in purifying CO and C_3H_6 emissions", *Applied Catalysis B: Environmental* **2019**, 244, 327.

First-Author Research Reports

3. "Single-atom-gold oxo-clusters prepared in alkaline solutions catalyze the heterogeneous methanol self-coupling reaction", *Nature Chemistry*, In Press.
4. "A common single-site Pt(II)-$O(OH)_x$- species stabilized by sodium on active and inert supports catalyzes the water-gas shift reaction", *Journal of the American Chemical Society* **2015**, 137, 3470.
5. "Catalytically active Au-$O(OH)_x$- species stabilized by alkali ions on zeolites and mesoporous oxides", *Science* **2014**, 346, 1498.
6. "Atomically dispersed Au-$(OH)_x$ species bound on titania catalyze the low-temperature water-gas shift reaction", *Journal of the American Chemical Society* **2013**, 135, 3768.
7. "Effects of CO_2 and steam on Ba/Ce-based NO_x storage reduction catalysts during lean aging", *Journal of Catalysis* **2010**, 271, 228.
8. "Pd-support interaction-defined selective redox activities in Pd−$Ce_7Zr_{0.3}O_2$−Al_2O_3 model three-way catalysts", *Journal of Physical Chemistry C* **2009**, 113, 12778.

Keynote Talks

9. Aftertreatment PGM Catalysts with Atomic Precision for Optimal Reaction Efficiency, *2018 CLEERS Workshop*, 9/18/2018.
10. Atomically Dispersed Precious Metal Species on Various Oxide Supports for Catalytic Hydrogen Upgrading and Emission Control, *Microscopy and Microanalysis 2016 Meeting*, 07/27/2016.

581794 Molecular Simulations of Interfacial Dynamics in Biological Systems

Research Experience:

My background is in Chemical Engineering and my research expertise is in Computational Biophysics and Molecular Biology. The focus of my work is to understand the molecular interactions that govern membrane dynamics in terms of lipid-lipid and lipid-protein interactions. The need for accurate lipid composition in membrane modeling has become more apparent in recent years, especially in the context of protein-lipid interaction for cell homeostasis or molecular mechanisms of disease. During my graduate training, I used molecular dynamics simulations to study complex membranes models of eukaryotic cells and their interaction with peripheral membrane proteins. My systems were the first to model eukaryotic membranes with at least 7 lipid species at the all-atom level; these studies showed the diversity of lipid components does affect membrane properties that in turn modify membrane-protein interactions. The common topic among my graduate research projects was the study of molecular interactions at biological interfaces; however, I also explore the behavior of water-oil interfaces in the presence of small alcohols that disrupt said interface. I had the opportunity to work and publish with experienced scientists not only in the computational field, but also experimentalists both in the U.S. as well as international collaborations.

During my postdoctoral training, I am applying my expertise on cell membrane modeling to study the dynamics of assembly and membrane remodeling of viruses from the molecular point of view. I am quantifying protein-protein interactions on the surface of the plasma membrane and predicting how membrane lipids reorganize in response to those interactions. I am expanding my simulation skills to advanced techniques such as enhanced sampling and coarse-grain modeling. The rich background diversity of my current group has indeed enhanced my postdoctoral experience by giving me the opportunity to learn from experts in coarse-graining, machine learning, and the developers of enhance sampling techniques such as advanced metadynamics simulations.

Postdoctoral Projects:

1. "Multi-scale modeling of protein binding and aggregation of the matrix (MA) domain of HIV-1. Elucidating the role of lipidated tails in peripheral membrane proteins on membrane dynamics and lipid sorting."
2. "Next-generation coarse-grain modeling of relevant biomembrane lipids. Developing models for Cholesterol and charged lipids."

Advisor: Prof. Gregory A. Voth, Department of Chemistry, University of Chicago, Chicago, IL

PhD Dissertation:

"Computational studies of membrane models and their interaction with a peripheral protein in yeast, & disruption of the water-oil interface by a hydrotrope."

Advisor: Prof. Jeffery B. Klauda, Chemical and Biomolecular Engineering Department, University of Maryland, College Park, MD.

Successful Proposals: Wrote and directed a proposal for XSEDE Educational Allocation. Assisted in writing and gathering preliminary results for NIH, XSEDE, and Anton/Anton2 (Pittsburgh Supercomputing Center) applications.

Research Interests:

Protein-lipid interactions are key for cell homeostasis and growth, and are involved in key steps of disease progression. Over the past ten years there has been more interest on the dynamics of peripheral membrane proteins with membrane lipids. Molecular dynamics (MD) simulations generate a trajectory based on the thermodynamics and forces that act on a given system. MD is a powerful tool to understand the behavior of a system, interpret experimental data with observations at the atomic/molecular level, and suggest or eliminate experimental targets. Increasing computational power allows us to simulate protein-membrane interactions at relevant length and time scales to understand the inner workings of a given protein and propose mechanisms of action of cellular processes.

As a faculty member I plan to lead a computational research group in molecular biophysics, promote participation of women and minorities in science, encourage interdisciplinary research stablishing strong collaborations with experimentalists, and strengthen existing connections with international scientists to develop international collaborations. The specific research topics in my lab will revolve around membrane modeling, protein-protein and protein-membrane interactions in the context of multi-scale study of liver disease, and molecular biology of onco-viruses. In the future, I would like to incorporate machine learning into the design of simulation experiments and data analysis.

Teaching Interests:

During my years at the University of Maryland as a graduate student, I worked with and mentored both high school and undergraduate students as well as served as a teaching assistant in multiple engineering classes over 4 terms. I am invested in becoming a good mentor and role model for my students; I have participated in seminars and workshops on course design and student-centered teaching. During my postdoctoral tenure, I completed the Certificate for College Teaching at the University of Chicago, for which I took classes in pedagogy, and design and taught a hands-on workshop in molecular dynamics along with a journal club geared towards incoming graduate students in the Biophysical Sciences. In particular, I would like to teach thermodynamics, statistics, introduction to chemical engineering, and design elective courses that introduce and promote computational methods in molecular biology.

Selected Publications (Citations: 683. Google h-index: 7)**:**

1. **Monje-Galvan, V.**; Warburton, L; Klauda, J.B. Setting-up all-atom molecular dynamics simulations to study the interactions of peripheral membrane proteins with model lipid bilayers in *Methods in Molecular Biology Series. Intracellular Lipid Transport.* Guillaume Drin, Ed. Springer, (2019)
2. Leonard, A.N.; Wang, E.; **Monje-Galvan, V.**; Klauda, J.B. Developing and testing of lipid force fields with applications to modeling cellular membranes. *Rev.* DOI: 10.1021/acs.chemrev/8b00384 (2019)
3. **Monje-Galvan, V**.; Klauda, J.B. Preferred binding mechanism of Osh4's amphipathic lipid-packing sensor motif, insights from molecular dynamics. *Phys. Chem. B.*, 122(42): 9713-9723 (2018)
4. **Monje-Galvan, V.**; Klauda, J.B. 2016. Peripheral Membrane Proteins: Tying the Knot between Experiment and Computation. *BBA: Biomembranes,* 1858: 1584-1593 (2016).
5. **Monje-Galvan, V**.; Klauda, J.B. 2015. Modelling Yeast Organelle Membranes and How Lipid Diversity Influences Bilayer Properties. 54(45), 6852-6861 (2015).
6. Wu, E.L.; Cheng, X.; Jo, S.; Rui, H.; Song, K.C.; Davila-Contreras, E.M.; Qi, Y.; Lee, J.; **Monje-Galvan, V**.; Venable, R.M.; Klauda, J.B.; Im, W. CHARMM-GUI Membrane Builder toward Realistic Biological Membrane Simulations. *J. Comput. Chem.* 35(27), 1997-2004 (2014).

581921 Reproduce - Remediation and Production Using Computational and Electrochemical Approaches

Summary

Over my career as a graduate student, research scientist, business development professional and lecturer, I have been involved in projects that are at the heart of global megatrends (urbanization and climate change) and how science and technology can ease the strain of these megatrends on resources (water availability, energy demand and food). My interest in a faculty position stems from the belief that this role is where solutions can be truly initiated through primary research and developed via collaborations with public and private entities. This interest is supported by experience in computational modeling, electrochemical conversion and polymer synthesis in both academic and industrial settings (see research interests for past, current and future work). This interest in finding solutions is balanced by providing adequate mentorship to the next generation of scientists through education, advising and professional development (see teaching interests for philosophy and proposed courses) and generating the excitement to work on these solutions. This is the foundation upon which my future lab – REPRODUCE – will be built.

Research Interests:

I have specific research experience in solid oxide fuel cells, electro-catalysis of nitrogenous compounds (ammonia and urea), phenolic resin synthesis, composite fabrication and nutrient recovery. These projects were executed using a variety of tools including density functional theory, flowsheet modeling and wet chemistry. As a future faculty member, my research will explore biomass remediation and phosphorous recovery. I will briefly highlight these topics below and look forward to a discussion at the conference. Agencies that would be interested in funding these potential projects include Department of Energy – Bioenergy Technologies Office, National Science Foundation, Department of Energy – Office of Science, and Department of Agriculture – National Institute of Food and Agriculture. I have co-written proposals to these agencies previously and I am familiar with each agency's process.

Biomass Remediation: Plant biomass – a complex mix of components – is broadly sub-divided into cellulose, hemicellulose and lignin macromolecules. The lignin fraction is the most underutilized biomass component and this macromolecule is largely made up of aromatic and aliphatic sub-units linked together through direct bonds on the aromatic ring, carbon-oxygen-carbon linkages between aromatic rings and phenol-type linkages. Aromatic compounds are molecules/building blocks produced commercially for use in various applications including adhesives (phenol-formaldehyde resins), foams (polyurethanes) and propulsion (20% v/v of jet fuel) and these compounds are typically generated from crude oil. Lignin represents an opportunity for generating these aromatic compounds from a renewable source, but introduces several challenges including a heterogeneous molecular structure due to biomass source and separation techniques as well as highly oxygenated functional groups leading to low energy content and downstream transportation issues (pipeline corrosion). Projects I am interested in developing in this area include: (1) predictive modeling of heterogeneous electrocatalytic lignin depolymerization, (2) electrocatalyst validation and optimization via in situ experimental analyses and (3) macromolecular synthesis using lignin-based building blocks.

Phosphorous Recovery: Phosphorus is an essential mineral for animal nutrition, and plays key roles in development and maintenance of skeletal tissue, acid-base balance, energy utilization and other factors. Animals
deficient in this macronutrient can suffer from reduced economic performance and resistance to disease (e.g., reduced egg yield in hens, lower weight in cattle and hogs). This nutrient is also an essential element for plant growth and has to be provided to soil via application of inorganic fertilizer or manure. This practice improves soil quality by adding important nutrients, with Concentrated Animal Feeding Operations (CAFOs) being the largest users of land-applied manure as a management practice for animal waste. Although CAFOs are subject to state and federal regulations requiring approved Manure Management Plans, nutrient runoff is still a widespread issue. Notwithstanding this nutrient's critical importance,

phosphorous (along with nitrogen) from animal manure is a primary contaminant of surface and groundwater leading to harmful algal blooms. Therefore, the reduction of P and N present in manure and the recovery of these nutrient species could serve as a viable approach to reducing environmental impact of runoff while also providing provide an additional source of revenue for farm operations. Typical methods for P recovery as a solid use chemicals such as sodium hydroxide and magnesium chloride, which raise the cost of the recovery process. An alternative approach would use electrochemical oxygen reduction to generate the basic environment required for P recovery as solid struvite. Projects I am interested in developing in this area include: (1) thermodynamics and kinetics of electrochemical P recovery via anode dissolution, (2) process design and optimization for farm-scale phosphorous recovery and (3) comparative techno-economic analyses of phosphorous recovery using wet chemistry and electrochemistry

Teaching Interests:

In the past, I have worked as a college tutor (over 8 years) and an MCAT instructor (1 year). I am currently teaching a "Principles of Engineering Materials" Course at Ohio University and have been the lecturer for this class since Fall 2017, with an average class size of 48 students and average instructor evaluations of 3.6 out of 4.

My teaching philosophy is centered on the idea that "*Learning is initiated by listening, furthered by interrogation and crystallized by doing.*" This is an approach that I apply to mentoring, coursework and professional development. Although most of the learning process is based on the student's actions, the teacher is the initiator and eventual catalyst for this learning process. Therefore, a teacher that is approachable, available and reachable coupled with a conducive environment that supports the different learning methods and backgrounds of each student are crucial elements of the process (see Diversity and Inclusion section below). This should clearly come across to the students in the classroom setting, small group sessions and/or in 1-on-1 conversations. Establishing a conducive environment at the beginning of the relationship (teacher-student, mentor-mentee or advisor-advisee) and maintaining this environment during the relationship allows the students to establish trust and encourages their participation in the learning process. At that point, I can then begin to challenge them in my role as teacher/mentor/advisor to take further responsibility in the questioning and doing phase of the learning process.

In the future, I would be interested in teaching classes that my training (as a chemical engineer and chemist) has prepared me for including: Mass and Energy Balances, Materials Science and Engineering, Chemical Reaction Engineering, Thermodynamics, Kinetics, Transport Phenomena, Engineering Mathematics, General Inorganic Chemistry, Physical Chemistry and General Physics. In addition, I look forward to creating elective courses based on my research interests including Computational Materials Modeling, Molecular Thermodynamics, Microkinetics, Electrochemical Engineering Principles etc. In creating these elective courses, I will focus on unique course objectives so as not duplicate other coursework's efforts.

Diversity and Inclusion (Brief Comment):

As a potential faculty member and a former international student, I tend to draw on my experience when I think about diversity and inclusion and the impact of my new environment on me (adapting) as well as my impact on that new environment (influencing). The history of physical sciences has shown that under-representation in the field is not always due to a lack of capable participants, but a leaky pipeline that does not often prioritize a combination of recruitment, integration and retention. Ensuring diversity and inclusion in STEM begins with creating multiple entry-points into the field (recruitment/engagement). Some of the avenues to accomplish this are through outreach at national conferences as well as engagement with junior chapters of minority profesional organizations such as the National Society of Black Engineers. The necessary next step after recruitment is furnishing individuals with the tools required to succeed in the new environment (integration). This could be a combination of creating avenues for student engagement with faculty (outside of lecture hours) and professional development opportunities offering training in soft-skills. Finally, an often-forgotten element of inclusive excellence is the "art of the follow-up" to ensure progress (retention). This is perhaps the most difficult aspect due to the dynamics of career decisions - students changing majors, faculty members changing schools etc. My role will be to foster an environment that prioritizes this inclusive excellence via the 3 elements I mentioned combined with my personal experiences to strengthen and grow the REPRODUCE lab.

Publications

1. Estejab, D.A. Daramola and G.G. Botte, Water Research 77, pp I33-I45 (2015). DOI: 10.1016/j.watres.2015.03.013
2. G.G. Botte, D.A. Daramola and M. Muthuvel, "9.14 Preparative Electrochemistry for Organic Synthesis" In Comprehensive Organic Synthesis II (Second Edition), G.A., Molander; P. Knochel, Eds., Elsevier: Amsterdam, 351 – 389 (2014). DOI: 10.1016/B978-0-08-097742-3.00940-X
3. D.A. Daramola and G.G. Botte, Journal of Colloid and Interface Science 402, pp 204-214 (2013). DOI: 10.1016/j.jcis.2013.03.067
4. D.A. Daramola and G.G. Botte, Computational and Theoretical Chemistry 989, pp 7 – 17 (2012). DOI: 10.1016/j.comptc.2012.02.032
5. D.A. Daramola, D. Singh and G.G. Botte, Journal of Physical Chemistry A 114, pp 11513 – 11521 (2010). DOI: 10.1021/jp105159t
6. D.A. Daramola, M. Muthuvel and G.G. Botte, Journal of Physical Chemistry B 114, pp 9323 – 9329 (2010). DOI: 10.1021/jp9077135

Pending Publications

1. E. Grossman, D.A. Daramola and G.G. Botte, Comparing B3LYP and B97 dispersion-corrected functionals for Studying Adsorption and Vibrational Spectra in Nitrogen Reduction.
2. Y. Al Majali, C. T. Chirume, D.A. Daramola, K.S. Kappagantula and J.P. Trembly, Coal Filler Based Thermoplastic Composites as Construction Materials: A New Sustainable End-Use Application.

581318 From Biosynthesis to Human Health: Harnessing Biomolecules with Multiscale Quantum Mechanics–Molecular Mechanics Simulations

From Biosynthesis to Human Health: Harnessing Biomolecules with Multiscale Quantum Mechanics–Molecular Mechanics Simulations

Research Interests:

Computational engineering of new reactions, proteins, and enzymes presents a holy-grail challenge in both academia and industry. The advancement of computing hardware and algorithm has largely boosted the integration of experiment and computation to innovate biomolecular research. Despite significant progresses, accurate computation of mechanisms, rates, and dynamics for biomolecules are still difficult, because of the large size of the systems and extensive computational sampling that is required to achieve convergence. Multiscale quantum mechanics/molecular mechanics (QM/MM) methods have been proposed to balance accuracy and efficiency through the division of a complex system into a portion where a QM treatment is deemed essential and the rest where an MM treatment is suitable. Enabled by graphics processing unit (GPU)-acceleration, a QM/MM calculation can now incorporate a large-scale electronic structure where 500-1000 atoms can be treated quantum mechanically. This opens up the possibilities of uncovering the quantum mechanical nature of biomolecules that are synthetically useful and biologically important.

My research interests involve the development and utilization of large-scale QM/MM simulation tools to investigate the mechanisms, catalytic origins, and dynamics of enzymatic reactions, revealing how solvent molecules, active site amino acids, and greater protein environment dictate enzyme catalytic actions. Specifically, I will focus on 1) pericyclases that construct complex macrocycles in biosynthesis, 2) glycyl radical enzymes that mediate metabolism in the human gut microbiota, and 3) lasso peptides that exhibits superior thermal stability and potential anti-microbial activity. These studies will elucidate the biomolecular functions from multiple temporal and spatial scales, informing new metrics and descriptors for computational engineering of biomolecules for synthetic and pharmaceutical uses.

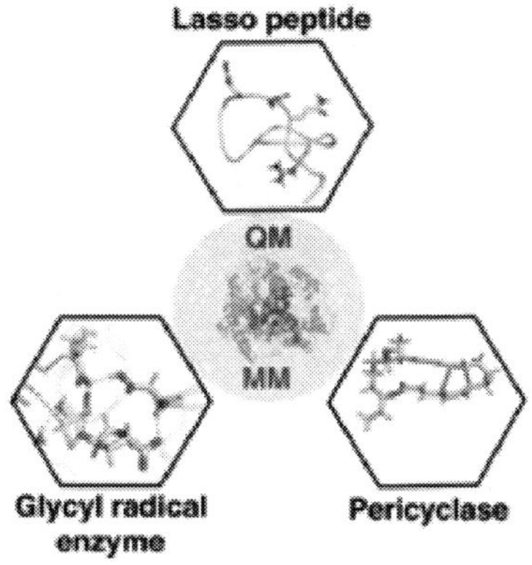

Teaching Interests:

My teaching interests lie in the general chemical engineering core classes, including Thermodynamics, Kinetics, and Physical Chemistry. My vision is to help students wield the weapon of thermodynamics and statistical mechanics to achieve fundamental understandings of chemical phenomena, and to conquer cutting-edge challenges in energy, biosciences, materials, and medicines. I majored in chemistry at college and computational chemistry in graduate school. These programs involve a heavy load of classes that overlap largely with the chemical engineering programs. As a graduate student at UCLA, I served as a teaching assistant (TA) for the freshman-level general chemistry laboratories between 2014 and 2016. I also served as a teaching assistant for the graduate-level Computational Chemistry in 2015 and Quantum Mechanics in 2016. These experiences endowed me with valuable intuitions regarding how to better organize hard-core class materials for students with different science and engineer backgrounds. Collectively, my academic background and teaching experiences will make me well positioned to take on the teaching duties in chemical engineering departments.

568104 Designing the Structure and Function of Soft, Complex Materials with Computational Modeling

Research Interests:

Soft, complex materials like colloidal dispersions and polymer solutions are widespread in many fields of science and engineering, including biotechnology, consumer products, and advanced materials. Because of their weak internal interactions, soft materials must be carefully engineered to obtain the desired properties under the conditions in which they are processed and deployed; these conditions are often far from equilibrium. **A key outstanding technological challenge is to relate the controllable microscopic characteristics of a soft material and its processing conditions to its emergent macroscopic properties (e.g., self-assembled structure, rheology).** This relationship can then be inverted to design materials tailored for a specific application. Computer simulations are powerful tools for tackling this problem; they provide detailed microscopic information that can be used to develop predictive models, and they can be used within an inverse-design framework to rapidly determine optimal candidate materials that can be tested in collaborative experiments.

During my graduate research (with Prof. Thanos Panagiotopoulous, Princeton University), I developed multiscale models to study the coupled relationship between structure and dynamics in mixtures of colloids and polymers driven out of equilibrium. I demonstrated how the microscopic interactions in flowing colloidal suspensions affect the solute spatial distribution and long-time dynamics in confinement, which are relevant to enhanced oil recovery, lab-on-a-chip devices, and membrane separations. I also showed how drying controls self-assembly in thin films, including colloidal crystallization and demixing, with applications in the fabrication of functional coatings and polymer nanocomposites. As a postdoctoral fellow (with Prof. Tom Truskett, University of Texas at Austin), I am applying my foundation in computational modeling to new research areas with a focus on design—namely, reconfigurable nanoparticle gels with targeted optical and rheological properties and porous polymer membranes for water treatment—in close collaboration with experimental research groups.

Building on these prior experiences, my future research program will initially be focused in three principal areas:

1. **Designing multilayer coatings by controlled evaporation-induced assembly.** There are currently no theoretical models to quantitatively predict the internal structure of mixtures in drying thin films. Such models can be exploited to discover and design the interactions needed to efficiently deposit functional, multilayer coatings with a single-step process.
2. **Manipulating nanoscale soft materials with flow.** Nanofluidic and microfluidic devices are promising technologies for the continuous separations needed to purify water or to collect rare cell markers in the blood for diseases like HIV and cancer, but most prior design efforts for these devices have been guided primarily by intuition. Mesoscale coarse-grained simulations are a powerful tool to optimize the separations of soft particles using tunable parameters like the device geometry and novel engineered surfaces.
3. **Robust, multiscale inverse design of self-assembling materials.** Colloidal particles arranged in specific lattices with long-ranged order are good candidates for materials with tunable optical activity, and self-assembly is a promising high-throughput strategy for fabricating them. However, it is a long-standing challenge to efficiently find the appropriate chemistries and conditions that will give the interactions that promote self-assembly into a target structure and not its competitors. Multiscale simulations coupled to a robust inverse-design approach can rapidly find these conditions subject to physical constraints of the experiments.

The aim of all three research projects is to make predictions for optimal designs that can be tested experimentally and translated to realizable technologies.

Teaching Interests:

I believe teaching and mentorship should be integrated into all interactions with students, including in coursework, in research, and with non-traditional learners through community outreach.

I promote an active learning environment that engages students in their coursework with applied examples taken from everyday life and the chemical engineering practice. I use small-group discussions and scaffolded problem-solving to enhance traditional lecture methods, and I solicit anonymous feedback throughout a course to adapt and differentiate its content. As a teaching assistant, I have delivered lectures, created assessments, and run interactive learning sessions for both a graduate-level thermodynamics course (15-30 students) and an undergraduate-level material balance course (50-180 students). In my office hours and lessons, I strove to create a safe and inclusive space where all students were comfortable and able to learn. In future, I have particular interest in teaching thermodynamics and statistical mechanics, transport, and fluid mechanics courses, but I am comfortable teaching all core courses in the chemical engineering curriculum. I additionally hope to develop project-based elective courses in molecular simulation techniques and sustainable scientific software development that can be tailored to both the undergraduate and graduate levels.

In my research group, I will foster a mutually supportive, mentorship-focused, inclusive environment that promotes participation by researchers at all levels of training and experience, especially from historically underrepresented groups in engineering and the computational sciences. My experience so far has strongly evidenced that the best research is done through collaboration of diverse viewpoints. I have had rewarding experiences mentoring undergraduate students through domestic and international programs, leading to multiple peer-reviewed publications. Based on this and the pivotal impact undergraduate research has had in my own career, I will actively involve undergraduate students in my own research group through semester-projects and summer research programs. The development of open-source scientific software, which underlies my research, can act as a gateway for these students that will also build a valuable foundation in a high-demand skill. Mentorship also extends beyond the university; I have engaged in community outreach activities geared towards elementary-school-aged children (e.g., hands-on simple experiments at the library, science fair), and I will continue to develop and participate in these activities in future.

582827 Computational Protein Design Using Optimization Programs and Force-Field Calculations

Research Interests:

Computational protein design enables us to generate a focused library of "designer" proteins with a desired phenotype. Computational techniques typically aid to scan thousands of amino acid type and spatial orientation combinations in reasonable time as a lever for directed evolution of proteins up a fitness landscape along the desired property – under polynomial type. These in turn, serve as precedents for experimental validation. Herein, we have laid out mixed-integer linear programming (MILP) as a method of efficiently performing more than five simultaneous amino acid substitution on a polypeptide backbone to tailor its functionality. Our work is divided into three main paradigms: (a) computational redesign of channel proteins for precise sub-nm tuning of pore size and chemistry for desired bioseparations, (b) altering substrate specificity of enzymes by exploring novel scaffolds and binding chemistry, and (c) *de novo* design of variable antibody domains targeted against a specific antigen epitope. First, monodispersed angstrom-size pores embedded in a suitable matrix are promising for highly selective membrane-based separations. They can provide substantial energy savings in water treatment and small molecule bioseparations. Such membrane proteins (primarily aquaporins) are commonplace in biological membranes but difficult to implement in synthetic industrial membranes due to their modest and non-tunable selectivity. PoreDesigner, a computational design workflow for the redesign of the robust beta-barrel Outer Membrane Protein F as a scaffold targeting of any specified pore diameter (spanning 3–10 Å), internal geometry and chemistry. PoreDesigner uses a mixed-integer linear program to optimally place long side - chain hydrophobic amino acids at the pore constriction region that yield a smaller and more hydrophobic pore by maximizing the interaction energy between the pore wall and the permeating water wire. This lays a promising foundation for using biomimetic membrane materials as viable filtration assemblies for performing precise angstrom-scale separations. Secondly, Iterative Protein Redesign and Optimization Suite of programs (IPRO) uses a MILP to optimally alter binding pocket residues of an enzyme to alter substrate affinity. A new algorithm has been devised that besides predicting amino acid substitutions, also predicts the possibility of introducing an insertion or deletion at the binding pocket to affect substrate recognition – thus using protein length as a lever for directed evolution. Finally, we use OptMAVEn-2.0 algorithm to rapidly design highly humanized antibody variable fragments targeted against an antigen epitope protein. This serves as a one-stop solution for designing antibody-mediated therapeutics during disease outbreaks upon experimental assessment of the best designs using immune-sorbent assays. Overall, in all cases the "designed" proteins are selected and ranked depending on thermodynamics of interactions by computing stability scores or interaction energy terms using force-field calculations that account for non-covalent forces (electrostatics, van der Waals, and solvation effects).

Teaching Interests:

While an integrative research and education proposal is always a priority of NSF CAREER program, developing a platform to maximize the public benefits of any proposed work can be tricky. To this end, advances in information technology provide a growing set of tools that can be applied to get the message out to a wide range of audience. To this end, I will pursue two different objectives: (a) Prepare workshops to engage structure biology mentorship to middle school, high school and undergraduate students, and (b) Introduce a graduate level coursework titled "Optimization in Structural Biology" to provide a relevant knowhow of linear algebra with tantamount interfacing with relevant biochemistry and chemistry.

For the first, I am already a part of NSF-CAREER proposal of Dr.Rajib Saha, Asst. Professor, department of Chemical Engineering, University of Nebraska, Lincoln and will have partial support for organizing "fun-camps" for young individuals from middle and high school which will use "Lego" and "flexible polymers", and hands on computer simulations as models to provide visual understanding of protein-protein interaction, lock and key hypothesis of enzymes and protein folding dynamics. For undergraduate students, I will be able to arrange visits to the headquarters of "Applied Biomimetic Inc." where Dr. Mariusz Grzelekaowski -the Chief Technical Officer, has expressed interest in showing how they express and purify the designed channel proteins (from PoreDesigner work – where we have an agreement over our Intellectual Property) and use them for various state-of-the art bioseparations. Additionally, I plan to organize popular lecture series where each student from my lab will provide short "layman's" description of their research dring these camps.

Regarding the coursework - I have an Online Teaching Certificate from World Campus Penn State (OSL-2050) which has trained me how to frame coursework material compliant with university policies. I think, currently there exist no single coursework which talks about (a) optimization programs, (b) structural biology, and (c) force-field based molecular mechanics under the same umbrella. This course work will provide students engaged in experimental pursuits a clear understanding of the computational paradigm, and computational students a deeper dive into how to write algorithms that could be employed for *in silico* protein evolution for any desired biochemical objective. The course will have two guest lectures (a) one by an experimental professor and (b) one by another computational scientist, who will tie the course content to real-world projects they have performed thus providing insights about systems that they have worked on.

582019 Data-Driven Modeling and Control of Batch and Batch-like Processes

Research Interests: Process Modeling and Control, Performance Monitoring, System Identification, Time-frequency Analysis and Forecasting, Machine learning and Causality Analysis

The key motivation that drives my research is the development of methods and strategies for the design of advanced process control systems that (i) account for the complex process dynamics including the non-availability of measurements throughout the process dynamics, (ii) ensure an efficient operation of processes by having relevant model-based control algorithms, (iii) ensure optimal operation of the process by having data-driven process monitoring tools, and (iv) could detect faults in the process as they occur and determine its root cause. My research interests are in the general area of process systems engineering, with emphasis on the development of modeling, control, monitoring and fault detection tools for chemical engineering processes using data-driven approaches. The central objective of my research is the development and implementation of data-driven advanced process control solutions for different chemical engineering applications. By bringing together tools from control and systems theory, chemical engineering and data analytics, we have been able to develop and implement practical solutions to important process control problems including (a) modeling and control of batch chemical engineering processes, (b) analysis and design of controllers, and (c) process network reconstruction using causality analysis.

Postdoctoral Project: "Development of data-driven modeling and control solutions for chemical engineering processes"

Under supervision of Prof. Prashant Mhaskar, McMaster Advanced Control Consortium (MACC), Department of Chemical Engineering, McMaster University

PhD Dissertation: "Data-Driven Modeling and Control of Batch and Batch-Like Processes"

Under supervision of Prof. Prashant Mhaskar, McMaster Advanced Control Consortium (MACC), Department of Chemical Engineering, McMaster University

Master's Dissertation: "Causality Analysis for Topology Reconstruction and Interaction Assessment"

Under supervision of Prof. Arun K. Tangirala, Indian Institute of Technology Madras

Research Experience: I am currently a postdoctoral fellow working on data-driven modeling and control strategies for chemical engineering systems. My current research projects are in collaborations with McMaster Manufacturing Research Institute (MMRI) and Linde, where I have developed novel data-based modeling and control approaches and is an extension of my doctoral research (e.g., [1-3,5-7]). My research contributions include the development of a framework to utilize big data aspects for modeling and control of the batch particulate process using a dynamic subspace identification based approach coupled with a static product quality model; optimal synthesis of a hydrogen plant startup process; data-driven closed-loop control of a uni-axial rotational molding process. In my masters' research, I focused on causality analysis and made some significant contributions to the problem of causal network reconstruction and interaction assessment from process data (e.g., [4]). The central objective of my research is the development and implementation of data-centric advanced process control algorithms.

Teaching Interests: Process control, Modern control theory, System identification, Time-series analysis

I view teaching as an integral and very important part of my overall academic activity. I plan to actively participate in the enhancement of the entire undergraduate and graduate curriculum. Given my interdisciplinary background with B.Tech in Electronics and Instrumentation Engineering and Masters' and PhD in Chemical Engineering, and my experience as a teaching assistant in several undergraduate and graduate courses at Indian Institute of Technology Madras and McMaster University, I feel confident that I can effectively teach any courses in the undergraduate and graduate Chemical Engineering curriculum.

The focus in my teaching would be on developing a curiosity for the subject and enabling students to apply the concepts learned in the classroom to practical problems through careful design of course projects. As a teacher, I believe in the transfer of fundamental subject knowledge to students in such a way that it cultivates their critical thinking skills. I will strive to instill a sense of curiosity in my students that will challenge them to go beyond the typical course requirement. Further, I will value the individual backgrounds and experiences of my students and create a learning environment that's fair for everyone. I will strive to nurture an environment that will encourage them to seek areas that excite them, as true learning occurs best when it is most meaningful. Further, I will adapt my classroom practices to continuously improve the effectiveness of my teaching based on feedback.

Future Directions: Building upon the training and research experiences that I had during my studies and postdoctoral fellowship; I would like to continue contributing to the systems engineering field. Some of the main directions that I plan to pursue in my future research activities include (a) development of machine learning approaches suitable for process control, (b) development of performance monitoring/decision support tools for large-scale plants, (c) development of hybrid modeling techniques for process control applications, and (d) study the problem of causal inference and its application to process control and monitoring.

Selected Publications (1 book, 6 published peer-reviewed journal articles, 5 conference papers):

[1] Gomes F. P. C., **Garg, A.**, Mhaskar, P. and Thompson M. R. (2019). Data-Driven Advanced in Manufacturing for Batch Polymer Processing Using Multivariate Nondestructive Monitoring, Ind. Eng. Chem. Res. 58 (23), 9940-9951.

[2] **Garg, A.**, Gomes, F. P. C., Mhaskar, P. and Thompson M. R. (2018). Model Predictive Control of Uni-Axial Rotational Molding Process, Comp. & Chem. Eng. 121, 306-316.

[3] **Garg, A.**and Mhaskar, P. (2018). Utilizing Big Data for Batch Process Modeling and Control, Comp. & Chem. Eng. 119, 228-236.

[4] **Garg, A.**and Tangirala, A. K. (2018). Metrics for Interaction Assessment using Causality Analysis, Ind. Eng. Chem. Res., 57(3), 967-979.

[5] **Garg, A.**and Mhaskar, P. (2017). Subspace Identification Based Modeling and Control of Batch Particulate Processes, Ind. Eng. Chem. Res., 56(26) 7491-7502.

[6] **Garg, A.**, Corbett, B., Mhaskar, P., Hu, G., & Flores-Cerrillo, J. (2017). Subspace-Based Model Identification of a Hydrogen Plant Startup Dynamics. Comp. & Chem. Eng., 106, 183-190.

Books:

[7] Mhaskar, P., **Garg, A**. and Corbett, B. (2019). Modeling and Control of Batch Processes. Springer International Publishing. http://doi.org/10.1007/978-3-030-04140-3

582022 Optimization and Control of Chemical Process Systems Under Uncertainty

Research Interests:

Future research interests:

Motivation: Optimization models for engineering systems often contain uncertain model parameters due to uncertainties in market forces and the environment, use of surrogate models, and difficulty in measuring parameters accurately. Optimal solutions to models that simply ignore the uncertainty in model parameters can be economically worthless and even disastrous in safety-critical applications. Rigorously accounting for uncertainties in process engineering models is particularly challenging due to their nonlinear and combinatorial nature. The goal of my research group will be to develop scalable solutions to optimization and control problems under uncertainty that possess theoretical performance guarantees. In particular, our focus will revolve around the following broad research themes.

Optimization and control of dynamic systems under uncertainty: There are several applications of this model class in process engineering, including bioprocess optimization under parametric and implementation-related uncertainties, optimal experimental design for inferring parameters in constrained dynamic systems, and model predictive control of dynamic systems subject to stochastic disturbances. Existing approaches typically work with an explicit discretization of the dynamic system, which may lead to inaccurate solutions and increased computation times. Practical optimization approaches usually do not provide rigorous optimality guarantees as well. Our goal will be to develop efficient and implementable optimization schemes that leverage state-of-the-art methods for simulating dynamic systems while simultaneously providing theoretical performance guarantees. In particular, we will consider optimization problems with embedded dynamic systems and chance-constraints/robust constraints to ensure reliable system operation, and develop local optimization methods and decomposition algorithms for solving these problems.

Modeling and optimizing the resilience of systems: A resilient system is one that can anticipate and prepare for changing conditions, and withstand and recover rapidly from disruptive events. Increasing uncertainty surrounding high-impact rare events such as climate-related disruptions and the ever-growing threat of cyberattacks has led to renewed interest in modeling and optimizing the resilience of critical infrastructure systems such as the electric power grid. Another application that has witnessed a recent surge in interest is optimization of the resilience of chemical plants to disruptive events. Although there isn't a single agreed upon resilience metric for such systems, the associated models typically include several challenging features from the viewpoint of optimization, including discrete decisions to model resilience-enhancing alternatives, multiple objectives corresponding to different stakeholders, different models of uncertain constraints (such as probabilistic and robust constraints), and multi-stage and multi-level optimization frameworks. We will develop application-specific resilience metrics based on existing literature and by working with domain experts, and design decomposition approaches for tackling the resulting large-scale stochastic optimization models.

Sustainable process design and operation: The problem of sustainable design and operation of engineering processes involves several challenges including multiple economic and environmental objectives, multiple time-scales and multi-level modeling frameworks, the lack of clearly defined system boundaries, and uncertainty in and unavailability of model data. We will consider a multi-objective stochastic optimization model for sustainable design and operation of specific chemical processes under probabilistic constraints on the environmental objectives. Tailored decomposition approaches will be developed to solve the resulting large-scale stochastic optimization model.

Summary of past research:

Doctoral research: I developed a decomposition algorithm for solving a broad class of two-stage stochastic mixed-integer nonlinear programs, and implemented it in a general-purpose software for solving this class of problems. The software, which also includes implementations of other state-of-the-art algorithms, consists of about 100,000 lines of C++ code, and also includes a test library that has been used by other members of the process systems engineering community. I also investigated the rate of convergence of deterministic branch-and-bound algorithms for constrained optimization, and used this theory to understand when branch-and-bound algorithms can mitigate the undesirable cluster problem in constrained global optimization.

Postdoctoral research: I designed a local optimization algorithm for solving chance-constrained nonlinear programs that uses techniques from the machine learning community. I have also been working on investigating the theory behind a purely data-driven stochastic optimization approach, designing a local optimization method for solving a broad class of stochastic optimization problems, constructing better optimal power flow models in the face of renewable energy uncertainty and designing approaches to solve these models, and exploring the design of resilience metrics for the power grid.

Teaching Interests:

I am passionate about teaching, and have actively sought out teaching opportunities. During my undergraduate studies, I worked as a Mathematics Olympiad trainer at an academy in India for groups of twenty students. One trainee was among the forty students nationwide selected to attend the International Mathematics Olympiad Training Camp in 2010. At MIT, I was a teaching assistant for the graduate Chemical Reactor Engineering course for which I helped develop computational exercises. I am also passionate about creating freely accessible educational resources and disseminating them to disadvantaged students. At IIT Madras, I was a volunteer for the Science Activities team of the National Services Scheme where I designed and demonstrated science experiments in underprivileged schools. While at MIT, I also recorded video lectures on discrete math as part of a team that generated free online content for the entrance exam for the Indian Institutes of Technology. Our effort was supported by the MIT Office of Digital Learning, and a couple of my video lectures were featured on MIT OpenCourseWare.

My main teaching objective will be to instill critical-thinking and problem-solving skills in students. My teaching philosophy is to use lots of carefully-designed working examples to build intuition, to have clearly-specified learning objectives for each class and to continually assess students' proficiency with respect to these objectives through in-class exercises, and to develop assignments and projects by drawing on applications in other chemical engineering subdisciplines (e.g., an exercise on sparse computational linear algebra based on sensitivity analysis of a concrete large-scale kinetic model).

I am particularly interested in teaching undergraduate and graduate-level courses on numerical analysis, optimization, control, and systems engineering. I am also interested in teaching advanced graduate-level courses on nonconvex optimization and stochastic optimization.

Selected Publications:

Dissertation (2018): Algorithms, analysis and software for the global optimization of two-stage stochastic programs, Massachusetts Institute of Technology (includes two working papers).

Kannan and P. I. Barton (2018). Convergence-order analysis of branch-and-bound algorithms for constrained problems. Journal of Global Optimization, 71(4), pp. 753-813.

Kannan and P. I. Barton (2017). The cluster problem in constrained global optimization. Journal of Global Optimization, 69(3), pp. 629-676.

Kannan and A. K. Tangirala (2014). Correntropy-based partial directed coherence for testing multivariate Granger causality in nonlinear processes. Physical Review E, 89(6), 062144.

Selected Working Papers:

Kannan, L. Roald, and J. Luedtke. Stochastic DC Optimal Power Flow With Reserve Saturation.

Kannan and J. Luedtke. A stochastic approximation method for chance-constrained nonlinear programs. Preprint at https://arxiv.org/abs/1812.07066

Kannan and P. I. Barton. GOSSIP: decomposition software for the Global Optimization of nonconvex two-Stage Stochastic mixed-Integer nonlinear Programs.

Kannan and P. I. Barton. A modified Lagrangian relaxation algorithm for two-stage stochastic mixed-integer nonlinear programs.

Selected Honors:

George M. Keller Graduate Fellowship, Dept. of Chemical Engineering, MIT, 2012.

Oil and Natural Gas Corporation Scholarship, India, 2012.

Reliance Heat Transfer Pvt. Ltd. Award for academic excellence, IIT Madras, 2012.

Institute Merit Award, IIT Madras, 2010 and 2011.

582124 Computational Design and Development of Advanced Catalytic Materials

Research Interests:

The central goal of my research is to computational discover, design and develop advanced functional materials for energy applications, *e.g.* electrocatalysis. The multiscale computational tools *e.g.* density functional theory, *ab initio* molecular dynamics and many body perturbation theory (GW/BSE) and high-throughput computational method *e.g.* machine learning will be integrated together to achieve the goal. Specific interested objectives are unconventional atomic structures that have great potentials in future electrochemical catalysis. I aim to develop the hybridized nano-structures combing the advantages of homogenous and heterogeneous catalysts with expected functions based on computational modeling. Engineering approaches from both intrinsic and extrinsic factors such as defect, doping, alloying and applying mechanical deformations will be applied to tune the materials towards desired properties. The high-throughput and machine learning models will be integrated to accelerate the advanced catalysts design process.

Teaching Interests:

My philosophy of teaching is to create a heuristic teaching environment. I can teach materials related courses, including thermodynamics, mechanics of materials, electronic structure of materials, solid state physics as undergraduate level courses. I also can provide courses related to advanced computational method such as density functional theory based simulations to teach the students helpful skills for their research. Depending on my research progress, I am highly interested in opening an interdisciplinary course on machine learning in materials science, which is a thrusting and cutting-edge topic in the materials science area.

582142 Studying Soft Materials in- and out-of-Equilibrium Using Analytical and Numerical Field Theories

Research Interests: Field theories have played an important role in polymer science, ever since the development of modern polymer physics as we know it. Powerful analytical techniques such as the random phase approximation, developed during those early years, still find great utility today despite vast improvements in computational power and efficiency which have understandably shifted much of the focus of polymer and soft materials research into the simulation domain. There is little doubt that state-of-the-art research in these areas, field-theoretic or otherwise, will continue to be largely computational in nature. However, it is important not to rely too heavily on computer simulations; one should also seek elegant and deep insights into the phenomena under study. For this latter objective, the classic analytical techniques, taught to us by Sam Edwards, Pierre-Gilles de Gennes and other pioneers of polymer field theory, are indispensable. The most significant conceptual advances are often made when theory and simulation work in concert.

My vision for a research program is one in which analytical and numerical field-theoretic techniques are used in tandem to address fundamental and applied problems in soft materials research. Through my Ph.D. and post-doctoral work, I have developed expertise in the construction and manipulation of classical statistical field theories, of both the equilibrium and dynamical kind, including the development of new field-theoretic techniques. This expertise will be leveraged in my research group to tackle new and challenging problems while ensuring that the analytical techniques of the polymer field theory tradition are handed down to the next generation of polymer and soft materials theorists. My particular research interests are in the structure and dynamics of complex fluids in which fluctuations/correlations play an important role. Examples include charged and polarizable systems such as (poly)electrolytes, (polymeric) ionic liquids, and salt-doped polymers. I am also broadly interested in the development of dynamical field theories for polymers that can incorporate electrostatic, hydrodynamic and entanglement effects. Such field theories could be an important step toward a computational framework in which phase diagrams that relate industrial melt processing variables to final morphologies are achievable.

Teaching Interests: For many years as a Ph.D. student, I developed and ran mandatory tutorial sessions as a teaching assistant for undergraduate physics courses, which were essentially mini-lectures designed to complement the main course lectures. The responsibilities included independently preparing lecture notes and quizzes. Through that experience I developed all of the essential skills required for a course lecturer. My background in Physics makes me particularly suitable for courses in Thermal Physics/Thermodynamics, Statistical Mechanics, Transport, Reaction Kinetics, Electromagnetism.

582190 QM-Based Multiscale Simulations for Applications in Electrocatalysis, Interfacial Chemistry, and Energetic Materials

Research Interests:

There is an economic and environmental imperative to design and discover new materials. Computational science has become an essential partner to experiment for understanding the fundamental physical and chemical properties of materials. Recent dramatic developments in first-principles quantum mechanics (QM) methods have enabled accurate predictions for the properties of various systems. However, such QM methods are limited to several hundreds of atoms for dozens of picoseconds (ps) (10^{-12} seconds). This capacity may grow by an order of magnitude over the next few years, but pure QM methods will remain far from meeting the requirement to inform the continuum degrees of freedom (e.g. solvent effect in electrochemical systems), which require millions of atoms (10^6) and microseconds (10^{-3} seconds) of simulations. Thus, there is an enormous gap between the scale of current QM methods and that of the real applications. My research is to **develop and validate new computation tools and software for multiscale reactive simulations of materials and processes** and then to **apply these tools and methods to develop new generations of materials designed to have improved performance**. My research is motivated toward a diverse set of applications. In particular, I am interested to apply these methods to electrochemical processes, catalytic reactions, interfacial chemistry, and materials under extreme conditions.

Postdoctoral Studies:

Advisor: **Prof. William A. Goddard III**, Caltech

Projects:

- Development of the quantum mechanics-based polarizable reactive force field (**RexPoN**) for multiscale simulations of materials and systems
- Development of polarizable charge equilibration (**PQEq**) model for predicting accurate electrostatic interactions in liquid, solids, and interfaces
- Prediction of structures and properties of green energetic materials from density functional theory (DFT) and reactive molecular dynamics

Co-authored Proposals:

- DOE, DE-SC0014607, funded (09/2015-09/2019)
- ONR, N00014-16-1-2059, funded (12/2015-11/2018)
- DARPA, 140D6319C0019, funded (3/2019-11/2019)

PhD Studies:

Advisors: Profs. Muhammad Sahimi and Theodore T. Tsotsis, University of Southern California

Dissertation: A process-based molecular model of silicon carbide (SiC) nano-porous membranes

Research Experience:

As a postdoctoral fellow, I developed the methodologies for constructing a very accurate reactive force field (FF), **RexPoN**, capable of describing complex reaction dynamics at the accuracy of ab initio CCSD QM (better than DFT). RexPoN retains high level QM accuracy for multiscale simulations of materials and energy systems with millions of atoms needed to make breakthroughs in the design of novel complex materials.

The application of RexPoN FF to water resulted in incredibly accurate properties of water (e.g. melting point to 0.2K accuracy). It showed that the lifetime of strong hydrogen bonds is ~ 90 femtoseconds at 298K, which was confirmed experimentally. In addition, it explained the puzzling hydrogen bonding network in water (a hundred years old problem) and the possible origin of anomalies in supercooled water.

To predict accurate electrostatic interactions in materials I developed the **Polarizable Charge Equilibration (PQEq)** method which uses atomic sized charges (not point charges) which are allowed to polarize every time step leading the accuracy of QM as dipoles are brought up to various molecules. I have validated the accuracy of PQEq for nearly all elements of periodic table including main groups and transition metals.

In addition, I have built on the accuracy for RexPoN to develop a new generation of hybrid computational framework, **RexPoN embedded QM, or REQM**, aimed at practical simulations on reactive systems with 100,000 to a million atoms while retaining the 0.05 eV accuracy of QM. In REQM the whole system (200,000 atoms for a 10 nm nanoparticle) is described with RexPoN, except for an embedded region of 200-300 atoms that uses QM to describe bond breaking processes. REQM can allow practical full solvent studies of electrocatalysis where QM is used only for atoms directly involved in bond breaking. This will extend full solvent studies to practical electrocatalysis on nanoparticles and nanowires. Also it will enable direct comparison to operando experiments for validation.

Future Plans:

I. New Generation of Catalysts for Electrochemical Water Splitting: Electrocatalytic water splitting driven by renewable energy input to produce clean H_2 is an essential component of the future energy portfolio. The challenge in developing dramatically improved catalysts is to develop deep understanding of the catalytic mechanisms. QM calculations of mechanism using explicit solvent have been validated by experiment, but the best electrocatalysts are nanoparticles and nanowires with 100,000 to millions of atoms. To make progress the QM accuracy has to be extended to such sizes.

REQM will enable this, allowing direct comparison to operando experiments and providing the means to use theory to guide the improvements in the catalyst. I propose to develop new generation of bifunctional nonprecious electrocatalysts to drive both HER and oxygen evolution reaction (OER) under the same conditions, useful for a wide pH range. My goal is to achieve at least one order of magnitude higher HER turnover frequency per catalyst site.

I anticipate funding from NSF (DMR, CBET, and CDS&E) and DOE. This work will be the focus of 50% of my efforts.

II. Thermodynamics and Adsorption Mechanism of Ions at Aqueous Interfaces: The adsorption and distribution of ions at the aqueous interfaces (solid and vapor) play a key role in the performance of many physical, chemical, atmospheric, and biological processes. The traditional view has ignored the presence of explicit ions at the liquid/vapor interface. However, recent experimental works using second order nonlinear optical techniques have proved that certain ions are enhanced at the interface.

Despite some theoretical efforts, clear understanding of the behavior of ions at the interface has remained challenging. I aim to change this situation. My group will utilize RexPoN FF and two-phase thermodynamics method (developed by Goddard group) to understand the thermodynamics and adsorption mechanism of the ions at the graphene/water interface due to its importance in many systems such as lithium-ion batteries and porous membranes for filtering, transistors, and supercapacitors. Our results will help experimentalist for effective engineering of the aqueous interface for potentially game-changing properties.

I anticipate funding from NSF (DMR, CBET, and CDS&E) and DOE. This work will be the focus of 30% of my efforts.

III. Structures, Properties, and Performance of Materials under Extreme Conditions: QM is providing increasingly accurate descriptions of materials under ordinary conditions of temperature and pressure. My interest is to extend these methods to the no-man's region of extremely high temperature and pressure/stress gradients involving complex reaction kinetics, combustion, and explosions. I want to develop techniques to predict reliably the constitutive properties and reaction kinetics processes that control stability, performance, and sensitivity of high energy materials (EMs). I want to do this prior to synthesis and characterization and discover the best possible new EMs to help guide the experimentalists to focus only on the most promising systems. Short term challenge is to design insensitive EMs (due to risks in their use) without any loss in their performance. Long term is to describe combustion, chemical vapor deposition, and atomic layer deposition.

One goal is to utilize and validate RexPoN FF and REQM to develop a new generation of two-dimensional (2D) composite EMs with desirable insensitivity and much higher performance than the conventional EMs (e.g. HMX and RDX).

To ensure maximum impact, I will establish collaboration with the experimentalists. I anticipate funding from DoD, ONR, AFOSR, and perhaps from DOE. This work will be the focus of 20% of my efforts.

Teaching Interests:

I have experience teaching and mentoring students in the classroom and in the laboratory. As a teaching assistant I have given lectures and tutorial sessions in Transport Phenomena (i.e. Mass Transfer, Heat Transfer, and Viscous Flow), Reactor Design and Kinetics, and Advanced Engineering Mathematics. Also I have assisted in courses on molecular and materials simulations and quantum mechanics. During my postdoctoral studies I have mentored more than 10 postdocs, graduate, undergraduate, and high schools students for a variety of different projects. I am enthusiastic about the opportunity to teach undergraduate and graduate courses including Quantum Mechanics, Computer Simulation and Modeling, Transport Phenomena, Chemical Reaction Engineering, Engineering Thermodynamics, and Advanced Engineering Mathematics. I would like to offer at least one but possibly two courses related to multiscale atomistic simulations of materials and systems.

22 peer-reviewed publications (11 first author).

Selected Publications:

1. **Naserifar** and W. A. Goddard III, "Anomalies in Supercooled Water at ~230 K Arise from a 1D Polymer to 2D Network Topological Transformation", *Proceedings of the National Academy of Sciences* (2019), *Under Review*
2. **Naserifar**, J. J. Oppenheim, H. Yang, T. Zhou, S. Zybin and W. A. Goddard III, "Development of Accurate Nonbond Potentials based on Periodic Quantum Mechanics Calculations for Molecular Simulations of Materials and Systems", *The Journal of Chemical Physics* (2019), *Under Review*
3. **Naserifar** and W. A. Goddard III, "Liquid Water is a Dynamic Polydisperse Branched Polymer", *Proceedings of the National Academy of Sciences*, 116 (6), 1998-2003 (2019)
4. **Naserifar** and W. A. Goddard III, "The Quantum Mechanics-based Polarizable Force Field for Water Simulations", *The Journal of Chemical Physics 149*, 174502 (2018)
5. **Naserifar**, D. J. Brooks, W. A. Goddard III, and V. Cvicek, "Polarizable Charge Equilibration Model for Predicting Accurate Electrostatic Interactions in Molecules and Solids", *The Journal of Chemical Physics* 146, 124117 (2017).
6. **Naserifar**, S. Zybin, C. Ye and William A. Goddard, III, "Prediction of Structures and Properties of 2,4,6-Triamino-1,3,5-Triazine-1,3,5-Trioxide (MTO) and 2,4,6-Trinitro-1,3,5-Triazine-1,3,5-Trioxide (MTO3N) Green Energetic Materials from DFT and ReaxFF Molecular Modeling", *Journal of Materials Chemistry A* 4, 1264-1276 (2016)
7. **Naserifar**, W. A. Goddard, III, M. Sahimi and T. T. Tsotsis, "First Principles-based Multiparadigm, Multiscale Strategy for Simulating Complex Materials Processes with Applications to Amorphous SiC Films", *The Journal of Chemical Physics* 142, 174703 (2015)
8. **Naserifar**, T. T. Tsotsis, W. A. Goddard, III, and M. Sahimi, "Toward a Process-based Molecular Model of SiC Membranes: III. Prediction of Transport and Separation of Binary Gaseous Mixtures based on the Atomistic Reactive Force Field", *Journal of Membrane Science* 473, 85-93 (2015)
9. Jaramillo-Botero, **S. Naserifar**, and W. A. Goddard, III, "A General Multi-objective Force Field Optimization Framework and its Application to the Design of Reactive Force Fields for Silicon

Carbide", *Journal of Chemical Theory and Computation* 10, 1426–1439 (2014)

10. **Naserifar**, L. Liu, W. A. Goddard, III, T. T. Tsotsis, and M. Sahimi, "Toward a Process-Based Molecular Model of SiC Membranes. I. Development of a Reactive Force Field", *The Journal of Physical Chemistry C* 117, 3308–3319 (2013)
11. **Naserifar**, W. A. Goddard, III, L. Liu, T. T. Tsotsis, and M. Sahimi, "Toward a Process-Based Molecular Model of SiC Membranes. II. Reactive Dynamics Simulation of the Pyrolysis of Polymer Precursor to Form Amorphous SiC", *The Journal of Physical Chemistry C* 117, 3320–3332 (2013)

581728 Computational Heterogeneous Catalysis for Energy Storage and Conversion

Research Interests:

I obtained my Ph.D. in Materials Science and Engineering at University of North Texas (UNT) in May 2017. After graduation, I worked as a postdoctoral researcher with Prof. Jens K. Nørskov in SUNCAT Center for Interface Sciences and Catalysis at Stanford University. In Sep. 2018, I moved to the University of Notre Dame to join Prof. William F. Schneider in the Department of Chemical and Biomolecular Engineering. My interdisciplinary background will compliment the existing strength of chemical engineering research of the department, and will open pathways towards materials chemistry.

I have been conducting independent computational research on heterogeneous catalysis for clean-energy applications such as fuel cells and metal-air batteries using high-performance computing and machine learning. I also worked on kinetics and mechanism of Cu-zeolite for deNOx reaction. My future research interests are combining computational and experimental (in collaboration) methods to design and search novel catalysts for next-generation energy storage and conversion, and using machine learning to accelerate the searching process.

1. Kinetic modeling and solvation effect of carbon-based nanomaterials as bifunctional oxygen evolution and reduction electrocatalysts.

2. High-throughput screening and machine learning algorithm of metal oxides as novel oxygen electrochemical catalysts.

3. Synergistic effect on heteroatom doped metal phosphide for efficient hydrogen evolution.

4. Mass transport and oxygen permeation in PEM fuel cells.

Teaching Interests:

Based on my research experience and the experience of being a teaching assistant, I am able to teach core courses in Chemical Engineering such as physical chemistry, thermodynamics, heterogeneous catalysis, transport phenomena, electronic structures. In addition, I can also develop new advanced courses according to my research area such as computational materials chemistry, electrochemistry, machine learning in catalyst discovery, energy storage and conversion.

I served as a teaching assistant for 2 classes (Bonding and crystallography, and electronic properties of materials), and a guest lecturer for 1 class (fundamentals of materials science) when I was a Ph.D. student at UNT. The most challenging part I found was to specialize course materials for students with different scientific backgrounds. After I took the teaching training at Stanford University when I was doing my postdoc, I learned a systematic teaching strategy including how to control the class atmosphere, how to lead the class, how to get positive feedback and how to improve my course quality. Those trainings benefit my teaching ability a lot that I have more capability to generate a creative, communicative, diverse class environment.

I have helped advising junior graduate students in our labs, and I have formally advised 2 high school students from Texas Academy of Mathematics & Science while at UNT, both of whom got admitted by top-tier research universities. One of the most challenging aspects of mentoring has been identifying what kind of communication works best with each student. Another challenge is to identify how frequently to interact with a student; some students benefit from more hands-on tutorials than others. My approach has been to make myself readily available to students and make it clear that if they need something when I am available. To make sure that students are on track, I also encourage frequent casual meetings and occasional written proposals for new ideas and directions. I have found that giving thorough feedback on a student's proposal is one of the best ways to steer a project. I also effectively interact with students from underrepresented minorities to enhance the diversity in the field.

581588 Computational Crystallization and Assembly of Polymers and Soft Matter

Research Interests:

Complex structural and dynamical behaviors can lead to various assembled structures, and in turn govern the final material properties of polymers and soft matter. To predict the structure-property relations for these soft materials, a fundamental understanding of these behaviors at molecular level is essential. My future research will focus on predicting the assembly and properties of a variety of soft materials, including polymers, liquid crystals, colloids, surfactants and peptides, from their molecular structures and processing conditions using statistical mechanics theories and multi-scale simulations.

During my graduate training at the Pennsylvania State University, I studied the properties and behaviors of conjugated polymers from theoretical perspectives under the guidance from Professors Enrique Gomez and Scott Milner. By combining multi-scale simulations, ranging from density functional theory to coarse grained molecular dynamics simulations, with statistical mechanics theories such as self-consistent field theory, we developed various methods to predict material parameters, such as persistence length, Maier-Saupe parameter and Flory-Huggins χ from chemical structures for conjugated polymers. We have also demonstrated that orientational order of polymer backbones can impose molecular weight dependence on charge transport for conjugated polymers.

As a postdoctoral fellow at University of Michigan, I work with Professor Ronald Larson on colloidal assembly and polyethylene (PE) crystallization. We developed a theoretical tool to predict the effective interactions between colloidal particles induced by telechelic polymers that can form loops on a single particle or bridges between two colloids. The effective interactions can be then used in a novel Brownian dynamics method to predict the shear rheology of these colloid/polymer solutions. We have also applied atomistic simulations and analytical theories to predict the effects of flows and side-chain branching on PE crystallization. Combining our crystallization simulations with rheology modeling, we aim at predicting the crystallization behaviors and the resulting morphologies and properties for blown PE films.

Moving forward, my independent research will focus on bulk and interfacial crystallization and assembly of soft materials. Using multi-scale simulations and analytical theories, we aim at predicting assembly pathways, and final morphologies and properties from molecular structures and processing conditions for model materials. We will also seek collaborations with experimentalists to validate our computational approaches and obtain macroscopic material properties that are difficult to measure in simulations. Using both simulations and experimental material properties as training data, we can then apply artificial neural networks to estimate structure-property relations for rational design of novel high-performance soft materials. My initial research directions are:

1. Polymer crystallization at interfaces and in flows;

2. Hierarchical assembly of peptides and peptide amphiphiles;

3. Conjugated polyelectrolyte and doped conjugated polymers in bulk and at interfaces.

Teaching Interests:

Over years, I have been taught, trained and advised by excellent teachers and supervisors, from whom I have learned that clarity in teaching and student engagement are the keys to deliver high quality education. The clarity in presentation echoes the logic nature of science, and thus can greatly improve the efficiency in the knowledge delivery process. The participation and actual practice from students are indispensable and learning is most effective when students are highly motivated. From my own learning and previous mentoring experience, both the joy of discovery and credited achievements are positive incentives that can drive students to improve themselves constantly. My goal is to incorporate this philosophy in teaching in order to train the next generations of excellent chemical engineers.

I have put my teaching philosophy in practice and observed successful outcomes. At the Pennsylvania State University, I mentored an undergraduate student in an independent research project and, at the end, he published a peer-reviewed paper as a first author. At the University of Michigan, I instructed junior Ph.D. students on molecular simulations and polymer physics. In addition to mentoring, I also worked as a teaching assistant in an undergraduate fluid mechanics class and as a guest lecturer on Monte Carlo simulations for an undergraduate elective course.

Given my educational background (B.S. and Ph.D in chemical engineering) and experience, I feel confident in teaching several chemical engineering courses at both undergraduate and graduate levels. For undergraduate courses, I am interested in teaching mass and energy balances, thermodynamics, reaction kinetics, mass and heat transfer, and fluid mechanics. Among the graduate level courses, I would like to teach statistical mechanics and polymer physics.

Selected Publications:

12. "Modeling inter-colloidal interactions induced by adsorption of mobile telechelic polymers onto particle surfaces", Zhang, W.; Travitz, A.; Larson, R.G., Macromolecules, in press.

11. "A metastable nematic precursor accelerates polyethylene oligomer crystallization as determined by atomistic simulations and self-consistent field theory", Zhang, W.; Larson, R.G.,Journal of Chemical Physics, in press.

10. "Thermal fluctuations lead to cumulative disorder and enhance charge transport in conjugated polymers", Zhang, W.; Bombile, J.H.; Weisen, A.R.; Xie, R.; Colby, R.H.; Janik, M.J.; Milner, S.T.; Gomez, E.D., Macromolecular Rapid Communications, in press.

9. "Tension-induced nematic phase separation in bidisperse homopolymer melts", Zhang, W.; Larson, R.G., ACS Central Science, 2018, 4, 1545-1550.

8. "Direct all-atom molecular dynamics simulations of the effects of short chain branching on polyethylene oligomer crystal nucleation", Zhang, W.; Larson, R.G., Macromolecules, 2018,51, 4762-4769.

7. "Nematic order imposes molecular weight effect on charge transport in conjugated polymers", Zhang, W.; Milner, S.T.; Gomez, E.D., ACS Central Science, 2018, 4, 413-421.

6. "Predicting Flory-Huggins χ from simulations", Zhang, W.; Gomez, E.D.; Milner, S.T., Physical Review Letters, 2017, 119, 017801.

5. "Using surface-induced ordering to probe the isotropic-to-nematic transition for semiflexible polymers", Zhang, W.; Gomez, E.D.; Milner, S.T., Soft Matter, 2016, 12, 6141-6147.

4. "Predicting the Flory-Huggins χ parameter for polymers with stiffness mismatch from molecular dynamics simulations", Kozuch, D.J.; Zhang, W.; Milner, S.T., Polymers, 2016, 8, 241.

3. "Surface induced alignment for semiflexible polymers", Zhang, W.; Gomez, E.D.; Milner, S.T., Macromolecules, 2016, 49, 963-971.

2. "Predicting nematic phases of semiflexible polymers", Zhang, W.; Gomez, E.D.; Milner, S.T., Macromolecules, 2015, 48, 1454-1462.

1. "Predicting chain dimensions of semiflexible polymers from dihedral potentials", Zhang, W.; Gomez, E.D.; Milner, S.T., Macromolecules, 2014, 47, 6453-6461.

581632 Computational Modeling of Chemical Interactions at Interfaces for Environmental Applications

Research Interests:

Chemical reactions at interfaces are important in both natural environments and environmental applications. For example, crystal growth and dissolution reactions at mineral/water interfaces have been applied in environmental remediation, material synthesis, and carbon sequestration. Another good example is use of nanoparticles to remove contaminants, where nanoparticle surface chemistry can be modified to target chemicals. Experimental methods are limited by temporal and spatial resolution, resulting in ambiguous data interpretation and a lack of fundamental understanding of mechanisms underlying these reactions. These knowledge gaps can be addressed with computational chemistry tools. Atomistic simulations can provide molecular scale insights into the reactions at interfaces and guide theoretical models for predicting reaction thermodynamics and kinetics. In my research, I propose to use state-of-the-art simulations to gain molecular insights into chemical interactions at solid-liquid interfaces. Building on my previous experience, I intend to pursue three research thrusts that investigate crystal growth via different pathways at the mineral-water interface and the computational design of micellar-enhanced ultrafiltration (MEUF) techniques.

The topic of thrust 1 is **atomistic understanding of the growth mechanism of metal oxides.** A molecular-level understanding of metal oxide growth mechanisms can help solve diverse environmental problems including aquifer chemistry, removal of heavy metals, treatment of acid mine drainage, and treatment of nuclear waste. Aluminum hydroxides, the major components of the aluminum ore bauxite, will be chosen as a model system for this study. By using rare-event sampling techniques and reactive force fields, I will map the free energy landscapes for monomer and oligomer attachment and predict rates for aluminum hydroxides growth. For validation of our theoretical predictions, I will seek opportunities for collaborations with experimentalists in atomic force microscopy (AFM) measurements of growth velocities and transmission electron microscopy (TEM) observations. The mechanistic insights gained in this study will also shed light on the growth mechanism for layer-structured silicates.

The topic of thrust 2 is**quantification and control of solvent effects in oriented attachment.** Nanoparticle-nanoparticle aggregation, or oriented attachment (OA), has been shown to occur during the growth of multiple natural materials. Recent work revealed the critical role of solvent in the form of a hydration force that prevents particle coalescence at close separations while enabling the co-alignment of particles. A comprehensive knowledge of the underlying mechanisms behind the hydration force is fundamental to understanding nanoparticle-based aggregation. I will utilize multi-scale computational efforts (MD, QM, and MC methods) to reveal the effects of surface and solution chemistry on the magnitude and sign of hydration force. I also plan to seek collaborations with experimentalists who can perform force measurement using AFM. This thrust has a potential transformative impact by tweaking the identified factors to accurately control the inter-particle forces and thus optimizing the conditions for particle-particle aggregation.

The topic of thrust 3 is **computational design of micellar-enhanced ultrafiltration (MEUF) techniques.** MEUF is a versatile wastewater treatment technique that removes dissolved ions and molecules by ultrafiltration membranes using surfactants. Although a large number of surfactants are available in the design space, only a limited number of them have been adopted in MEUF. Therefore, it is helpful to develop an efficient computational method for predicting optimal surfactant types to avoid extensive and cost-prohibitive experimental efforts. I will use a combined method of a continuum model (COSMO-RS) and MD simulations to select optimal surfactants for removing organic contaminants from wastewater. This work will provide potential surfactants and experimental variables that can be tested and verified in the experiments. Therefore, I seek to collaborate with experimentalists in measuring partition coefficients of contaminants between micelles and water and adsorption isotherms. Given the extensive amount of data produced from simulations, I also seek to collaborate with machine learning experts to facilitate faster predictions.

Teaching Interests:

I am interested in teaching a graduate-level course on atomistic simulations as well as an undergraduate-level course on interfacial chemistry and am able to more broadly teach subjects related to environmental chemistry and thermodynamics.

581510 Modeling and Optimization for Process System Engineering Problems

Research Interests:

I am interested in using mathematical modeling and optimization to solve practical chemical engineering problems. Mathematical Optimization is the selection of the optimal elements from many, possibly infinite number of, alternatives, based on a proposed model for the process to be studied, which will provide timely and intelligent aids in decision-making. Over the past few decades of study, optimization has been proved to improve many process performances, including but not limited to: product quality, process robustness, environment sustainability, capital and operational cost, etc. To use optimization properly for solving chemical engineering problems, one should (i) model the process effectively and accurately enough, and (ii) solve the models in a timely fashion.

Previously, I studied the modeling and optimization in three process system engineering problems, including (i) inventory routing, (ii) production scheduling, and (iii) process characterization for the pharmaceutical industry. In the future, I plan to study how to solve the online scheduling and inventory routing problem through the perspectives of finding the "better" solution from the pool of symmetric solutions. I also plan to combine the knowledge of optimization and design of experiments together, which will not only provide the optimal solution, but also offer statistical insights. Additionally, I want to study how to use the modeling method called DRSM, which I have been working on, together with model predictive control. These works will be very useful because of their direct industrial relevance. These problems are also very interesting academically because they are closely related to different areas of process system engineering community, with potential for follow-up and additional work by many other researchers.

Teaching Interests:

I am very interested in teaching courses on (i) process synthesis, design and optimization, (ii) process dynamics and control, and (iii) design of experiments and statistics. First, I have a solid background on process synthesis, mathematical optimization, design of experiments, and artificial intelligence. Therefore, I will love to teach how to analyze and synthesize the chemical processes, as well as how to design and optimize them using the modern optimization or artificial intelligence tools (e.g., integer programming, neural networks, etc.); additionally, design of experiments will also be a very tool for many chemical engineering students to include in such a course. Second, I have studied process dynamics and control, including the advanced model predictive control (MPC), in both undergraduate and graduate years; I have also assisted the teaching of such a course for many semesters (including guest-lecturing). Thus, I would love to teach students how to study the dynamics of process, and how to control the dynamic processes, using the important ideas such as feedback and close-loop; moreover, topics such as MPC, the comparison of knowledge-driven and data-driven models will be interesting to introduce in this course. Finally, design of experiments are not offered in many schools, but it presents a very useful statistics tool for researchers in almost any area of the broad chemical and biological engineering discipline. I am open to teaching other courses too.

Overall, I wish I can lead and guide the students to learn the knowledge, as well as to obtain the tools for self-education, because learning should be a lifelong process, especially considering the rapid growth of technology.

Previous Research Topics and Results:

First, inventory routing problem (IRP) is a supply chain optimization problem that arises in a wide range of manufacturing sectors, including industrial gases and petrochemicals. To improve the efficiency of their supply chains, vendors in a wide range of manufacturing sectors switch to vendor-managed inventory (VMI) practices. VMI refers to an agreement between a vendor and a customer in which the latter allows the vendor to choose the timing and volume of deliveries, while the former agrees to ensure that the customer does not run out of product. A switch to such a practice has a distinct advantage because the vendor can combine deliveries to make more efficient use of the resources and can therefore reduce the distribution cost. However, executing VMI is non-trivial, which requires the solution of IRP, a difficult combinatorial problem. We have conducted studies for IRP in the following three aspects: propose mixed-integer programming models that handle different features and constraints that appear in IRP, especially, driver constraints (1); develop solution methods that can solve IRP efficiently (2); and study how reoptimization should be carried out for IRP (3).

Second, production scheduling has been widely studied, because it is a complex and practical problem that needs to be addressed for many chemical manufacturers. In the face of new information and uncertainties, online scheduling is required, and terminal constraints should be added in the model. However, how to write the terminal constraints is not apparent. The traditional approach would require that the terminal inventory level of each material is required to be greater than a lead-time-based threshold, but this method does not exploit the relationships of inventory levels among materials. Considering this issue, we propose new types of terminal constraints on inventory levels for different network structures (4). The proposed terminal constraints lead to better closed-loop solutions in two aspects, compared to the traditional approach: they prevent stockout and save inventory holding cost. Additionally, we proposed different formulations, and compared their theoretical tightness and computational performances for modeling the sequence-dependent changeovers (5).

Third, Dynamic Response Surface Modeling (DRSM) is a recently proposed data-driven method that models time-resolved outputs, and it can be used for the classical Design of Experiments (DoE) and the generalized Design of Dynamic Experiments (DoDE). Compared to the classical approaches, DoDE and DRSM methodologies allow time-varying inputs, accommodate time-resolved output measurements, and model their interrelationship. Because DRSM quantifies the effects of experimental conditions on the process outputs over time, it can enable more convenient and comprehensive process optimization. We apply the DRSM model in a pharmaceutical problem to identify stoichiometries and obtain the optimal operating condition. The optimization is aimed to maximize the length of an operational window, which is defined based on the species concentrations. Because DRSM provides the models of the concentrations under different conditions and at different times, it is very convenient and useful to apply it for this optimization problem. Moreover, we also incorporate two types of robust constraints in the optimization to explicitly handle the uncertainty in the operating condition as well as in the DRSM model.

Reference

- Dong, Y., Pinto, J.M., Sundaramoorthy, A., and Maravelias, C.T. MIP Model for Inventory Routing in Industrial Gases Supply Chain. *Industrial & Engineering Chemistry Research*, 2014, 53, 17214-17225.
- Dong, Y., Maravelias, C.T., Pinto, J.M., and Sundaramoorthy, A. Solution Methods for Vehicle-based Inventory Routing Problems. *Computers & Chemical Engineering*, 2017, 101, 259-278.
- Dong, Y., Maravelias, C.T., and Jerome, N.F. Reoptimization and Policy Analysis for Maritime Inventory Routing under uncertainty. *Optimization & Engineering*, 2018, doi: 10.1007/s11081-018-9383-8.
- Dong, Y. and Maravelias, C.T. Terminal Constraints for Single-stage Multi-product Scheduling Problems. In *Foundations of Computer Aided Process Operations (FOCAPO)*, Tucson, AZ, January 8-12 2017.
- Velez, S., Dong, Y., and Maravelias, C.T. Changeover Formulations for Discrete-Time Mixed-integer Programming Scheduling Models. *European Journal of Operational Research*. 2017, doi: 10.1016/j.ejor.2017.01.004.
- Dong, Y., Georgakis, C., Mustakis, J., Hawkins, J. M., Lu, H., Wang, K., McMullen, J. P., Grosser, S. T., Stone, K., Constrained Version of the Dynamic Response Surface Methodology for Modeling Pharmaceutical Reactions. *Industrial & Engineering Chemistry Research*. **2019,** doi: 10.1021/acs.iecr.9b00731.

582329 Predictive Descriptors for the Targeted Synthesis of Solid-State Materials

Research Interests:

Paramount to meeting the increased demand for renewable and efficient energy sources is the discovery and design of next-generation solid-state materials, which are the active components of countless energy storage and generation devices including batteries, fuel cells, photovoltaics, and heterogeneous catalysts. First-principles electronic structure calculations utilizing density functional theory (DFT) have emerged as a standard tool in the computational characterization of materials and can be used in a high-throughput manner to reliably predict the thermodynamic stability of new materials with targeted properties. However, thermodynamic stability alone does not determine synthesizability and does little to indicate how the material of interest should be synthesized—e.g., what precursors, temperatures, pressures, atmospheres, etc. lead to this material being realized in the laboratory. This challenge is made especially daunting by the vast diversity of chemistries, compositions, and structures that can be realized in the solid state. My research group will approach this challenge by a combined theoretical and data-driven approach. High-throughput DFT calculations allow for the rapid generation of thermodynamic and kinetic data for bulk structures and interfaces. By parsing these calculations with the help of sophisticated statistical approaches, we will pursue so-called descriptors that enable us to relate geometric, thermodynamic, and electronic signatures in a specified target or precursor material to the most likely synthesis pathway that these materials will undergo. Ultimately, our work will accelerate the realization of new technologically relevant materials by providing fundamental understanding as to how synthesis proceeds in the solid-state and predictive tools that rapidly convey that understanding through accurate high-throughput predictions.

Teaching Interests:

My desire to pursue a tenure-track faculty position was cemented in my third year of graduate school when I worked as a teaching assistant for graduate Analytical Methods in Chemical Engineering and began mentoring undergraduate research assistants. I've found many parallels between teaching and mentoring students, especially the critical nature of drawing clear connections between the material that is being presented and how that material manifests in "real world" applications that are tangible to a Chemical Engineer. Performing fundamental research in the area of solid-state synthesis uniquely positions me to interact directly with each of the pillars of Chemical Engineering – thermodynamics (e.g., phase diagrams), kinetics (e.g., ion diffusion), and transport (e.g., within a reactive particle). My expertise in data science, quantum mechanics, and solid-state chemistry also presents an opportunity for me to develop courses that are typically outside the norm for a Chemical Engineering curriculum but have significant utility in exposing undergraduate or graduate students to disciplines that are relevant to both industrial and academic careers.

581838 Understanding Multimodal Interactions through Deep Neural Networks and Statistical Mechanics

Research Interests:

I am a fourth year PhD candidate in the Chemical and Biological Engineering department at RPI with an interest in combining MD simulations with AI/machine learning techniques to facilitate the design of proteins and multimodal surfaces for a range of applications. I am fascinated with the fundamental problem of understanding how water mediates complex interactions between surfaces containing multiple modes of interaction (e.g. hydrophobic or charge interactions). Despite the fact that these "multimodal" interactions play a central role in many biological processes, our fundamental understanding of the principles governing these interactions remains limited. In recent years, the advent of GPU computing has resulted in dramatic improvements in the computational speed of MD simulations. Additionally, the field of AI has experienced a renaissance, with neural networks being successfully employed for a variety of problems including reading text, classifying images, and even recently folding proteins. Thus, a new field combining the wealth of data arising from rapid MD simulations and the AI tools to mine this data is emerging. As a junior faculty member, my aim will be to leverage these powerful tools to address the fundamental question of how the structure of multimodal surfaces governs their interactions.

Teaching Interests:

As a graduate student, I have enjoyed the process of mentoring students at different levels. I have mentored three high school students through the process of formulating a research question, performing molecular dynamics simulations, and ultimately presenting their data at either a poster session or in a group meeting. I have also created more complex projects for three talented undergraduate students. In this case, because the students were more advanced, I guided them to learn the basic academic skills required as well as to learn to become partially independent researchers. Watching these students learn to internalize their projects to the point where they are excited to work independently and propose new ideas has been extremely rewarding. As a senior graduate student, I have recently played the role of mentoring first and second year graduate students through this process as well, requiring me to teach the creative skill of being a truly independent researcher, while also requiring me to teach thermodynamics and statistical mechanics at a deeper level. As a professor, I look forward to mentoring students as well as teaching these related courses (thermodynamics, separations, or statistical mechanics) or any other core courses formally in a classroom setting. I feel that my teaching experience has been personally rewarding and has helped me strengthen my own knowledge in these fields.

Future Directions:

There are two main areas of work that I would like to expand upon from my PhD research:

1. **A Neural Network Approach to Predicting Patch-Based Protein Surface Properties**
   Understanding the physical properties associated with protein surface patches is important for a wide variety of applications including protein solubility (important for protein crystallization or formulation of biologic therapeutics) and protein-x interactions (important for drug discover, protein-protein interactions, and protein surface interactions). Researchers have used variational autoencoders to evaluate the similarity of small molecules in latent space in order to predict molecular properties. We have recently developed an approach for decomposing protein structures (from PDBs) into protein surface patches (represented using a voxel representation with channels being used for atomic occupancies) and using a variational autoencoder with 3D convolutional layers to encode these patches into a latent space representation. In this representation, we can observe relationships between protein surface patches that capture classical heuristics (i.e. hydrophobic or charge classifications) as well as identify unexpected relationships. In the future, I plan to build upon this work to identify minimal sets of protein patches from which important classes of proteins can be reconstructed. I will then perform molecular simulations to quantify

important parameters such as solvation (free energy of cavity formation) and protein-ligand interactions for this minimal set and use this data to predict these important parameters for the relevant classes of proteins. An advantage of this approach is that this latent space representation can also suggest mutations to proteins that might significantly alter the behavior of a given patch (with the caveat that that mutation may change the protein structure in unexpected ways).

2. **Designing Patterned Surfaces through Self-Assembly of Small Molecules**
   In my PhD, I have studied designing complex multimodal surface (including charged, hydrophobic, and hydrogen bonding characteristics for the application of chromatographic materials for protein separation. In addition to this application, there are many other applications for which designing patterned surfaces is important (design of biomaterials, biosensors, etc.). We have recently identified (using molecular dynamics simulations) a class of chromatographic ligands which self-assemble into patterned surfaces containing concentrated patches of hydrophobicity and charge on the nanometer length scale. The principles underlying this phenomenon are closely related to the self-assembly observed in surface-grafted block copolymer systems but differ importantly in that the ligands are much shorter and contain small chemical moieties instead of "blocks" resulting in shallow pattern formation instead of the formation of different phases. In the future, I will build upon this work to develop a statistical mechanical model for predicting the surface structure that an individual ligand will self-assemble into. Through this, I will map out the design space of surfaces which can be synthesized in this manner. Additionally, I will collaborate with experimentalists to evaluate important surface structures experimentally. This work has important implications for the design of new chromatographic materials and for the design of functionalized surfaces in general.

581797 Mechanistic Electrochemistry Driving Energy Science: From Catalysts to Batteries

Vision

Fundamental electrochemical science and engineering are at the heart of the push towards a sustainable energy ecosystem, from revolutionizing transportation with electric vehicle batteries, to agriculture with electro-catalyzing ammonia synthesis. With my formal training in the electrochemistry of transition metal-based systems for catalysis and energy storage, I am interested in working on fundamental scientific challenges aimed at hastening these grand-scale transitions.

Research Experience

Postdoctoral Work: Suppression of Interfacial Degradation in High-Energy Lithium-Ion Batteries
As a postdoctoral scholar in Prof. Bryan McCloskey's research group in the Chemical Engineering department at UC Berkeley, I'm studying fundamental degradation processes at the cathode-electrolyte interface in advanced Li-ion batteries. Cathode materials with high specific charge capacities at high voltages are needed to significantly push the energy density of Li-ion batteries, which will serve to lower the associated cost per unit of energy. While there are several cathode materials that meet this requirement, such as the layered Li-rich NMC (Ni, Mn, Co) oxide with specific capacities greater than 250 mAh/g, degradation processes at high voltages (> 4.5 V vs. $Li^{+/0}$) severely affect cycle life and overall battery performance. Specifically, the decomposition of the organic electrolyte at the cathode surface causes CO_2 release with the concomitant deposition of insulating films on the cathode surface. Cathode surface impurities such as Li_2CO_3 also decompose to produce CO_2. In addition, the participation of the lattice oxide atoms in charge compensation ("O-redox") during charging (Li^+ removal), results in surface O_2 release above 4.5 V. The evolution of gases such as CO_2 and O_2 is therefore indicative of interfacial degradation at high voltages. Electrochemical interfaces in batteries are, however, buried and therefore difficult to mechanistically probe.

1. Quantification of High-Voltage Interfacial Degradation: I have utilized online electrochemical mass spectrometry (OEMS) to quantify processes leading to interfacial degradation in high-energy Li-ion cathode materials, as a function of state-of-charge of the metal oxide. Ex-situ acid titrations of charged electrodes with isotopically labelled cathode particles and gas analysis with OEMS quantify the peroxo-like species generated during O-redox. The studies revealed a surface-to-bulk propagation of these peroxo-like O-species with increasing state-of-charge.
2. Interface Engineering to Suppress Degradation: With the tools in hand to quantify interfacial degradation manifesting in gas release, I have pursued methods to "reactively" passivate the electrode surfaces against such degradation, i.e. by utilizing a surface chemical reaction to produce a uniform passivation layer. These can be classified as:
    1. Additive: Deposition of electro-inactive coating layers on the cathode surface by atomic layer deposition (ALD). I found that uniform layers of early transition metal oxides such as MoO_x and WO_x are effective in suppressing high-voltage gas release leading to superior long-term cycling performance.
    2. Subtractive: I found that a controlled acidic treatment of the metal oxide surface to produce a few nm thick Li-deficient reaction layer on the surface leads to a near-complete suppression of gas release at high voltages, with no loss of extractable capacity, along with significantly improved long term cycling and rate performance.

The electroanalytical studies of the cathode materials in their pristine and surface-treated forms were supplemented with electron microscopy, X-ray diffraction, inductively coupled plasma mass spectrometry as well as surface and bulk-sensitive spectroscopic techniques to provide a generalizable model for interface passivation. I also extended these studies to other cathode materials such as nanoscale-$LiCoO_2$,

Li_3RuO_4 and the Ni-rich NCM oxides.

PhD work: Mapping Free Energies to Screen Molecular Electrocatalysts on a Computer
Density functional theory based computational methods have traditionally been used in molecular catalysis to *post-facto* study the thermodynamics and kinetics of catalysts with extensive experimental benchmarking. In my PhD work at Stanford University, advised by Prof. Christopher E. D. Chidsey and co-advised by Prof. Robert M. Waymouth, I explored turning this workflow around, that is, using state-of-the-art density functional theory for *a priori* molecular electrocatalyst design. Specifically, for the competing two-electron reductions of CO_2 to CO and formate, and of protons to H_2, I demonstrated how the use of two thermodynamic descriptors, viz. the free energies of two key catalytic intermediates, streamlined the screen for promising catalyst candidates from a large library of transition metals and organic ligands. The predictions from the *in silico* screen were validated by subsequent experimental synthesis and electrochemical studies of an iron complex that was found to be active towards CO_2 and proton reduction at the expected reduction potential.

This work, however, was the culmination of several experimental and computational endeavors to understand the electrocatalytic reactivity of transition metal complexes towards CO_2 reduction at low driving forces, some of which included a) the unraveling of a new pattern of CO_2 activation by singly reduced Ru-complexes, b) engineering single step two-electron reductions in first row transition metal complexes to minimize overpotentials associated with electron transfer, c) highlighting a key deactivation step when using transition metal hydrides as alcohol-oxidation catalysts in fuel cells.

Prior Research (MSc And BSc)

I studied the disparity in the low temperature microwave spectrum of the benzene-H_2O dimer and the benzene-H_2S dimer using the van Vleck contact transformation to the Vibrational-Rotational-Translational (VRT) Hamiltonian (IISc Bangalore). I used hyper-spectral imaging and electron microscopy to study the growth of higher order gold nanostructures in *L. acidophilus* bacteria (IIT Madras).

Research Interests:

I am broadly interested in studying fundamental electrochemical processes in energy storage materials. My research laboratory would strive to answer two scientific questions: (a) how to engineer cooperativity between transition metals and p-block elements to achieve multi-electron redox at low driving forces; and (b) how to rationally passivate electrode-electrolyte interfaces in high-voltage batteries.

The insights derived from this endeavor will aid materials' design for electrochemical energy conversion and storage processes including high-energy batteries and electrocatalytic fuel syntheses. The proposed research will involve a multi-pronged experimental effort involving synthetic, electrochemical, spectroscopic and analytical tools, aided by synergistic collaborations.

Teaching Experience

I served as the head teaching assistant for a first-of-its-kind Electrochemical Measurements laboratory course with Prof. Chidsey at Stanford University, designed for upper-class chemistry majors and graduate students. As part of curriculum development for this course, I designed and assembled a custom 'open-box' potentiostat with tunable circuit elements connected to a data acquisition device controlled by MATLAB™, with the aim of helping the students develop a circuit-level understanding of electrochemical measurements.

Apart from being a teaching assistant (TA) for several lecture and laboratory-based undergraduate chemistry courses at Stanford University, I also served as a TA trainer for first-year graduate students taking on teaching roles in the chemistry department.

My lecturing experience includes a three-part lecture series titled 'Discovering the World of Batteries' for the Science Circle High School Program at Stanford University, and a guest lecture on 'Density Functional

Theory for Organometallic Chemists' as part of the graduate level Advanced Inorganic Chemistry course at Stanford University.

Apart from teaching in the formal settings outlined above, I particularly enjoy mentoring younger graduate students, undergraduates and high school students towards their scientific advancement.

Teaching Interests:

I am interested in teaching lecture-based as well laboratory-based courses related to energy science and engineering, such as Electrochemistry, Inorganic Chemistry, Solid-State Chemistry, Thermodynamics, Kinetics, Quantum Chemistry, Electrochemical Energy Storage and Conversion. I am also interested in teaching courses that broadly cover topics related to renewable energy conversion and storage technologies.

Publications

1. Ramakrishnan, Moretti, Chidsey "Mapping Free Energy Regimes in Electrocatalytic Reductions with Transition Metal Complexes", *Chem. Sci.* **2019** ASAP (doi: 10.1039/C9SC01766F)
2. McLoughlin, Waldie, Ramakrishnan, Waymouth "Protonation of a Cobalt Phenylazopyridine Complex at the Ligand Yields a Proton, Hydride, and Hydrogen Atom Transfer Reagent", *J. Am. Chem. Soc.* **2018**, 140(41), 13233-13241.
3. Ramakrishnan, Chidsey "Initiation of Electrochemical Reduction of CO_2 by a Singly Reduced Ruthenium(II) Bipyridine Complex", *Inorg. Chem.* **2017**, 56(14), 8326-8333.
4. Waldie, Ramakrishnan, Kim, Maclaren, Chidsey, Waymouth "Multi-Electron Transfer at Cobalt: Influence of the Phenylazopyridine Ligand" *J. Am. Chem. Soc.* **2017**, 139(12), 4540-4550.
5. Ramakrishnan, Chakraborty, Brennessel, Jones, Chidsey **"**Rapid Oxidative Hydrogen Evolution from a Family of Square–Planar Nickel Hydride Complexes" *Chem. Sci.* **2016**, 7, 117-127.
6. Ramakrishnan, Waldie, Warnke, de Crisci, Batista, Waymouth, Chidsey "Experimental and Theoretical Study of CO_2 Insertion into Ruthenium Hydride Complexes" *Inorg. Chem.* **2016**, 55(4), 1623-1632.

Publications (in preparation)

7. Ramakrishnan, Park, Wu, Yang, McCloskey "Suppression of Interfacial Outgassing and Improved Cycling in High Energy Li-rich NMC Cathodes enabled by an Acid-treated, Li-deficient Surface".
8. Ramakrishnan, Asundi, Park, Bent, McCloskey "Effective Interface Passivation of Li, Mn-rich high energy cathodes with Atomic Layer Deposited Molybdenum Oxide".
9. Ramakrishnan, Hu, Cabana, McCloskey "Improved Interfacial Stability with Gradient Al_2O_3 Coatings on $LiCoO_2$ Nanoplates".
10. Li, Ramakrishnan, McCloskey, Cabana "Definition of Redox Centers in Reactions of Lithium-Intercalation in Li_3RuO_4Polymorphs".

582152 Heterogeneous Electrocatalysis: Developing Strategies to Engineer Industrially Relevant Catalysts from Fundamental Activity Trends

Research Interests:

Heterogeneous electrocatalysis is an established field with growing applications in renewable energy. For example, hydrogen production and consumption are a carbon neutral routes towards energy storage and conversion to electricity (via water electrolysis and hydrogen fuel cells, respectively). *My research interests focus on understanding and engineering material surfaces during electrocatalysis as to provide a feedback loop for the design of more highly active and earth abundant catalysts for industrial electrochemical reactors.*

Postdoctoral Project:

Development and characterization of active and stable electrocatalysts for the oxygen reduction reaction as part of a project with the Toyota Research Institute to understand the operando active site structure and composition.

Supervisor: Professor T. F. Jaramillo

Chemical Engineering Department at Stanford University

Fall 2017 - *Current*

PhD Dissertation:

***Title*: Fundamentals and Industrial Applications: Understanding First Row Transition Metal (Oxy)Hydroxides as Oxygen Evolution Reaction Catalysts**

Supervisor: Professor S. W. Boettcher

Chemistry Department at the University of Oregon

Fall 2012 – Spring 2017

Research Experience:

My studies have focused on describing and understanding fundamental catalyst activity and stability trends in lab-scale electrochemical devices, using coupled operando electrochemical-synchrotron techniques, and in industrial electrochemical reactors to identify strategies for activity and stability enhancement. During my PhD, at the University of Oregon, my research centered on studying first-row transition metal (oxy)hydroxides for the oxygen evolution reaction (half reaction in water splitting for hydrogen production) in alkaline media. Specifically, the goal of this research was to understand the role of Fe in highly active Ni-Fe catalysts and use this knowledge to design catalysts for industrial anion exchange membrane electrolyzer systems. This project was part of an academic-industry collaboration with Proton Onsite (Nel Hydrogen), an onsite gas (H_2, etc.) generation company. My postdoctoral work at Stanford University is a collaboration of experimental, theory, and machine learning research that focuses on studying the oxygen reduction reaction, half reaction in hydrogen fuel cells that limits total efficiency, and is part of an effort funded in-part by the Toyota Research Institute. This project focuses on designing tools to predict new non-platinum and acid-stable catalysts. Materials systems we have focused on includes metal-organic-frameworks and transition metal nitrides, oxides, and carbides. These studies focus on understanding the active site composition and structure in operando and have been greatly aided by in suit studies at the Stanford Synchrotron Radiation Lightsource. *Broadly, my research has been motivated by the principle that a better understanding of the active site during catalysis will inform the design of a more active industrial-scale catalyst.*

During this research I have gained invaluable skills in electrochemistry, synthesis, material science, and chemical engineering (to name a few). Furthermore, I am experienced in developing and promoting collaborations with researchers skilled in theoretical modeling (e.g. density functional theory), machine learning, and advanced material characterization (e.g. beamline scientist at synchrotrons).

Teaching Interests:

I am a committed educator with a passion for teaching and have developed and taught curricula that focus on evidence-based teaching and active learning. As a co-instructor and co-course designer I implemented these strategies into a small pilot course, research immersion class, that was an alternative to the General Chemistry laboratory course at the University of Oregon. I believe that good mentorship and teaching can be the difference between someone becoming a successful researcher or not pursing science and I am also committed to promoting diversity and inclusion and have participated in several mentoring programs focused on supporting women in science (Program: *Association of Women in Science Mentoring*) and diversity on a broader scale (Program: *Someone like Me*).

Future Direction:

One critical challenge today is comparing fundamental electrocatalyst studies to industrially relevant systems due to differences in corrosion, oxidation, and/or structure of the catalyst during operation in the different systems. Due to this incomparability, few new-non precious metal electrocatalysts have successfully transitioned into industrial systems. My vision is to develop a world-class research group that bridges the gap between fundamental electrocatalysis research and industrially-relevant devices and provide new strategies for energy storage and conversion technologies. **My research group** *would design and develop characterization techniques to facilitate the successful transition of catalyst materials from fundamental research laboratories to practical devices.*

Selected Publications: [17 published, 6 first author, 1 patent submitted, a few highlighted below]

Note: I have also published using the name Michaela S. Burke

Kreider, M.E.; **Stevens, M. B**.; Gallo, A.; Abroshan, H.; Back, S.; Siahrostami, S.; Norskov, Y. K.; King, L. A.; Jaramillo, T. F. Oxidized Surface Layer on Transition metal Nitrides: Active Catalysts for the Oxygen Reduction Reaction. U.S. Patent Application No. 16/420,890, Unpublished (filing date 05/23/2019).

Stevens, M. B.*; Enman, L. J.*; Korkus, E. H.; Zaffran, J.; Trang, C. D. M.; Asbury, J.; Kast, M. G.; Caspary Toroker, M.; Boettcher, S. W. Ternary Ni-Co-Fe Oxyhydroxide Oxygen Evolution Catalysts: Intrinsic Activity Trends, Electrical Conductivity, and Electronic Band Structure. *Nano Res.* **2019**, *12*, 1-8.

*These authors contributed equally

Xu, D.; **Stevens, M.B.**; Cosby, M.; Oener, S. Z.; Smith, A.; Enman, L. J.; Ayers, K. E.; Capuano, C. B.; Renner, J.; Danilovic, N.; Li, Y.; Wang, H.; Zhang, Q.; Boettcher, S. W. Earth-Abundant Oxygen Electrocatalysts for Alkaline Anion Exchange Membrane Water Electrolysis: Effects of Catalyst Conductivity and Comparison with Performance in Three-Electrode Cells. *ACS Catal.* **2019**, *9*, 7-15.

Enman, L. J.; **Stevens, M. B.;** Dahan, M. H.; Nellist, M.; Caspary Toroker, M.; Boettcher, S. W.; Operando X-Ray Absorption Spectroscopy Shows Fe Oxidation is Concurrent with Oxygen Evolution in Cobalt-Iron (Oxy)hydroxide Electrocatalysts, *Angew. Chem. Int. Ed.* **2018**, *57*, 1-6. [Highlighted in *Nature Catalysis*: https://www.nature.com/articles/s41929-018-0154-x]

Stevens, M. B.; Trang, C. D. M.; Enman, L. J.; Deng, J.; Boettcher, S. W., Reactive Fe-sites in Ni/Fe (oxy)hydroxide are responsible for exceptional oxygen electrocatalysis activity. *J. Am. Chem. Soc.* **2017,** *139*, 11361-11364. [Times cited: > 80]

Stevens, M. B.; Enman, L. J.; Batchellor, A. S.; Cosby, M. R.; Vise, A. E.; Trang, C. D. M.; Boettcher, S. W. Measurement Techniques for the Study of Thin Film Heterogeneous Water Oxidation Electrocatalysts. *Chem. Mater.* **2017**, *29*, 120–140. [Times cited: > 110]

Burke, M. S.; Enman, L. J.; Batchellor, A.; Zou, S.; Boettcher, S. W. Oxygen Evolution Reaction Electrocatalysis on Transition Metal Oxides and (Oxy)hydroxides: Activity Trends and Design Principles. *Chem. Mater.* **2015**, 27, 7549-7558. [Times cited: > 340]

Burke, M. S.; Zou, S.; Enman, L. J.; Kellon, J. E.; Gabor, C. A.; Pledger, E.; Boettcher, S. W. Revised Oxygen Evolution Reaction Activity Trends for First-Row Transition Metal (Oxy)hydroxides in Alkaline Media. *J. Phys. Chem Lett.* **2015,** *6, 3737-3742.* [Times cited: > 170]

Burke, M. S.; Kast, M. G.; Trotochaud, L.; Smith, A.; Boettcher, S. W. Cobalt-Iron (Oxy)hydroxide Oxygen Evolution Electrocatalysts: The Role of Structure and Composition on Activity, Stability, and Mechanism. *J. Am. Chem. Soc.* **2015,** *137, 3638-3648.* [Times cited: > 550]

580939 Corrosion Mitigation and Wastewater Treatment

Research Interests:

- Study of performance of volatile corrosion inhibitors for prevention of *top of the line corrosion and* black powder formation *in oil and* export *gas pipelines.*
- Study of adsorption of corrosion inhibitors using *in-situ* electrochemical/atomic force spectroscopy.
- Mitigation of scale formation using green inhibitors.
- Evaluation of an electrochemical approach to recover nitrogen (N) and phosphorus (P) from Concentrated Animal Feeding Operations (CAFO) waste as a solid slow release fertilizer product.

Successfully Completed Research Proposals:

- *TLC-Mitigation JIP* funded by five industrial oil and gas companies with budget of $600K, 2017-2020

- *Understanding and using corrosion inhibitors project* funded by Total with budget of $690K, 2019-2021.

- *Reduction of Nutrient Runoff from CAFO Waste Land*, the Ohio Water Development Authority (2016 R&D grant).

- *Black powder formation project* funded by Petroleum Institute of Abu Dhabi with budget of $900K, 2015-2016.

Research Experience:

- ***Institute for Corrosion and Multiphase Technology (ICMT)***

Supervisors: Dr. Srdjan Nesic and Dr. Marc Singer

My research focuses on methods to reduce and prevent corrosion and scaling in oil and gas pipelines, in particular, mitigation of Top-of-the-Line Corrosion and black powder formation using volatile corrosion inhibitor. Some results of this project have been reported in the US patent application (US20170233637A1). In addition, I am using atomic force microscopy (AFM) to study adsorbed morphologies of corrosion inhibitors on metal surfaces. The objective of this work is to provide a better understanding of the inhibition mechanism by coupling *in situ* electrochemical (EC) methods with AFM.

- ***Institute for Sustainable Energy and the Environment (ISEE)***

Supervisor: Dr. Jason Trembly

My research focuses on developing methods to recover nutrients from CAFO waste using electrochemical techniques. My research goal is to demonstrate that it is possible to use the electrochemical processes to recover N and P from CAFO waste as a solid slow release fertilizer product ($MgNH_4PO_4{\cdot}6H_2O$). Results of this project have been published in Journal of the Electrochemical Society.

- ***Laboratory Interfaces and Electrochemical Systems, La Sorbonne University, France***

PhD advisor: Dr. Bernard Tribollet

My PhD work focused on the development of a new green inhibitor, Paronychia Argentea extract for reducing calcium carbonate scale formation on metallic surfaces. One of the outcomes of my PhD research was to *understand* the mechanism of formation of calcium carbonate layers. The second outcome of my research was to study performance of scaling inhibitors to mitigate formation of calcium carbonate. In addition, I used the Comsol Multiphysics® for 2D and 3D modeling of calcium carbonate deposits and mass transfer at the interface solid/solution. The results of my research led to a US Patent: US 9234166 B2, 2016.

Teaching Interests:

Aside from research, I am passionate about teaching. My first opportunity to teach students was as a laboratory instructor for pharmaceutical engineering at the Bejaia University (2008-2009) when I was completing my MS degree. During the fall 2017, I taught an undergraduate level course "Principles of Engineering Materials" at the College of Engineering, Ohio University. This class is taken by students from different majors: civil, chemical, mechanical engineering and industrial and system engineering. I have trained and mentored undergraduate and graduate students in planning, assigning and scheduling fall test matrix, reviewing results to ensure quality standards.

Future Directions:

To create the next generation of materials and technologies to address critical challenges in environment sustainability, petroleum engineering, advanced manufacturing and durability of engineering materials, we need to create new methodologies and research partnerships that will address these challenges. Therefore, my work will involve understanding of materials and theirs interactions with environment at the nanoscale level. I would like to unite the knowledge that I obtained from the topics that I worked on during my Ph.D. and postdoc: scale and corrosion inhibition and recovery of nutrient from animal waste water.

As faculty I would like to work in the area of electrochemical corrosion diagnosis and control of structure in aggressive environment, and use the electrochemical impedance spectroscopy to study the performance of corrosion inhibitor, the coatings, and characterize metallic and advanced materials. The overall objective is to use my expertise in corrosion and electrochemistry to develop practical mitigation methods involving corrosion inhibitors and coating that can protect metals form corrosion.

Atomic force spectroscopy (AFM) has proved to be a powerful tool for investigating the surface structure, the changes in surface morphology of materials. My ambition is to use in-situ AFM couple with electrochemical technic to study for example the effects of the addition of nanoparticles and nanoreinforcements to cement and identity the electronic and mechanical proprieties of nanoengineered and nanocomposites materials. As well, study the structural proprieties of nanomaterial.

I would like also to use my expertise in AFM, electrochemistry and water treatment to seek new opportunities for collaborative, multidisciplinary research projects. In particular, I hope to expand my involvement in international research programs.

I would like to work on projects funded by agencies, such as the Department of Energy and the Department of Transportation / PHMSA (Pipeline and Hazardous Materials Safety Administration). Such efforts would necessarily entail a better understanding of the processes involved in applying for and maintaining grants from Federal bodies. With these grants, I can work on projects "looking into the future" and those are not directly applicable to the industry but are at more fundamental level. In addition, national recruitment should also be encouraged and I propose to set up an NSF Research Experience for Undergraduates (REU) within the related research institute.

I look forward to officially supervising graduate students, leading or participating in PhD and MS degree committees, as well as contributing to the administrative functions of the department.

Patents

1. Nesic, M. Singer, F. Farelas, T. N. Vu, D. Young, Z. Belarbi, "Methods for Inhibiting Metal Corrosion", US 20170233637A1, 2017
2. Gamby, Z. Belarbi, L. Chaal, F. Chaussec, L. Makhloufi, B. Saidani, B. Tribollet, B. Sotta, "Manufacturing process for a vegetal matter-based anti-lime scale product ", US 9234166 B2, 2016

Publication in peer-reviewed journals

1. Z. Belarbi, J. Dominguez Olivo, F. Farelas, D. Young, M. Singer, S. Nesic; submitted to Corrosion

journal (2019)
2. Z. Belarbi, J. P. Trembly, Journal of the Electrochemical Scociety, 165 (2018) (13) E685-E693
3. Belarbi, T. N. Vu, F. Farelas, D. Young, M. Singer, S. Nesic; Corrosion 73 (7) (2017)892-899
4. Belarbi, F. Farelas, M. Singer, S. Nesic; Corrosion 72 (10) (2016)1300-1310
5. Belarbi, J. Gamby, L. Makhloufi, B. Tribollet; Electrochimica Acta 189 (2016) 118-127
6. Belarbi, J.Gamby, L. Makhloufi, B. Sotta, B. Tribollet; Journal of Crystal growth 386 (2014) 208-214
7. Belarbi, J. Gamby, L. Makhloufi, B. Tribollet; Electrochimica Acta 109 (2013) 623-629

Conference Papers

1. Belarbi, J. Dominguez Olivo, F. Farelas, D. Young, M. Singer, S. Nesic; CORROSION/2019, paper no. 13184 (Nashville, Tennessee, NACE, 2019)
2. M. Sani, Z. Belarbi, B. Brown, S. Nesic; CORROSION/2019, paper no. 13026 (Nashville, Tennessee, NACE, 2019)
3. Singer, F. Farelas, Z. Belarbi, S. Punpruk, A. Jenkins; CORROSION/2019, paper no. 13303 (Nashville, Tennessee, NACE, 2019)
4. Belarbi, F. Farelas, D. Young, M. Singer; CORROSION/2018, paper no. 10823 (Phoenix, Arizona, NACE, 2018)
5. Belarbi, D. Young, N. Nesic, R. Nogueira, M. Singer; CORROSION/2018, paper no. 10962 (Phoenix, Arizona, NACE, 2018)
6. Belarbi, T. N. Vu, F. Farelas, D. Young, M. Singer; CORROSION/2017, paper no. 9096 (New Orleans, Louisiana, NACE, 2017)
7. Belarbi, B. Brown, M. Singer; CORROSION/2017, paper no. 9290 (New Orleans, Louisiana, NACE, 2017)
8. Belarbi, F. Farelas, M. Singer, S. Nešić; CORROSION/2016, paper no. 7274 (Vancouver, BC: NACE, 2016)
9. Belarbi, J. Gamby, B. Sotta, L. Makhloufi, F. Chaussec, B. Tribollet; 21st Day Information Water, 4 - 6 November, 2014. Poitiers, France
10. Belarbi, J. Gamby, B. Tribollet, L. Makhloufi; 20th Day Information Water, 25 - 27 September, 2012. Poitiers, France

572800 Nanoscale Design of Electrocatalysts for CO_2-to-Fuels Powered Using Renewable Electricity

Research Interests: I am passionate about renewable energy technology and I want to develop my research program to study the fundamental mechanisms of electrocatalytic materials for the sustainable production of commodity chemicals.

On the cathodic side, how can we achieve high-value chemical products, such as CO_2 to C4+? What is a suitable anodic reaction that can generate value added products, other than oxygen, and maximize the overall energy efficiency at the same time? To answer these questions, my research program will focus on a combination of computational chemistry, electrochemical experiments and in-situ characterizations.

Teaching Interests: with a B.S. degree in chemical engineering and a number of years working as a R&D engineer, I am comfortable teaching any chemical engineering courses in an undergraduate curriculum. I am particularly interested in chemical kinetics, transport phenomena and material&energy balances.

My Ph.D. training is in electroanalytical methods, thus I am particularly suited to teach an electrochemical engineering method class. This class will cover topics such as electrochemical thermodynamics, kinetics and catalysis. We will apply the fundamental principles of electrochemistry to understand engineering applications such as fuel cell, battery and corrosion. The difficulty of this class is suitable for an upper-level undergraduate student or a graduate student.

566228 Reinforced Anion Exchange Membrane (AEM) Separators Based on Triblock Copolymers for Electrode-Decoupled Redox Flow Batteries (RFBs)

Shrihari Sankarasubramanian

Research Scientist

Department of Energy, Environmental and Chemical Engineering, Washington University in St. Louis, St. Louis, MO

Research motivation: Energy consumption is one of the barometers of economic and human well-being and the equitable development of humankind requires sustainable energy equity. Achieving this overarching goal requires us to address issues such as the intermittency of ecologically non-disruptive renewable energy technologies (solar and wind), decarbonize the transportation sector, and find alternative chemical pathways for large scale, energy intensive industrial processes. *I aim to help address these key issues by leveraging my training and experience in multi-scale theoretical and experimental investigations of electrochemical systems.*

Postdoctoral project – *"Reinforced anion exchange membrane (AEM) Separators Based on Triblock Copolymers for Electrode-decoupled redox flow batteries (RFBs)"*

Supervisor: Prof. Vijay K. Ramani, Department of Energy, Environmental and Chemical Engineering, Washington University in St. Louis

PhD dissertation – *"Investigation of the oxygen reduction reaction at the lithium-oxygen cell cathode"*

Supervisor: Prof. Jai Prakash, Department of Chemical and Biological Engineering, Illinois Institute of Technology

Research Interests:

My research journey began with my undergraduate work on continuum scale modeling of the capacity fade mechanisms in lithium-ion batteries with Prof. Balaji Krishnamurthy. An internship at Tata Power Solar, where I assisted in formulating a new silicon texturing bath composition (resulting in a ~1% improvement in final cell efficiency), whetted my appetite for translational research.

During my PhD research with Prof. Prakash at Illinois Institute of Technology, I: 1) evaluated novel lithium-ion cell chemistries for use in sub-zero temperatures (in conjunction with Hydro-Quebec), 2) performed the first ever accelerating rate calorimetry study of the thermal safety of lithium-sulfur cells (again with Hydro-Quebec), and 3) performed a multi-year theoretical and experimental investigation of catalysts, solvents and kinetics of the oxygen reduction reaction (ORR) in lithium-oxygen cells in collaboration with Toyota Motors. Over the course of these projects, I extensively employed density functional theory (DFT) simulations, continuum scale kinetic modeling, and classical electrochemical and electroanalytical experimental techniques.

As an integral part of my postdoctoral research project, I am investigating anion exchange membranes for use in electrode-decoupled redox flow batteries (RFBs). Herein, I am applying my knowledge of electrochemistry and transport phenomena to understand the transport of ions across a functionalized membrane. The final goal of this project, funded by ARPA-E, is to develop selective membranes that allow for the deployment of low-cost, electrode-decoupled RFBs such as the Fe-Cr system.

Future research directions: *I plan to leverage my training and experience in multi-scale theoretical and experimental investigations of electrochemical systems to carry out transformative, translational research into next generation, beyond lithium-ion batteries, grid scale energy storage using redox flow batteries,* *and the electro-synthesis of bulk and fine chemicals.*

Project 1: *Metal- (metal superoxide) batteries* – Lithium-oxygen batteries have been the holy grail of next generation battery systems due to their unmatched theoretical specific energy (3458 Wh kg^{-1}). Despite years of extensive work on catalysts, redox mediators and ever-more-stable electrolytes, a practical lithium-oxygen battery has not been realized. As part of this project I will examine sodium- and potassium-oxygen batteries whose reduction products are superoxides instead of the peroxides, allowing me to circumvent the columbic efficiency and degradation problems of lithium-oxygen batteries. I will combine insights from electrochemical kinetics studies with oxygen impermeable membranes and water tolerant ionic liquids to demonstrate long life alkali metal- (metal superoxide) batteries. *The end goal will be the development of a true metal-ambient air battery.*

Project 2: *Low-cost, grid scale energy storage* – The adoption of intermittent renewable energy sources calls for the large-scale deployment of energy storage solutions to maintain grid reliability. The DOE estimates that $100/kWh is the cost at which storage solutions become cost competitive. This cost target precludes both the Li-ion battery and the all-vanadium redox-flow battery. I will investigate alternative, low cost, earth abundant chemistries for redox-flow batteries. The use of catalytic inhibitors to suppress the hydrogen evolution and oxygen evolution reactions, in conjunction with electrolyte systems that increase the nominal redox potential of the redox couple of choice, will pave the way for >2V RFBs utilizing couples such as Cr-Ce and provide an unparalleled cost advantage compared to existing technologies. Further, I will develop highly perm-selective, high anion transference number anion-exchange membranes for this system both in-house and in collaboration with polymer scientists and material chemists.

Project 3: *Electrochemistry in extreme environments* –

Fundamental investigation of CO_2 and O_2 electrochemistry on Mars– Electrochemical processes under conditions encountered in Mars pose several interesting fundamental questions. The low temperatures increase the half-life of reaction intermediates (for example by decreasing the rate of the superoxide disproportionation mechanism) and severely diminishes the rates of transport and kinetic processes. This can lead to an increase in free radical degradation of electrochemical cell components and also decrease the selectivity of catalysts. The linearity of Arrhenius and Eyring relationships may be non-applicable in and we may need to reevaluate our understanding of solvation under these conditions. Thus, I propose to undertake a comprehensive electrochemical study of the major soluble components of the Martian regolith, eventually moving on to regolith simulants like JSC-1.

Development of a CO_2-brine electrolyzer- The CO_2-brine electrolyzer which reduces CO_2 on the cathode side while producing oxygen by water oxidation on the anode side is an elegant, unitized solution to both capture CO_2 and produce valuable industrial O_2 simultaneously. This system is inefficient at the system level because of the low concentrations of CO_2 available in the Earth's atmosphere and the associated costs of concentrating it to useable levels. But using this under Martian conditions, where the atmosphere is 96% CO_2 changes the economic value proposition and makes it an enabling technology for eventual Martian exploration and settlement. I will investigate the fundamental electrochemistry of CO_2 electroreduction under the near-neutral conditions found in the regolith simulant electrolyte, develop catalysts and eventually demonstrate such an electrolyzer system.

Teaching Interests:

The experience I have gained assisting in the teaching of 9 courses across the spectrum of ChE, from the sophomore level to the advanced PhD level, has prepared me well to serve as an instructor teaching the traditional chemical engineering core. I have served as Assistant to instructor and Support to instructor respectively for courses on Electrochemical Engineering and Energy Conversion and Storage at Washington University, where I have been actively involved in class planning and laboratory sessions in addition to traditional teaching assistant roles. During my PhD, I served as a Teaching assistant for 7 courses including thermodynamics, transport phenomena, undergraduate laboratory and electives in energy technology. The experience of assisting in the teaching of thermodynamics as an introductory first course, an intermediate course for 1st year master's students and as an advanced course for PhD students provided invaluable insights into the nuances of balancing depth and clarity without sacrificing technical accuracy. The mentor-mentee relationships I formed with high school, undergraduate and graduate students in the lab redoubled and renewed my commitment to excellence in teaching and training. I look forward to bringing this passion to my future role as a ChE faculty member.

Invention and technology translation (1 patent issued, 3 disclosed): I have always wanted to see my work changing society in a reasonable timeframe. My PhD work on Pd and Pt alloy catalysts for lithium-oxygen batteries with Toyota resulted in a patent and is the basis of their continued research in this field. My on-going postdoctoral work on reinforced membranes for flow batteries has already resulted in two $50,000 Leadership in Entrepreneurial Acceleration Program (LEAP) grants from Washington University to serve as a pre-seed fund for a start-up. My future independent research career will maintain that same strong practical, inventive and translational emphasis.

Selected Publications (15 published, 5 under review)

1. **Sankarasubramanian,** V. Ramani, Dimethyl sulfoxide -based electrolytes for high current potassium-oxygen batteries, *Journal of Physical Chemistry C,* **122** (34), 19319 (2018).
2. Wang, J. Parrondo, C. He, **S. Sankarasubramanian**, V. Ramani, Microscale-Bipolar-Interface-Enabled pH Gradients in Electrochemical Devices, *Nature Energy* (2019) DOI: 10.1038/s41560-019-0330-5.
3. **Sankarasubramanian,** N. Singh, F. Mizuno, J. Prakash, *Ab initio* investigation of the Oxygen Reduction Reaction activity on noble metal (Pt, Au, Pd), Pt_3M (M=Fe, Co, Ni, Cu) and Pd_3M (M=Fe, Co, Ni, Cu) alloy surfaces, for Li-O_2 cells, *Journal of Power Sources,* **319**, 202-209 (2016).
4. **Sankarasubramanian,** J. Seo, F. Mizuno, N. Singh, K. Takechi, J. Prakash, Enhancement of oxygen reduction reaction rate by addition of water to an oxidatively stable ionic liquid electrolyte for lithium-air cells, *Electrochemistry Communications,* **73,** 55-57 (2016).
5. **Sankarasubramanian,** J. Seo, F. Mizuno, N. Singh, J. Prakash, Elucidating the Oxygen reduction reaction kinetics and the origins of the anomalous Tafel behavior at the Li-O_2 cell cathode, *Journal of Physical Chemistry C*, **121,** 4789-4798 (2017).
6. Zhang, J. Parrondo, **S. Sankarasubramanian,** V. Ramani, Detection of Reactive Oxygen Species in AEM Fuel Cells using *in situ* Fluorescence Spectroscopy, *ChemSusChem*, **10**, 3056-3062 (2017).

581970 Developing Electrochemical-Based Predictive Battery Health Monitoring for Future Battery Management Systems

Developing Electrochemical-based Predictive Battery Health Monitoring for Future Battery Management Systems

Saeed Khaleghi Rahimian, Technical Lead of Modeling Team, Battery Technology Group, SERES (formerly SF Motors), Santa Clara, CA

Research Interests:

Primary Research: Develop High-fidelity Electrochemical-based Battery Life Models

Due to global concerns about the environmental impacts of fossil fuels, there has been a significant growth in Electric Vehicles (EVs) and Energy Storage Systems (ESSs) market. Li-ion batteries due to high energy and power densities are the dominant technology in EVs and they have a huge potential for use in off-grid power supply systems in combination with renewable photovoltaics and wind power. However, these batteries are prone to unexpected failures and safety issues. A battery management system (BMS) is a key element in EVs and the smart grid to make the batteries operate within the proper voltage and temperature interval, guarantee the safety and prolong their service life. To fulfill this goal, the BMS requires an accurate estimation of the battery state of charge (SOC), state of power (SOP) and state of health (SOH). The accurate and reliable online battery state estimation is a challenging issue since these states are not measurable using readily available sensors. The SOH, which depends on the battery capacity and power fade, predicts the number of times the battery can be charged and discharged before its life is terminated, thus it can help prevent catastrophic hazards and premature failure. To optimize the battery performance while maintaining its safety and reliability, a physics-based battery life model is significantly more beneficial compared to the popular data-driven or semi-empirical equivalent circuit models. Nonetheless, due to highly complex nature of degradation mechanisms in Li-ion cells [1], there is a considerable lack of high-fidelity life models that can predict the cell behavior under a variety of cycling and storage conditions.

Past/Current Works:

I am currently leading a battery modeling group to build and validate full-order/reduced-order electrochemical-based life models for commercial Li-ion cells with Ni-rich cathodes and Si/graphite composite anodes. We initially incorporated the Li plating side reaction on anode particles into the well-known Newman electrochemical model to study the impact of battery pack cooling strategy on nonuniform Li plating under fast charge conditions [2]. As part of the offline SOH estimation, implemented in the existing BMS, we built a control-oriented physics-based calendar life model by modeling the anode Solid Electrolyte Interphase (SEI) formation [3] as well as a cathode parasitic side reaction [4]. The model was able to predict the cell capacity fade and resistance rise data in a wide range of SOC and temperature storage conditions [5]. I am also working on a novel online SOH estimation approach that can track the loss of electrodes' active materials and Li^+ inventory by fitting the cell open circuit voltage at a few SOCs [6].

Future Direction

As an assistant professor, I would like to continue working on physics-based life modeling of Li-ion batteries to develop an advanced SOH estimation algorithm for BMS future generations. To achieve this goal, my plan is to incorporate the other critical aging phenomena such as active materials loss, electrodes' microstructure changes and electrolyte decomposition with gas generation into the life model. This life model is crucial to determine the battery dominant degradation mechanisms during its lifetime. Thus, it would help us to optimize the operation conditions (e.g. fast charge) to enhance the battery durability while avoiding catastrophic failures. The model can also be integrated with a simplified battery pack thermal simulation to improve the battery thermal management system. The other application of the model is to predict the cell remaining useful life, which can be greatly useful for decision making on second-life of EV batteries [7]. This type of modeling would be computationally expensive; however, the simulation can be run remotely with the advanced cloud technology [8].

Secondary Research: Develop Efficient Numerical Global Optimization Methods

Global optimization (GO) possesses a vast application in different fields of science and engineering [9]. GO aims at finding the best solution or a set of local solutions with high quality of a problem subject to a set of constraints which define a feasible region. Homotopy continuation methods are promising GO techniques that extend the convergence domain of locally convergent methods (e.g. Newton-Raphson) to seek all roots of a nonlinear system of equations [10].

Past/Current Works:

In parallel to my electrochemical engineering research, I also worked on a new numerical technique that significantly enhanced the conventional homotopy continuation methods in terms of finding all solutions of a single (multiple) nonlinear equation(s) [11-12]. Recently, I came across a great mathematical paper by Decker and Keller [13], which was the motivation to improve the method robustness by applying the appropriate numerical continuation method [14]. However, the computation time increased dramatically with the number of equations ($\sim 2^n$, n= number of equations) and thus, it was limited to small systems ($n<10$). By using a novel idea, I was able to reduce the computation time significantly ($\sim n^2$), that allowed me to solve large systems up to 30 equations [15].

Future Direction

I have a strong interest to apply the new homotopy method to solve the nonlinear systems of equations arising in physics, chemistry and biological sciences. In particular, I would like to find all stationary points of potential energy functions, which play a crucial role in determining the rates of chemical and physical transformations. My next plan is to combine the new method with a locally convergent optimization routines (e.g. Nelder-Mead) to locate the global optimal solution(s) of objective functions with several local optimums. I am highly ambitious to develop a novel GO method to locate the global minima of potential energy functions, which is extremely useful in prediction of the structure of proteins, clusters, crystals and biomolecules. This would help to accelerate the new material discovery [16] and the structure-based computer-aided drug design [17].

Teaching Interests:

I believe a major part of my role as a teacher of engineering students is to help them prepare to be successful professional engineers. A student embarking on a career path, whether in academia or industry, should have a comprehensive knowledge base, a wide range of problem-solving skills, and the ability to communicate and collaborate with others. This involves not only imparting knowledge directly related to course materials, but also fostering the development of secondary skills which will help them approach real-world problems.

My teaching interests are quite broad at both the undergraduate and graduate levels. At the undergraduate level, I am interested in teaching Reaction Kinetics, Transport Phenomena, Thermodynamics, Numerical Methods in Chemical Engineering, and Introduction to Electrochemical Energy Systems. At the graduate level, I am primarily interested in teaching courses in Advanced Reaction Engineering, Advanced Numerical Methods applied in Chemical Engineering, and Electrochemical Energy Systems Modeling and Optimizations.

References

[1] M. Palacín et al., Why do batteries fail?, *Science*, 351 (2016).

[2] M. Forouzan et al., Non-Uniform Li Plating Behavior of Li-Ion Cells in a Battery Pack with Bottom,

Symmetric Top-Bottom, or Side Cooling Thermal Management Systems during Fast Charging, *235th ECS Meeting* (2019).

[3] M. Pinson et al., Theory of SEI Formation in Rechargeable Batteries: Capacity Fade, Accelerated Aging and Lifetime Prediction, *J. Electrochem. Soc.*,160 (2013).

[4] J. Kim et al., Prospect and Reality of Ni-Rich Cathode for Commercialization, Adv. Energy Mater. (2017).

[5] S. Khaleghi Rahimian et al., A Reduced-Order Physics-Based Calendar Life Model for Li-Ion Cells, submitted to *AIChE Annual Meeting* (2019).

[6] S. Khaleghi Rahimian et al., A Novel Online State of Health Estimation of Li-Ion Cells, submitted to *AIChE Annual Meeting* (2019).

[7] M. Baumann et al., Could-connected Battery Management for Decision Making on Second-Life of Electric Vehicle Batteries, *13th International Conference on Ecological Vehicles and Renewable Energies* (2018).

[8] T. Kim et al., Cloud-Based Battery Condition Monitoring and Fault Diagnosis Platform for Large-Scale Lithium-Ion Battery Energy Storage Systems, *Energies,* 11 (2018).

[9] C. Floudas et al., A Review of Recent Advances in Global Optimization, *J Glob Optim*, 49 (2009).

[10] M. Kuno et al., Computing All Real Solutions to Systems of Nonlinear Equations with a Global Fixed-point Homotopy, *Ind. Eng. Chem. Res.,* 27 (1988).

[11] S. Khaleghi Rahimian, et al., A New Homotopy for Seeking all Real Roots of a Nonlinear Equation, Comp. & Chem. Eng., 35 (2011).

[12] S. Khaleghi Rahimian et al., A Robust Homotopy Continuation Method for Seeking All Real Roots of Unconstrained Systems of Nonlinear Algebraic and Transcendental Equations, *Ind. Eng. Chem. Res.*, 50 (2011).

[13] D. Decker et al., Multiple Limit Point Bifurcation, *J. Mathematical Analysis and Applications*, 75 (1980).

[14] S. Khaleghi Rahimian et al., A Global Homotopy Continuation Algorithm to Solve Systems of Nonlinear Equations, submitted to *Journal of Computational and Applied Mathematics.*

[15] S. Khaleghi Rahimian et al., Finding All Solutions of Systems of Nonlinear Equations with an Efficient Global Homotopy Continuation Method, submitted to *AIChE Annual Meeting* (2019).

[16] A. Pulido et al., Functional Materials Discovery using Energy–Structure–Function Maps, *Nature*, 543 (2017).

[17] G. Sliwoski et al., Computational Methods in Drug Discovery, *Pharmacological Reviews*, 66 (2013).

581958 Charge Storage and Transport in Electrochemical Science: From Bulk to Interfaces

Research Interests:

Seeking clean energy is one of the most urgent problems humanity is facing. As far as cleanness and efficiency are concerned, electrochemical technologies for energy storage and conversion can offer reliable and sustainable solutions in this endeavor.

My research interests focus on fundamental understanding of electrochemical processes, which essentially refers to thermodynamics, kinetics, and transport of materials. As electrode materials typically are required to convey both electrons and ions, their thermodynamics are driven by the electrochemical potentials of charge species. Such a framework holds not only for bulk phases but also at the interface of materials. By performing experimental characterizations and theoretical calculations, I have comprehensively studied the storage phenomena for energy materials. To elucidate the storage mechanisms, I developed a generalized model that describes the fundamental relation between the storage capacity and the component activity [1]. The developed model is in a good agreement with experimental evidences for metal storage (Li and Ag) and for gas storage (H_2) [2,3].

Equally important to thermodynamics is kinetics. It determines the charging speed of energy storage devices, e.g. batteries, or the power density of energy conversion devices, e.g. fuel cells. I am particularly interested in studying kinetic problems at heterointerfaces. A recent work showed that a prototype composite consisting of a superionic conductor and an electronic conductor exhibits an unprecedentedly fast kinetics in Ag storage. The rate characterized by chemical diffusion exceeds any other solids at room temperature, and even exceeds the rate of NaCl in liquid water [4]. The finding of this study may open a new class of composite materials that allow for ultrafast energy delivery.

Aside from energy storage and conversion technologies, I am also interested in irreversible electrochemical processes, e.g. corrosion. My previous works investigated the mechanism of carbon corrosion induced by molten salts, a critical reaction governing the practical efficiency of direct carbon fuel cells. The *in situ* observation showed that the carbon corrosion kinetics is strongly correlated with the wetting behavior of molten salts [5]. This work together with a modeling study were later adopted by a fuel cell company to optimize the design of electrode reactors.

Selected publications:

[1] C.-C. Chen and J. Maier, Space charge storage in composites: thermodynamics, ***Physical Chemistry Chemical Physics***, 19, 6379-6396 (2017)

[2] C.-C. Chen and J. Maier, Decoupling electron and ion storage and the path from interfacial storage to artificial electrodes, ***Nature Energy***, 3, 102–108 (2018)

[3] L. Fu, C.-C. Chen, et al., Thermodynamics of lithium storage at abrupt junctions: Modeling and experimental evidence, ***Physical Review Letters***, 112, 208301, 1–5 (2014)

[4] C.-C. Chen, L. Fu, and J. Maier, Synergistic, ultrafast mass storage and removal in artificial mixed conductors, ***Nature***, 536, 159–164 (2016)

[5] C.-C. Chen, et al., Wetting behavior of carbon in molten carbonate, ***Journal of The Electrochemical Society***, 159, D597–D604 (2012)

Teaching Interests:

I have broad interests in teaching the Chemical Engineering coursework, including thermodynamics, transport, reaction kinetics, interface science, and electrochemistry. At Stanford University, I served as a guest lecturer in a renewable energy class. My goal was to teach thermodynamics, kinetics, and transport phenomena in energy technologies. To this end, I developed an instructive set of materials and structured the lecture in a manner that the students can clearly see the connection between the fundamental knowledge and the real-world applications. During my PhD study, I also mentored several graduate students working on different projects.

582017 Electrocatalysts and Gas Diffusion Electrodes for Vapor-Phase Carbon Dioxide Electrolyzers and Hydrogen Fuel Cells

Postdoctoral Research:

Sept. 2019 – Present, *Stanford University, Chemical Engineering, CA USA*

Advisor: Thomas Jaramillo, Fritz Prinz

Projects:

(1) Development of electrocatalysts and gas diffusion electrodes for vapor-fed carbon dioxide reduction electrolyzers.

(2) Development of gas diffusion electrodes for hydrogen fuel cells using atomic layer deposition.

Ph.D. Research:

2011-2017, *University of Waterloo, Chemical Engineering, ON Canada*

Advisor: Zhongwei Chen

Thesis: Development of novel nanostructured bi-functional catalysts and design of advanced electrodes for rechargeable metal-air batteries

Funding Proposals:

- (Co-writer) Agency: *Ontario Centres for Excellence (OCE)*, Title: Advanced Rechargeable Aqueous Lithium-ion Zinc Flow Battery
- (Co-writer) Agency: *Natural Sciences and Engineering Research Council of Canada (NSERC) – Strategic Project Grants*, Title: Ionic Liquids as Electrolytes for Supercapacitors and Metal-air Batteries
- (Co-writer) Agency: *Natural Sciences and Engineering Research Council of Canada (NSERC) - CRD*, Title: Production Scale-up and Optimization of Core-Corona Structured Bifunctional Catalysts for Rechargeable Zinc-Air Battery

Research Interest:

My broad research interest includes development of electrochemical energy systems and understanding the factors that govern their performance. My two current thrusts are (1) *development of catalyst and gas diffusion electrode systems for* ***vapor-fed carbon dioxide electrolyzers*** *for energy and fuel generation*, and (2) *precious and non-precious metal catalyst design and testing for* ***hydrogen fuel cells***.

To perform the above research in a systematic manner and gain as much insight as possible, I plan to divide my research into following multi-disciplines: (1) Understanding phenomena at gas diffusion electrodes, (2) electrode engineering, (3) device integration, (4) catalyst/electrode characterization, (5) novel electrode designs and fabrication, and (6) in-situ, -operando studies.

Teaching Interest:

I am prepared to teach both fundamental undergraduate and advanced graduate level courses in the field of chemical engineering, and materials sciences and engineering. Based on my undergraduate training in Nanotechnology Engineering (U. of Waterloo), I plan to design new courses with emphasis on multi-disciplinary science, such as (1) structures and properties of nanomaterials, (2) physicochemical characterization of nanomaterials, and (3) materials for energy applications.

Publications *(5 selected out of 42)*:

- **Lee, D. U.**, Li, J., Park, M. G., Seo, M. H., Ahn, W., Stadelmann, I., Ricardez-Sandoval L., Chen, Z. W. (2017) *Self-Assembly of Spinel Nanocrystals into Mesoporous Spheres as Bifunctionally Active Oxygen Reduction and Evolution Electrocatalysts*. ChemSusChem. 10 (10): 2258-2266
- **Lee, D. U.**, Xu, P., Cano, Z., Kashkooli, A. G., Park, M. G., and Chen, Z. W. (2016) *Recent Progress and Perspective on Bi-functional Oxygen Electrocatalysts for Advanced Rechargeable Metal-air Batteries*. Journal of Materials Chemistry A. 4: 7107-7134
- **Lee, D. U.**, Fu, J., Park, M. G., Liu, H., Kashkooli, A. G., and Chen, Z. W. (2016) *Self-Assembled* $NiO/Ni(OH)_2$ *Nanoflakes as Active Material for High-Power and High-Energy Hybrid Rechargeable Battery*. Nano Letters. 16 (3): 1794–1802
- **Lee, D. U.**, Choi, J.-Y., Feng, K., Park, H. W., and Chen, Z. W. (2013) *Advanced Extremely Durable 3 Dimensional Bi functional Air Electrode for Rechargeable Zinc Air Batteries*. Advanced Energy Materials. 4 (6): 1301389
- **Lee, D. U.**, Kim, B., and Chen, Z. W. (2013) *One pot synthesis of a mesoporous* $NiCo_2O_4$ *nanoplatelet and graphene hybrid and its oxygen reduction and evolution activities as an efficient bi-functional electrocatalyst*. Journal of Materials Chemistry A. 1 (6): 4754-4762

Patents *(2 selected out of 4)*:

- **Lee, D. U.**, and Chen, Z. W. (2016) *Single layer air electrode and processes for the production thereof.* US 2016/0322645 A1
- **Lee, D. U.**, Park, H. W., Feng, K., and Chen, Z. W. (2012) *Advanced Bi-functional Electrode Based on Metal Oxide Nanowires and its Method of Growth on Metal Support.* US patent serial #: 61/963,877

582109 Combining Advanced Materials and Manufacturing and Electrochemical Engineering for Building a Sustainable Future

Research Interests:

Our planet – including our international community – must address pressing global issues such as climate change, food and water quality, and growing energy demand. Chemical engineering, including reactor design and catalysis development, are crucial to meeting these challenges, including efficient production of fuels and chemical precursors and removal of toxins and pollutants from water and food sources. Advanced electrochemical energy storage, separation and conversion devices, such as batteries, electroactive membranes, and catalytic reactors, are promising approaches to addressing these concerns using renewable energy sources. Effective engineering of these devices is complicated by limitations on the control of the electrochemical cell component morphologies across scales, thereby hindering our ability to unlock the full potential of these technologies.

My research program will focus on pioneering potential solutions to this problem by using advanced materials development and additive manufacturing techniques. These approaches can be used to design and control structures from the sub-micron through the centimeter length scales. Furthermore, the use of advanced manufacturing will allow my lab to explore novel catalyst and reactor designs that would be otherwise unattainable through other manufacturing approaches. These electrochemical studies can allow us to design experiments to accurately answer fundamental questions of the influence of environment, operating conditions and fluid dynamics on the intrinsic properties of new materials. Coupled with automatic design methodology, we can develop a systematic path toward rational design of electrochemical devices to address some of the most pressing needs of the planet.

Teaching Interests:

My objective as an educator is to build a framework and culture for not only learning, but creative and critical thinking, innovative application of the concepts conveyed, and widespread service to others. When I teach, my objective is for my learners to gain understanding, apply knowledge to new areas, and utilize learning as a tool for aiding society. As instructors, we are uniquely positioned to equip the engineers of tomorrow with the technical foundation they need to be able to evaluate, assess and address the daunting challenges of energy, water, food, and climate facing our planet. Furthermore, I believe as leaders in education, we are called to not only design courses and curriculum that engage students from a plethora of backgrounds and experiences, but to also impart a culture of service within all students. By combining strong engineering principles with a mentality that focuses on aiding others, we enable students to become the best engineers and leaders and to embody a paradigm of progress and service.

581873 Design of Electrochemical Biosensing Strategies Toward High-Sensitivity, Non-Fouling, Multiplex Point-of-Care Systems

Abstract: Electrochemical biosensors are some of the most well-established and reliable technologies for point-of-care sensors, with the electrochemical glucose sensor setting the gold standard for such devices. Their continued success can be attributed to the simplicity and low cost of the instrumentation required. However, limitations and challenges remain for electrochemical biosensor, including detection sensitivity, matrix effect and multiplex sensing. My doctoral study focused on the development of novel biosensing strategies to solve these limitations, producing a rapid, high-sensitivity, non-fouling, multiplex, universal biosensing system for point-of-care detection of nucleic acids, proteins and small molecules. The developed biosensing strategies involved the application of multidisciplinary areas, including synthetic biology, biochemistry, immunology and electrochemistry.

Keywords: synthetic biology, bioengineering, electrochemistry, biosensor, nucleic acid displacement reaction, peptide displacement reaction, CRISPR.

Research Interests: electrochemistry biological devices, synthetic biology network, electrochemically controlled CRISPR.

Teaching Interests: biochemistry, electrochemistry/electrochemical engineering, thermodynamics, biotechnology.

588415 High-energy Aqueous Li-ion Battery

4th Year as Assistant Research Scientist

Research Interests:

Non-aqueous electrolyte in state-of-the-art Li-ion batteries (LIBs) give rise to severe safety and environmental concerns, as they exhibit intrinsic drawbacks, including flammability, toxicity, and high sensitivity toward moisture. Aqueous batteries hold promise to address this issue, but are often maligned because of their low energy densities due to the narrow electrochemical stability window of water (1.23 V). Recently, we addressed this challenge and made ground-breaking efforts by expanding the ESW of water to >3.0 V when first introducing the concept of the "water-in-salt" electrolyte class (WiSE), based on highly or superconcentrated lithium salts (>21 mol/kg LiTFSI). After that, I dedicated last 4 years on fundamental understanding of this unique electrolyte system, including solvation structure, ionic transportation behavior, surface reaction behavior, etc. Based on this, I conducted a series of researches to push boundary of WiSE much further:

1. Highly reversible Li-sulfur chemistry was built on the nature of thermodynamically phase-separation between TFSI and polysulfide anions. High capacity sulfur anode left the energy density of aqueous battery from 80 Wh/kg up to 200 Wh/kg (material level). [Paper 4]
2. A flexible and wearable 3V-class aqueous symmetrical lithium-ion battery is developed using a single $LiVPO_4F$ material as both cathode and anode in a "water-in-salt" gel polymer electrolyte, showing the high-safety and eco-friendly features of aqueous electrolyte. [Paper 3]
3. The strong hydrophobic passivation layer was developed to minimizes the competitive water reduction during interphase formation in WiSE, enabling a 4V-class aqueous Li-ion batteries with high efficiency and reversibility. Such aqueous Li-ion batteries offer energy densities approaching very closely to those of non-aqueous Li-ion batteries (~300 Wh/kg, material level), but without the safety concern of the latter. [Paper 2]
4. A brand new "halogen conversion-intercalation" cathode chemistry inside graphite was developed, based on a densely packed stage-I graphite intercalation compound, C3.5[Br0.5Cl0.5], forming reversibly in WiSE. By coupling it with a passivated graphite anode, this high capacity cathode blazed a trail for high-energy (>400 Wh/kg, material level) aqueous batteries. [Paper 1]

Successful Proposal and Grant: DoD Independent Research and Development (IR&D) Funds, subcontract from Johns Hopkins Applied Physics Lab (JHU/APL) ($48,000 USD), 2018

Leading Research Projects:

1. "High Energy Aqueous Lithium Ion Batteries", DoE Advanced Research Projects Agency-Energy (ARPA-E)
2. Phantom Lord PLA "Improve the maturity of the 4.0 V aqueous flexible Li-ion battery technology", subcontract from US Army Research Laboratory
3. Phantom Lord PLR "Improve the maturity for the Gel Polymer Electrolyte
4. (GPE) flexible aqueous battery from lab based to fieldable", subcontract from US Army Research Laboratory
5. "Robust, Flexible, Aqueous Polymer Electrolyte Based Li-ion Batteries for Size and Weight Reduction", ORION funded by Johns Hopkins Applied Physics Lab

All under supervision of Prof. Chunsheng Wang, Department of Chemical and Biomolecular Engineering, University of Maryland

Research Experience:

My academic career path has covered multiple areas of material science and engineering, including electron structure and transportation in thermoelectric materials, photo-induced minority carrier injection in photovaltaic materials, and electron-ion charge transfer in electrochemical materials, *ect*. These crossed experiences always give me a broader vision to design desired function material systems. My formal training also includes a systematic physics study, which makes me be passionate for electrochemistry – most quantifiable chemistry science and engineering. Hence, I joined Prof. Chunsheng Wang's group and devoted myself to advanced battery research. Meanwhile, projects that I worked on were always close collaborations with large material characterization facilities: Argonne National Laboratory, Brookhaven national laboratory and NIST. As a result of these collaborations I have acquired invaluable and versatile experience in many research areas.

Teaching Experience and Interests:

Aside from my research career, I also have tried to seize any opportunity to gain teaching experience. I TAed "solid-state physics" course during my undergraduate period at Nanjing University. I guest-lectured graduate lecture "Advanced Fuel Cells and Batteries" in Department of Chemical and Biomolecular Engineering, University of Maryland. Meanwhile, I was very active in research mentorship and have advised 8 diverse students to-date, including 5 graduate students. As faculty, I am looking forward to teaching introductory and advanced courses on topics of physical chemistry, electrochemistry and solid physics and chemistry. I am especially interested in developing coursework that designs, synthesis, characterization, fabrication and application of materials for energy storage or harvest devices.

Future Direction:

As faculty I would like to continue developing advanced energy storage devices, which will mostly be focused on next-generation rechargeable batteries. Large-scale application of clean energy urgently demands new energy storage systems with higher energy density, higher safety and low cost. I believe innovate materials and fabrication strategies should be developed to build up new electrode chemistries to meet these requirements. I will also delicate to the fundamental understanding of electrolyte systems, which will give us huge advantages to manipulate their electrochemical stability and transport property.

Meanwhile, based on my previous research experiences of energy harvest materials, I will try to explore more possibility of devices or systems combining both energy harvest and storage. In many successful systems, especially electrochemical ones, they basically share the same principle or mechanism, but with different outputs or applications. Therefore, in the future, I foresee myself working on building more innovate systems, or potentially opening up new areas.

Selected Publications:

Papers: **(citation: 4973; h-index: 33)**

1. **Yang**, J. Chen, X. Ji, T. P. Pollard, X. Lü, C. Sun, S. Hou, Q. Liu, C. Liu, T. Qing, Y. Wang, O. Borodin, Y. Ren, K. Xu, C. Wang, Aqueous Li-ion Battery Enabled by Halogen Conversion-Intercalation Chemistry in Graphite, ***Nature***, 2019, 569, 245–250.
2. **Yang**, J. Chen, T. Qing, J. Chen, X. Fan, W. Sun, A. v. Cresce, M. S. Ding, M. A. Schroeder, N. Eidson, C. Wang, K. Xu, 4.0 V Aqueous Li-ion Batteries, ***Joule*** (Cell press), 2017, 1, 122–132.
3. **Yang**, X. Ji, X. Fan, T. Gao, L. Suo, F. Wang, W. Sun, J. Chen, L. Chen, F. Han, L. Miao, K. Xu, K. Gerasopoulos, C. Wang, Flexible Aqueous Li-ion Battery with High Energy and Power Densities, ***Advanced Materials***, 2017, *29*, 1701972.
4. **Yang**, L. Suo, O. Borodin, F. Wang, W. Sun, T. Gao, X. Fan, S. Hou, Z. Ma, K. Amine, K. Xu, C. Wang, Unique aqueous Li-ion/sulfur chemistry with high energy density and reversibility, ***Proceedings of the National Academy of Sciences***, 2017, *114*, 6197-6202 (Direct Submission).
5. **Yang**, Z. Wang, T. Lin, H. Yin, X. Lv, D. Wan, T. Xu, C. Zheng, J. Lin, F. Huang, X. Xie, M. Jiang, Core-Shell Nanostructured "Black" Rutile Titania as Excellent Catalyst for Hydrogen Production Enhanced by Sulfur Doping, ***Journal of the American Chemical Society***, 2013, *135*, 17831–17838.
6. Lin, I-W. Chen, F. Liu, **C. Yang**, H. Bi, F. Xu, F. Huang, Nitrogen-doped mesoporous carbon of extraordinary capacitance for electrochemical energy storage, ***Science***, *2015*, 350, 1508-1513.

Key words: energy storage, Li-ion battery, aqueous electrolyte, electrochemistry

582238 Advanced Materials for Efficient Energy Conversion Based on Spectroscopic and Mechanistic Study

Abstract:

Developing catalytic materials and device to enable efficient interconversion of chemical and electrical energies, *e.g.*, fuel cells and electrolyzers, is key to sustainable future. Despite the explosive growth of new materials, few has demonstrated satisfactory performance under realistic conditions in an MEA configuration. Guiding principles in rational design of electrocatalysts remain elusive. Developing advanced materials with enhanced catalytic performance, as well as manipulation of the interfaces between catalysts and membrane, is key to enable efficient energy conversion under realistic conditions in an MEA configuration. Through spectroscopic and fundamental mechanistic studies, the reaction pathway on catalysts can be monitored in MEA configuration, which provides insights for the rational design of nanostructured materials with well-defined structure/composition and manipulation of MEA interfaces to achieve high performance.

Research Interests:

Developing catalytic materials and device to enable efficient interconversion of chemical and electrical energies, *e.g.*, fuel cells and electrolyzers, is key to sustainable future. Guiding principles in rational design of electrocatalysts remain elusive. Despite the explosive growth of new materials, few has demonstrated satisfactory performance under realistic conditions in an MEA configuration. Developing advanced materials with enhanced catalytic performance, as well as manipulation of the interfaces between catalysts and membrane, is key to enable efficient energy conversion under realistic conditions in an MEA configuration.

In my PhD research, with Prof. Erkang Wang at Changchun Institute of Applied Chemistry, I developed a selective, active, and stable catalyst for oxygen reduction reaction (ORR), whose sluggish kinetics hinders the commercialization of fuel cells. By decreasing the size of silver nanoparticles to < 2 nm (2 to 5 atoms), the utilization efficiency of catalyst atoms was improved significantly, leading to superior performance as compared to the commercial Pt/C. During my postdoctoral research with Prof. Younan Xia at Georgia Institute of Technology, I developed a type of cost-effective Pt nanocages for ORR with enhanced catalytic activity and durability. Owing to the unique architectures of the synthesized nanoparticles, such as hollow interiors, porous walls, and well-defined facets, the Pt utilization efficiency and specific activity of Pt nanocages were significantly enhanced. During my postdoctoral research with Prof. Bingjun Xu and Prof. Yushan Yan at the University of Delaware, I successfully developed catalysts for electrochemical nitrogen reduction reaction (ENRR) by evaluating them under realistic conditions in an MEA configuration. Further, I have also developed expertise in operando surface enhanced infrared absorption spectroscopy (SEIRAS) to quantify the surface pH changes within the electrochemical double layer using CO_2 reduction reaction as a model system.

In my future research, I plan to develop advanced materials for efficient energy conversion including fuel cell and electrolysis by taking advantage of my unique background in material science, electrolysis, and spectroscopy. Through spectroscopic and fundamental mechanistic studies, the reaction pathway on catalysts can be monitored in MEA configuration, which provides insights for the rational design of nanostructured materials with well-defined structure/composition and manipulation of MEA interfaces to achieve high performance. For example, ORR as the cathode reaction has attracted intensive research interests because of its rate-determining role in fuel cells. Numerous nanomaterials have been demonstrated as catalysts for ORR with enhanced activity and durability in rotating disk electrode (RDE) setup. However, few of them show good performance in fuel cells due to the differences between RDE and MEA configurations. Through operando spectroscopic monitoring of ORR in both RDE and MEA configurations, the interfaces between catalysts and ionomer/membrane can be manipulated, which enables the discovery of advanced materials with enhanced catalytic performance in MEA. With extensive experience in the synthesis of nanomaterials and strong expertise in MEA and operando SEIRAS, I am uniquely suited to undertake these challenging research efforts.

Teaching Interests:

I have extensive experience working with undergraduate and graduate students during my stay at Georgia Institute of Technology and the University of Delaware, and enjoyed teaching them material syntheses, electrochemical measurements, and SEIRAS. I am interested in teaching most of the core Chemical Engineering courses, including thermodynamics and kinetics. I also plan to develop elective courses close to my research interest such as, spectroscopy, catalysis, material sciences, nanoscale fabrication, and electrochemistry.

Selected Publications:

1. Nash, J.*; **Yang, X.***; Anibal, J.; Wang, J.; Yan, Y; Xu, B. Electrochemical Nitrogen Reduction Reaction on Nobel Metal Catalysts in Proton and Hydroxide Exchange Membrane Electrolyzers. Journal of the Electrochemical Society **2017**, 164, F1712–F1716.
2. Dunwell, M.; **Yang, X.**; Setzler, B.; Anibal, J.; Yan, Y.; Xu, B. Examination of Near-Electrode Concentration Gradients and Kinetic Impacts on the Electrochemical Reduction of CO2 using Surface-Enhanced Infrared Spectroscopy. ACS Catalysis **2018**, 8, 3999–4008.
3. **Yang, X.***; Gilroy, K. D.*; Vara, M.; Zhao, M.; Zhou, S.; Xia, Y. Gold icosahedral nanocages: facile synthesis, optical properties, and fragmentation under ultrasonication. Chemical Physics Letters **2017**, 683, 613–618.
4. **Yang, X.**; Roling, L. T.; Vara, M.; Elnabawy, A. O.; Zhao, M.; Hood, Z. D.; Bao, S.; Xia, Y. Synthesis and characterization of Pt–Ag alloy nanocages with enhanced activity and durability towards oxygen reduction. Nano Letters **2016**, 16, 6644–6649.
5. Wang, X.; Figueroa-Cosme, L.; **Yang, X.**; Luo, M.; Liu, J.; Xie, Z.; Xia, Y. Pt-based icosahedral nanocages: Using a combination of {111} facets, twin defects, and ultrathin walls to greatly enhance their activity toward oxygen reduction. Nano Letters **2016**, 16, 1467–1471.
6. **Yang, X.**; Gan, L.; Zhu, C.; Lou, B.; Han, L.; Wang, J.; Wang, E. A dramatic platform for oxygen reduction reaction based on silver nanoclusters. Chemical Communications **2014**, 50, 234–
7. **Yang, X.***; Gan, L.*; Han, L.; Wang, E.; Wang, J. High-yield synthesis of silver nanoclusters protected by DNA monomers and DFT prediction of their photoluminescence properties. Angewandte Chemie International Edition **2013**, 52, 2022–2026.

582036 Fabrication of Conducting Polymers and Superconductors Using Chemical Vapor Deposition Methods for Energy and Electronic Application Devices

Research Interests:Â

Device Fabrication based on oxidative Chemical Vapor Deposition (oCVD) synthesize of Conducting Polymers (CPs) combines biocompatibility, electronic conductivity, optical transparency, and mechanical flexibility compatible with lightweight substrates. Due to these features oCVD CPs exhibit promising performance for next-generation devices in a wide range of applications including but not limited to electronic, optoelectronic, electrochemical, optochemical, drug release, tissue engineering, medical devices, sensors, soft robotics, corrosion protection layers, economical and distributed devices for the Internet of Things (IoT), thermoelectric devices, and energy storage and harvesting devices.[1, 2]

Nowadays, there is a significant implication of highly versatile electronic materials in energy harvesting and storage devices. Electronic materials are a vital building block for the advanced technologies as faster computing, and smaller device size is demanded to make our daily life more comfortable and productive. My research during Ph.D. and Postdoctoral span Superconductors, Semiconductors, and Conducting Polymers, which are the three main electronic materials, in both fabrication (using highly versatile roll-to-roll Metal Organic Chemical Vapor Deposition (MOCVD), and oCVD techniques), and energy/electronic device applications in macro and nanometer scales. Such an enriching experience provides me with a strong background to put my research direction in the field of electronic device fabrications.

During my postdoctoral research at the Massachusetts Institute of Technology (MIT), under the supervision of Prof. Karen Gleason, I am currently working on device fabrication based on the oCVD synthesis of conjugated conducting polymers (CPs). We developed a method to use volatile liquid oxidant such as vanadium oxytrichloride ($VOCl_3$) instead of common solid oxidants such as iron chloride ($FeCl_3$). $VOCl_3$ is a volatile liquid at room temperature, and its flow rate to the oCVD reactor is more controllable than its solid oxidants counterparts, which resulted in better control of the $VOCl_3$ surface concentration. In addition, the main advantage of using volatile liquid oxidant is that the final CPs thin films exploit directly in device fabrication without the need for any post-deposition rinsing step to remove unreacted oxidants and oxidation by-products. The resulting oCVD CPs properties such as conductivity can be tuned by controlling the $VOCl_3$ and monomer flow rates as well as the substrate temperature and other processing parameters. The outstanding results relates to exploiting CPs in perovskite solar cell device and flow battery were obtained and published in Science Advanced,[1] and submitted to Advanced Materials, respectively.

My Ph.D. research, focused on fabrication of Second-Generation High Temperature Superconductors (2G-HTS) by roll-to-roll MOCVD method. The 2G-HTS have a wide range of applications in power electronic devices. At the same time, I worked on developing the structure and texture analyses methods for 2G-HTS with outstanding current carrying capacity and mechanical strength on Ion Beam Assisted Deposition (IBAD)/MgO template (Figure 2).[3] As the outcome of my doctoral research, we obtained the world-record of the high critical current density value of 20 MA/cm^2 at 30 K, 3 T ($B||c$), and high pinning force value of 1000 GN/m^3 at 20 K in the REBCO coated conductors at the operating condition of wind turbines. This world-record of the high critical current density and high pinning force already mentioned and acknowledged in numerous publications by other researchers in the superconductivity community.

Electronic materials have provided motivation and inspiration for many innovations in modern technologies and energy harvesting and storage devices. There is a wide room for future modern device applications using Conducting Polymers (CPs) as biocompatible electronic components grown by highly versatile deposition technique like CVD and its derivative techniques such oCVD. As global energy demand continues growing, there is an urgent need for the development of clean, and renewable energy devices, which can reduce the dependence on fossil fuels. My research interests lie in electronic materials fabrication using novel vapor phase techniques (e.g., oCVD, MOCVD) for large-scale production of organic and inorganic nanostructures and implementation of these electronic materials into devices based on industrial demands and energy applications. Therefore, I plan to investigate several key points which are listed as three projects by focusing on the large-scale production by a versatile roll-to-roll CVD process and will be discussed at the meeting.

- Energy harvesting and storage devices (by focus on photovoltaic and electrochemical energy storage devices)
- Thermoelectric devices
- Integrating superconductors and conducting polymers in a flagship biocompatible electronic device

Teaching Interests:

I have had several enriching teaching and mentoring experiences during my graduate studies, and I view teaching as an essential and enjoyable part of being a faculty member. Like my research, my teaching experience includes both theoretical and practical courses, enabling me to comfortably step into teaching roles for theoretical as well as practical undergraduate and graduate courses. I served as a mentor of few undergrad and M.S students, during my Ph.D. and Postdoctoral studies and was serving as a teaching assistants of several courses.

During my Postdoctoral studies at MIT, I earned the Kaufman Teaching Certificate Program (KTCP) certificate as the evidence of a strong commitment to the teaching enterprise. During KTCP workshop sessions at MIT, I became completely familiar and got useful experience and knowledge relates to designing a course and constructing a syllabus, planning and facilitating a class session, interactive teaching and active learning, constructing effective problem sets and exam questions, and diversity.

(1) M. Heydari Gharahcheshmeh, M. M. Tavakoli, E. F. Gleason, M. T. Robinson, J. Kong, K. K. Gleason, Tuning, Optimization, and Perovskite Solar Cell Device Integration of Ultrathin Poly(3,4-ethylenedioxythiophene) (PEDOT) Films via a Single-Step All-Dry Process. Science Advances. In Press. (2019).

(2) M. Heydari Gharahcheshmeh, K. K. Gleason, Device Fabrication Based on Oxidative Chemical Vapor Deposition (oCVD) Synthesis of Conducting Polymers and Related Conjugated Organic Materials. Advanced Materials Interfaces. **6**, 1801564 (2019).

(3) M. Heydari Gharahcheshmeh, G. Majkic, E. Galstyan, A. Xu, M. Kochat, X-F. Li, V. Selvamanickam, Superconducting characteristics of REBCO coated conductors with different Zr content, IEEE Transaction on Applied Superconductivity. **29** (5), 6601105 (2019).

582052 Sustainability in Process Systems Engineering

Research Interests:

Environmental conservation, impact reduction, and economic efficiency are vital for the provision of manufactured goods, energy, food, shelter, transport, and almost all basic functions of society. Population growth, economic growth, and increasing requirements for the standard of living mean that more and better goods are in demand, which in turn require more natural resources and manufacturing activity. Such developments, if not carefully designed, can lead to resource depletion/degradation, more landfill waste, higher levels of pollutants, and increased environmental impacts.

My research interests focus on solving the aforementioned challenges through the development and use of PSE tools.

Research strand 1: Advancements in Sustainable Process Systems: Inspired by my Post-Doctorate work, in this strand I will examine the use of PSE tools such as supply chain optimization, process intensification, and data analysis to explore and give novel solutions to the fields of a) Circular Economy Systems Engineering, b) Food-Energy-Water Nexus and c) Renewable Energy Transition.

Research strand 2: Theoretical Advancements & Computational tools for large scale and hierarchical optimization: Inspired by my Ph.D. work, Strand 2 is focusing on the algorithmic development and mathematical tools & frameworks for large scale and/or hierarchical mixed integer optimization problems as they will be vital for the advancements in strand 1. Both deterministic and data-driven approaches will be explored for the development of reliable solution approaches for challenging large-scale problems with multiple decision makers.

Teaching Interests:

Being a teaching assistant in both Imperial College London and Texas A&M University I had the opportunity to assist or co-instruct a total of 6 different PSE classes, making me both competent and interested in designing and teaching different PSE courses including process analysis, design, modeling, optimization, and control for both undergraduate and graduate students. Furthermore, I am interested in teaching classes related to sustainable development such as “Environmental Engineering” and “Circular Economy Systems Engineering”.

I am especially interested in alternative methods of teaching and assessment that encourage critical thinking. Along with more traditional tools for teaching, such as lectures and tutorials, I am planning to enhance student engagement by active and team-based learning. Furthermore, I will be using projects (both group and individual) and presentations as alternative teaching and assessment methods.

The supervision of students is something that I really enjoy and look forward to doing in my academic career. During my Ph.D. and Postdoctorate, I had the opportunity to co-supervise 4 undergraduate students, 2 master students, and 5 Ph.D. students and I am finding this experience both enjoyable and gratifying.

582130 Nanostructured Materials for Next-Generation Lithium-Sulfur Batteries

Research Interests:

As interest in renewable energy grows and demand for electric vehicles strengthens, it is crucial for energy storage technology to improve beyond the currently used lithium-ion (Li-ion) batteries. Lithium-sulfur batteries (Li-S) are one of the most promising battery candidates to replace Li-ion batteries due to their high theoretical specific energy of 2510 W H kg^{-1}. In addition to their many advantages, Li-S batteries still suffer from a number of challenges including safety concerns regarding the presence of flammable liquid electrolytes and polysulfide transport. In my research group I will leverage my background in nanoscale materials and polymer synthesis to design and develop new nanostructured materials for Li-S cathode and separator. Design of tailored materials can solve fundamental Li-S challenges, improve safety. and lead to improved battery performance.

Graphene-based materials will also constitute another area of my research. As a highly conductive carbon with a unique 2-d structure, graphene has received intense academic interest and has promising applications in a large number of fields. In particular, graphene fibers can be produced graphene sheets using wet spinning to form an alternative type of carbon fiber. I will utilize graphene fibers in conjunction with modified cathode composition or as a coating on separators to further improve the Li-S battery performance. Graphene also has promise for biomedical applications, mainly due to its unique electrical and thermal conductivity. I plan to investigate the design of biosensors and drug delivery systems based on graphene and its derivatives. Further, due to multidisciplinary background which includes polymer synthesis, nanomaterials, biomaterials and material/chemical engineering, I am highly enthusiastic about collaborations with other researchers in industry and academia.

Research experience:

Postdoctoral projects

Gel-electrolyte Li-S batteries, Graphene-based materials, Joo Group, Cornell University, 2018-present.

My research background is a mixture of versatile disciplines including polymer synthesis, biomaterials, tissue engineering and material/chemical engineering. My current research is focused on Gel-electrolyte systems in Li-S batteries that have already overcome many challenges in Li-ion batteries. I develop novel gel systems, modify cathode materials and investigate novel compositions as anode materials in Li-S batteries. To further bring my study closer to practical applications in industry, I fabricate Li-S pouch cells in addition to coin cells to obtain high specific capacity Li-S in large sizes. Graphene-based materials is another area of my current research. I was able to develop a simple yet effective method to fabricate graphene fibers with excellent electrical, thermal and mechanical properties.

High internal phase emulsion, Biopolymer/protein interactions, Abbaspourrad Group, Cornell University, 2016-2017.

Prior joining the Robert Frederick Smith school of Chemical Engineering at Cornell University I worked on biopolymer/protein emulsions and their application in food industry. I also investigated protein/surfactant/biopolymer interaction in the Department of Food Science at Cornell University. During this time, I had the opportunity to visit a materials science lab in the Department of Materials Science and Engineering to work on living radical polymerization and emulsion polymerization.

Biodegradable photo-cured polymers in protein- delivery and tissue engineering, PPDDRL, Qatar University, 2013-2014.

As a postdoctoral fellow at Qatar University in collaboration with University of Bristol the main focus of my research was on biodegradable polymer synthesis, UV-crosslinking, and their thermal, mechanical and chemical characterization. Further, I opened a new route for research in that lab and defined projects based on electrospinning for the Master's and undergraduate students.

PhD dissertation

Biodegradable block copolymer nanoparticles suitable for drug delivery systems, Khoee Lab, University of Tehran, 2007-2012.

Pursuing my passion for polymer chemistry that I had been introduced to during my Master's project with photoactive poly amide-imides, I continued working in Khoee lab as a grad student and worked on block copolymer synthesis and biodegradable nanoparticles in drug delivery. The synthesized triblock-copolymers were able to self-assemble in aqueous environment and were further modified to a core-shell nanoparticle to improve the pH-sensitivity of the particles for proper targeted delivery of the loaded drug. Along with this project, I was involved in some side projects of electrospun fibers and biodegradation studies.

Selected Publications and Presentations:

1. **Zamani**, G. Shebert, R. Zhang, Y. Joo, "Effect of Gel electrolyte on polysulfide transport in Li-S batteries" (Meeting Abstract, The Electrochemical Society, Dallas, TX, May 2019)
2. **Zamani**, H. Ismaeil, M. Elrayess, W. Kafienah, H. Younes, "A Biodegradable Photo-Crosslinked Fibrous Elastomer for Cardiac Tissue Engineering Applications" (Meeting Abstract, International Conference on Biopolymers and Bioplastics, Baltimore, MA, March 2019).
3. AlAmer, **S. Zamani**, K. Fok[a], A. Satish and Y.L. Joo, "Preparation of expandable graphite and exfoliated graphene with a Taylor-Couette flow reactor" (under preparation to Advanced Functional Materials).

G.L. Shebert, **S. Zamani,** and Y.L. Joo, "End-of-discharge failure and recovery in Li-S batteries with liquid and gel electrolyte" (under preparation for Nano Energy).

1. Hassouna, **S. Zamani**, H. Younes, "Synthesis & Characterization of Tricarballylate Based Thermally crosslinked Elastomers for Drug Delivery & Tissue Engineering Applications" (Materials Science and Engineering C, **2018**, 93, 254-264).
2. **Zamani**, N.M. Malchion, M.J. Selig, A. Abbaspourrad, "Formation of shelf stable Pickering high internal phase emulsions (HIPE) through inclusion of whey protein microgels" (Food & Function, **2018**).
3. Abdallah, F. Jalali, **S. Zamani**, et.al., "Fabrication & Characterization of 3D Electrospun Biodegradable Nanofibers for Wound Dressing, Drug Delivery and Other Tissue Engineering Applications" (Pharmaceutical Nanotechnology, Volume 4, Number 3, September **2016**, pp. 191-201)
4. **Zamani**, S. Khoee, "Preparation of Core-Shell Chitosan/PCL-PEG triblock copolymer nanoparticles with ABA and BAB morphologies: effect of intraparticle interactions on physicochemical properties" (Polymer 53, **2012**, 5723-5736).

Teaching Interests:

Over the course of academic career my interest and contributions in teaching have grown. During my graduate studies I had the chance to serve as an instructor and teaching assistant for a large number of students in the courses of "General Chemistry" and "Organic Chemistry", respectively. Later in my postdoctoral fellowship at Qatar University, I was the co-lecturer of "Polymer synthesis and characterization" for graduate students.

More recently, in Spring 2019, I served as the instructor for the General Chemistry labs at State University of New York at Cortland. It was an amazing experience to interact with students who are taking their first steps into the huge world of chemistry. To strengthen my teaching skills, I participated in the Connecting Research Teaching Conference at Cornell University, which discussed high-impact teaching practices and techniques to improve student interaction.

Additionally, over the course of my career I have trained and mentored a number of undergraduate and graduate students. I enjoy passing on my experience and knowledge to the students, and it is especially rewarding when my help makes a big difference in their research.

As a faculty, I would be interested in teaching any course related to polymer chemistry and characterization, organic chemistry, materials science and biomaterials. Depending on the department, I would also like to develop new courses on biopolymers and biomaterials.

582135 Kinetics and Chemistry of Fast Pyrolysis of Carbon-Based Material Using Novels Reactors Approaches

Research Interests:

Fast pyrolysis of carbon-based material promises to be an effective and environmentally sustainable path to renewable energy and organic starting materials, i.e., a replacement for petroleum. In the fast pyrolysis process, the reactor is the performance-controlling process and a variety of alternatives have been considered at all scales: laboratory, pilot and pre-commercial/commercial. Yields and selectivity for pyrolysis is a function of numerous factors, including rates of heating and quenching (residence times for heterogeneous and homogeneous events) and condensed phase properties (such as the dynamic shrinkage of the materials). My Ph.D. thesis was focused on demonstrating a laboratory-scale fast pyrolysis technique (micro-particle micro-reactor system (MSMR)) using manufactured biomass microspheres. A unique single-particle (~10 μg) microreactor technology coupled with a millisecond response flame ionization detector was used to investigate the effects of relevant particle and process parameters and to capture the dynamics of real-time microscale single-particle pyrolysis for the first time. Another of my Postdoctoral research projects was working with a new experimental technique capable of elucidating the kinetics and chemical intermediates of biopolymer chemistry during the pyrolysis process. The technique of PHASR (Pulsed Heated Analysis of Solid Reactions) kinetics enables a millisecond temporal analysis of solid/melt phase reactions. Based on my research background and my focus during my Ph.D. and Postdoctoral studies, I would like to introduce a new field of research on the following topics during my initial years:

(1)- New Technique for Enhanced Oil Recovery: Institute Fast Pyrolysis.

(2)- Fundamentals of the Kinetics and Chemistry of Heavy Crude Oil (including asphaltene) Pyrolysis.

(3)- Fast Pyrolysis of Biomass for Production of Deoxygenated Biofuel without Catalysts or Hydrogen.

Postdoctoral Projects: "*Mechanism of xylose pyrolysis-NSF.*"

"*Methods and Mechanisms of Pyrolysis: Polyolefin Decomposition for a Circular Economy-ExxonMobil.*"

"*New Detector for Liquid chromatography (LC) analysis. Activated Research Company (Part-time).*"

Under the supervision of Prof. Paul Dauenhauer, Chemical Engineering and Materials Science, University of Minnesota

Ph.D. Dissertation: "*kinetics of biomass fast pyrolysis-NSF.*"

Under the supervision of Prof. Joseph J. Biernacki, Chemical Engineering Department, Tennessee Tech University.

Master Dissertation: "*Determination of Optimum Gas Injection Condition in EOR Process by Interfacial Tension Investigation.*"

Under the supervision of Prof. Shahab Ayatollahi, Chemical and Petroleum Engineering Department, Shiraz University.

Research Experience:

My research over the last ten years has focused on the relationship between the surface and interface of fluids for the purpose of enhanced oil recovery and asphaltene behavior in petroleum reservoirs under different conditions and kinetics, and the chemistry of Fast Pyrolysis of Carbon-Based Material.

In the beginning, my research was about the effect of temperature and pressure on the kinetics of asphaltene deposition during the capturing of CO_2 as a global gas environment. Then I started my thesis on investigating the effects of temperatures, pressures and composition on CO_2 solubility and miscibility by measuring interfacial tension during CO_2 caption.

After finalizing my master thesis, I became a member of a group that is conducting research on modifying the Parachor model, neural network, and molecular dynamics simulation by interfacial tension technique to measure the solubility of the global gas environment (such as CO_2) at different pressures and temperatures.

Since I am so interested in research and development, particularly in regards to the problematic relationship between population and global warming, I was searching for a project to help the environment and ultimately benefit all humankind. I was searching the internet, and, interestingly enough, I found the project titled "A multi-scale environmental and kinetic study on the pyrolysis of sustainable biomass feedstock" at Tennessee Tech University. During my Ph.D. research, a novel laboratory-scale fast pyrolysis technique is demonstrated using biomass microspheres and a unique single-particle micro-reactor technology. During my postdoctoral research, I had the opportunity to work under the supervision of Prof. Paul Dauenhauer and use unique isothermal techniques (PHASER reactor) to obtain the fundamental information of material during the pyrolysis process.

Teaching Interests:

I believe in the real and complex engineering and science environment, the principle key for solving sophisticated problems and engineering obstacles is creativity and how to learn the problem-solving process. During my first mathematics and chemical engineering design courses in university, I made a great effort to improve my creativity skills along with learning the problem-solving process. That became a fundamental motivation for me during graduate school. Therefore, as a teacher, I will strive to teach my students with these two essential factors to motivate them and provide them with extensive understanding. I believe it will encourage students learning challenging topics, guide them to recognize and completely understand the fundamental concepts in chemical engineering and build confidence in their abilities. I would like to make my students capable of analyzing and evaluating on their own to become an independent learner.

Chemical engineering is one of a fascinating major among all engineering disciplines with distinct theoretical and experimental aspects. From the theoretical point of view, strong knowledge and deep understanding of applied mathematics are ineluctable specifically at senior and graduate levels. Applied mathematics brings about great insight for the student to trust their intuition and problem-solving skills to assess and investigate widespread problems. In my Thermodynamic course, I encouraged students to focus on conceptual aspects of each problem by solving selected practical problems in class and providing alternative approaches to express different perspectives at each problem. This strategy of teaching will give rise to more motivation, satisfaction, and self-confidence.

Additionally, my experience as a teaching assistant in Thermodynamic course enabled me to realize the importance of practical and experimental aspects of chemical engineering by suggesting different methodologies for involving students in practical perspectives of materials and problems. First, I will present the theories for each section concisely and will spend more time on technical and practical examples. As an example, in Fluid mechanics course, I provided lab demonstration and placed students in groups to work with experimental setups and software to build their practical skills. Second, I will take advantage of visual demonstrations and graphs. Finally, I will involve my students in practical discussions and by offering them the opportunity to experience challenging concepts. During my teaching experience, I recognized that a proper evaluation method was to assign students homework problems which required comprehensive work throughout the semester. Homework assignments and examinations helped students demonstrate their knowledge and problem-solving skills and analyze solutions properly. In most chemical engineering courses, results can be reached by more than one method. I require my students to verify their analysis by computer simulation, as well. Moreover, individual and group assignments or final projects can help students acquire deeper and conceptual understanding. Projects should be flexible enough to serve different student as interests and to evaluate each student as skill and expertise. I take advantage of various types of exams and projects, such as problem-solving and take home exams. Additionally, I welcome feedback from students which allow me to evaluate my performance during a semester by conducting teacher evaluation survey in my class to assure what my students expect of me to improve my teaching skills and methods.

Teaching Experience:

Instructor

University of Applied Sciences and technology

1-Industrial security and safety

2-Thermodynamic

3-Fluid mechanics

Assistant Instructor

Shiraz University, Shiraz, Iran

Fluid and rock properties Lab: spring 2011- Fall 2014.

Willing to teach the following courses: (1) Unit Operations Laboratory, (2) Chemical Process Design.

Selected Publications:

Zolghadr, Ali, Matthew D. Kelley, Ghazal Sokhansefat, Masoud Moradian, Brianna Sullins, Tyler Ley, and Joseph J. Biernacki. "Biomass microspheres–A new method for characterization of biomass pyrolysis and shrinkage." *Bioresource technology* 273 (2019): 16-24.

Zolghadr, Ali, Joseph J. Biernacki, and Ronald J. Moore. "Biomass Fast Pyrolysis Using a Novel Microparticle Microreactor Approach: Effect of Particles Size, Biomass Type, and Temperature." *Energy & fuels* 33, no. 2 (2018): 1146-1156.

Haghbakhsh, Reza, **Ali Zolghadr,** Sona Raeissi, and Shahab Ayatollahi. "Investigation of volumetric fluid properties of (heptane+ hexadecane) at reservoir conditions." *Journal of Natural Gas Science and Engineering* 22 (2015): 377-394.

Zolghadr, Amin Reza, Mohammad Hadi Ghatee, and **Ali Zolghadr**. "Adsorption and orientation of ionic liquids and ionic surfactants at heptane/water interface." *The Journal of Physical Chemistry C* 118, no. 34 (2014): 19889-19903.

Zolghadr, Ali, Mehdi Escrochi, and Shahab Ayatollahi. "Temperature and composition effect on CO2 miscibility by interfacial tension measurement." *Journal of Chemical & Engineering Data* 58, no. 5 (2013): 1168-1175.

Zolghadr, Ali, Masoud Riazi, Mehdi Escrochi, and Shahab Ayatollahi. "Investigating the effects of temperature, pressure, and paraffin groups on the N2 miscibility in hydrocarbon liquids using the interfacial tension measurement method." *Industrial & Engineering Chemistry Research* 52, no. 29 (2013): 9851-9857.

Reviews:

Zolghadr A, C. Templeton, Biernacki, J. J. Biomass fast pyrolysis using a novel micro-sphere micro-reactor approach: Model-based interpretations. (Summited to the Energy & Fuels)

https://scholar.google.com/citations?user=5atMn1EAAAAJ&hl=en

All Since 2014

Citations	389	369
h-index	7	7
i10-index	7	7

582147 Energy Storage and Conversion with Organic Molecules and Advanced Porous Electrodes

Research Interests:

Enabling renewable energy technology is critical for reducing carbon emissions and mitigating the harmful effects of climate change. Broadly, my research has sought to understand and apply electrochemical engineering principles to develop and improve energy storage and conversion technologies, with the ultimate goal of making a clean energy future possible. By combining electrochemical characterization of materials, flow cell experiments, and multi-scale simulation, a deeper understanding of the interplay between mass transport and electrochemical kinetics can be gained within these devices.

As a faculty member, I plan to study the electrochemical properties of organic molecules, such as quinones, and their use in energy storage and conversion devices. Quinone electrochemistry is particularly interesting because their electrochemical properties can be tuned by attaching different functional groups to the molecule. This feature opens up a wide range of new redox couples, with applications in energy storage and carbon-dioxide sequestration. The effect of various substituents on the reduction/oxidation mechanism of these molecules is not well understood, but critical for identifying possible kinetic limitations and extending battery lifetime. Through a combination of half-cell electrochemical characterization, simulation, and full-cell experiments, my research will develop a fundamental understanding of reduction/oxidation pathways in these molecules. This knowledge will be harnessed both to improve the efficiency and lifetime of flow batteries and to explore new applications, such as electrochemical synthesis.

I am also interested in studying the role of porous electrode flow cell devices for energy storage and conversion applications. Flow batteries, fuel cells, and electrolyzers have immediate applications in decarbonizing our economy and rely on carefully engineered porous electrodes. The proper structure and design of a porous electrode for flow batteries, for example, is not well known, yet many electrochemical couples are so kinetically facile that mass transport within the electrode is the greatest hindrance to flow battery operation at high current. My research will use simulation and flow-cell experiments to establish connections between porous electrode structure, mass transport, and performance to enable high-current operation of electrochemical flow devices.

In summary, I plan to advance understanding of electrochemical kinetics and mass transport within porous electrodes to improve flow battery efficiency and lifetime, and apply the knowledge gained to energy conversion applications such as hydrogen production and electrochemical synthesis.

Teaching Interests:

I have served as a teaching assistant for two courses: Introduction to Solid-State Chemistry, a freshman-level general chemistry course for materials scientists, and Engineering Thermodynamics, a junior and senior level course for engineering students. I have taught in both traditional classroom and large-scale online format courses, with responsibilities including lecturing, course organization, problem set and exam design, grading, and running a laboratory class session. Furthermore, as a participant in the Harvard Departmental Teaching Fellow program, I took and later co-taught a pedagogy course for science and engineering, as well as participated in biweekly seminars on best teaching practices.

I am interested in applying active learning strategies in my classroom to improve student engagement and retention of information. This could take the form of low-stakes, multiple-choice questions followed by a short discussion between students sitting next to each other, which has been shown to help students arrive at the correct answer to a particular question. Reserving a few minutes at the end of each class period for students to reflect on and write about what they've learned and what still confuses them can also help encourage learning, and is useful feedback to the instructor as well. On an individual student basis, and as a research mentor, I ask a lot of questions, in an effort both to gauge the student's understanding of particular material and to help push them to understand new material.

Mathematical modeling is a powerful teaching tool, because it integrates programming, mathematics, and science classes. A course on mathematical methods would give students important transferable skills, like computer programming and data analysis, and encourage discovery learning by allowing them to test hypotheses on how a system would operate. A question like "What happens if I raise the temperature?", for example, moves from a lengthy experiment to a few keystrokes. Thus, one of the courses I plan to develop is a mathematical modeling in electrochemistry course, which would allow interested students to build models to tackle questions like these. Such a course could also build upon concepts covered in core chemical engineering classes including transport, kinetics, and thermodynamics.

Research Experience:

Postdoctoral Scholar, Lawrence Berkeley National Laboratory, 2017 – present

Advisor: Adam Weber

- Developed a multiphysics computational model of anion-exchange membrane fuel cells to understand the role of water transport and carbon dioxide contamination in cell performance.
- Developed anion-exchange membrane electrolyzer model and advised undergraduate researcher to illustrate effects of electrolyte composition.
- Simulated water transport and hydrogen crossover in small-scale, passive-water-management fuel cell systems, including sub-freezing operation.

Graduate Student, Applied Physics, Harvard School of Engineering and Applied Sciences, 2012 – 2017

PhD thesis: *Topics in the Development of Aqueous Quinone Flow Batteries*

Advisor: Michael Aziz

- Characterized electrochemical properties of several organic molecules for energy storage applications via cyclic voltammetry and rotating disk electrode studies.
- Combined simulation and electrochemical characterization to suggest a reduction mechanism for 2,6-dihydroxyanthraquinone. This information was later used to identify decomposition mechanisms.
- Demonstrated the effects of chemical substituents on operating voltage, efficiency, and lifetime of organic-molecule-based flow batteries by designing, machining, and operating lab-scale flow battery for charge/discharge cycling.
- Simulated electrochemical and fluid dynamic phenomena in flow battery with analytical and finite element computational techniques to understand the role of mass transport in flow battery performance.

Selected Publications:

M.R. Gerhardt, L.M. Pant, and A.Z. Weber. Along-the-Channel Impacts of Water Management and Carbon-Dioxide Contamination in Hydroxide-Exchange-Membrane Fuel Cells: A Modeling Study. *Journal of the Electrochemical Society*, 166(7):F3180-F3192, 2019.

M.R. Gerhardt, A.A. Wong, and M.J. Aziz. The Effect of Interdigitated Channel and Land Dimensions on Flow Cell Performance. *Journal of the Electrochemical Society*, 165(11):A2625-A2643, 2018.

Q. Chen, M.R. Gerhardt, and M.J. Aziz. Dissection of the Voltage Losses of an Acidic Quinone Redox Flow Battery. *Journal of the Electrochemical Society*, 164(6):A1126-A1132, 2017.

M.R. Gerhardt, L. Tong, R. Gómez-Bombarelli, Q. Chen, M.P. Marshak, C.J. Galvin, A. Aspuru-Guzik, R.G. Gordon, and M.J. Aziz. Anthraquinone Derivatives in Aqueous Flow Batteries. *Advanced Energy Materials*, 3:1601488, 2016.

Q. Chen, M.R. Gerhardt, L. Hartle, and M.J. Aziz. A Quinone-Bromide Flow Battery with 1 W/cm^2 Power Density. *Journal of the Electrochemical Society*, 163(1):A5010–A5013, 2015.

K. Lin, Q. Chen, M.R. Gerhardt, L. Tong, S.B. Kim, L. Eisenach, A.W. Valle, D. Hardee, R.G. Gordon, M.J. Aziz, and M.P. Marshak. Alkaline quinone flow battery. *Science*, 349(6255):1529–1532, 2015.

B. Huskinson, M.P. Marshak, C. Suh, S. Er, M.R. Gerhardt, C.J. Galvin, X. Chen, A. Aspuru-Guzik, R.G. Gordon, and M.J. Aziz. A metal-free organic–inorganic aqueous flow battery. *Nature*, 505(7482):195–198, 2014.

582168 Materials for Energy Storage Applications: Fundamental Insight for Rational Design and Development

Research Interests:

Environmental challenges and economic forces are reshaping the way we generate and consume energy on a global scale. To keep up with the accelerating adoption of electric vehicles, allow for grid scale energy storage, and meet the demands of future electronic applications, new materials for high energy density batteries must be developed. Lithium ion batteries are presently the dominant form of energy storage used in consumer electronics and, recently, electric vehicles. However, high costs have prevented widespread deployment of lithium ion batteries for applications other than portable electronics, and the safety hazards of exothermic reactions associated with traditional materials during cell failure remain to be addressed. Therefore, strategies to enhance the mechanical and chemical stability of advanced electrode materials are key to their successful application in commercial batteries.

My future research will use insight gained from the study of charge transport processes and interfacial reactions/interactions to inform the design of new materials for energy storage applications.

Overall, the goal of my work will be to identify and understand molecular phenomena that can be utilized to design and develop materials to specifically address challenges related to enabling advanced battery chemistries. These new materials will pave the way for low cost and safe energy storage technology by stabilizing high energy density battery chemistries, replacing flammable liquid carbonate electrolytes, and enabling new form factors for batteries in advanced electronic applications. Using spectroscopic and electrochemical techniques and in situ characterization tools my research will answer questions regarding the solvation structure of ions in polymer and small molecule electrolytes for fast and selective lithium ion transport and the kinetics and mechanisms of electrochemical reactions related to advanced battery chemistries. Furthermore, I plan to utilize tools for automated experimentation and high throughput analysis to accelerate the materials discovery process. This research program builds on my combined expertise in materials synthesis, characterization of kinetic and transport properties, and electrochemical device fabrication gained through my graduate and postdoctoral research.

Graduate Research with Prof. Zhenan Bao at Stanford University

During my PhD studies I designed and synthesized new self-healing polymers and elastomers with a focus on specialized mechanical properties for improving the stability of lithium ion batteries (1). Silicon and lithium metal can store ten times more charge than the graphite used in commercial electrodes; however, silicon, due to its brittle nature and large volume expansion during lithiation, has poor cycling stability, while lithium metal suffers from significant side reactions, poor quality deposition, and the potential to form hazardous dendrites. Using a supramolecular self-healing polymer as the binder for a silicon microparticle electrode or coating on the lithium metal anode, I demonstrated improved cycling stability of these high capacity materials. As a binder, the self-healing polymer enables the electrode to heal after fracture of the silicon, and by identifying an optimal relaxation time to maximize cycling stability I found that the success of this self-healing electrode concept relies on a balance of stress relaxation and stiffness in the binder (2). Moreover, the self-healing polymer coating on lithium metal enabled dendrite free deposition at current densities as high as 5 mA cm^{-2} (3). To understand this effect, I studied several carefully selected polymers with varied chemical and mechanical properties as coatings on the Li metal anode. Thermodynamic analysis of the lithium nucleation process identified the polymer dielectric constant and surface energy as two key descriptors of the polymer coating on lithium deposition (4). Furthermore, I explored the design and synthesis of elastic polymer networks that contained a combination of covalent and dynamic hydrogen bonding crosslinks for enhanced stress dissipation and improved toughness. I used these dynamic hydrogen bonds to create a self-healing elastomer that enabled the *first* demonstration of a low potential, high capacity electrode for stretchable lithium ion batteries based on a graphitic carbon/silicon composite foam electrode (5). Further development of this class of tough elastomers led to the creation of a highly resilient lithium ion conducting material able to elastically absorb as much energy as Kevlar based electrolyte materials while conducting lithium ions at the same order of magnitude as commercial separators (6).

Postdoctoral Research with Prof. Yang Shao-Horn at MIT

For my postdoctoral research, I am working in two areas related to understanding the mechanisms of electrochemical instability and dynamics of ion transport in small molecule and polymer electrolytes. First, I am using in situ characterization techniques to study the electrolyte decomposition reactions at the lithium metal interface in order to clearly identify the components of the solid electrolyte interphase (SEI), which strongly influences the cycling stability of lithium metal batteries, and the mechanism of its formation. The structure and composition of the SEI on lithium metal is poorly understood and there is great need to expand knowledge in this area to provide guidance for the development of new electrolytes and additives. Second, I am working as part of a large combined synthetic and computational effort to design and discover polymer electrolytes for Li-ion batteries. To aid in this work, I have led the development of an automated platform for fully integrated electrolyte formulation and testing. This high throughput tool allows for the accumulation of extremely large, high quality data sets required to enable machine learning methodologies for the computer-generated prediction and optimization of polymer electrolytes.

Teaching Interests:

As a teacher and mentor, I aim to provide students with the skills to clearly identify questions that interest them and the tools to answer those questions. During my PhD at Stanford University I was a teaching assistant for Polymer Chemistry and Micro and Nanoscale Fabrication Engineering where I gained experience preparing course materials and guiding student projects. In the laboratory, I have formally mentored 1 high school, 4 undergraduate, 2 Master's, and 3 PhD rotation students at various times during my PhD and postdoctoral work. Additionally, I have worked with various programs at Stanford and MIT to promote improved access to higher education among students from underrepresented minority groups. With my undergraduate and graduate background in chemical engineering, I am able to teach core chemical engineering courses at any level, and I am particularly interested in teaching courses where I can combine chemical engineering principles with a molecular perspective. I am also excited about teaching or developing lecture and laboratory based courses in polymer sciences and electrochemistry.

Selected Publications:

1. **Lopez, J.;** Mackanic, D.G.; Cui, Y.; Bao, Z. Designing Polymers for Advanced Battery Chemistries. *Nat. Rev. Mater.* **2019,** 4 (5), 312–330
2. **Lopez, J.;** Chen, Z.; Wang, C.; Andrews, S. C.; Cui, Y.; Bao, Z. The Effects of Cross-Linking in a Supramolecular Binder on Cycle Life in Silicon Microparticle Anodes. *ACS Appl. Mater. Interfaces* **2016**, *8* (3), 2318–2324.
3. Zheng, G.*; Wang, C.*; Pei, A.*; **Lopez, J.*;** Shi, F.; Chen, Z.; Sendek, A. D.; Lee, H.-W.; Lu, Z.; Schneider, H.; Safont-Sempere, M. M.; Chu, S.; Bao, Z.; Cui, Y. High Performance Lithium Metal Anode with a Soft and Flowable Polymer Coating. *ACS Energy Lett.* **2016**, *1* (6), 1247–1255. *equal contribution
4. **Lopez, J.;** Pei, A.; Oh, J. Y.; Wang, G.-J. N.; Cui, Y.; Bao, Z. Effects of Polymer Coatings on Electrodeposited Lithium Metal. *J. Am. Chem. Soc.* **2018,** *140* (37), 11735–11744.
5. Sun, Y.*; **Lopez, J.*;** Lee, H.-W.; Liu, N.; Zheng, G.; Wu, C.-L.; Sun, J.; Liu, W.; Chung, J. W.; Bao, Z.; Cui, Y. A Stretchable Graphitic Carbon/Si Anode Enabled by Conformal Coating of a Self-Healing Elastic Polymer. *Adv. Mater.* **2016**. *28* (12), 2455-2461. *equal contribution
6. **Lopez, J.;** Sun, Y.; Mackanic, D. G.; Lee, M.; Foudeh, A. M.; Song, M.-S.; Cui, Y.; Bao, Z. A Dual-Crosslinking Design for Resilient Lithium-Ion Conductors. *Adv. Mater.* **2018**, *30* (43), 1804142.

579913 Realizing CO_2 Utilization and Carbon-Neutral Energy Conversion in Practical Scenarios --- Mining the Air, Extracting Industrial Wastes, One-Pass Conversion, and Scale-Out

Research Interests:

My special inter-disciplinary fundamental and experimental ingenuities in electrochemistry, catalysis, microfluidics, system integration and techno-economics have aligned me with the knowledge base and holistic view to begin bridging the gap between the current technology and a chemical engineering CO_2 mitigating process that can be considered stable and economical. So far, most studies in electrochemical CO_2 utilization have been relying on high-purity feed gases and electrolytes, operated in ideally conditioned small apparatus. The throughputs and costs could not fulfil the low-carbon economy roadmap set out by authorities. To cope with the target by delivering realistic CO_2 value chain enabling technologies, my primary research focuses as a Lecturer are: 1. the technological advancement of CO2 electrolyzers through the construction and assessment of physically-scaled reactors; 2. the realization of electrochemical processes driven by practical sources, for instances, intermittent renewables as the electricity input, industrial effluents as the feed gas, and industrial waste liquids as the electrolytes; 3. the development of effective structural engineering strategies for reactive sites with near-unity one-pass CO_2 conversion efficiency so that it could be readily used in industrial carbon sources. Up to now, few efforts have looked to understand the reactions and dynamics taking place within a reactor larger than 1 cm^2, and even greater engineering challenges and opportunities will arise in practical scenarios.

This research will also look to address a yet unanswered question within the field: what will a carbon-neutral energy conversion cycle look like? Answering these questions will require expertise across several academic disciplines. To aid in my lab's research I will collaborate with an international network of researchers that I have built up, particularly subject-matter experts in various electrolyzer components. Funding for the project and collaborations will be sought from various government and private sector sources, some applications of which are currently underway. My strong networking capabilities and inter-sectoral visions will consolidate the connections with the industry, enabling grants with larger consortium in the future and offering more career possibilities for our graduates. The proposed research is also well-positioned within the faculty team in the department, functioning between fundamental catalyst research and system-integration.

Teaching Interests:

I was a teacher of science subjects in front of a large classroom in a private school prior to my undergraduate study. Since my postgraduate program, my teaching competences have been further advanced by structured teaching curriculum, generic skill workshops and training-on-work. I served as a teaching assistant for three undergraduate/postgraduate courses in chemical/mechanical engineering stream (*Energy Conversion 1-3, Thermofluids 1-2, Heat Transfer 3-4, Engineering Mechanics, Product Design and Development*). In each of these courses I was responsible for a wide range of tasks including leading and teaching tutorials, guiding laboratory experiments and planning course materials, in addition to compiling and grading problem sets. Therefore, I am confident in my teaching competences for the bachelor's and master's programs in chemical or mechanical engineering.

I am also willing to serve as a cohort mentor for students to help consolidate their fundamental base, foster cross-border knowledge acquisition and develop out-of-box thinking. The most effective mentoring style, gained from the past experiences, is to encourage achievements whilst acknowledge mistakes. Therefore, I will ensure weekly-based communication with each mentee and recalibrate the project schedule to maximize what students could earn. All mentees will be granted diverse conceptual and technical trainings according to individual interest and passion (e.g. fundamental or practical, theoretical or experimental, chemical or engineering). These are believed to boost their internal enthusiasm and satisfaction, maintaining a life-time awareness of science attitude.

Looking to the future, I am eager to pioneer a new course that provides interested students with i) a well-rounded knowledge of carbon-neutral energy resources and usage, and ii) the skills to assess the efficiency, potential and emissions of existing and new technologies. In this course, the art and science revolving a carbon-neutral energy conversion cycle will be taught through four modules: CO_2 emission, CO_2 capture, CO_2 utilization and storage, and fuel cell. Current challenges for renewables related to fossil fuel crisis and CO_2 concentration will be introduced. Pros and cons of different CO_2 capture and sequestration strategies will be discussed. CO_2 price fluctuation and safety concerns would lead students to alternative solutions, such as CO_2 utilization. This session will be the core module of this course and include photochemistry, electrochemistry, chemical synthesis etc. Last but not least, fuel cell technology, as a green way to release the energy in fuels into electricity and the final step of the energy conversion loop, will be elaborated. Course materials with sufficient case studies will be well prepared based on classical textbooks, recent reports from the International Panel on Climate Change, BP's Statistical Report on World Energy as well as scientific articles. As this course is closely related to my study, my up-to-dated research outcomes will be readily accessible to our students and transformed into teaching, tutorial and training materials. A combination of laboratory and industry challenges will be posed to students who will then be introduced to different effective solutions. After following a series of clues, students will then be asked to submit reports and give presentations on methods they would use to unravel the mysteries. I will establish an internal website to preserve and pass on the knowledge gained to future generations of graduate students.

580792 Advancing the Development of Fuel Flexible Combustion, Compact Energy Systems, and the Sustainability Analysis Methods

Research Interests:

A. Fuel Flexible Combustion

The purpose of the proposed research is to harness blue whirl combustion as a fuel-flexible means of extracting energy from a wide array of liquid fuels for energy & power applications by the U.S. Army. The PIs will investigate the physics underpinning blue whirl combustion through experiments and modeling. This knowledge will serve as the foundation for understanding the scaling and stability of efficient blue whirl combustion across a broader range of liquid fuels than can be accomplished by existing technology. Consequently, this will extend the range and endurance of autonomous systems in multi-domain operations, improve energy security, and reduce energy logistic costs. *The overarching goal of this study is to advance the development of fuel flexible combustion technology by understanding the physical and chemical aspects of the blue whirl phenomenon. One* specific task of study include:

(i) The PIs will examine the ability of the blue whirl phenomenon to combust a wide range of model fuels representing biomass and petroleum derived fuels. Additionally, PIs will investigate the role of chemical functionality and molecular structure on the blue whirl combustion efficiency and emission characteristics.

B. Compact Energy Systems

The development of compact energy generation systems that use locally available a low-energy density biomass such as switchgrass and wood can minimize warfighter and robotic autonomous platforms exposure to threats by reducing the reliance on consolidated forward-operating bases and subsequent lines of communication. The intended research project study the potential of production of electrical energy using complex pyrolysis vapors as a fuel source to one type of fuel cell. Further, the possibility of switching between one type of fuel cell to another type of fuel cell to maximize the power density through elimination of coke on the anode will be investigated. To study the performance of fuel cell operating with pyrolysis vapors as a fuel source, the current density versus voltage and current density versus power density curves will be generated. These curves further used to determine optimal operating conditions such as temperature of pyrolysis process as well as fuel cell. The anode surface of the fuel cell will be investigated using any one or more of the following analytical techniques: The scanning electron micrograph (SEM), transmission electron microscopy (TEM), X-ray diffraction (XRD), FT-IR, X-ray photoelectron spectroscopy (XPS) and the In-situ HT-Raman spectroscopy.

C. Sustainability Analysis Methods

The standard techno-economic analysis (TEA) approach in literature for evaluating new biomass upgrading technologies involves estimating the process costs of converting biomass like wood to chemicals and fuels and compare these costs with that of incumbent petroleum technologies. However, the results of TEA can not be used to inform designing of new policies that are aimed to penalize environmentally detrimental technologies as the TEA analysis cannot account social costs/benefits of the production of chemicals and fuels from renewable and non-renewable feedstocks. ***This shortcoming of the TEA analysis can be overcome by analyzing new biomass conversion technologies with the new proposed method of integrated TEA and ecological life cycle assessment (E-LCA).***

The application of the new method of integrated TEA and E-LCA for evaluating renewable energy technologies consists of three steps: First step consists of expanding the process boundary to include eco-system services and apply life cycle assessment framework to determine the marginal change in the quality or quantity of the selected ecosystem services; The second step is to use the ecosystem valuation techniques to quantify the costs/benefits of marginal change in the quality or quantity of the selected ecosystem services; and the final step is to combine the social costs/benefits with process costs. Our research group will ***study the sustainability of making renewable fuels via algae and pyrolysis*** *technologies using the new integrated TEA and E-LCA method.*

Teaching Interests:

I have observed while I was mentoring undergraduate students as well as co-supervising Masters students that students try to memorize scientific concepts. I also noticed among these students that there is a fear of learning new knowledge as well as difficulty with communicating their work properly. Having desire to teach how to learn science has motivated me to purse career as a faculty member. As faculty member, I will help students to develop their creative abilities and critical analysis skills through class assignments. I will put effort to improve student's communication skills through assigning team projects as well as asking them to give presentations very often. In my lectures, I will include application of engineering concepts in studying other fields of science to encourage students to learn new fields and to promote interdisciplinary understanding.

My course assignments and exams will be a blend of qualitative and quantitative questions. I believe that solving qualitative questions related to scientific concepts will encourage students to deconstruct scientific concepts. Such deconstruction will enable students to understand scientific concepts thoroughly. Solving assigned quantitative problems with the theoretical understanding will provide an opportunity for students to improve their problem solving skills. Such improvement will help students to succeed in their selected career paths. I use classroom chalkboard while I am deriving any equations and solving any quantitative problems, as it will provide enough time for students to think and understand what I am teaching. At other times, I use PowerPoint visuals for teaching course material as it will improve the students focus.

I frequently quiz students during every class to know how well students are interacting with the class material. If I find out students are finding difficult to understand my course material, I will analyze the problem and try to overcome it as soon as possible. Additionally, I will not hesitate to ask advice from the department faculty to improve my teaching. I will consider remarks made by students in their evaluation forms and I will put effort to improve my quality of teaching if needed. I try to understand classroom dynamics in a few weeks after the beginning of a semester and I will adjust my pace accordingly that majority of students in the class will follow my lectures.

557584 Electrochemical Engineering for Energy, Environment and Separation Applications

Research Interests:

Electrochemistry, which connects electricity with redox chemical reaction, provides novel and exciting opportunities for engineering solutions to some of the grand challenges faced by our human society. My research will primarily be focused on engineering electrochemical materials, devices and system for energy and separation applications, with an emphasis on the below technologies.

Energy:

- Li-ion battery
- Next generation batteries system (Li metal, solid state, multi-valent)
- Enhanced oil recovery

Separation:

- Shock electrodialysis for water desalination and treatment, nanoparticles, colloid separation etc.
- Faradic ion adsorption and separation

My method will primarily be experiment. For some projects, I will also leverage the large amount of experiment data and use data analytics to optimize and design electrochemical systems. Also, I collaborate closely with computational scientists and theorists for modeling at different length scales.

Teaching Interests:

I will develop a graduate course: Electrochemical Engineering for Energy, Environment and Separation based on my research experience.

I can teach the following undergraduate course based on my education and training.

Thermodynamics, kinetics, transport phenomena, fluid mechanics, system and control engineering.

582828 Illuminating Solid-Water Interfacial Processes at Nanoscale for Sustainable Environmental Remediation and Resource Recovery

Research Experience:

My research has focused on diverse redox reactions and subsequent crystal nucleation and growth of nanoparticles occurring at solid-water interfaces in nature and engineered application systems. The overarching goal is to understand unconsidered crucial roles that abiotic and biotic sources play in redox reactions of transition metals and to develop sustainable chemistry for water treatment and energy applications. Through my Ph.D. and postdoctoral researches, I have elucidated the crucial effect of interfacial processes on the nucleation of geological (Mn oxides) and engineering materials (Li dendrite) using *in situ/ex situ* synchrotron-based X-ray techniques and atomic force microscopy (AFM). By applying photo- and electrochemical redox reactions, I have obtained fundamental understanding of redox reactions and (trans)formation of nanoparticles, all of which occur at solid-solution interfaces.

The diverse routes for redox reactions and nucleation have inspired me to learn more about how nature forms different mineral phases via abiotic and biotic redox reactions and subsequent nucleation, and how I can use this understanding to develop better engineered applications, sustainable chemistry, and sustainable engineering. To embody the curiosity and enthusiasm, I am interested in pursuing the following projects.

Research Interests:

With growing populations and improved standards of living in the world, we are facing daunting challenges in environmental sustainability: sustainably supplying food, water, and energy and designing a future without pollution or waste[1]. Based on the ubiquitous presence and substantial surface energy (~100 $J\ g^{-1}$) of minerals and nanoparticles in nature and engineered systems, understanding the interfacial processes at the molecular level is significant to treat contaminants and to design sustainable systems [3]. With a strong fundamental background and techniques of interfacial processes (e.g., *in situ/ex situ* synchrotron-based X-ray and AFM), my interdisciplinary research aims to understand local, regional, and global aspects of elemental cycles and impacts of pollution at the molecular level in natural and engineered systems. The fundamental understanding of solid-water interfaces at the molecular level will enable my research group not only to design an improved system for the mitigation of pollution and waste, but also to recover resources from waste, all of which occur at the interface between solid and water. My research will focus on the three major projects:

Project 1: Establishing a new understanding of the interfacial processes of metal oxides and emerging contaminants in nature and engineered aquatic systems.

Understanding of the physico-chemical nature of the interfacial processes still remains elusive because of extreme difficulties in analyzing such interfaces in their relevant environments (e.g., *in situ*), as they evolve (e.g., in real-time), and lack of knowledge at the molecular-scale resolution. This project will explain the fate of metal oxides (or nanoparticles) at mineral-water interfaces, and important roles in controlling the redox reaction and structure of natural minerals under environmentally relevant conditions. Specifically, by developing *in situ* characterization platform (such as synchrotron-based X-ray absorption/scattering (e.g., XAS and SAXS/WAXS) and electrochemical-AFM), my group will not only visualize how Mn oxide nanoparticles form and react with contaminants depending on redox conditions at nanoscale interface, but also elucidate a new Eh-pH diagram (Pourbaix diagram) obtained from nanoscale analyses.

Project 2: Sustainable synthesis of nanomaterials and developing re-activation system of the nanomaterials in contaminant removal and energy catalysis.

Designing a system is the most critical stage in determining the types and amounts of wastes or pollutions. While Mn(III/IV) oxides are very useful environmental and energy catalysts, due to its high reactivity and redox capacity, previous studies have shown that external heat and pressure are necessary to control the structure and oxidation state of Mn oxides. Also, the decay of reactivity occurs because of the reduction of Mn(III/IV) to Mn(II) in engineering applications systems. Using the design of green chemistry (reducing extra chemicals and external energy), we will synthesize varied structures of Mn(III/IV) oxides, and develop the re-activation mechanism of them in the energy and environmental applications.

Project 3: Mining valuable elements from waste: A low-cost green approach for Lithium-ion battery recycling.

The popularity of electric vehicles (EV) starts to grow explosively. This substitution of fossil-fueled vehicles to EV will generate multimillion-metric-tons of used Li-ion batteries. This situation recently drives researchers to recycle valuable materials in the used Li-ion batteries. However, current recycling pathways are energy-intensive or environmentally-harmful processes (e.g., high-temperature melting-and-extraction, chemical-extraction using strong acids). My research group will aim to develop sustainable pathways to dissemble cathodes, containing valuable metals (e.g., Co, Mn, Ni) using biotic processes (e.g., microbes), and selectively uptake the metals using (photo)electrochemistry. This project will bring a cost-effective and sustainable processes of Li-ion battery recycling.

Teaching Experience:

My experiences in lecture and teaching assistant for graduate and undergraduate are transport phenomena, kinetics and thermodynamics, which are my teaching interests as well, in environmental and engineered systems. In aquatic chemistry and geochemistry classes, I gave lectures for kinetics and thermodynamics in environmental systems. Also, as a teaching assistant in thermodynamic course of chemical engineering, I leaded problem-solving classes for undergraduate students biweekly. I also had experiences in laboratory courses of general chemistry lab and environmental engineering lab as a teaching assistant. Specially, I was an instructor of general chemistry lab for 2 years in my master degree. The experiences helped for me to learn how to make students motivated and teach them efficiently.

As a research mentor, I have mentored for 3 graduate students, 3 undergraduate students, and 1 high school student. In particular, I have learned how it is joyful to be a mentor for a beginner by working with one of the undergraduate students for 2 years. This experience was basically learning process for me about how advisors teach graduate students. Providing motivations for a conducting project and careful advices made her as the most outstanding undergraduate researcher in her class. She became 1st author in a peer-reviewed journal (crystal growth & design) with me, and got research awards from ACS and Washington University in St. Louis. In this way, I have learned how to teach new subjects to students and research beginners and the importance of knowing your audience and presenting information clearly.

Teaching Interests:

Due to my background in aquatic chemistry, kinetics, and thermodynamics in environmental and chemical engineering systems, I would be particularly well-suited and interested in teaching kinetics and thermodynamics, and aquatic chemistry for both undergraduates and graduates. Also, my most favorite course, which I got an award at collegiate competition hosted by the Korean Institute of Chemical Engineers, is transport phenomena which including fluid dynamics, heat transfer, and mass transfer. I can develop the course at the view point and issues of environmental and energy systems.

Together with the development of the pre-existing courses, I will also add new curriculum of the department, environmental nanochemistry. This course involves the study of nanochemistry at various environmental interfaces, focusing on colloid, nanoparticles, and surface reactions. The aim of this course is to help students attain a better understanding of how nanoscience could potentially lead to better water treatments, more effective contaminated-site remediation, or new energy alternatives.

Successful Proposals:

Funding Proposals

Photochemically-Induced Nucleation and Growth of Manganese Oxides at Environmental Interfaces (NSF

CHE-1610728), Contribution: developed first idea and results, and made a draft. ($ 402,000.00)

Selected Synchrotron National Facility Beamtime Proposals (20 proposals in total)

Haesung Jung, and Yuanzhi Tang, Effects of metal impurities on the redox property and structure of layered manganese oxides, NSLS II 6BM (National Synchrotron Light Source II, Brookhaven National Laboratory) beamtime proposal (2019-2): XAS experiments.

Young-Shin Jun, **Haesung Jung**, Byeongdu Lee, and Doyoon Kim, Photochemically-assisted Fast Abiotic Oxidation of Mn^{2+} (aq), and Nucleation and Growth of δ-MnO_2 at Environmental Interfaces, *APS 12ID-B (Advanced Photon Source, Argonne National Lab) beamtime proposal* (2017-1): *In situ* time-resolved GISAXS measurements.

Selected Publications:

Haesung Jung, Byeongdu Lee, Miklos Lengyel, Richard Axelbaum, Jeeyoung Yoo, Youn Sang Kim, and Young-Shin Jun, "Nanoscale *in situ* Detection of Nucleation and Growth of Li Electrodeposition at Various Current Densities", *Journal of Materials Chemistry A*, 2018, 6, 4629–4635.

Haesung Jung, Tandeep Chadha, Yujia Min, Pratim Biswas, and Young-Shin Jun, "Photochemically-assisted Synthesis of Birnessite Nanosheets and Alteration of Its Structure in the Presence of Pyrophosphate", *ACS Sustainable Chemistry & Engineering*, 2017, 5, 10624–10632.

Yue Hui[†], **Haesung Jung**[†], Doyoon Kim, and Young-Shin Jun, "Kinetics of α-MnOOH Nanoparticle Formation through Enzymatically-catalyzed Bio-mineralization inside Apoferritin", *Crystal Growth & Design*, 2017, 17, 5675–5683. [†]The authors contributed equally.

Haesung Jung, Tandeep Chadha, Doyoon Kim, Pratim Biswas, and Young-Shin Jun, "Photochemically-assisted Fast Abiotic Oxidation of Manganese and Formation of δ–MnO_2 Nanosheets in Nitrate Solution", *Chemical Communications,* 2017, 53, 4445–4448.

Haesung Jung and Young-Shin Jun, "Ionic Strength-controlled Mn (Hydr)oxide Nanoparticle Nucleation on Quartz: Effect of Aqueous $Mn(OH)_2$", *Environmental Science & Technology,* 2016, 50, 105–113.

Selected Awards:

Graduate Student Research Award in Environmental Chemistry Division, American Chemical Society, 2018

Graduate Student Research Award in Department of Energy, Environmental & Chemical Engineering, Washington University in St. Louis, 2017

581831 Winning the Energy Dilemma: Carbon-Free Energy and Environmental Sustainability

Winning the Energy Dilemma:

Carbon-Free Energy and Environmental Sustainability

Simona Liguori – Department of Chemical Engineering, Worcester Polytechnic Institute, Worcester, MA

Assistant Research Professor – *Worcester Polytechnic Institute* 2018 – present

Assistant Research Professor – *Colorado School of Mines* 2016 – 2018

Physical Science Research Associate – *Stanford University* 2014 – 2016

Post Doc – *Institute on Membrane Technology (National Lab-Italy)* 2012 – 2014

Motivation and Research Interests:

The generation of carbon-free electricity and the production of low-carbon fuels are urgent needs that must be addressed as the atmospheric concentration of carbon dioxide (CO_2) continues to rapidly increase. The proposed solutions for mitigating the effects of climate change are the deployment of low-carbon fuels and improvements to the efficiency of current technologies for energy conversion.

Carbon-free hydrogen is viewed as a promising long-term option for tackling CO_2 emissions from distributed sources, as well as generating electricity and energy as needed. Electrolytic processes such as water electrolysis and photo-reduction combined with photovoltaic solar and/or wind energy can be utilized to produce 'carbon-free' hydrogen. However, these technologies suffer from various drawbacks such as the high electric energy requirements, low efficiency, and very high hydrogen production cost compared to the conventional methane steam reforming process. Furthermore, in some cases, significant land area is required to deploy these technologies.

One of the primary goals of my research is to advance the science of *metallic membrane reactors* for catalytic reactions to produce various chemicals and feedstocks via non-traditional pathways. The proposed technology presents a unique method to generate hydrogen through a catalytic membrane configuration using hydrogen selective materials. Membrane reactors make it feasible to combine the production and separation steps in one unit. This leads to lower operating temperatures and pressures, higher yields, and lower capital and operating expenses. The produced CO_2 can be easily compressed and captured for geologic sequestration. The combination of these state-of-the-art technologies will make it possible to produce carbon-free hydrogen.

The concept of membrane reactors can also be applied to produce ammonia using a nitrogen-selective membrane. The central premise of this research is to physically separate the energy-intensive nitrogen dissociation from the serial exothermic hydrogenation reactions (using an atomic nitrogen-permeable catalytic membrane). In doing so, separate catalysts and operating pressures can be utilized for each step in the catalytic cycle.

The efforts proposed herein leverage intensified reaction/separation engineering to synthesize next generation ultra-selective metallic membranes. This will overcome the perceived barriers associate with the traditional technologies by increasing the hydrogen and ammonia production and eliminating the CO_2 emissions through the use of catalytic membrane reactors.

Another area of my research interests focuses on fostering scientific curiosity through negative emission technologies, particularly mineral carbonation technologies. In mineral carbonation, the CO_2 contained in the atmosphere chemically reacts with calcium- and/or magnesium-containing minerals to form stable carbonate materials. Mineral carbonation presents several fundamental engineering challenges. This includes developing methods to effectively increase the carbonation rate of these materials while intensifying the system to exploit the exothermic nature of these reactions. Additionally, these efforts must be performed within environmental constraints, such as minimal energy and material losses.

In summary, my research interests are focused on:

Developing the science of metallic membrane reactors, as well as synthesizing and testing new materials for membranes and catalyst to

- Produce carbon-free hydrogen, and capture CO_2
- Decouple the ammonia reaction into two-half steps realized across a reactive membrane to reduce operating pressure, increase yields, and decrease capital and operating expenditures

Furthering Negative Emission Technologies by

- Improving mineral carbonation pathways to remove CO_2 from the atmosphere and permanently store it as a carbonate material.

Teaching Experience and Teaching Interests:

I have taken any opportunity to share my knowledge and experience with others to promote science. To achieve this goal, I worked as a Teaching Assistant (TA) for Inorganic Chemistry courses during my post-doctoral appointment in Italy. Additionally, I was the co-instructor of a Carbon Capture course at Colorado School of Mines. During these appointments, I realized that teaching could be both challenging and rewarding. Teaching requires extensive planning, organizing materials, prioritizing ideas, interacting with students, learning to monitor and adjust to the atmosphere of a classroom, tailoring instructions for students with varying levels of ability, and learning how to accomplish goals that sometimes seem mutually exclusive. All this while keeping the attention of a class and meeting the expectations of administrators and peers. Nevertheless, I am always enthusiastic and passionate when I have the opportunity to share my knowledge with students. Especially, when I can engage them in scientific activities. In addition to formal methods of teaching, I have had the opportunity to serve as mentor for three undergraduate students in Italy and two PhD students at Stanford. I also currently mentor a PhD student at Worcester Polytechnic Institute. During these years, I have given all the help I could to ensure my mentees success in both academic and professional settings.

Over these years I have developed a growing interest for chemical engineering subjects, including fluid mechanics, transport phenomena, thermodynamics, energy conversion processes, and chemical reaction and kinetics. As a future faculty, I intend to combine my research experience with newly developed techniques for interactive learning to create a positive and engaging classroom atmosphere where everyone feels included and empowered to pursue their educational goals.

Research Experience:

My research experience spans multiple topics, including chemical reactors and reaction analysis, reaction kinetics, heat and mass transport processes, reactive diffusion through catalysts, gases separation and membranes.

For my PhD research, I focused on both experimental and simulation investigations into the reforming reactions of bio-fuels in both membrane reactors and fixed-bed reactors for hydrogen production. After completing my PhD, I continued my research in membrane science and technology as a Post-Doctoral Research Fellow at the Institute on Membrane Technology in Italy, as a Physical Science Research Associate at Stanford University, and as a Research Assistant Professor at Colorado School of Mines. During these years, I have published numerous peer-reviewed articles and registered one scientific patent with the U.S. patent office on an alternative, novel method to produce ammonia via a nitrogen-selective membrane reactor.

Additionally, I applied years of my research experience in the areas of membrane technology, hydrogen production, and CO_2 mitigation to set up two new research laboratories at Colorado School of Mines and Worcester Polytechnic Institute. As a Research Assistant Professor, I currently work on synthesizing new membranes and developing new catalysts for ammonia and hydrogen production to mitigate CO_2 emissions.

Future Direction:

As a faculty member I would like to continue applying membrane science and technology to different areas of engineering, as well as industrial applications. I believe the interdisciplinary research activities that I performed in the areas of chemical engineering and environmental science have equipped me with the essential tools to advance my research toward mitigating the greenhouse gas emissions, such as CO_2, by producing lower-carbon fuels.

Three of the areas that I have found myself very fascinated about are the development of organic membranes for gas separation processes, metal organic frameworks (MOFs) with the purpose of carbon capture, and, finally, improving the mineral carbonation mechanism with the aim of permanent removal of CO_2 from the atmosphere.

Selected Publications:

1. **S. Liguori (corresponding Author)**, K. Lee, J. Wilcox, "Innovative N_2-selective metallic membranes for the potential use of CO_2 capture", *J. Membr. Sci.*, 585 **(2019)** 52-59

2. Z Zhang, **S Liguori**, TF Fuerst, JD Way, CA Wolden, "Efficient ammonia decomposition in a catalytic membrane reactor to enable hydrogen storage and utilization", *ACS Sust. Chem. Eng.*, 7 (**2019**) 5975-5985

3. K Kian, C Woodall, J Wilcox, **S. Liguori (corresponding Author)**, "Performance of Pd-based membranes and effects of various gas mixtures on H_2 permeation", *Environments* 5 (**2018**) 128

4. B. Anzelmo, J. Wilcox, **S. Liguori (Corresponding Author)**, "Hydrogen production via natural gas steam reforming in a Pd-Au membrane reactor. Investigation of reaction temperature and GHSV effects and long-term stability", *J. Membr. Sci*, 565 (**2018**) 25-32

5. B. Anzelmo, J. Wilcox, **S. Liguori (Corresponding Author)**, "Hydrogen production via natural gas steam reforming in a Pd-Au membrane reactor. Comparison between methane and natural gas steam reforming reactions", *J. Membr. Sci*, 568 (**2018**) 113-120

6. **S. Liguori**, J. Wilcox, "Membrane considerations and plant design for post-combustion CO_2 capture", Ch. 14, In Current Trends and Future Developments on (Bio-) Membranes, A. Basile, E. Favvas (Eds.), (**2018**) Elsevier, ISBN 9780128136454

7. J Wilcox, **S. Liguori** "Ammonia synthesis using nitrogen-selective membrane" – (**2018**), *US Patent No 15/707,007*

8. B. Anzelmo, J. Wilcox**, S. Liguori (Corresponding Author)**, "Natural gas steam reforming reaction at low temperature and pressure conditions for hydrogen production via Pd/PSS membrane reactor", *J. Membr. Sci.*, 522 (**2017**) 343-350.

9. **S. Liguori**, J. Wilcox, "Silica membranes application for Carbon Dioxide separation", Ch. 11, In Current Trends and Future Developments on (Bio-) Membranes, A. Basile, K. Ghasemzadeh (Eds.), (**2017**) Elsevier, pp. 265-294, ISBN 9780444638663

10. J. Wilcox, P.C. Psarras, **S. Liguori**, "Assessment of reasonable opportunities for direct air capture", *Env. Res. Lett.*, 12 (**2017**) 065001. (**It was invited**)

11. P. Psarras, H. Krutka, M. Fajardy, **S. Liguori**, S. Zhang, N. Mac Dowell, J. Wilcox, "Slicing the pie: how big could Carbon Dioxide removal be?", *WIREs Energy and Environment*, 6 (**2017**) e253

12. **S. Liguori**, K. Kian, N. Buggy, B. Anzelmo, J. Wilcox, "Opportunities and challenges for hydrogen production/separation via metallic membranes", *submitted*, *Progress in Energy and Combustion Science*

13. L. Giugliano, **S. Liguori**, K. Kian, J. Wilcox, T Bandhauer, "Technoeconomic analysis of steam generation with Carbon Dioxide capture via steam methane reforming in a membrane reactor", *submitted*, *Environmental Science & Technology*

14. J. Wilcox, P.C. Psarras, H. Pilorge **S. Liguori**, J. He, M. Yuan, C Woodall, K. Kian, L. Pierpoint, J. Jurewicz, M. Lucas, R. Jacobson, N. Deich, "Strategy for jumpstarting modular DAC using low-temperature heat", *submitted*, *Proceedings of the National Academy of Sciences.*

581960 Organic Molecular Electrocatalysts for Energy-Water Nexus and Beyond

Xi Yin, *Ph.D.*

Los Alamos National Laboratory, Materials Physics & Applications,
Los Alamos, New Mexico 87545

Office Phone: (505) 667-3238; E-mail: xiyin@lanl.gov
Personal website: https://sites.google.com/view/xiyin

Highlights

- Expertise in advanced energy materials, electrochemistry and computational modeling.
- Invented the first organic molecular electrocatalyst (OMEC) for energy-water nexus.
- Initiated and led a $990k three-year project (FY19-FY21).
- First-authored 11 publications, 2 covers, 17 total, citations > 550, total if >160, h-index

Expertise

(i) Materials chemistry, coordination chemistry, and inorganic chemistry; (ii) molecular simulation and computational modeling; (iii) physical chemistry, electrochemistry, and electrocatalysis; (iv) organic molecular electrocatalysis; (v) energy conversion and storage, polymer electrolyte fuel cells; (vi) machine learning and data mining for materials design.

Education

- Ph. D., Chemical Engineering (advisor: Prof. Hong Yang), University of Illinois at Urbana-Champaign (2012-2015)
- M. Sc., Materials Science, University of Rochester
- M. Sc. and B.Sc., Materials Science and Engineering, Tsinghua University

Employment

- Research Scientist/Project Leader, Los Alamos National Laboratory/New Mexico Consortium, Los Alamos, NM (2018-present)
- Postdoctoral Research Associate (mentor: Dr. Piotr Zelenay), Los Alamos National Laboratory, Los Alamos, NM (2015-2018)

Awards & Honors

- MRS Graduate Student Award Silver Medalist, Materials Research Society (2015)
- SCS Science Image Challenge Finalist, UIUC School of Chemical Science (2014)

Summary of Previous Works

Throughout my research, I have sought to combine experimental studies with computational modeling and theory to accelerate discoveries that will have extensive impact on society. My dissertation work at the University of Illinois at Urbana-Champaign focused on a fundamental understanding of the nucleation and growth of metal nanocrystals for catalysis. Different from most observation-based studies, my work emphasized the chemistry of metal-ligand complexation during pre-nucleation and designed ligand environments to tune nucleation and growth *via* thermodynamic and kinetic leverages. By employing density functional theory (DFT) calculations and reaction kinetic models, I developed new theories on nonclassical nanocrystal growth. These works led to the worldâ€™s first atomically thin palladium nanosheets and a quantitative understanding of the roles of surfactant ligands in nanocrystal synthesis for electrocatalysis, gas sensing, and biological applications.

My postdoctoral work at Los Alamos National Laboratory has focused on the development of platinum group metal-free (PGM-free) catalysts for fuel cell cathodes, driven by the need to replace scarce and expensive PGM materials with compounds based on earth-abundant elements, such as C, N, Fe, and other transition metals. Through molecular-level design of catalyst templates, kinetic study of catalyst degradation, and optimization of the electrode structure, my work has covered a broad spectrum of fuel cell research to boost the performance of state-of-the-art PGM-free catalysts.

While expanding my research portfolio from precious metal-based materials to carbon-based electrocatalysts, I noticed that most electrocatalysts are built on metal-centered active sites that have been electronically modified *via* metallic, ionic, or covalent bonds. The activity improvement of conventional catalysts may intrinsically be limited by the possible electronic structures of these metal sites. To overcome this limitation, I conceived of an idea to use organic molecules as metal-free electrocatalysts, which are attractive due to their highly tailorable geometry, tunable electronic structures, and chemical robustness. With these factors in mind, I designed and modeled a metal-free N-containing molecule as a bifunctional catalyst for H_2O_2 production and thehydrogen evolution reaction. ***This discovery is the first demonstration of high catalytic activity from an inexpensive, small organic molecule in a practical polymer electrolyte membrane (PEM) device for crucial electrochemical reactions.*** Building upon these findings, I plan to establish a world-leading program that aims to develop an entirely new class of electrocatalysts based on rationally designed organic molecules for clean energy and environmental applications, which goes beyond conventional electrocatalysts that rely on precious metal or transition metal elements.

Research Interests:

My future research aims to unravel the structure-activity relationships in novel ***Organic Molecular Electrocatalysts*** (**OMECs**) and apply that fundamental knowledge to the molecular-level design of next-generation electrocatalysts for renewable energy, environmental and computing applications. Different from conventional metal-containing electrocatalysts, OMECs are solely based on the earth abundant non-metal elements, *e.g.*, C, N, H, etc. The covalent bonding in organic structures ensures the robustness of OMECs in harsh electrochemical environment. The highly-tailorable structures of OMECs will allow the tuning of electrochemical functions in a large organic chemical space. Although millions of organic compounds have been designed and synthesized, no systematic works have been done to fully explore the electrocatalytic activities of the metal-free organic molecules.

My proposed research will encompass three main thrust areas: (1) distributed electrochemical production of hydrogen peroxide (H_2O_2) for water treatment; (2) electrochemical hydrogen production and energy storage; and (3) carbon dioxide (CO_2) reduction into high-value liquid fuel. The first thrust aims to address the increasing need for distributed on-site production of H_2O_2 as a water disinfection agent for use in remote areas and disaster mitigation during the disruption of regular infrastructure services. The second and third trusts will provide solutions for the integration of hydrogen facilities with smart grids, as well as the production of clean fuels.

Teaching Interests:

My research background is in chemical engineering and materials science, with expertise in electrochemistry, materials chemistry, computational modeling and machine learning. My research experience allows me to confidently teach the following courses:

- Thermodynamics of Chemical Processes
- Chemical Reaction Engineering
- Electrochemical Engineering and Fuel Cell
- Electrochemistry
- Computational Design of Energy Materials
- Computational Electrocatalysis
- Fundamentals of Materials Science and Engineering

- Reaction Kinetics
- Synthetic Nanomaterials
- Materials Chemistry

Proposal Writing Experience

1. "Field-deployable electrochemical device for decontamination and water treatment" to LANL Laboratory Directed Research and Development, Exploratory Research (LDRD-ER), **Co-PI**, 2019
2. (**Awarded, $990K**) "Organic molecular electrocatalysts for hydrogen evolution reaction" to LANL Laboratory Directed Research and Development, Exploratory Research (LDRD-ER), **Co-PI**, 2018
3. (**Accepted**) "Activity of organic molecular electrocatalysts for hydrogen evolution reaction" to LANL Institutional Computing (IC) Program for Projects in Computational Science and Engineering, **1st Co-Investigator**, 2018
4. "Center for earth-abundant materials for carbon-free fuels conversion" to DOE Office of Basic Energy Sciences (BES) the Energy Frontier Research Center program (EFRC), **contributor**, 2018
5. "High performance direct liquid feed fuel cells powered by dimethyl ether" to DOE Office of Energy Efficiency and Renewable Energy (EERE), **contributor**, 2018
6. "Metal-free molecular catalysts for hydrogen evolution reaction" to LANL Laboratory Directed Research and Development, Exploratory Research (LDRD-ER), **Co-PI**, 2017
7. "A new class of metalorganic structure containing infinite one-dimensional metal chain" to National Science Foundation, **contributor**, 2013

Publications (11 first/co-first authored publications, 3 cover designs, 17 total, citations > 600, total if > 165, h index 9, i10-index 9 from google scholar)

1. **X. Yin**, H. T. Chung, U. Martinez, L. Lin, K. Artyushkova, P. Zelenay, "PGM-free ORR catalysts designed by templating PANI-type polymers containing functional groups with high affinity to iron", ***Journal of the Electrochemical Society***, **2019**, 166 (7), F3240-F3245
2. U. Martinez, S. K. Babu, E. F. Holby, H. T. Chung, **X. Yin**, P. Zelenay, "Progress in the development of Fe-based PGM-free electrocatalysts for oxygen reduction reaction", ***Advanced Materials***, **2019**, 1806545 (**Back Cover Design**)
3. **X. Yin**, P. Zelenay, "Kinetic models for the degradation mechanism of PGM-free ORR catalysts", ***ECS Transactions***, **2018**, 85, 85 (13), 1239–1250
4. **X. Yin**, M. Shi, K. S. Kwok, D. L. Gray, J. A. Bertke, H. Yang, "Dish-like higher-ordered nanostructures of palladium through metal ion-ligand complexation", ***Nano Research***, **2018**, 11, 3442–3452 (**Cover Design**)
5. **X. Yin**, M. Shi, J. Wu, D. L. Gray, J. A. Bertke, H. Yang, "Quantitative analysis of different formation modes of platinum nanocrystals controlled by ligand chemistry", ***Nano Letters***, **2017**, 17 (10), 6146–6150
6. **X. Yin***, L. Lin*, H. T. Chung, S. K. Babu, U. Martinez, G. M. Purdy, P. Zelenay, "Effects of MEA fabrication and ionomer composition on fuel cell performance of PGM-free ORR catalyst", ***ECS Transactions***, **2017**, 77 (11), 1273–1281 (*Equal contributors.)
7. J. Kim, P.-C. Shih, K.-C. Tsao, Y.-T. Pan, **X. Yin**, C. J. Sun, H. Yang, "High-performance pyrochlore-type yttrium ruthenate electrocatalyst for oxygen evolution reaction in acidic media", ***Journal of the American Chemical Society***, **2017**, 139 (34), 12076–12083
8. P.-T. Pan, J. Wu, **X. Yin**, H. Yang, "*In situ* ETEM study of composition redistribution in Pt-Ni octahedral catalysts for electrochemical reduction of oxygen", ***AIChE Journal***, **2016**, 62 (2), 399–407
9. **X. Yin**, J. Wu, P. P. Li, J., H. Yang, "Self-heating approach to the fast production of uniform metal nanotructures", ***ChemNanoMat***, **2016**, 2 (1), 37–41 (**Front Cover Design**)
10. **X. Yin**, X. H. Liu, Y.-T. Pan, K. A. Walsh, H. Yang, "Hanoi tower-like multilayered ultrathin palladium nanosheets", ***Nano Letters***, **2014**, 14 (12), 7188–7194
11. **X. Yin**, S. A. Warren, Y.-T. Pan, K.-C. Tsao, D. L. Gray, J. Bertke, H. Yang, "A new motif for infinite metal atom wires", ***Angewandte Chemie International Edition***, **2014**, 53 (51), 14087–14091
12. P.-T. Pan*, **X. Yin***, K. S. Kwok, H. Yang, "Higher-ordered nanostructures of flexible palladium two-dimensional nanosheets for fast hydrogen sensing", ***Nano Letters***, **2014**, 14 (10), 5953–5959

(***Equal contributors.**)

13. J. Kim, **X. Yin**, K.-C. Tsao, S. Fang, H. Yang, "$Ca_2Mn_2O_5$ as oxygen-deficient perovskite electrocatalyst for oxygen evolution reaction", ***Journal of the American Chemical Society***, **2014**, 136 (42), 14646–14649
14. J. Wu*, M. Shi*, **X. Yin**, H. Yang, "Enhanced oxidation reduction stability of (111) surface dominant $Pt_3Ni@Pt_3Pd$ core-shell nanoparticle catalysts", ***ChemSusChem***, **2013**, 6 (10), 1888–1892 (*Equal contributors.)
15. J. Wu, P. Li, Y.-T. Pan, S. Warren, **X. Yin**, H. Yang, "Surface lattice-engineered bimetallic nanoparticles and their catalytic properties", ***Chemical Society Reviews***, **2012**, 41, 8066–8074
16. **X. Yin**, K. X. Chen, H. P. Zhou, X. S. Ning, "Combustion synthesis of Ti_3SiC_2/TiC composites from elemental powders under high gravity conditions", ***Journal of the American Ceramic Society***, **2010**, 93 (8), 2182–2187.
17. **X. Yin**, K. X. Chen, H. P. Zhou, X. S. Ning, "Combustion synthesis of Ti_3SiC_2 from Ti/Si/C powders under high gravity conditions", ***Rare Metal Materials and Engineering***, **2008**, 37 (9), 1606–1609.

Manuscripts under Review/in Preparation

1. **X. Yin**, L. Lin, U. Martinez, P. Zelenay, "2,2'-Dipyridylamine as heterogeneous organic molecular electrocatalyst for two-electron oxygen reduction reaction in acid media", submitted, **2019** (**Preprint**, DOI: 10.26434/chemrxiv.8283962).
2. **X. Yin**, L. Lin, H. T. Chung, U. Martinez, A. M. Baker, S. Maurya, P. Zelenay, "2,2'-Dipyridylamine as organic molecular electrocatalyst for hydrogen evolution reaction in acidic electrolytes", submitted, **2019** (**Preprint**, DOI: 10.26434/chemrxiv.8273627)
3. **X. Yin**, H. Gao, "Prediction of proton reduction potential at six-membered N-heterocycles using machine learning", **2019**, in preparation
4. L. Lin, **X. Yin**, H. Chung, U. Martinez, K. Moore, P. Zelenay, "Hierarchically porous PGM-free ORR catalysts for high performance polymer electrolyte fuel cells", **2019**, in preparation

Patent

1. **X. Yin**, P. Zelenay, "Organic molecular electrocatalysts for hydrogen evolution reaction and hydrogen peroxide production", U.S. Provisional Patent, 2019

Selected Conference Presentations (One invited talk, coauthored 7 DOE Annual Merit Review presentations)

1. **X. Yin**, E. Holby, P. Zelenay, "Data mining of structure-activity relationship in organic molecular electrocatalysts for hydrogen evolution reaction", ***235th ECS Meeting***, Dallas, TX, 2019
2. **(Invited) X. Yin**, U. Martinez, S. Komini Babu, H. T. Chung, G. M. Purdy, P. Zelenay, "Kinetic insight into the degradation mechanism of PGM-free ORR catalysts", ***233rd ECS Meeting***, Seattle, WA, 2018
3. **X. Yin**, L. Lin, H. T. Chung, U. Martinez, A. Baker, S. Maurya, P. Zelenay, "Organic molecular catalyst for hydrogen evolution reaction", ***232nd ECS Meeting***, National Harbor, MD, 2017
4. **X. Yin**, L. Lin, U Martinez, H. T. Chung, P. Zelenay, "Organic molecular catalyst for electrochemical production of hydrogen peroxide", ***232nd ECS Meeting***, National Harbor, MD, 2017
5. **X. Yin**, L. Lin, H. T. Chung, S. Komini Babu, U Martinez, G. M. Purdy, P. Zelenay, "Effects of porosity and ionomer composition on fuel cell performance of PGM-free ORR catalysts", ***231st ECS Meeting***, New Orleans, LA, 2017
6. **X. Yin**, H. T. Chung, U. Martinez, L. Lin, P. Zelenay, "Magnetic purification of PGM-free catalysts", ***PRiME 2016/230th ECS Meeting***, Honolulu, HI, 2016
7. **X. Yin**, H. T. Chung, U. Martinez, J. H. Dumont, G. M. Purdy, P. Zelenay, "Non-PGM ORR catalysts prepared from polyaniline-type polymers with strong affinity to iron", ***229th ECS Meeting***, San Diego, CA, 2016

8. **X. Yin**, J. Wu, Y.-T. Pan, H. Yang, "Quantitative analysis of kinetically controlled growth of metal nanoparticles based on ligand-metal ion interactions", ***2015 MRS Spring Meeting***, San Francisco, CA, 2015
9. **X. Yin**, X. Liu, Y.-T. Pan, K. Walsh, H. Yang, "Hanoi tower-like multilayered 2D Pd nanosheets and the molecular level understanding of the formation mechanisms", ***2015 MRS Spring Meeting***, San Francisco, CA, 2015
10. **X. Yin**, J. Wu, P. Li, H. Yang, "Formation of nanoparticles in self-regulated reaction systems", ***246th ACS National Meeting***, Indianapolis, IN, 2013
11. **X. Yin**, J. Wu, W. Zhou, H. Yang, "Development of facet-controlled Pt and Pt alloy nanocatalysts for oxygen reduction reaction", ***246th ACS National Meeting***, Indianapolis, IN, 2013

Journal Reviewer

- *ACS Catalysis, Nano Energy, Nano Letters, Carbon, Fuel Cells, ACS Applied Materials & Interfaces, ACS Applied Energy Materials, Journal of Materials Chemistry A (Outstanding Reviewer in 2018), Journal of Electroanalytical Chemistry, Nanotechnology, The Journal of Physical Chemistry, Journal of Nanoparticle Research, etc.*

580901 Energy Conversion and Storage

Research Interests:

The depleting fossil fuels unprecedentedly motivate the development of new energy technologies, nowadays. It is highly desirable to produce energy from renewable sources, to storage energy and then to work efficiently and cleanly. To implement the new energy system at a large scale, the new energy technologies must be competitive with current fossil fuel energy system. With this requirement in mind, Kang Group focuses on the fundamental understanding of catalytic processes that are involved in energy conversion and storage. We address the challenge through a three-step-approach: we first investigate the structure-property correlation on a well-defined surface (e.g. single crystal); second, we construct highly controlled nano-scale model materials to verify the validity of structure-property correlation that is observed on well-defined systems; and finally we implement the knowledge learned from our well-defined and nano-sized models into the real word – to produce the catalytic materials that are potentially usable in industry. Currently, our research interests cover following topics:

- Hydrogen economy.

The so-called "hydrogen economy" is a proposed new energy system in which the hydrogen is the energy carrier. The hydrogen comes from the renewable energy powered hydrogen production which is ideally water splitting. The hydrogen is consumed by the end-users through a device called fuel cell. We are closely working on the catalytic processes associated with the hydrogen economy, including hydrogen oxidation reaction (HOR), oxygen reduction reaction (ORR), hydrogen evolution reaction (HER), and oxygen evolution reaction (OER).

Representative publications:

1. Huang, L.L.; Chen, P.; Liu, M.H.; Fu, X.B.; Gordiichuk, P.; Yu, Y.N.; Wolverton, C.; **Kang, Y.J.***; Mirkin, C.A.* Catalyst design by scanning probe block copolymer lithography ***Natl. Acad. Sci. USA*, 2018,** 115 (15), 3764-3769
2. Qu, G.X.; Zhou, Y.; Wu, T.; Zhao, G.; Li, F.; **Kang, Y.J***.; Xu, C.* Phosphorized MXene-phase Molybdenum Carbide as Earth-Abundant Hydrogen Evolution Electrocatalyst ***ACS Appl. Energy Mater.* 2018**, 1 (12), 7206-7212
3. **Kang, Y.J.**; Yang P.D.; Markovic, N.M.; Stamenkovic, V.R. Shaping electrocatalysis through tailored nanomaterials ***Nano Today*, 2016**, 11 (5), 587-600
4. Chen, C.†; **Kang, Y.J.**†(**†co-first authors**); Huo, Z.; Zhu, Z.; Huang, W.Y.; Xin, H.L.; Snyder, J.D.; Li, D.G.; Herron, J.A.; Mavrikakis, M.; Chi, M.F.; More, K.L.; Li, Y.D.; Markovic, N.M.; Somorjai, G.A.; Yang, P.D.; Stamenkovic, V.R. Highly Crystalline Multimetallic Nanoframes with Three-Dimensional Electrocatalytic Surfaces ***Science*, 2014,** 343 (6177), 1339

- Chemical upgrade of environment-abundant small molecules.

While the renewable energy sources may solve the global energy problem, it is unable to directly provide chemical products that are currently from petro-industry. Our solution is to chemically upgrade the small molecules that are abundant in our environment, including CO and CO_2.

Representative publication:

1. Luc, W.; Fu, X.B.; Shi, J.J.; Lv, J.J.; Jouny, M.; Ko, B.H.; Xu, Y.B.; Tu, Q.; Hu, X.B.; Wu, J.S.; Yue, Q.; Liu, Y.Y.; Jiao, F.*; **Kang, Y.J.*** Two-Dimensional Copper Nanosheets for Electrochemical Reduction of Carbon Monoxide to Acetate ***Nature Catal.*, 2019,** 2, 423-430 (cover featured)

- Nitrogen reduction for ammonia synthesis.

Ammonia is a key chemical for our industries and daily life, however, the ammonia synthesis is a high energy consumption industry and environment-unfriendly. Nitrogen reduction (NRR) to produce ammonia attracts great attention recently, however, the reported ammonia production rate (~10^{-8} mol cm^{-2} h^{-1}, usually) is not even close to any practical application. Our recent revolutionary result increases the ammonia production rate by 3-4 orders of magnitude (to be published), making NRR a promising route for ammonia production. In the other hand, we are also pursuing the alternative route other than NRR to convert N_2 in the air to ammonia.

- Energy storage

Among many new technologies toward energy storage, Kang Group focuses on Li metal batteries and Li-S batteries for their high theoretic capacity and foreseeable future.

Representative publications:

1. Xu, Y.; Li, T.; Wang, L.P.; **Kang, Y.J.*** Interlayered Dendrite-free Lithium Plating for High-performance Lithium Metal Batteries ***Adv. Mater.,*** **2019** doi.org/10.1002/adma.201901662
2. Xu, Y.; Wang, L.P.; Jia, W.; Yu, Y.N.; Zhang, R.Y.; Li, T.; Fu, X.B.; Niu, X.B.; Li, J.Z.*; **Kang, Y.J.*** Three-dimensional carbon material as stable host for dendrite-free lithium metal anodes ***Electrochimica Acta*, 2019** 301, 251-257

Teaching Interests:

Any subject that is related to energy conversion and storage, including but not limited to: transport, thermodynamics, reaction and reaction engineering, catalysis, materials science and engineering, materials chemisty and etc.

577591 Efficient Carbon Modification for Sustainable Food/Energy/Water Nexus

Research Interests: Bio-processes

- Bioenergy and Biofuels (Biodiesel and Biochar)
- Activation of Carbonaceous Graphitic Structures
- Wastewater Treatment
- CO_2 Capture
- Sono-Physics (acoustic-based bio-processes)
- Computational Fluid Dynamic (CFD) simulation
- Techno Economic Analysis (TEA)

Teaching Interests: All Chemical Engineering Courses

Abstract

The sustainable food / energy / water (FEW) nexus is arguably the leading grand challenge facing mankind. Our critical reviews of physical modifications (*Reviews in Chemical Engineering*, 2019. in press, doi.org/10.1515/revce-2017-0113) and chemical modifications (*Reviews in Chemical Engineering*, 2019. in press, doi.org/10.1515/revce-2018-0003) of biochar, in addition to our experimental results, suggest that *acoustic, photochemical* and *plasma* treatments, in selected reaction environments and conditions, are capable of inducing either structural or functional group changes on carbonaceous materials. These tunable, low energy treatments have potential to play a major role in FEW systems through a number of transformative tracks.

Our transdisciplinary team revealed that single-staged *ultrasound* and *photochemical* treatments of biochar in H_2O with dissolved CO_2 results in *fixation of C* from CO_2 on biochar, *fixation of H* from water on biochar, mineral leaching by water including minerals detrimental to gasification, increase in biochar's *heating value* (due to the 3 processes stated above), and increase in biochar's *internal surface area* (*AIChE Journal*, 2014. **60**(3):1054-1065). These synergisms seem to be tunable by feedstock, reactants stoichiometry and reaction conditions in pyrolysis, and treatments (*Fuel*, 2019. **235**:1131-1145). Carbon and hydrogen fixations seem to be connected to the formation of H_2, CO, formic acid, formaldehyde, and associated radicals during sonolysis of aqueous CO_2. Similar to ultrasound waves, *non-thermal plasma* can split water vapor and CO_2 to excited chemicals and fuels including methanol, H_2 and CO (*Current Opinion in Green & Sustainable Chemistry*, 2017. **3**:45-49). The presence of carbon in these plasma systems has not been explored. However, treatment of biochar in non-thermal chlorine plasma yields a high adsorption capacity of elemental mercury in flue gas due to the creation of Cl-active sites (*Chemical Engineering Journal,* 2018. **331C**:536-544).

In this seminar, we will discuss the potential routes of the observed synergisms in *mitigation of climate change through CO_2 capture/recycle, energy production through advanced gasification, Remediation of global carbon cycle* (*Fuel*, 2018. **225**:287-298) & *water resource* (*Ultrasonics Sonochemistry*, 2019, **51**, 20-30,). These treatments can also open new routes for tackling challenges such as *water desalination* of through functionalized defected graphene membrane. We are exploring the potential of employing these low-temperature biochar modification technologies on improving the soil productivity through water- and nutrient-holding capacity, and pH adjustment. Moreover, the presence of essential *small molecules*, such as H_2, CO, formaldehyde, etc., produced during the treatments are likely to have other applications in FEW nexus and climate change mitigation.

553974 Multiscale Approach to Future Energy Science and Engineering

Research Interests:

My academic career path covers a variety of topics in energy science and engineering. My academic experiences in both Chemical Engineering Departments and Petroleum Engineering Departments provide me unique chances of conducting research at multiple scales of interests. My research in Petroleum Engineering Departments focuses on computational simulation of macroscopic transport processes from continuum viewpoints. Such approach strives to strike a balance between accuracy and feasibility, and thus are promising for industrial applications. My work in Chemical Engineering Departments, however, tackles problems from more microscopic perspectives and aims to elucidate the fundamental sciences at molecular levels or even below. From these experiences, I have acquired invaluable and versatile skills in various research topics including soft matter, complex fluids, multiphase transport, surface chemistry, molecular modeling, process simulation, and machine learning, with their applications in tackling various pressing issues in conventional and renewable energy fields. My research has already found several industrial applications for fuel transportation and biofuel production. For more information about my research, please visit my website at

https://sites.google.com/view/yingdalu/

Future Direction:

My diverse research experiences have led me to believe that future energy research can greatly benefit from an improved elucidation of the underlying physics involved at more microscopic levels. Thus, I envision my future research will employ a multiscale approach that combines perspectives from the disciplines of both Chemical Engineering and Petroleum Engineering to tackle the pressing issues in modern energy science and engineering. This type of research maintains the edges of current approaches used in the field, but blends with more microscopic perspectives with the aid of modern computational modeling techniques such as molecular modeling. It also incorporates the recent boom in artificial intelligence and data processing techniques such as machine learning. This represents long-term and highly interdisciplinary research that allows my research group to make valuable contributions to modern energy science and engineering. Specific research topics include:

a. Computational Guided Design of Functional Molecules for Controlling n-Alkane Crystallization and Gelation

(Research themes: Soft matter, Computational Chemistry, Material Science)

b. Reliable Prediction of Multiphase Flow Features by Machine Learning

(Research themes: Data Science, Multiphase Flow)

c. Numerical Simulation of n-Alkane Crystallization and Gelation in Multiphase Flow

(Research themes: Computational Physics, Multiphase Transport)

d. Multiscale Molecular Simulation of Bulk and Interfacial Behaviors of Asphaltenes

(Research theme: Molecular Simulation, Self-assembly, Interfacial Science)

Teaching Experiences:

I am an experienced instructor and dedicated mentor. I have supervised 2 Ph.D. and 2 M.S. students to successfully complete their degrees, and am currently mentoring 1 Ph.D. student, 2 M.S. students, and 2 undergraduate students. Furthermore, I have instructed 5 foundational courses, 1 laboratory course, and 1 senior design course across different departments in different countries. I value an interdisciplinary approach to teaching Chemical Engineering course, and emphasize improving students' self-learning, critical thinking, and creative thinking skills.

Teaching Interests:

I enjoy teaching foundational chemical engineering courses including fluid dynamics, thermodynamics, transport, and laboratory, as well as advanced topics with vast applications in energy science and engineering, including rheology, colloid science, computational fluid dynamic simulation, molecular simulations. I am also interested in developing general skill improvement courses to enhance students' problem solving, critical thinking, creative thinking and troubleshooting skills and to ease the transition from college to workplace.

Selected Publications (16 articles and 230+ citations):

First-author Articles:

1. **Lu, Y.**; Nariman Fathi Najafabadi and Abbas Firoozabadi. Effect of Low-Concentration of 1-Pentanol on the Wettability of Petroleum Fluid–Brine–Rock Systems. ***Langmuir*** 2019, 35 (12), 4263-4269

2. **Lu, Y.**; Nariman Fathi Najafabadi and Abbas Firoozabadi. Effect of Temperature on Wettability of Oil/Brine/Rock Systems. ***Energy & Fuels*** 2017, 31 (5), 4989-499

3. **Lu, Y.** and Savage, P. E. Supercritical Water Gasification of Lipid-extracted Hydrochar to Recover Energy and Nutrients. ***The Journal of Supercritical Fluids*** 2015, 99, 88-94

4. **Lu, Y.**; Levine, R. B. and Savage, P. E. Fatty Acids for Nutraceuticals and Biofuels from Hydrothermal Carbonization of Microalgae. ***Industrial & Engineering Chemistry Research*** 2015, 54, 4066-4071

5. **Lu, Y.**; Huang, Z.; Hoffmann, R.; Amundsen, L. and Fogler, H. S. The Counterintuitive Effects of Oil Flow Rate on Wax Deposition. ***Energy & Fuels*** 2012, 26, 4091-4097

Corresponding-author articles:

1. Liu, H.; Zhang, J. and **Lu, Y.*** Yielding characterization of waxy gels by energy dissipation. ***Rheologica Acta*** 2018, 57(6), 473-480

2. Liu, H.; **Lu, Y.*** and Zhang, J. A Comprehensive Investigation of the Viscoelasticity and Time-dependent Yielding Transition of Waxy Crude Oils. ***Journal of Rheology*** 2018, 62(2), 527-541

3. Li, Y.; Han, S.; **Lu, Y.*** and Zhang, J. Influence of Asphaltene Polarity on Crystallization and Gelation of Waxy Oils. ***Energy & Fuels*** 2018, 32, 1491-1497

4. Ma, C.; **Lu, Y.***; Chen, C.; Feng, K.; Li, Z.; Wang, X. and Zhang, J. Electrical Treatment of Waxy Crude Oil to Improve Its Cold Flowability. ***Industrial & Engineering Chemistry Research*** 2017, 56 (38), 10920-10928

External Links:

1. Research website: https://sites.google.com/view/yingdalu/

2. Google Scholar: https://scholar.google.com/citations?user=xdpCAR4AAAAJ&hl=en

579263 Biopolymer Encapsulated Lipase for Biodiesel Production

Abstarct: Immobilized encapsulated lipase was taken in stirred tank batch bioreactor (STIBR) and packed bed bioreactor (PBBR). The studies were carried out in a batch mode of operation and various process parameters were optimized for biodiesel production. HPLC was used for analyzing the biodiesel. The optimum conditions for processing palm oil in a stirred tank immobilized bioreactor (STIBR) were 30oC, 72 h reaction time and 23.7 x g relative centrifugal force. Similarly, the optimal conditions for processing palm oil in a (PBBR) were 1.5 ml/min and 264 h reaction time. STIBR shown conversion up to 100% and the PBR has shown conversion up to 82%. Since the STIBR has higher conversion rate, the kinetic parameters Km and Vmax were evaluated and found to be 600 mol.m-3 and 0.84 mol.m-3min-1. The kinetic parameter values were substituted into Michaelis–Menten empirical equation and the batch time was found to be same as experimental value of 72 h. The encapsulated lipase retained 82% relative conversion after 5 cycles of reuse. Based on the experimentation and the results, it is concluded that biodiesel production using encapsulated lipase in an immobilized bioreactor may open new vistas for the scaleup studies of this technology in near future.

Research Interests:

Dr.Ravindra Pogaku research proposal:

My overarching research interests revolve around green chemical engineering, particularly, the integrated approach between chemical and bioprocess engineering. I apply principles of chemical engineering and apply for bioprocesses to answer the how, who, where and how fast questions of bio-waste can be converted to bio-wealth. More specifically, I am interested in examining the direct and intimate relationship between the chemical and biochemical activity of the bioprocesses, the chemical process engineering signature of the activity and how that signature may translate between heterogeneous environments thus, enhancing our understanding of the current and past chemical and bioprocesses that drive our planet earth to the sustainable future.

Introduction:

Green chemical engineering and bioprocess engineering are the offshoot of chemical engineering. The integration of the chemical and bioprocess engineering knowledge can be utilized for exploiting natural resources on the planet earth. The abundance of available lignocellulose waste on earth can be tapped into rich source of bio-wealth by transforming the bio-waste into commercial processes and products for the welfare of mankind. The complex chemical and bioprocess reactions involved during the conversion of lignocellulose biomass are controlled by accelerated interactions between the bio-catalysts. The pathways or mechanisms that drive these processes, while ultimately thermodynamically favored, are often carried out and exploited from micro to macro scale. The results of the processes change the chemical and biochemical environment, often on a time-scale much more rapid than abiotic chemistry could explain. For example, the oxygenation of the atmosphere and the accumulation of fixed carbon and nitrogen are direct results of microbial processes. My research interests center on examining the intimately linked interactions between the chemical engineering principles and application for bioprocesses. I strive to use my novel training skills as both a chemical engineer and a bioprocess engineer, including knowledge of food engineering, environmental engineering and microbial enhanced oil recovery, to make deep insight studies on green engineering in the form of integrated bio-refinery chemicals and bioenergy. These acquired skills directly apply to the central questions of bioprocesses, *who* (metabolic and biochemical path ways), *what process* and *where* (biochemical processes, natural lignocellulose waste) and *how fast* are these processes going (in nature).

My focus is on the green chemical engineering by integrating chemical and bioprocesses. Both the concepts are essential for sustainable life and because of their many useful products, the resultant intermediate and end compounds can be used in benign environments, in highly economical way, with less energy intensity and equity to maintain the dynamic equilibrium of nature. Collaborations with other researchers within and outside of the department are also more easily forged in this small-scale and pilot scale research setting.

Research settings:

I envision my laboratory to have integrated base of chemical, bioprocess and computational machine learning framework for diverse and heterogeneous systems. I have worked in the following fields during my 35 years of teaching and research career: I envision my laboratory to have integrated base of chemical, bioprocess and computational machine learning framework for diverse and heterogeneous systems. I have worked in the following fields during my 35 years of teaching and research career. I have generated 2 million USD research and consultancy funding while working in Universities in India and Malaysia.

I would continue to generate funds nationally and internationally as I have wide network with academia, research and industry. I would propose to get funding from international institutions with collaborative research and consultancy activities. The conceptualization of my research proposal is shown in figure 1.

Figure 1: Conceptualization of Research proposal

Further, I will develop an internationally known research programs in the area, and will collaborate with researchers in Engineering. I shall submit research proposals to get funds from NSF, DOE and Industries in USA and also international funding. I would envision a Symbiotic research model as shown in figure 2.

Figure 2: Symbiotic Research model:

Research Approach:

There is an abundance of non-edible biomass to replace about one third of the world's petroleum resources. In recent days, bio catalysts are used in a wide variety of industrial applications to accelerate the rates of chemical, biochemical transformations. My research approach would be to strive to combine the chemical engineering for traditional bioprocess to address the who, what, where, how and how fast the reactions transform, are the questions that forms the basis of green engineering. There are many advantages to using an interdisciplinary approach with interactive role of the learned people, such as described below and different techniques. First, each technique approaches the question from a different perspective, different set of assumptions and the results may integrate over different scales for better productivity in enhancing green engineering knowledge for the sustainable society.

Student- Scholar Interactive role:

I have worked with and mentored students (high school, undergraduate and graduate), technicians, industry employees, postdoctoral researchers and visiting scientists. I would envision a role for researchers at each stage in my laboratory. Particularly, for the most junior, the high school and undergraduate students I prefer to employ them in small, term-length projects, where techniques and fundamentals are heavily emphasized. As these students gain experience and technical know-how, they can be given more freedom and creativity in the laboratory to have hands-on experience. Graduate students initially (first year or two) will be trained on both technique and research philosophy, gaining greater independence as experience grows. I visualize Master's degree students as trained skilled researchers, but they ought to have well targeted projects, that are mainly driven by my research interests. I anticipate Ph.D. students to have a greater role in the development of the project rationale, methodology and perhaps funding effort. The goal is to foster Ph.D. scholars' development into independent and talented researchers to apply their acquired knowledge for the better society. Postdoctoral researchers join my lab to either gain experience in cutting edge technologies, adding to their professional toolbox. While I will strive to include postdoctoral funding in my research proposals, I also believe it is important for young Ph.D. level researchers to attempt to establish their own funding resources. This benefits the Postdoctoral scholars to excel in establishing well distinguished laboratories. The postdoctoral period is also a time to enhance mentoring skills by interacting closely with graduate students, perhaps on a joint project, in the laboratory. I also foresee the role of visiting scientists from national and international institutions for collaborative research linkages and submission of joint research proposals for funding.

Future Directions:

In green engineering systems, I am interested in asking: Where does commercial exploitation of microbes occur in the bioprocesses for the welfare of humankind? What are the potential integrated chemical and bioprocesses and rates of their rapid conversions for integrated bio-refinery and bioenergy products? How do other allied fields, such as environment, food, enhanced oil recovery, interact with green chemical engineering? What technologies may enhance the quality of life for living creatures to establish dynamic equilibrium between man and society?

Research Expertise:

My research activities interface between green chemical engineering and bioprocess engineering. My research team worked on research questions involving bio-catalysis, bio integrated refinery, bio-derived materials, bio-mining, bio-wealth from waste, bio fuels, bio-lubricants, bioremediation and bio encapsulation technology, microbial enhanced oil recovery, food Engineering.

Research Interests:

Currently, my research focus is on bio refinery for bio based economy, materials to produce renewable transportation fuels, heterogeneous catalysis and bio-chemicals production.

- Kinetic studies on biomass conversion in gas and liquid phase
- Understanding and improving the stability of solid catalysts in fluids
- Engineering for biofuel and bio refinery from renewable sources
- Application of Machine learning higher throughput for developing new catalysts and catalytic

formulations

- Synthesizing of supports for bimetallic catalyst reactions
- Rational design and development of bimetallic catalyst

Current research at University of South Carolina:

The production of hydrocarbon fuels and chemicals from renewable biomass resource has become more pressing in recent decades. The biggest challenge in biomass conversion is to develop active, selective and stable catalysts for particular applications. The objective of this research is to optimize catalytic performance for hydro deoxygenation (HDO) and hydrogenation reactions by enhancing the stability of the support, tuning metal particle size, and controlling surface composition. Application of modern computational technique such as Machine learning with High Throughput Experimentation (ML-HTE) for catalyst discovery is actively pursued in our research group.

Conclusion:

I have developed a unique skillset that encompasses traditional chemical engineering, bioprocess engineering and machine learning for establishing green engineering to harness sustainable living. I am excited about using this skillset and training to investigate how the chemical and bioprocess world effects and drives the green engineering for bio based economy. I believe the combination of techniques, knowledge and skills from a range of disciplines is necessary to address these questions in the green engineering. My background and training bring this approach and skillset together for making better world.

Teaching Interests:

Statement of Educational Philosophy- Ravindra POGAKU -PhD; (P. Eng.)

"Teaching" is amongst the noblest and prestigious professions where one is involved in transforming the future of the country. My father was a teacher and my first academic teaching experience began in my tenth grade math classroom, when my father asked me to coach the students on a one-on-one basis in our rural residential area. My extrovert nature and zeal towards helping other students by tutoring activities, has slowly carved the "teacher" in me. I firmly believe that the best way to learn is to teach and continue to learn all through the life. In addition to faculty input, and discussions with peers, my students' responses are the best source for improving my teaching techniques which are evolving on a continuous basis.

I am fortunate to have associated with academicians of various national and international institutions. I believe mentors make a difference in one's life. I will always be grateful for the many positive influences that eminent teachers brought in selecting teaching as my profession, and helping me in shaping my career. From time to time, I have transformed myself and reached a position to inspire budding chemical engineering professionals, to achieve their ambition in this competitive world of knowledge and thereby help them to unravel secrets of nature.

My educational philosophy is based upon answering four key questions including (a) why I want to teach, (b) what I teach, (c) how I teach, and (d) how I measure my effectiveness.

The most important reasons that I want to teach are applying my skills and knowledge to enhance science and engineering knowledge for sustainable society. I believe sharing my fervor for engineering science to trigger students' curiosity and appreciation of the natural world, and encouraging students to be their personal best by becoming critical, creative thinkers and interested in learning. An important goal of mine is to positively influence students by showing them how they are valued individuals who should always strive for self-improvement and that everyone should be respected, recognized for their work, and encouraged to share ideas in the diverse world.

With respect to what I teach, my goal to show students that science is for everyone and that fun and creativity are important aspects of science whether one is doing a "backyard" experiment or working as a research scientist. I feel that it is vital to let students know that they are stakeholders in their education, it is vital for them to develop solid problem solving skills, and they should never stop being curious or imaginative in all that they do.

Concerning how I teach, I strive to use a variety of instructional methods to actively engage students regardless of learning style, background, or aptitude. There is value in integrating technology in the classroom since most people's home, work, and social lives are touched in some way in our digital society. Critical to lesson planning are setting goals that include knowing what students should know and be able to do at the lesson's end, what essential questions are being asked, and what big ideas should be understood. In order for the lesson to be a meaningful learning experience, it must be in harmony with academic standards as well as four critical components known as the Four Pillars of Instruction including (a) curriculum requirements, (b) engaging instruction, (c) genuine assessments, and (d) management practices that facilitate gaining and maintain student cooperation.

In addressing how I measure and assess my performance, my objectives are to continuously work at improving my teaching skills through self-reflection, professional development courses, workshops, and conferences, consulting with my teaching colleagues and school administrators, and seeking feedback from students and their caregivers.

My value as an educator and success in the classroom depend upon knowing as much as possible about each student which will enable me to collaborate with them to maximize their learning and allow me to create a safe and welcoming classroom. In getting to know each student, I have three central interests which are (a) identifying their individual learning styles, aptitudes, interests, and special needs, (b) how they solve problems, handle conflicts, and what about school interests them most and least, (c) what types of support do they have at home and what they and their caregivers expect of me as a teacher.

Treating students as unique people and individual learners facilitates their ability to be cooperative and collaborative learners, to identify and solve problems, and to be personally accountable for their actions.

I will teach undergraduate and graduate courses, supervise undergraduate, graduate, and postdoctoral researchers and establish network with other colleagues and interdisciplinary departments.

Ravindra Pogaku

583260 Functionalized Materials for Sustainable Energy Applications: Structure and Reactivity

Research Interests:

Past and Current Research:

Recent advances in material synthesis enable the controllable design and construction of nanostructured model materials for applications in important fields such as catalysis, energy storage and conversion, optoelectronics, gas separation, biomedical diagnosis, etc. The fundamental basis for the optimization of synthetic methodology and development of material structure/performance relationships are still a matter of great interest.

During my Ph.D. research, I studied two-dimensional (2D) nanosheets obtained by delamination of bulk layered materials, which exhibit novel physical and chemical properties, different from their bulk precursors, that are responsible for their excellent performance in energy storage. However, producing atomic thickness 2D nanosheets for large-scale commercial applications remain unrealistic in large scale due to the fragmentation, morphological damage, and re-aggregation of the detached nanosheets associated with the exfoliation process. I developed a novel exfoliation method to achieved single layer 2D nanosheets with large lateral size for energy storage and conversion devices with enhanced electrochemical performance. Through a combination of experimental and theoretical studies, I demonstrated that intercalation of solvated ions weakens the interaction between host layer and interlayer species and promote the total delamination of bulk precursor into single layer 2D nanosheets.

In addition, I studied nanoporous materials and well-defined nanoparticles for emission control, particularly for CO oxidation reaction under moisture. In practical applications, a large proportion of water plays a completely different role in catalytic reactions in presence/absence of precious metals. Water could be either a devastating catalyst poison on transition metal oxides or a promoter on supported precious metal catalysts. However, the high cost of precious metals limits its large-scale commercial applications. Therefore, I developed functionalized materials with a significant reduction in the amount the precious metals by modifying the physical and chemical properties of the materials. The resulting functionalized materials show enhanced water-tolerance and catalytic activity. Through in situ spectroscopic monitoring of reaction, I found that water directly participates in the reaction by formation of active intermediates at metal-support interfaces and lowers the activation barrier of the reaction. I have also utilized these synthetic strategies during my postdoctoral research. Currently, I am working as a postdoc at the University of Oklahoma with Prof. Daniel E. Resasco. We have advanced synthetic methodology–structure–reactivity relationship through comprehensive studies on reaction kinetics and theoretical calculations. We have demonstrated that water can play an unexpected promotional role in some biomass upgrading reactions. In liquid phase reactions, the presence of water changes the kinetics of C-C bond forming reactions by exerting remote-C=O bond polarization from the surface site to the reactant via water bridges. I have demonstrated that this phenomenon only occurs with solvent molecules that are able to create H-bonds, but not with aprotic (polar) solvents.

Future Research:

My future research plans focus on understanding and controlling the interaction between molecules and solid materials within various scales to engineer novel functionalized materials for sustainable energy applications, such as fuel cells and C1 conversion. For example, C1 building blocks, originating from fossil (e.g. natural gas, shale gas) or renewable (e.g. biomass, organic waste) sources, can utilize catalytic process to generate valuable products, which may help alleviating environmental problems and increasing the widespread use of clean energy. Conventionally, C1 molecules are firstly converted to active species for cascade catalytic reactions to form value-added components. These two steps are separated. Developing a multifunctional catalyst for coupling reaction in one-pot is an energy-efficient catalytic route to directly convert C1 molecules into high value compounds. I plan to design multicomponent architectures of functionalized materials with controllable catalytic activity and selectivity on basis of the well-studied surface composition and electronic/geometric structure of catalyst, reaction kinetics and theoretical calculations. I plan to develop a general principle in rational design of structured functional materials for sustainable fuel and chemical production through a variety of chemical transformations.

Teaching Interests:

I have extensive teaching experience with undergraduate and graduate students throughout my academic career, including teaching them material synthesis, characterization and catalytic performance measurements both in the classroom and laboratory. I am interested in teaching any core course in undergraduate chemical engineering, such as Introduction to Chemical Engineering, Mass and Energy Balances, Thermodynamics, Reaction Kinetics, Chemical Reaction Engineering, and Unit Operation Laboratory. Also, I would be interested in developing elective graduate courses such as nanomaterials design for electrocatalysis applications.

581888 Sustainable Conversion Processes and Reactor Design to Produce Fuels and Chemicals

The objective of this abstract is to briefly present my academic background, future research aspiration, research experience and teaching interest. I was awarded PhD in Chemical Engineering from University of Alberta, Canada in 2016, obtained MESc in Chemical Engineering from Western University, Canada in 2011, and BSc (Engg.) in Chemical Engineering and Polymer Science from Shahjalal University of Science and Technology with distinction securing highest CGPA among 52 students in 2005.

Research Interests:

My future research aspiration falls within the strategic target areas of *Energy and Reaction Engineering*. In next five years, my research activities will focus towards: (i) conversion processes to produce fuels and chemicals from alternative sources, (ii) application of microfluidics to develop sustainable manufacturing processes, (iii) engineering aspect of liquid phase oxidation, and (iv) fouling and sludge formation of fuels during storage and transportation. My research will able to develop alternative and sustainable processes that have potential to meet the future demand of fuels, biofuels, petrochemicals and/or pharmaceuticals while mitigating environmental impact.

- Sustainable conversion processes will use to produce fuels and chemicals from alternative resources such as biomass, coal, bitumen, and waste materials. Some of these materials are rich in oxygen (for example, Canadian coal has about 20 wt % O), moisture (e.g. coal and biomass) and sulfur (oilsands bitumen, for instance, has about 5% S). Oxidative or oxidative hydrothermal dissolution to depolymerize the alternative sources has great prospect to recover chemicals (such as organic acids, phenol) from O-rich materials and to produce fuel having higher heating value and lower sulfur content. Both the catalytic and non-catalytic routes will be investigated by using air and H_2O_2 as a green and economic oxygen source. Catalysts will be synthesized by wet impregnation method. Fuel will also be produced from the recovered organic acids via catalytic decarboxylation. The depolymerization of cellulose to glucose by oxidation-hydrolysis will also be considered. Glucose can then be used to produce renewable fuels.
- Microfluidic reactor will employ for better understanding of the industrially important chemical processes to develop a sustainable manufacturing process. For instance, ibuprofen can be produced only in three minutes in microfluidic reactor whereas traditional complex routes require many steps and produce huge wastes (E-factor: 20–100). Oxidation of benzyl alcohol to benzaldehyde can also conduct easily in microfluidic reactor instead of traditional catalytic route that produce huge wastes. Moreover, alcohol and ketone production will be manipulated by using different level of oxygen in a reactor (e.g. BUSS loop reactor) that would be beneficial for pilot/industrial scale. Micro/Nano droplet formation using hydrocarbons, alcohol and H_2O_2 will also be investigated to develop sustainable oxidation.
- Engineering aspects of liquid phase oxidation will be explored measuring *in situ* oxygen level in different organic liquids at different experimental conditions. A model will be developed to predict and validate the oxygen transport data for different compound classes based on the elemental component and the structures. In addition, work will be extended to monitor the transport of other gases into the hydrocarbons. Moreover, there is scope to develop more sustainable HIOXY patches to monitor *in situ* oxygen level at wide temperature (currently available patches can only work at 50 °C) and hydrocarbon ranges.
- Part of my PhD work ended up indicating the role of heterocyclic compounds in bitumen weathering, fouling and sludge formation during transportation of fuel. However, very little information is available in literature despite considerable industrial interest. I envision working in the field.

Research Experience:

As a postdoctoral fellow in the department of Chemical and Materials Engineering, University of Alberta, currently, I am involved with multiple research projects including but not limited to oxidative coal dissolution, multiphase reaction engineering in a microfluidic reactor, micro/nano droplet formation for sustainable oxidation, and heavy oil upgrading. During my PhD at University of Alberta, I worked on a multidisciplinary project and the main accomplishments were: (i) identified the main challenges (such as viscosity increase) of oxidative bitumen upgrading, (ii) identified the main compound classes responsible for free radical addition reactions (potentially caused viscosity increase) with an explanation based on their reaction pathways, (iii) demonstrated by applying a microfluidic reactor how oxygen availability can be used to control oxidation selectivity independent of conversion by manipulating reactor hydrodynamics, (iv) developed a method to monitor *in situ* oxygen transport in liquid hydrocarbon by using an oxygen phase fluorimeter and modified the current theoretically predicted threshold for the oxidation in the film based on the Hatta-number. Following the *in situ* method developed to monitor the liquid phase oxygen during my PhD study, I also worked on the oxygen transport in different liquid hydrocarbons at Shahjalal University of Science and Technology in collaboration with University of Alberta. The objectives were to generate more engineering experimental data that could be used for reactor design and operation. A manuscript is in preparation based on the experimental data.

In addition, I conducted experimental research in biodiesel production from solid sludge of wastewater treatment during my MESc study at Western University. I am advising one graduate student of Chemical and Materials Engineering Department of University of Alberta. I also supervised six undergraduate students during my work at University of Alberta, Western University and Shahjalal University of Science and Technology. My research has been resulted in eleven research articles in reputable journals such as Renewable & Sustainable Energy Reviews, Fuels & Energy, Fuels, Industrial and Engineering Chemistry Research, Reaction Chemistry and Engineering, and in the proceedings of eight conferences including prestigious conference ACS. These contributions have made positive impacts within the research community, with 261 citations to date (Google scholar h-index: 6; June 29, 2019). Recognizing my contribution in fuels and petrochemicals research area, I received Gerald J Maier/NOVA Chemicals Corporation Graduate Recruitment Award (CAD 5000) at University of Alberta. I also received Julie Lassonde Scholarship (CAD 10,000) and GTA Outstanding Research Contribution Scholarship (CAD 500) at Western University in recognition of biodiesel production from wastewater sludge.

I have extensive hands-on experience in many modern analytical tools and experimental techniques. I have synthesized copper oxide supported on activated carbon catalyst (CuO/C) and used for the decarboxylation of the model acids and organic acids extracted from coal via oxidative dissolution of coal as a part of my postdoctoral project. Moreover, I prepared and characterized heterogeneous acid and base catalysts and zeolite catalysts and applied in biodiesel production from wastewater sludge. I gained extensive experiences in liquid phase oxidation during my PhD research. These experiences will facilitate conducting research that I envisioned.

I believe effective collaboration and pursuing funding play a significant role in accelerating academic research and innovation. I have been involved in various research and engineering projects with industrial partners, i.e. Cenovus FCCL Ltd, Mancal Incorporation, Suncor Energy, Institute for Oil Sands Innovation (IOSI) and 3P Technologies. I have learned how to effectively communicate with potential industrial partners to find a "sweet spot" for collaboration and pursuing funding. Furthermore, I assisted my current and former advisors in the preparation of several funding applications and research proposals for the federal agencies such as NSERC, MITACS and provincial agencies such as AITF. I believe that given my drive and motivation as reflected by my research experience, I will be able to link with existing research expertise in my future work place with industrial partners, external research agencies, and other universities (especially with University of Alberta, Canada) to accelerate my research.

Teaching Interests:

My teaching interests include but are not limited to different undergraduate courses on Chemical Process Analysis, Fluid Dynamics, Heat Transfer, Mass Transfer, Statistical Process/Statistics for Engineers, Chemical Reaction Engineering, Petroleum Refining, Natural Gas Processing, Petrochemical Technology, and Fuel and Energy. I am interested in teaching graduate courses on Energy and Environment, Green Energy and Engineering, Instrumental Analyses for Engineers, Liquid Phase Oxidation, Petroleum

Refining, and Advanced Chemical Engineering Topic (such as Thermal Analysis-Fundamentals and Applications, and Rheology- Flow Properties of Non-Newtonian Fluids). For advanced graduate courses, I plan to develop course materials and lectures that will motivate the graduate students to learn advanced fundamental concepts as well as to stimulate their critical thinking skills that would be important for cutting-edge research and innovative engineering practice.

I have a strong commitment to facilitate and promote the equity, diversity, and inclusion. I believe my five years of teaching experience at the university level, strong academic records and ten years of research experience in the field of conversion processes to produce fuels and chemicals by using free radical chemistry and multiphase reaction engineering would allow me to perform successfully in this position.

581635 Sustainable Energy Production from Renewable and Fossil Fuels

Research Interests:

As the world energy consumption will increase by 48% between 2012 and 2040, projected by U.S. Energy Information Administration (EIA) in International Energy Outlook 2017 (IEO2017), our society is faced with the essential challenge to fulfill our energy needs in a most efficient, economic and environment friendly manner. My research interests synergistically bind conventional and alternative energy together and will give me the unique opportunity to pursue externally funded research as a faculty member on both sides of the energy resource spectrum. Overall **my research aims to develop simple, robust and effective engineering solutions to produce sustainable energy from both renewable and fossil fuels.** All of these researches will be done by collaboration with different industries and by using facilities at various national labs.

Future Research Plans

My research will be mainly focus on the following research areas,

1. **Second generation Biofuel production from biomass:** To make biorefineries profitable and not to compete with food resources it is very important to produce second generation biofuels from non-food crops essentially lignocellulosic biomass deriving from woods, crop residues or organic waste. The main limitation for processing lignocellulosic biomass is that it requires additional step to separate main components (cellulose, hemicellulose and lignin) before processing followed by high temperature treatment to get bio-oil which is the main starting material to generate biofuels and other biorefinery products. To overcome these limitations, I have two goals for this project. (1) Develop solvent that could potentially dissolve/disperse all the lignocellulosic components so that the whole biomass could be treated simultaneously in a single reactor (2) Replace the high temperature thermal treating step by economical photocatalysis step using mesoporous titania-silica as a photocatalyst.
2. **Mesoporous Silicon/Carbon anode electrodes for high-performance Lithium-Ion batteries:** Various drawbacks of commercially available graphite anode including low theoretical specific capacity (372 mAh g^{-1}) and poor rate capability limit the performance of Lithium-Ion (Li-Ion) batteries as an energy storage material. Research is ongoing to replace graphite with silicon (Si) a negative electrode material due to its high theoretical capacity (4200 mAh g^{-1}) for the production of next generation Lithium-Ion batteries with high energy density and high-power performance. My goal for this project is to develop novel mesoporous Si electrode using inexpensive surfactant templated mesoporous silica thin film as the starting material. The pore orientation and pore size will be tuned by templating surfactant structure and by substrate surface modification. This novel design will help to release the developed stress and strain during volume expansion and maintain the structure stability for the anode electrode in Li-Ion batteries.
3. **Formulation of environment friendly hydraulic fracturing fluids for shale gas reservoir:** In hydraulic fracturing, the selection of suitable water based fracturing fluid additives is the most important step to recover natural gas effectively and environment friendly from low permeable shale reservoirs. The current challenges with conventional polymer based fracturing fluid additives are their hazardous nature, poor reservoir conductivity, formation rock damage, additional chemical breaker requirements, incomplete cleanup, reducing porosity and permeability of proppant pack, sensitive to high temperature/salinity in the reservoir. My goal for this project is to develop environment friendly nanoparticle-modified viscoelastic mixed surfactant hydraulic fracturing fluids with enhanced proppant carrying capacity, high temperature/salinity tolerance and pH adjustable viscoelastic property for improved natural gas recovery.
4. **Shale flowback water treatment using novel Forward Osmosis membrane:** Economical and sustainable water management in natural gas production via hydraulic fracturing from

unconventional shale formations is currently the biggest challenge for Oil &Gas industry. Millions of gallons of fresh water along with proppants and chemical additives are used every time to fracture a new well and after fracturing completion the gas industry has to deal with the large amount of recovered or the flow back water. The flowback water contains high amount of total dissolved and suspended solids, corrosion and scale products, production chemicals and bacteria which make it unsuitable for reuse both internally and externally before proper treatment. For this research project my objective is to develop highly selective, robust and low-fouling nano-composite membrane for forward osmosis (FO) to handle large amounts of flowback water with minimum pre/post-treatment. The FO membrane will be prepared by layer-by-layer assembly using porous silica nanoparticles incorporated in polymeric membranes.

Research Experience

1. Postdoctoral Fellow, Chemical & Environmental Engineering, **University of Cincinnati 2018-Current**, Project: "**Development and application of novel colloidal systems for drug delivery and skin model systems**", Advisor Dr. Yoonjee Park (Chem & Env Eng. Dept.)
2. Postdoctoral Research Fellow, Civil & Environmental Engineering, **University of Michigan 2015-2018**, Project: "**Study the impact of surfactant additives used in hydraulic fracturing on the extent of water entrapment and natural gas production in shale gas reservoirs**", Advisor Dr. Brian Ellis (Civil & Env Eng. Dept.)
3. PhD in Chemical Engineering, **University of Kentucky 2010-2015**, Thesis:" **Fundamental studies of surfactant templated metal oxide materials synthesis and transformation for adsorption and energy applications**", Advisor Dr. Stephen E. Rankin (Chem & Mat. Eng. Dept.)

Successful Research Grant/Fellowship

1. Dow Sustainability Postdoctoral Fellowship, 2-year Research & Development Grant 2016

Peer Reviewed Publications (selected)

1. **Das, S.**; Adeoye, J.; Dhiman, I.; Bilheux, H.; Ellis, B. "Imbibition of Mixed-Charge Surfactant Fluids in Shale Fractures". **Energy & Fuels, 2019, 33, 4, 2839-2847**
2. **Das, S.**; Oldham, E.; Lehmler, H.; Knutson, B.; Rankin, S. "Tuning the position of head groups by surfactant design in mixed micelles of cationic and carbohydrate surfactants". **Journal of Colloid and Interface Science, 2018, 512, 428-438**
3. **Das, S.**; Xu, W.; Lehmler, H.; Knutson, B.; Rankin. S. "Inverted Micelle-in-Micelle Configuration in Cationic/Carbohydrate Surfactant Mixtures". **ChemPhysChem, 2017, 18(1), 79-86**
4. **Das, S**.; Nagpure, S.; Garlapalli, R.; Islam, S.; Strzalka, J.; Rankin, S. "Pore Orientation Effects on the Kinetics of Mesostructure Loss in Surfactant Templated Titania Thin FIlms". **Physical Chemistry Chemical Physics, 2016, 18, 2896-2905**
5. Zhou, L.*; **Das, S.***; Ellis, B. "Effect of surfactant adsorption on the wettability alteration of gas-bearing shales". **Environmental Engineering Science, 2016, 33, 10, 766-777** * *Equal contribution*
6. Das, S.; Wu, Q.; Garlapalli, R.; Nagpure, S.; Strzalka, J.; Jiang, Z.; Rankin, S."In-Situ GISAXS studies of Orientation Effects on the Transformation Mechanisms in Mesoporous Titania Thin Films". Journal of Physical Chemistry C, 2014, 118, 968-976

Successful Grant of Time at National Lab Facilities (selected)

1. User proposal (IPTS-22133) "Nanostructure change of light-activatable drug-encapsulated liposome by laser irradiation and colloidal stability", Extended Q-Range Small Angle Neutron Scattering (EQ-SANS): Beam line BL-6, Spallation Neutron Source (SNS), Oak Ridge National Lab (ORNL) 2019
2. User proposal (IPTS-22139) "Molecular Interaction Between Surfactant from Body Wash and Skin Stratum Corneum Surface", Neutron Reflectometry: Beam line BL-4B, Spallation Neutron Source (SNS), Oak Ridge National Lab (ORNL) 2019
3. User proposal (Proposal ID: 22902) "Pores Accessibility of Mixed Surfactant Fracturing Fluids", Small Angle Neutron Scattering (SANS): Beam line NG-B-30-m SANS, Center for Neutron Research (NCNR), National Institute of Standards and Technology (NIST) 2017

4. User proposal (IPTS-15695) "Effect of Surfactant Adsorption on Water Imbibition in Organic-rich Shale: Neutron Imaging Study", Neutron Imaging: Beam line CG-1D, Spallation Neutron Source (SNS), Oak Ridge National Lab (ORNL) 2016
5. General user proposal (GUP-32929) "Pore Orientation Effects on the Kinetics of Mesostructure Loss in Surfactant Templated Titania Thin Films", GISAXS/ GIWAXS: Beam line 8-ID-E, Advanced Photon Source (APS), Argonne National Lab (ANL) 2013

Teaching Interests:

My philosophy of teaching is to create an environment for students where the skills and knowledge learned from the class will eventually be used in practical fields. I believe that the most significant learning occurs in situations that are both meaningful and realistic. My goal as an educator is to inspire students to pursue their academic interests by learning to ask the right questions, not memorize the right formulas. I view teaching as a privilege and hope that my efforts will have a positive influence on my students as my past professors have had on mine. My previous teaching experience as teaching assistant at University of Kentucky and my postdoctoral course on college science teaching at University of Michigan will help me to become a successful educator. I am confident that I would be able to teach any course in the Chemical & Materials engineering curriculum, and am interested in teaching the following courses: (1) Engineering thermodynamics, (2) Fluid dynamics, (3) Transport phenomena, (4) Separation processes and (5) Reaction engineering. Apart of that I am also interested in developing elective courses on (1) Colloid & Interfacial science (2) Sol-gel chemistry & applications.

581194 A Structured Approach to the Design and Optimization of Sustainable Energy Systems

Research Interests:

My research interests lie in the design and optimization of energy systems, at scales from the individual unit to the entire supply chain. My passion is to develop and improve sustainable energy sources to reduce the world's reliance on fossil fuels. I plan on pursuing projects in this sector to improve the performance, economic viability, and efficiency of these systems as applied to alternative energy sources to provide a more sustainable energy future.

Proposed Research:

As a faculty researcher, I seek to apply a systems perspective to the plant design process, with the goal of facilitating the design and development of sustainable energy production. My vision is to begin at the equipment level, focusing on features that affect the power, stability, cost, and environmental impact of a unit. With success, I will expand the ideas to whole-plant design.

It can be difficult to know *a priori* how design decisions will impact the overall performance of a unit operation. One approach is to develop detailed theoretical models that enable all design decisions to be considered in simulation, but this is time consuming and computationally intensive.

As an alternative, this work will explore the creation of a higher order framework of system components to assist in understanding how design decisions interact to impact performance. The framework will combine component data, a mass and energy flow network, and constraints from reduced order physics models to estimate the impact of a design on the performance of the final system. System optimization will then push the framework to suggest a design which meets the desired conditions, such as maximum power or minimum cost, given environmental restrictions.

As a complement to this investigation, an experimental program will provide a testbed to assess and refine the theoretical developments. In particular, a research program for designing flow batteries with potential for large energy storage will be undertaken. Flow batteries are a promising alternative for energy storage that offers great potential for scalability and flexibility in application. The testbed will be used to gather component impact data, validate the design framework, and explore methods of increasing the power density and longevity of flow batteries through flow field design. This will flex the capabilities of the design framework, with the goal of boosting flow battery utility and viability as alternative energy storage.

Initially, experiments will identify the impact of key design decisions on the flow field to better understand how individual design characteristics affect the power density and longevity tradeoff. It will be important to complement this experimental work with electrochemical and fluid dynamics modelling of the reactants in the flow field to understand and predict where there may be dead spots or areas of low flow. This will improve the design process and identify ways that the designs may be changed to improve stability and performance, for both the short and long term.

With success in these early efforts, research will expand to applying the framework to improve the robustness of energy supply chain networks. While supply chain network optimization offers a reliable method for finding solutions that are feasible and maximized (or minimized) for some objective, it does have a weakness that small differences in starting data can result in significant differences in the form of the optimized result.

The challenge is that similar values of the objective function can potentially correspond to several different supply chain frameworks, so a small difference in initial conditions can produce very different recommendations for the final design. While this is not always a concern, it can be problematic, for example, in high investment situations. Restrictions imposed by local zoning can also skew results when a portion of a supply chain moves in or out of that area.

Envisioned is a supply chain optimization framework that considers these and other practical constraints in the formulation of the design. For example, can a locational consistency serve as a soft constraint within the framework? And can it be introduced without the need for significant user effort? If successful, the result will be a robust network optimization framework.

Previous Work:

My postdoctoral research has focused on system design and optimization as applied to energy and process systems. This research has explored a range of applications for these techniques, starting with supply chain network optimization for biomass to produce bioproducts. The work developed techniques to incorporate spatially distributed, inherently variable data into the optimization without losing information to aggregation. This systems research extended into design of systems, as well, with a project focused on the design and life cycle analysis of a system to produce an alternative green process for a currently petroleum-based polymer. The vein of research continues with work on the system design and operation for a gigawatt scale electrolyzer system for optimal use of variable wind power. Ultimately, the unifying theme of my work is the optimal use of sustainable resources.

My PhD research focused on design of the gas distributor of polymer electrolyte membrane (PEM) fuel cell systems to increase the system power density. The main focus of the work was on interdigitated gas distributors, which are an alternative to the conventional, parallel-channel gas distributors used in many fuel cell system designs. In interdigitated gas distributors, reactant flow is forced through the gas diffusion medium of the cell, bringing reactants closer to the catalyst layer and forcing liquid water out of the gas diffusion medium, thus reducing the diffusion distance. My research explored how gas distributor design parameters, including channel length, width, depth, and GDL type could be improved to optimize the performance characteristics of the fuel cell system, such as system power density, stability and water handling. The research sought to understand the importance and effect of these design parameters for fuel cell systems with interdigitated gas distributors using the conventional design as a baseline.

Teaching Interests:

A comprehensive engineering education produces graduates who are renaissance persons of engineering whose consideration of the world engages with their trade in transformative ways. Their education should resemble a pyramid: built on a broad, strong base of knowledge to foster versatile problem-solving skill, narrowing in scope as the curriculum culminates in a peak of field-specific training. Above all, graduates should be prepared for the challenges they will face, able to identify with scientific rigor their questions, ideas, and solutions, and capable of communicating these to diverse audiences. Effective teaching is driven by a fundamental desire for student success, both academically and in life.

I have had the privilege of serving as both mentor and teaching assistant at two universities in two countries, informing my development as a teacher in the classrooms where I have been the student. With each new responsibility, I have developed my own guiding ethos enriched by the perspectives of colleagues and students from many backgrounds. To educate effectively in a faculty role is to maintain a healthy coexistence of dichotomous ideas. Over-reliance on the theoretical leaves students lacking experience in its specific application — on the practical, lacking perspective in its broad application. Focusing on one point obsessively always comes at the expense of another: research furthers the bounds of knowledge and keeps the professor abreast of recent scholarship, where teaching accessibly provides students the cognitive tools to improve and motivates research. Across disciplines, reliance on feedback to inform introspective self-reflection is key to unlocking the ultimate test of teaching quality: whether students are provided a productive educational environment. Ultimately, a neglected balance leaves students potentially discouraged by the learning process: the *'how'* and *'why'* of the material is lost.

My experiences have well prepared me to actively maintain the equilibrium these tenets demand. My pedagogy will introduce foundational concepts with unique examples: some highly theoretical, others, far more experiential and hinging on applied example to current issues. Thorough preparation for teaching duties leaves time for the one-on-one interactions with students which inform mindful selection of class materials. Outside of the classroom, my research can only benefit from this investment, rejuvenated by the opportunity to interact with students from diverse backgrounds and think in a new paradigm. This approach will frame curricula in relation to other fields, clearly evidencing that topics in engineering do not stand alone, but interlock.

Catering to these vital balances is made more achievable where a cooperative faculty environment is driven by a central goal: to bolster student success across course levels. I look forward to opportunities to grow in my teaching voice. Shakespeare is credited with regarding the world as a stage. I think it's a classroom.

581883 A Novel Route for the Flexible Preparation of Hydrocarbon Jet Fuels and Valuable Chemicals from Biomass-Based Platform Chemicals: A Case of Using 2, 3-Butanediol As a Feedstock

With the reduction of fossil energy reserves, increasing concerns of environmental pollutions and greenhouse gas emissions, considerable efforts and processes have been made for the production of bio-based chemicals and fuels to reduce the reliance on fossil resources. 2, 3-butanediol (BDO) is one of the most promising platform bio-based chemicals, which can be produced by fermentation of sugar, lignocellulosic wastes, glycerol and syngas.

A process employing acid catalysis dehydration, C-C coupling technics, and hydrodeoxygenation is being developed to convert BDO to methyl ethyl ketone (MEK), methyl-propanol (MPA), (2-ethyl-2, 4, 5-trimethyl-1, 3-dioxolane, ETMD), butadiene, and jet fuel. First, the products of BDO dehydration is completely detected, identified and characterized, and the mechanism is completely investigated. Next, the kinetic model of BDO dehydration is established to control the selectivity of variable products according to the marketing demand. Besides widely applications of MEK, the ETMD formed by ketalization of MEK and BDO, was reported as a promising next generation fuel additive. Then, by undergoing the aldol condensation and hydrodeoxygenation, MEK and MPA can be further converted to C8~C12 linear and branched hydrocarbons that meet the requirements of jet fuels. The aldol condensation kinetics are also investigated to make molecular size distributions in the final hydrocarbons controllable.

Research Interests: biomass conversion, biomass utilization, biofuel

Teaching Interests: biomass utilization,natural product synthesis

585027 Engineering Multifunctional Nanomaterials for Energy and the Environment

Research Interests:

I am a materials scientist with extensive research experience in nanomaterials for applications in the areas of energy and the environment. As an Assistant Professor, I am eager to bring in my multidisciplinary engineering expertise, my strength in building collaborations across multiple departments, and my proficiency in writing funding proposals. As an expert in the synthesis and characterization of nanomaterials, with a Ph.D. in Materials Science from CentraleSupelec and Ecole Normale Superieure (Paris-Saclay, France), I believe I am an excellent candidate for a faculty position in a leading institution. I can contribute to original and strong core programs in materials science and engineering, centered on composite materials, energy storage, water harvesting and treatment for a sustainable environment. I have built a unique background in fundamental and applied research at Princeton, in synergy with leading materials industry, and have acquired extensive knowledge and research experience in the area of graphene and carbon nanotubes for diverse applications. At Princeton, I have contributed to fundamental research on the chemical routes for synthesizing graphene at industrial scale. My contributions have resulted in the production of graphene-reinforced rubber composites that are currently being commercialized by the world's largest graphene manufacturer Vorbeck Materials (a Princeton University spin-off), where I have held a joint researcher position. Additionally, for the manufacturer Reliance Industries Ltd, I have demonstrated the effect of functionalized graphene in automobile tires to achieve low rolling resistance. My research on Li-ion batteries examines the multifunctional potential of graphene and other materials such as polymers and ceramics to extend their performance. Initially in the framework of an ARPA-E project, and later through partnership and full funding from a preeminent American electronics company, I have successfully demonstrated the role of graphene as a barrier to the formation of solid-electrolyte interphase at the anode level. This approach extends the life cycle of Li-ion batteries. Recently, I have been leading a research project on the development of organic and inorganic materials for atmospheric water harvesting. I have been synthesizing hygroscopic and biocompatible polymer- based solid desiccant materials that can readily absorb large quantities of water vapor.

Ultimately, freshwater is obtained using an isothermal regeneration cycle at low temperatures. In addition to successfully transferring technologies from lab to market, my research has

benefited important industries and resulted in the publication of 18 journal articles, 1 book chapter, 5 patents, and over 20 conference publications, with over 400 citations. As an Assistant

Professor, I plan to lead a multidisciplinary research group that will investigate the processing and application of nanomaterials and be at the forefront in this domain.

Teaching Interests:

As an Assistant Professor, one of my objectives will be to create a classroom composed of curious and engaging students, who will be the driving force for deep interest in the material taught. I aim to build a classroom environment where students across diverse backgrounds can be incorporated to build a harmonious education platform. I advocate teaching a course from different angles, different perspectives, so as to attempt to capture or appeal to the different learning styles or backgrounds of the various students. My ultimate goal is to train my students to be critical thinkers capable of raising pertinent scientific questions, and more importantly to propose answers, to innovate, and to advance science. I would be delighted to teach a variety of courses such as Structure and Properties of Materials, Synthesis and Applications of Polymers, or Energy Storage Systems. My passion and experience in materials science, fundamental and applied, inspire me to motivate and mentor the next generation of students in their careers.

587795 Probing Hydrogen activation to solve current energy challenges

PhD Research Summary

PI: Prof. Lars Grabow, University of Houston

Keywords: Catalysis, Hydrogen purification, Gold, Density Functional theory, Infrared Spectroscopy, Metal-Support Interface.

Renewable hydrogen (H_2) is touted to be the replacement for carbon based fuels in an effort to reduce greenhouse gas emissions. The current H_2 production route through steam reforming of natural gas followed by water-gas shift treatment, however, indirectly releases CO_2 and accounts for 3-5% of our global energy use. Moreover, the CO content in the effluent H_2 stream from the water-gas shift reactor must be lowered for H_2 applications that are sensitive to CO poisoning, for instance, fuel cells.

My research focuses on purifying hydrogen through preferential oxidation (PROX) of CO with O_2 in hydrogen-rich streams, which is a more energy efficient solution than the currently used CO methanation approach. As catalytic material I am studying supported gold nanoparticles, which is motivated by Haruta's discovery of their high CO oxidation activity at low temperature. The toolset of my research encompasses density functional theory, FTIR spectroscopy and kinetic experiments to study CO and H_2 oxidation on model Au/TiO_2 catalysts. Water was shown by our group to enhance CO oxidation rates through O_2 activation to active OOH^* intermediate at the metal-support interface (MSI). Moreover, we also suggested that the MSI is the active site for heterolytic H_2 activation over Au/TiO_2. This conclusion is corroborated by DFT calculations and H-D exchange experiments coupled with FTIR spectroscopy. While water promotes O_2 activation, we found it to poison the MSI sites and kinetically hinder the H_2 activation. Thus, water plays two distinct roles in enhances the selectivity for CO oxidation during PROX.

One of the applications of hydrogen produced through the steam reforming of methane is to selectively hydrogenate alkynes in polyolefin feedstock in the plastics industry. The alkyne impurities in the feed must be reduced to prevent deactivation of the downstream polymerization catalyst. Au catalysts have great potential as selective hydrogenation catalysts because they do not over-hydrogenate both the alkynes and the alkene feed into alkanes. The selectivity of Au originates from the weak binding of alkenes to Au, thus preventing further hydrogenation to alkanes. Using my Au/TiO_2 computational model, I have explored the active sites for H_2 activation to hydrogenate alkynes. My DFT model predicted that the alkyne hydrogenation requires negatively charged hydrogens which can be produced on the surface of the gold nanoparticle in the form of gold hydrides. In a particle size study we found that the turnover frequencies (TOFs) were identical for different particle sizes when they were based on surface sites for 1-octyne hydrogenation. Overall, computational predictions and kinetic experiments suggest that hydrides on surface gold sites are active for alkyne hydrogenation. We believe that our detailed insights into the hydrogenation kinetics over supported gold catalysts will contribute to further improve selective hydrogenation catalysis in industrially relevant processes.

In the broadest terms, my work will showcase the importance of studying H_2 activation and the application of novel gold catalysts to engineer novel solutions for the current challenges in hydrogen based industrial processes.

Research Experience: As a graduate research assistant, I have used density functional theory (VASP), FTIR spectroscopy and kinetic experiments to study the application of supported gold catalysts for hydrogen purification, selective hydrogenation of alkynes and natural gas upgrading through selective methane oxidation. I have had the opportunity to perform experiments for a year at our collaborator, Prof.

Bert Chandler's lab at Trinity University in San Antonio. I have discovered a novel H_2 activation pathway on Au/TiO_2 and came up with an explanation for enhanced PROX selectivity upon adding water to the feed stream. During my M.Tech. research, I used DFT (Quantum ESPRESSO) to study the effects of doping foreign metals in cerium oxide and the application of Ru-substituted CeO_2 for CO_2 methanation.

Areas of Interest for Postdoc: As a postdoc, I want to learn and apply new theoretical methods and experimental techniques to improve catalysts and materials in clean energy applications. I want to explore the area of machine learning for new material discovery and I want to gain experience in the application of tight binding method for large scale systems. I want to gain hands on experience in synthesizing catalysts and would love to learn new techniques such as mass and NMR spectroscopy. In addition to diversifying my technical skill set, I would love any opportunities to collaborate.

Selected Publications:

1.) Whittaker, T.;† **Kumar, K. B. S.**;† Peterson, C.; Pollock, M. N.; Grabow, L. C.; Chandler, B. D. "H_2 Oxidation over Supported Au Nanoparticle Catalysts: Evidence for Heterolytic H_2 Activation at the Metal-Support Interface", J. Am. Chem. Soc, 2018, 140, 16469-16487. († : equal contribution)

2.) **Kumar, K. B. S.**; Deshpande, P. A. "On identification of labile oxygen in ceria-based solid solutions: Which oxygen leaves the lattice?", J. Phys. Chem. C, 2015, 119, 8692–8702.

3.) Bruno, J. E.; **Kumar, K.B. S.**; Dwarica, N. S.; Hüther, A.; Chen, Z.; Guzman IV, C. S.; Hand, E. R.; Moore, W. C.; Rioux, R. M.; Grabow, L. C.; Chandler, B. D. "On the Limited Role of Electronic Support Effects in Selective Alkyne Hydrogenation: A Kinetic Study of Au/MO_x Catalysts Prepared from Oleylamine-Capped Colloidal Nanoparticles", ChemCatChem, 2019, 11, 1650-1664 (featured in "A Decade of Catalysis brought to you by ChemPubSoc Europe".

588023 Utilizing bioenergy for production of energy and mitigation of climate change

Research Interests: Bioenergy with carbon capture and storage; Thermochemical conversion processes; Chemical looping; Heterogeneous catalysis; Waste management; Process modeling, integration, and optimization

Post-doctoral Projects

1) Gasification and pyrolysis of garden waste and agro residue for thermal applications Funded by: Tata Centre for Technology and Design (TCTD) IIT Bombay in collaboration with TCTD MIT; Supervisor: Prof. Sanjay Mahajani, Department of Chemical Engineering, IITB

2) Developing low-cost biomass to power systems using engine for tar reduction

Department of Mechanical Engineering, MIT; Prof. John Heywood's research group with Dr. Leslie Bromberg (Funded by: Yanmar, Japan)

3) Current project: Decentralized biomass torrefaction systems for biochar production

Department of Mechanical Engineering, MIT (Funded by: J-WAFS) Supervisor: Prof. Ahmed Ghoniem

Research Experience

During PhD, I worked on an alternative chemical looping based process to produce feedstock for ammonia and nitric acid plants. The project was sponsored by Orica Mining Services, Australia, and the work primarily involved studying reaction of ammonia with different metal oxides, thermodynamic analysis, investigating reaction mechanism, and Aspen simulations. Later, in postdoc at TCTD, we developed an improved pelletization and gasification process, and patented a new fixed bed downdraft gasifier design for high ash feedstock. I worked on TGA, bench-top reactor, and pilot gasifier (40 kg batch capacity) primarily to investigate gasification performance, effect of operating parameters, and the role of ash on char reactivity. I also studied co-gasification of high ash coal and high ash biomass in the designed gasifier analyzing the effect of inorganic content on the synergistic performance. At MIT, my first project involved optimizing the induced downdraft gasifier and connecting it to the internal combustion engine that was investigated for its ability to crack tar in the producer gas. Presently, I am working on torrefaction of different biomass (e.g. pine shavings, hay, rice husk, wood chips, coconut shells) to produce biochar for improving crop yield and enhancing soil quality. We developed an auger- based moving bed torrefaction reactor where the reaction temperature is governed by controlling air-biomass ratio and the residence time is governed by controlling the speed of the motor operated auger. We defined an index of torrefaction to measure the severity of the process and proposed it as a parameter to guide the operation protocol of the reactor for a specific biomass based on application requirement. I am currently working on developing a control system for the reactor and carrying out a comparative life cycle analysis for non-oxidative and oxidative torrefaction.

Teaching Experience

1. Teaching Assistant in Department of Energy Science and Engineering, IIT Bombay (1 course each semester during PhD): a) Energy management; b) Process Integration; c) Non-conventional energy sources; d) Electrochemistry; e) Thermodynamics
2. Demonstrator for Mass Transfer (Separation) lab for undergraduate students at Department of Chemical Engineering, Monash University, Australia.
3. Lab instructor for Process Control at Department of Chemical Engineering, BITS Pilani.
4. Completed KTCP Summer Teachers' Training Program at MIT (2019), and Professional Practice courses at BITS Pilani (2010-11).

As a post-doctoral fellow at Tata Centre, I took several sessions on different topics such as stakeholder analysis, business plan preparation, waste management, to name a few. These sessions used to be organized once a month for about 25 Masters students. I prefer teaching the courses such as Introduction to Chemical Engineering, Thermodynamics, Chemical Process Safety, Thermodynamics, Reaction Engineering, and other courses related to energy and environment.

Future Direction

I want to contribute to resolve the issues of energy security, environmental emissions, and waste management using chemical engineering knowledge through experimental and theoretical research in the areas of bioenergy, alternative fuels, and sustainability.

1) Bioenergy with Carbon Capture and Storage (BECCS)

BECCS is viewed as a promising negative emissions technology because the carbon dioxide released after conversion of biomass is captured and stored, and the new growing biomass consumes CO2 already in the atmosphere. However, there is no successful BECCS plant at commercial scale with rare exceptions, which means there are still several challenges that need to be addressed. A single large sized (500 MW) coal-fired or biomass co-fired power station typically emits 2 to 3 million tons of CO2 per year. I will work on mapping biomass resource availability at regional and global level; investigating supply demand chain of biomass & BECCS technology scenarios; evaluating the environmental and economic impacts of BECCS; and identifying the potential challenges & opportunities of large scale deployment of BECCS. Recently, U.S. Department of Energy (DOE) announced over $79 million in funding for bioenergy research and development including biofuels, bio-products, and biopower.

Biochar: Production, Applications, and Impact

Since the technology of BECCS has been considered to achieve CO2 emissions target set by Paris Agreement, understanding the role of biochar in reducing carbon emissions and providing energy has become an important area. Primary processes for production of biochar includes pyrolysis, gasification, and torrefaction. I will do the experimental studies on producing different grades of biochar using each of these processes at different scales starting with TGA and bench- top reactor. I also plan to investigate the synthesis of different magnetic biochars (Ca/Mg/Fe loaded) and their effectiveness in removing different contaminants from waste water. On application side, I will investigate carbon accounting and sustainability assessment of biochar chain using Life Cycle Assessment. On implementation and commercialization side, I will work with end-users (e.g. farmers) to acquire and use biochar, create demonstration plots, and develop training methods and materials to teach growers to implement biochar strategy effectively.

2) Chemical looping

Chemical looping is an innovative approach assisting CCS as it enables an almost pure CO2 gas to be produced, which can then be relatively easily stored without any further major processing. The process uses an oxygen carrier that is cycled between a fuel and an air reactor to provide oxygen for partial (Chemical looping gasification (CLG)) or complete (Chemical looping combustion (CLC)) conversion of the feedstock such as biomass, avoiding requirement of air separation unit. The relatively high quantity of volatile matter, and low sulfur and ash content of biomass makes it better feedstock over coal in chemical looping process. Two main reasons for the technology not being commercialized are oxygen carrier (OC) issues (e.g. inadequate reactivity, recyclability, strength, attrition resistance, oxygen-carrying capacity) and plant complexity especially the solids-circulating systems and heat transfer. I will start with work on developing efficient oxygen carriers from cheaper metal oxides, ilmenite, industrial waste, and bimetallic oxides; and then would investigate them for their gasification and combustion performance.

Publications

1. Thengane, S.K., Hoadley, A., Bhattacharya, S., Mitra, S., Bandyopadhyay, S. (2014). Cost- benefit analysis of different hydrogen production technologies using AHP and Fuzzy AHP. *International Journal of Hydrogen Energy*, 39

2. Thengane, S.K., Hoadley, A., Bhattacharya, S., Mitra, S., Bandyopadhyay, S. (2016). Exergy efficiency improvement in hydrogen production process by recovery of chemical versus thermal energy. *Clean Technologies and Environmental Policy*, 18 (5)
3. Thengane, S.K., Hoadley, A., Bhattacharya, S., Mitra, S., Bandyopadhyay, S. (2016). An alternative process for nitric oxide and hydrogen production using metal oxides. *Chemical Engineering Research and Design*, 112
4. Thengane, S.K., Hoadley, A., Bhattacharya, S., Mitra, S., Bandyopadhyay, S. (2017). Thermodynamic evaluation of chemical looping based nitric oxide and hydrogen production. *Chemical Engineering Research and Design*, 132
5. Siddiqui, H., Thengane, S.K., Sharma, S., Mahajani, S.M. (2018). Revamping downdraft gasifier to minimize clinker formation for high-ash garden waste as feedstock. *Bioresource Technology* 266 (*Equal first author*)
6. Gupta, A., Thengane, S.K., Mahajani, S.M. (2018). CO2 gasification of char from lignocellulosic garden waste: Experimental and kinetic study. *Bioresource Technology*, 263
7. Thengane, S.K. (2018). Assessment of different technologies for managing yard waste using analytic hierarchy process. *Process Integ. and Opti. for Sustainability*, 3(2)
8. Thengane, S.K., Tan, R.R., Foo, D.C.Y., Bandyopadhyay, S. (2019). A pinch-based approach for targeting of carbon capture, utilization, and storage (CCUS) systems. *Industrial & Engineering Chemistry Research* 58 (8)
9. Thengane, S.K., Gupta, A., Mahajani, S.M. (2019). Co-gasification of high ash biomass and high ash coal in downdraft gasifier. *Bioresource Technology*, 273
10. Kung, K., Thengane, S.K., Shanbhogue, S., Ghoniem, A. (2019). Parametric analysis of torrefaction reactor operating under oxygen-lean conditions. *Energy*, 181
11. Gupta, A., Thengane, S.K., Mahajani, S.M. (2019). Kinetics of pyrolysis and gasification of cotton stalk in the Central parts of India. *Fuel* (*accepted*)
12. Kung, K., Thengane, S.K., Ghoniem, A. (2019). Functional mapping of torrefied product characteristics for biofuel and chemical production. *Journal of Cleaner Production* (*revised*)
13. Thengane, S.K., Bandyopadhyay, S. (2019). Biochar Mines: Panacea to Climate Change and Energy Crisis? *Clean Technologies and Environmental Policy* (*under review*)

588413 Producing cleaner fuels using waste and renewable resources

Profile and short bio:

I am a Chemical Engineer with a strong background in materials synthesis and characterization, and an expertise in polymers, carbon materials and nanotechnology. After completing my B.Sc. in Chemical Engineering (National Technical University of Athens, Greece, 2006) and my M.Sc. in Materials Science and Technology (Aristotle University, Greece, 2008), I worked in the private sector before joining the United Arab Emirates University (U.A.E.U.) in 2014 in order to pursue a PhD in Chemical Engineering under the supervision of Prof. I. Zuburtikudis. My PhD research is focused on the development of tailor-made adsorbents based on inexpensive renewable and/or waste resources for the desulfurization of commercial fuels. In parallel, I worked in the U.A.E.U. as a Teaching Assistant for four semesters in various undergraduate courses (thermodynamics, heat and mass transfer, reactor design and polymer engineering among others). After my graduation in May 2020, I am seeking to join a research-focused university as an assistant professor or as a postdoctoral associate.

Research interests:

Sustainability has emerged as an issue of immense importance that will determine the prosperity of future generations. Due to the gradual depletion of natural resources, especially of the fossil fuels, a significant part of the research and development in academia and in the industry is shifting towards the **utilization of waste or renewable resources for the development of new, advanced engineering materials**. Moreover, the increasing demand for **cleaner fuels** and the stringent regulations of commercial fuel specifications have driven the research of alternative methods to upgrade the conventional industrial **desulfurization** technologies.

These challenges have inspired my PhD research, in which I am developing **tailor-made adsorbents** using waste and renewable resources for the desulfurization of commercial fuels. Particularly, I am producing **carbon nanofibers** based on the **electrospinning** of **lignin** and **recycled PET**. Lignin is a very abundant natural polymer, the potential of which is still rather unexploited. On the other hand, poly(ethylene terephthalate) (PET) is one of the most widely consumed commodity polymers, therefore, it would be desirable to develop alternative uses of waste PET.

Carbon nanofibers (CNFs) are a one-dimensional form of carbon which are projected to be at the forefront of emerging technological advances, especially in applications related to energy storage, adsorption and catalysis. CNFs have a very high surface area which minimizes any limitations related to mass transfer. Moreover, a significant advantage of carbon nanofibers lies in the feasibility to be easily functionalized or doped with metals or heteroatoms. Therefore, in my research I present a road-map for the production of an advanced engineering material from low- cost resources. The method of preparation involves the **electrospinning** of the raw materials and the carbonization of the precursor nanofibers. Throughout my research I have used a **Design-of- Experiments** statistical methodology to optimize the materials preparation. By adjusting the process conditions, I have shown that the **textural characteristics** of the carbon nanofibers can be customized for a particular application. In my research, I have proven that these materials are efficient in the adsorption of refractory **sulfur compounds** from commercial fuels, such as alkyl- substituted benzothiophenes. Currently, I am concerned with the issues of **selectivity** and **scale-up**.

Besides the main topic of my PhD thesis, I am also working on a related side-project, in which I am developing activated carbon from **asphaltenes**. This involves the extraction of asphaltenes from bitumen, their stabilization, carbonization and activation using different methods, and their application for the desulfurization of fuels, as well.

Keywords: carbon nanofibers, lignin, PET, electrospinning, adsorption, desulfurization, asphaltenes

Future research plans:

I am very interested in the design of advanced materials and nanomaterials for the development of next-generation industrial processes, such as advanced separations and catalysis. Of particular importance for me is the connection between structure and properties, which determines the efficiency of a certain process. Moreover, I find the concepts of sustainability and circular economy very appealing, and in this context I would be very interested in working on the utilization of waste and/or renewable resources, on the advancement of cleaner energy and on the production of chemicals and polymers from biomass.

Teaching interests:

During my PhD studies in the U.A.E.U., I worked as a Teaching Assistant for four semesters in various undergraduate courses (including material and energy balances, thermodynamics, heat and mass transfer, reactor design and polymer engineering among others). As a Teaching Assistant I was responsible for the grading of tests and assignments and for the teaching of tutorials and exercises.

As a faculty member, I would like to teach undergraduate core subjects of the Chemical Engineering major, including reaction engineering, physical chemistry, heat and mass transfer and polymer science and engineering. Furthermore, at a graduate level I would be interested in organizing and teaching a course of Design and Analysis of Experiments and also a course related to Nanomanufacturing and Nanoengineering.

Publications in peer reviewed journals *(Orcid ID: 0000-0001-6303-8944)*:

1. **Svinterikos, E.,** Zuburtikudis, I., Al-Marzouqi, M., "Fabricating carbon nanofibers from a lignin/r-PET blend: the synergy of mass ratio with the average fiber diameter", *Appl. Nanosci.,* Under Review, October 2019
2. **Svinterikos, E.,** Zuburtikudis, I., Al-Marzouqi, M., The nanoscale dimension determines the carbonization outcome of electrospun lignin/recycled-PET fibers, *Chem. Eng. Sci.* 202 (2019) 26–35
3. **Svinterikos, E.,** Zuburtikudis, I., Al-Marzouqi, M., Carbon nanomaterials for the adsorptive desulfurization of fuels, *J. Nanotechnol.* (2019), Article ID 2809867
4. Al Haddad, Z.A., **Svinterikos, E.,** Zuburtikudis, I., Designing electrospun nanocomposite poly(vinylidene fluoride) mats with tunable wettability, *Colloids Surf. A* 523 (2017) 81–90
5. **Svinterikos, E.,** Zuburtikudis, I., Tailor-Made electrospun nanofibers of biowaste lignin/recycled poly(ethylene terephthalate), *J. Polym. and the Environ.* 25 (2017) 465–478
6. **Svinterikos, E.,** Zuburtikudis, I., Carbon nanofibers from renewable bioresources (lignin) and a recycled commodity polymer [poly(ethylene terephthalate)], J. *Appl. Polym. Sci.* 133 (2016) 43936
7. Tsimpliaraki A., **Svinterikos S.,** Panayiotou S, Marras S., Zuburtikudis I., Papadopoulou L., PBSA/layered silicate nanocomposites and the role of nanoclay on their electrospun fibrous structure, *Int. J. Nanomanufacturing* 6 (2010), 291-299
8. Tsimpliaraki A., **Svinterikos S.,** Zuburtikudis I., Marras S., Panayiotou C., Nanofibrous Structure of Nonwoven Mats of Electrospun Biodegradable Polymer Nanocomposites—A Design of Experiments (DoE) Study, *Ind. Eng. Chem. Res.* 48 (2009), 4365–4374

Presentations at international conferences:

1. **Svinterikos E.,** Al Marzouqi M., Zuburtikudis I., The nanoscale effect in the carbonization of electrospun lignin/PET fibers, *ANNIC 2018*, 22-24/10, Berlin, Germany
2. **Svinterikos E.,** Al Marzouqi M., Soliman A.M., Zuburtikudis I., Tunable fabrication of carbon nanofibers from renewable and waste resources, *2017 AICHE Annual Meeting*, 29/10 –03/11, Minneapolis, USA
3. **Svinterikos E.,** Zuburtikudis I., Tunable fabrication of carbon nanofibers based on the electrospinning of biowaste lignin and recycled PET, *Carbon 2016*, 10-15/07/2016, Penn State, USA
4. **Svinterikos E.,** Zuburtikudis I., A review on self-healing coatings based on nanocontainers for active corrosion protection, *NACE UAE Corrosion 2015*, 12-14 May, Abu Dhabi

Successful research proposals:

"Utilization of waste and inexpensive resources for the development of tailor-made adsorbents" funded by the Emirates Center for Energy and Environment Research with budget of ~$100K, 2018 – 2020

581954 Active Soft Materials: Data Driven Study to Understand, Control, and Design Bio-Soft-Materials

Research Interests:

Most biological structures are inherently soft and noisy. Soft matter physics holds the key to unlocking several important problems that impact human health, including identification of cancerous cells, and the development of antibiotic-free means to address biofilm infection. Furthermore, studies of such systems inspire engineered advanced materials; these structures pervade personal care, pharmaceuticals, and food products. Soft matter surrounds us in many forms- gels, emulsions, polymers, granular and glassy materials. These systems feature heterogeneous and complex microscopic structures that drive their unique mechanical, rheological, and optical properties. The challenge in identifying, predicting and controlling their properties lies in the weak interactions between their building units and their extreme sensitivity to external stimuli. In this data-driven era, the need for reliable data on complex, highly correlated systems with multiple responsive (living!) structures in paramount. In my research program, I will "*study physical aspects of these structures and their complex interactions and dynamics by utilizing state of art experimental tools that provide statistically reliable data*".

My research program will focus on the bio-soft materials particularly cells and cellular communities and their interactions with surroundings. From the perspective of human health, my research will help to find solutions to challenging issues such as high throughput identification of cancer cells and means to mitigate bacterial infection by predicting the response of cells to external stimuli. From the perspective of material science, the outcome of my research will help to design bioinspired materials with particular properties by predicting required microscopic structures.

Background and Future Plans

My background as an experimentalist at the interface of mechanical-chemical engineering and biophysics enables me to approach challenging problems in active soft matter including living cells within a statistical physics framework. I am adept at methods to provide much needed perspective on highly correlated collective dynamics in systems far from equilibrium, including living cells. I specialize in extracting information in highly challenging settings to understand relationships that govern structure formation, dynamics and properties on the micro- and nano-scale and across large domains.

For example, I am currently working to develop high throughput assays to identify cancer cells based on a nano-rheology technique to measure changes in the rheology and stiffness of the cell membrane. This project required that we develop means to follow, in real time, the motion of gold nanorods adhered to different segments of the cells. I have developed a nanoscale rheology technique based on angle resolved dark field microscopy that tracks translation and 3d reorientation of nanoscale probes with high spatial and temporal resolution [1]. This newly developed technique is adaptable to measure mechanical properties of other complex soft materials. We are already applying this method to study the complex rheology of polymer systems, wormlike micelles, and vesicles. This field is wide open for innovation and will form one "basis vector" of my independent program.

I have recently developed a collaboration in Penn Dental Medicine to study biofilm formation in the context of dental caries, a disease that strikes impoverished populations with devastating effects world-wide. To control and remove biofilms we need to better understand their mechanical structure and response to challenges. This project requires that we measure the change in the highly heterogeneous, spatially and temporally complex rheology of biofilms on the microscale when they are exposed to treatment. I approach this challenge by using advanced particle tracking and microrheology to deduce mechanical properties of the biofilm. The method we will develop here are going to applicable to other biofilm related disease including devise related infection such as implants or catheter or non-devise related such as cystic fibrosis.

I am interested as well in the fundamental aspects of active matter- I am working to develop the concepts of active interfaces as a means to remotely enhance transport in microscale and in confined settings. The idea is to arrange and control synthetic and biological self-propelled colloids in a two-dimensional setting to induce enhanced fluctuation and mixing in a three-dimensional fluid. Such active interfaces can potentially introduce a new class of materials such as active Pickering emulsions and revolutionize industrial techniques such as emulsion stabilization and oil recovery. I also study the phase transition, jamming, and glassy dynamics of complex interfaces, such asprotein-, surfactant-, and polymer-laden interfaces. These studies have been performed on a new apparatus that I developed to simultaneously measure surface tension and interfacial rheology.

Active soft materials are often formed by dense populations of active species. To understand the dynamics of individual species and properties of whole systems we need to investigate them in a dense setting but with single unit resolution. For example, in my Ph.D., I extended digital holographic microscopy to track thousands of bacteria in 3D with high spatial (sub-micron) and temporal resolutions [2] to investigate the effect of hydrodynamic interaction on bacterial attachment to solid surfaces [3-4].

In my independent research program, I will integrate my expertise in advanced imaging, microrheology, microfluidics, and microbiology to tackle emergent issues in human health, to design and control novel soft materials, and to improve our fundamental understanding of physics of soft materials.

Teaching Interests:

As an engineering scholar, I am eager to adopt modern teaching methods and hands-on experiences to excite the next generation of STEM leaders. I am especially interested in teaching courses in fluid mechanics, transport phenomena, rheology, and thermodynamics. With a background in chemical engineering, mechanical engineering, and biophysics, I would like to develop interdisciplinary courses such as statistical physics of soft materials and advanced optical techniques in biophysics and soft matter. I would like my teaching to reflect my passion for engineering and biology as well as research-based learning. I am also very interested in mentoring senior design projects that introduce undergraduates to department faculty and their research and to transfer graduate and higher education experiences to undergraduate students. My research program will provide a great platform for training students to integrate multidisciplinary approaches and perspectives to address questions at the interface of biology and engineering.

References:

[1] M. Molaei, E. Atefi, and J. C. Crocker. Nanoscale rheology and anisotropic diffusion using single gold nanorod probes. PRL, 120(11):118002, 2018.

[2] M. Molaei and J. Sheng. Imaging bacterial 3d motion using digital in-line holographic microscopy and correlation based de-noising algorithm. Opt Express , 22(26):32119{32137, 2014.

[3] M. Molaei, M. Barry, R. Stocker, and J. Sheng. Failed escape: Solid surfaces prevent tumbling of Escherichia coli . PRL , 113(6):068103, 2014.

[4] M. Molaei and J. Sheng. Succeed escape: Flow shear promotes tumbling of Escherichia coli near a solid surface. Sci Rep , 6:35290, 2016.

581992 Dynamics, Transport, and Self-Assembly in Flowing Polymeric Liquids

Research Interests:

My research interests lie in the area of dynamics of polymeric liquids and microscale flows. As the canonical example of non-newtonian fluids, polymeric systems have long served as a rich source of problems important both from a fundamental standpoint as well as engineering applications. While great progress has been made in our understanding of linear homopolymers, as polymer synthesis techniques rapidly progress, chemists are able to synthesize macromolecules with highly complex architectures, far beyond simple long chain branching, with unprecendented control over chemical details at the molecular level. Indeed, the current trend in polymer science towards developing functional materials relies more on novel molecular topology involving commodity polymers rather than synthesizing molecules with exotic monomers. Hierarchical self-assembly of such molecules, e.g., using block-copolymers is often leveraged for creating responsive materials whose properties can be manipulated on-the-fly with external stimuli. Presence of charges on polymer subchains leads to further complexity. Processing of such polymers presents unique challenges that require extensive research on their flow behavior and rheology, as our knowledge of linear chain behavior often do not translate to molecules with non-trivial topologies. I use theory along with computer simulations to gain insight into such problems, with an emphasis on out-of-equilibrium dynamics in flowing systems.

In this poster, first, I will present my work on elucidating the structure and dynamics of bottlebrush polymers. Bottlebrushes are highly branched macromolecules consisting of a large number of side chains densely grafted onto a central backbone chain, capable of flow-driven self-assembly to form hierarchical soft materials. I am investigating these systems in collaboration with experimental groups for potential application as photonic crystals with on-the-fly tunable properties via 3-D printing. Next, I will discuss my recent investigations on charge transport in flow batteries with redox-active polymers (redoxmers). This work is also in collaboration with other experimental groups, as part of the Joint Center for Energy Storage Research project, geared towards gaining a mechanistic understanding of redoxmer dynamics with the aim of improving the efficiency of flow batteries. Finally, I will discuss my proposed research projects for the near future that builds on my current expertise on polymeric liquids and fluid dynamics.

Teaching Interests:

I am willing to teach any of the core undergraduate level courses in chemical engineering. At the graduate level, I would prefer to teach fluid mechanics/transport and applied mathematics courses as these topics are closer to my areas of expertise and interest. I would also like to develop a new course based on a set of closely related topics centered around low Reynolds number hydrodynamics, Brownian motion, and possibly microscale transport. This course will be aimed at graduate students, and will specifically be useful for students working in micro or nanofluidics.

581736 Constitutive Modelling for Complex Fluids in Complex Flows

Research Interests:

From dry powders to entangled polymer melts, the processing of complex fluids at industrial scales is often limited by flow instabilities that are poorly understood on a fundamental level. In my research, I seek to address this problem by developing and applying improved constitutive models for complex fluids. For such systems, the coupling between flow and microstructure is generally complex, non-linear, and non-equilibrium, with underlying physics spanning many length-scales and time-scales. In my approach to the subject, I combine new ideas and insights with existing theories to produce a microscopically faithful continuum description of a fluid's response to deformation and flow. Thereafter, I use insights from physical intuition and numerical simulations to guide well-controlled simplification strategies (asymptotic expansions, pre-averaging approximations, galerkin expansions, moment expansions, etc.) until I reach a suitable balance between fidelity to the full model and computational tractability in complex flows. These simplified models are then applied to study flow instabilities observed in benchmark problems. Initial directions for future work will include polydisperse entangled polymers, dense granular suspensions, and dry fibers.

Selected Research Experience:

Constitutive modelling of wormlike micelles: Department of Applied Maths and Theoretical Physics, University of Cambridge (advised by Mike Cates)

My postdoctoral research focuses on developing and applying simplified constitutive models for the linear and non-linear rheology of wormlike micelles. Wormlike micelles are an important component of many consumer goods (hand-soap, shampoos, detergents, etc.) and industrial processes (e.g. advanced oil recovery). However, the rheology of wormlike micelles is notoriously sensitive to minute changes in formulation, and this creates problems for industrial-scale production. Improved constitutive models could reduce waste and down-time by helping to diagnose and correct problems with the formulation in real-time. By combining existing tube-based constitutive models (e.g. the Rolie Poly model) with a complete population-balance model for scission and reformation kinetics, we have produced a model that directly connects molecular-scale information to linear and non-linear rheological measurements. On some levels, this approach is conceptually similar to what has been done by others – however, by including previously-neglected physics we have upended several well-established results in the existing literature. For example, by incorporating the entire polydisperse molecular weight distribution in our population balances, we show that flow-induced breaking is not generally responsible for shear banding instabilities in wormlike micelles and that a Doi-Edwards type instability is more likely to be at play.

Constitutive modelling of polydisperse blends: Department of Chemical Engineering, University of Santa Barbara (advised by Gary Leal and Glenn Fredrickson)

In one component of my PhD studies, I developed a simplified constitutive model for polydisperse entangled linear polymer melts. Whereas monodisperse polymer melts have been a historical focal point of constitutive modelling efforts, industrially relevant polymers are highly polydisperse in their composition. It is well known that high polydispersity improves processability, but the precise mechanism by which this is achieved is not at all well-understood. By generalizing the 'double reptation' ansatz to non-linear modes of stress relaxation (e.g. chain retraction and convective constraint release), we have produced a simple constitutive model of polymer blends that retains a seemingly reasonable approximation for the effects of microscopic couplings between chains of differing lengths. When applying this model to complex flows, we find (as is seen in experiments) that adding a small fraction of short chains dramatically extends the window of stable processing conditions. This work has created new opportunities to understand how changes in melt formulation influence the processability of polymer melts.

Publications:

- **D. Peterson**, M. E. Cates, L. G. Leal, "Constitutive modelling of wormlike micelles", in prep.
- **D. Peterson**, C. Sasmal, V. Boudara, D. J. Read, L. G. Leal, "Nonlinear rheology predictions of bidisperse polymer blends in a complex flow", in prep.
- J. Gillessen, C. Ness, **J. D. Peterson**, H. J. Wilson, M. E. Cates, "A tensorial constitutive model for dense frictional suspensions", in prep.
- **D. Peterson**, G. H. Fredrickson, L. Gary Leal, "Does shear induced demixing resemble a thermodynamically driven instability?" *Journal of Rheology*, 63(2) pp. 335 – 359
- Boudara, **J. D. Peterson**, L. Gary Leal, D. J. Read, "Nonlinear rheology of polydisperse blends of entangled linear polymers: Rolie-Double-Poly models", *Journal of Rheology*, 63(1) pp. 71 – 91
- **D. Peterson**, M. E. Cromer, G. H. Fredrickson, L. G. Leal, "Shear banding predictions for the two-fluid Rolie-Poly model", Journal of Rheology 60(5) pp. 927 - 951

Recorded Talks (available online):

"Modelling polymer blends in flow," Joseph D. Peterson, Glenn H. Fredrickson, L. Gary Leal. Thesis defense (2018). https:/bit.ly/2IRJPmf

"Two-fluid models for polymer melts and solutions," Joseph D. Peterson, Glenn H. Fredrickson, L. Gary Leal. KITP dense suspensions workshop (2018). (talk begins at 1:00:58) https://bit.ly/2xe7OFl

Teaching Interests:

During my PhD Studies, I was awarded a departmental fellowship to be a co-instructor for an undergraduate chemical engineering course. Because my research is heavily math-oriented, I chose to co-teach the numerical methods class for chemical engineers. This was a great learning experience for me, and I look forward to future opportunities to engage with students in the classroom setting. Although I enjoyed teaching a subject relevant to my own research, I would also welcome the challenge of teaching other subjects as well: my background and training as a chemical engineer makes me qualified to teach any core subject in the discipline (thermo, transport, kinetics, process control, etc).

581407 Structure, Rheology and Processing of Complex Fluids Towards Scalable Manufacturing of Functional Devices

Research Interests:

There is growing interest in solution-processed manufacturing of functional devices through scalable (i.e. roll-to-roll) coating/printing technologies to new sectors such as energy, electronics and medicine, due to many potential benefits, including huge cost savings. Critical to the functionality of the manufactured device in many of these applications is achieving a desired pattern (1D, 2D or 3D) or a certain structure. Precise control of the microstructure of formulations and their processing behavior (in coating/printing and during processing) are critical to efficient fabrication of targeted pattern/structures. Material formulations are often complex in structure as they often contain several functional additives (i.e. particles, polymers and surfactants) and can exhibit a complex flow behavior. Current challenges in solution-processed manufacturing of many applications are the lack of clear understanding of the structure of functional formulations and the processing behavior of rheologically complex materials in various roll-to-roll coating/printing processes.

My PhD research focused on addressing fundamental problems in structure-rheology-processing relationships for roll-to-roll processing of complex fluids, with a focus on shear-thickening colloidal dispersions. The flow behavior of the shear-thickening particle dispersions in large amplitude oscillatory shear and extensional flows, and their dynamics in gravure printing and slot-die coating were studied. The studies provided new insights on the flow behavior and structure of shear-thickening dispersions, and that the fluid rheology can significantly influence the ink transfer and the coating stability in gravure printing and slot-die coating process, respectively.

My postdoctoral research at National Renewable Energy Laboratory (NREL) has been focused on investigating the structure and rheology of catalyst inks for solution-processed manufacturing of catalyst layers (CL) for proton-exchange-membrane fuel cells and electrolyzers. The catalyst ink formulations are a colloidal mixture of catalyst particles and ionomer in water-alcohol mixtures. The structure of the CL is found to strongly influence the electrochemical performance, however, how the ink microstructure and processing affects the evolution of CL structure is currently less clear. My research focused on understanding how the formulation components (ionomer-particle-solvent) interact and the impact the catalyst ink microstructure primarily using rheological tools, that has been little utilized, and USAXS in collaboration with Argonne National Laboratory (ANL). The studies revealed insights on the influence of ionomer and solvent (alcohol-ratio) on the agglomerated structure of different catalysts (based on structure and surface chemistry), and the implications on the processing and the evolution of CL microstructure.

My future research is motivated towards addressing the fundamental gaps in soft matter and non-Newtonian fluids dynamics that enable solution-processed manufacturing of functional devices. In the short-term, I hope to focus on fundamental problems that are critical for understanding the microstructure and processing behavior (in coating and drying processes) of catalyst ink formulations in the fabrication of catalyst layers for fuels cells, electrolyzers and other electrochemical technologies. The research problems span topics of the structure and rheology of colloid-polymer mixtures, and the dynamics of non-Newtonian fluids in coating and drying processes.

Teaching Interests:

Fluid Mechanics, Transport Phenomena, Rheology of Complex Fluids

Publications:

1. S Khandavalli, JP Rothstein Extensional rheology of shear-thickening fumed silica nanoparticles dispersed in an aqueous polyethylene oxide solution *J. Rheol.*, 2014, 58 (2), 411-431
2. S Khandavalli, JA Lee, M Pasquali, JP Rothstein The effect of shear-thickening on liquid transfer from an idealized gravure cell *J. Non-Newtonian Fluid Mech.*, 2015, 221, 55-65

3. S Khandavalli, JP Rothstein Large amplitude oscillatory shear rheology of three different shear-thickening particle dispersions *Rheol. Acta*, 2015, 54 (7), 601-618
4. S Khandavalli, JP Rothstein The effect of shear-thickening on the stability of slot-die coating *AIChE J.*, 2016 62 (12), 4536-4547
5. S Khandavalli, J Hendricks, C Clasen, JP Rothstein A comparison of linear and branched wormlike micelles using large amplitude oscillatory shear and orthogonal superposition rheology *J. Rheol.*, 2016, 60 (6), 1331-1346
6. S Khandavalli, JP Rothstein Ink transfer of non-Newtonian fluids from an idealized gravure cell: The effect of shear and extensional deformation *J. Non-Newtonian Fluid Mech.*, 2017, 243, 16-26
7. S Khandavalli, P Rogers, JP Rothstein Roll-to-roll fabrication of hierarchical superhydrophobic surfaces *Appl. Phys. Lett.*, 2018, 113 (4), 041601
8. S Khandavalli, JH Park, NN Kariuki, DJ Myers, JJ Stickel, K Hurst, KC Neyerlin, M Ulsh, SA Mauger Rheological Investigation on the Microstructure of Fuel Cell Catalyst Inks *ACS Appl. Mater. Interfaces, 2018,* 10 (50), 43610-43622
9. S Khandavalli, JH Park, NN Kariuki, S Zaccarine, S Pylypenko, DJ Myers, M Ulsh, SA Mauger Structure and Rheology of Catalyst Inks for Low-Temperature Proton Exchange Membrane Electrolyzers (*in preparation*)

565351 Complex and Biological Fluids

Research Interests: My research interest and expertise is in the emerging field of complex and biological fluids. The problems I am interested in also fall under the purview of soft and active matter. Many phenomena in nature involve a variety of forces that conspire together to show complex dynamics and even give rise to self-organisation, flocking and swarming. A large subset of these phenomena,for example, intracellular processes in living organisms, take place in a fluid medium where hydro-dynamic forces balance elastic, electric or magnetic forces. The main focus of my research has been to unravel and interpret these complex mechanisms using a variety of mathematical modelling and numerical techniques. Some questions that have fuelled my research work are: how do seemingly simple systems, living or synthetic, show self-organization or coherent motion at scales much larger than their individual size; how can we obtain macroscopic models from microscopic interactions at the individual level to explain such collective dynamics, how do microorganisms of increasing complexity perform locomotion in ingenious ways and how can we design synthetic microbots that mimic them. More recently, I have become interested in key questions in developmental cell biology that involve mechanical processes. Some of these questions are: what are the active forces arising from the motion of actin filaments or microtubules inside a cell that enables it to perform various important tasks, like migration,cell-division; how fluid-structure interaction due to ciliary motion inside oocytes enables organisms to have a left-right symmetry; how does tissue elongation occur due to cellular interactions. The common theme that triggers my interest in all these problems is the crucial role played by mechanics.

Complex Fluids: Complex fluids are found in abundance in both nature and industry. They comprise a class of fluids that are typically non-Newtonian in their behaviour. Examples include suspensions of colloidal particles, liquid drops or polymers suspended in a fluid. The application of thermal, magnetic or electric fields can alter the macrosopic properties of these fluids and render them non-Newtonian. My research work has focused on a part of this wide spectrum of complex fluids: multiphase flows and electro-rheological (ER) fluids. I will focus on a specific phenomena called as Quincke rotation, an instability under which dielectric particles or drops can spontaneously start rotating under the application of an electric field. Apart from having significance in colloidal hydrodynamics, this phenomena has seen a resurgence as it has proved to be an ideal system for making synthetic active matter.

Biological fluids: Locomotion at the microscale level has fascinated scientists since the last few decades. While larger organisms like fish and humans can take advantage of inertia to locomote, smaller organisms like
prokaryotes live in a world of low Reynolds numbers and cannot use inertia due to the well established Scallop theorem. In this section, I will focus on the hydrodynamics of bacteria and its interaction with the environment. Bacteria are found abundantly in nature and form an important part of the global ecosystem. It is crucial to understand their dynamics: how individual cells perform locomotion, how their colonies swarm and multiply eventually leading to formation of biofilm that has profound impact on food industry and understanding infectious diseases.

Teaching Interests: The core underlying philosophy that fuels my passion to teach and supervise students is to help them understand complex ideas that I have already understood and gained expertise in. My passion for teaching stems from the belief that it is a noble profession as it one of the best ways to contribute to a society's development and also have a sustainable positive influence on students that lasts a lifetime. Teaching also helps me in questioning and re-evaluating my own understanding of the subject. I have gained considerable teaching experience during my graduate and post-doctoral career spanning more than 8 years. I have designed and taught a course on fluid mechanics in the University of Illinois, Urbana-Champaign entirely on my own in the early stages of my graduate studies and served as a teaching assistant several times during the course of 3 years. During my postdoctoral career, I have supervised undergraduate students in Department of Applied Mathematics and Theoretical Physics, Cambridge and mentored an undergraduate student from Trinity College on a research project that I independently designed. Furthermore, I have also mentored PhD students in the biophysics/mechanics group on certain projects that have resulted in joint publications. I have also been involved in science festivals for public outreach held at the mathematics department in University of Cambridge.

583219 Microscale Transport Phenomena: Data Analytics and Fluid Dynamics

Research Interests

The Wang laboratory will develop analytic technologies in the domain of microscale transport phenomena that are relevant to environment, health and energy. Specifically, we are interested in (1) particle technologies in microfluidics relevant to biomedical research, (2) clean/conventional energy research using data science, and (3) polymer modeling for material design.

We will achieve these goals through a combination of state-of-art machine learning method and tractable mathematical modeling. Given the potential of expanding the level, number, and depth of University-Private Sector research partnership in the United States, our long term vision is to establish a research program that emphasizes Institution-Industry collaboration, such as environmental agencies, pharmaceuticals and Oil & Gas energy sectors.

Research Experience

Postdoctoral: data analytics in enhanced oil recovery & novel approach in fluid/particle science and technology (Advisor: Sangtae Kim, Chemical Engineering, Purdue, 08/2016-current)

The United States is currently the world's top oil producer, and the national policy "supports research and policy options to increase our domestic supply of oil while ensuring environmentally sustainable supplies domestically and abroad" (Department of Energy). An estimated two thirds of conventional crude oil in mature fields remains unproduced due to the physics of fluid flow. As such, new techniques are needed for chemically-enhanced oil recovery (EOR), in order to overcome the physical force holding hydrocarbons and turn the accumulations into oil reserves.

I have performed critical work on oil recovery, in collaboration with Pioneer Oil Company. One oil recovery method is to apply surfactants into injected water, but there is a lack of detailed information about the compounds and structure of the industrial surfactants. It is critically important to understand the phase behavior for the oil/water/surfactant systems. By applying machine learning, my goal is to form a microemulsion phase achieving the lowest interfacial tension, which increases the capillary number and dramatically recovers the remaining oil fraction within the pore. I treated the system as a multinomial classification problem, optimizing the system via support vector machine.

Another challenging and interesting research area which I am interested in is ellipsoidal microhydrodynamics. It has significant implications for research and industry as ellipsoidal nanoparticles are frequently encountered in various applications including those involving pharmaceuticals, foods and cosmetics.

Graduate: transport and mixing induced by swimming organisms and settling/rising particles (Advisor: Arezoo Ardekani, Mechanical Engineering, Notre Dame, 08/2011-05/2016)

The motion of microorganisms in the ocean is a crucial part of the effective management of ocean, coastal, and great lakes waters, "including increased understanding of potential responses to global environmental variability and climate change." as stated by Science and Technology Center for Microbial Oceanography Research and Education (C-MORE, Division of Biological Infrastructure, NSF). The influence of density stratification on the motion of a body is critical to understanding the stainability of the ocean environment, but is a complex problem which is only starting to be explored.

My past work has focused on small planktonic organisms that display unsteady or impulsive motion. They use contractile elements, cilia and flagella, and appendages to propel themselves, and their motion may affect ocean stratification. During my Ph.D. research, I have contributed to four projects funded by the National Science Foundation (NSF) that provided major advances in this area. I performed a significant study on sedimentation and swimming in ocean environments. In addition, I have studied the unbounded settling dynamics of a circular disk, a prototypical model of marine organisms, biomass, and pollutant particles. I examined three different environments and identified the transport phenomena of particles, bubbles, and marine organisms.

Teaching Interests

My teaching interests include the fundamental of momentum and mass transport and an elective focus on Microhydrodynamics. Thanks to my undergraduate and graduate training in fluid mechanics program, I have experience with core transport phenomenon, and heat and mass transport that are critical for chemical engineering students. I will leverage my graduate and postdoctoral training and provide students with an ability to intuitively understand how particle technologies can be applied to academic research and industrial settings. In addition, I will be actively seeking co-op opportunities for students by bridging the relationships between university and industries.

Teaching Experience

Mentoring: 2 undergraduates, 1 graduate student (Purdue, 2016-present)

Guest lecture, CHE37700 (Purdue, 2016 fall)

Tutor for 6th-9th graders participating in WIEP (For Your Imagination Program) event (Purdue, 2016)

Guest lecturer of Bowman Creek Project (Riley High School, South Bend, IN, 2013)

Selected publications (out of 13 total)

First-author contributions are underlined

Wang, S., Martin, C.P. and Kim, S., 2019. Improper integrals as a puzzle for creeping flow around an ellipsoid. Physics of Fluids, 31(2), p.021101. (Invited papers on transport phenomena in celebration of Prof. Robert Byron Bird's 95th birthday)

Wang, S., Ellett, K.M. and Ardekani, A.M., 2017. Assessing the utility of high-level CO2 storage and utilization resource estimates for CCS system modelling. Energy Procedia, 114, pp.4658-4665.

Wang, S. and Ardekani, A.M., 2015. Biogenic mixing induced by intermediate Reynolds number swimming in stratified fluids. Scientific reports, 5, p.17448. (Highlighted on Physics.org, Futurity, Geology Page, Purdue College of Engineering News Page, Purdue Research Computing Cluster)

Wang, S. and Ardekani, A.M., 2012. Unsteady swimming of small organisms. Journal of Fluid Mechanics, 702, pp.286-297.

Author

Shiyan Wang

Purdue University

581372 Numerical Simulation based Design of Point of Care Diagnostic Devices

Research Interests: Paper Microfluidics, Droplet dynamics, Computational fluid mechanics

Teaching Interests: Fluid Mechanics, Heat Transfer, Numerical methods for chemical engineers

DNA amplification refers to the production of multiple copies of a DNA sequence from a single copy. Polymerase chain reaction or PCR has been the regular choice for scientists in order to carry out DNA amplification reaction. However, the recent invention of the loop-mediated isothermal amplification (LAMP) method, a decade ago has given new impetus towards the development of point of care diagnostic tests based on amplification of DNA. In the recent past, LAMP has been carried out using various microfluidic technologies. The manufacturing of these devices is rigorous and expensive. Paper membranes have tremendous potentiality in replacing the existing technology as the manufacturing cost of a paper-based device is comparatively lesser and a paper based DNA amplification device would be relatively easy-to-use. However, this emerging technology is yet to be explored in details. The current work presents an insight into the CFD simulations for LAMP amplification reaction on a porous membrane. In addition, a mathematical model for LAMP amplification reaction was derived for the first time. The logistic equation and Richard's curve model were found to be the best fits. However, due to the significantly higher correlation coefficient for logistic equation, it was used in the simulation studies. The porous membrane used in the current work is a Nitrocellulose membrane (pore size values varied from 14.5 μm to 0.45 μm). Nitrocellulose membrane with a pore size of 0.45 μm is suitable for most protein blotting applications. Our major aim is to enhance the number of DNA copies via LAMP on the paper membrane. The mathematical model considered is a convection-diffusion-reaction model and the problem was solved using Finite Element method via COMSOL Multiphysics 5.0. The element size (e) varied between 2.4×10^{-4} (min) - 0.12 (max) and the maximum element growth rate was taken as 1.1. Triangular meshes were chosen for the entire domain except the boundaries (both the horizontal walls and the outlet) where boundary layers with a thickness adjustment factor of 0.5, were implemented. Results are demonstrated in terms of DNA amplicon production for various pore sizes (ε), aspect ratio of the geometry (A) and inlet concentration of the amplicons. Based on these simulations, a near optimal design was achieved based on the maximum production of DNA copies.

582129 Dynamics of Soft Material Systems Under Electric Fields

Research Group Vision and Impact:

I want to have an impact as a researcher in developing tools for controlling and processing soft material systems using external fields, with a thrust to electric fields. Soft material systems, for example, colloidal dispersions, emulsions and surfactant solutions are important in many chemical engineering processes like formulations design, enhanced oil recovery, and pharmaceutical product development. These systems have properties in between solids and liquids, and their structure can be controlled by forces of the magnitude of thermal fluctuations. The processing of these materials is not simple because flow and processing forces affect their structure and properties.

The response of soft materials scales as the square of the electric fields, unlike other traditional fields which produce a linear response. This means that the same effect can be achieved using a significantly smaller value of the electric field. However, despite tremendous potential, we lack a fundamental understanding of the effect of electric fields on soft materials. This limits the commercial applications of electric field driven technology for processing soft material systems. I want to use my background in electrohydrodynamics and field driven surfactant transport to develop electric fields as a robust tool to manipulate soft materials.

I will work in two different directions under this research theme: (1) uncover fundamental physics of these systems under deterministic electric fields; (2) develop stochastic models to predict system response under fluctuating fields. The first research direction is aimed to thoroughly understand the system under controlled environments. The second direction will provide insight into the system behavior under more realistic processing conditions, where disturbances in operating variables create stochastic fields. I will employ a combination of theoretical, computational and experimental techniques in my research group.

Research Experience:

I have expertise in both computational and experimental methods in the areas of fluid mechanics and interfacial science. During my Master's dissertation under Prof. Mahesh Tirumkudulu, I developed an experimental platform to measure the speed of crack propagation in thin films of drying colloidal dispersions. In my PhD under Prof. Lynn Walker and Prof. Aditya Khair, I worked on boundary integral computations and asymptotic theory to predict the deformation and breakup of drops under external electric fields. I implemented the computations for constant as well as stochastic external fields. I complemented the computations with experiments on the breakup of oil drops, and drops loaded with colloidal suspensions under electric fields. I also designed an experimental platform to quantify surfactant transport to oil-water interfaces under an electric field. I have a strong background in theoretical, computational and experimental tools, and I plan to combine these approaches in solving future research problems.

Research Interests:

My interest lies in determining the response of soft material systems to external fields (both deterministic and stochastic), in particular electric fields, and use this expertise to control their processing, and inform the design of tailored soft material systems. I hope to extend my work on surfactant transport under electric fields to quantify interfacial rheology as a function of electric fields. This will provide insight into the interfacial properties that are of significance in electrocoalescence of drops, and guide work on electric field assisted emulsification. I also hope to build collaborations with material scientists and chemists to design tailored materials that respond specifically to electric fields.

Another research area I want to delve into in the future is electrokinetic methods for water desalination. This technology uses electric current to remove ionic impurities in water by their adsorption to electrodes or ion-exchange membranes. Despite advances in electrode materials, and cell and membrane design, the lack of coherent models for charge/discharge mechanisms of electrodes, ion adsorption at electrodes and membrane transport equations limits the desalination efficiency of this method. I hope to use my background in electric field driven flows and surfactant adsorption to contribute to this area by elucidating mechanisms of transport and adsorption of ionic impurities in these systems.

PhD Dissertation:

Coupling electric fields and surfactants to quantify the mechanics of fluid-fluid interfaces, under the supervision of Prof. Lynn M. Walker and Prof. Aditya S. Khair, Department of Chemical Engineering, Carnegie Mellon University, Pittsburgh, PA.

Teaching Interests:

My core teaching interests are in the fundamental subjects of fluid mechanics, transport phenomena, mathematical and numerical methods, both at the undergraduate and graduate level. My goal in education will be to develop courses related to soft materials. More specifically, I want to include concepts of electrohydrodynamics in graduate level soft matter or fluid mechanics courses. I even plan to incorporate introductory concepts on soft materials in undergraduate courses to get students excited in this field.

I have had the opportunity to be a teaching assistant for six courses during my academic career. These include undergraduate courses on Heat and Mass Transfer, and Chemical Reaction Engineering, as well as graduate courses on Interfacial Waves, Advanced Mathematical Methods, and Mathematical Modeling of Chemical Engineering Processes. The primary responsibility was to hold frequent office hours, but I have had the opportunity to give lectures and develop questions for exams in some of the courses.

583903 Investigating Yielding Dynamics and Fracture Mechanics in Soft Materials

Research Interests:

My vision statement for academic oriented research is addressing relevant chemical engineering problems in non-newtonian fluid mechanics to unveil underlying physics by designing unique experiments. My research interests are aligned towards probing microstructure, capturing yielding dynamics and fracture mechanics in soft mater. Rheometry will be used as a pivotal tool to investigate both microscopic and macroscopic properties of unique complex materials like colloidal coacervates, polymeric gels and bio-materials. I intend to exploit existing well known techniques like light scattering to complement rheometry, gaining insights into yielding dynamics and fracture mechanics. Moreover, advanced rheometry techniques like orthogonal superposition rheometry will be employed to investigate crosslinking gels under flow. This will give fundamental insights into how flow affects the crosslinking mechanics during gelation process, which is very pertinent problem chemical process industries.

Teaching Interests:

My teaching philosophy is to develop an efficient active learning atmosphere for the students, with emphasis on critical thinking and problem solving. My undergraduate and master's degree background in Chemical Engineering enables me to seamlessly teach all basic Chemical Engineering subjects. I have special interests in teaching undergraduate courses like 'Chemical Process Principles', 'Transport Processes', and graduate courses 'Polymer Science' and 'Suspension Rheology'. I intend to design new courses targeting unique experimental design to solve complex flow related problems. I was selected for 'Preparing the Professoriate 2017-2018' Program, one-year training program at NCSU, where I got an opportunity to closely follow a mentor and improve my teaching skills in an undergraduate classroom.

587485 Understanding the hierarchy of structure, dynamics, and interactions in soft matter

Research Interests

In soft matter systems, the structure, dynamics, and interactions between foundational building blocks lead to complex material responses. Engineers have exploited the interplay of these properties to design materials with similar compositions but with drastically different properties, such as processable polymer composites that are as strong as metal or swollen polymer gels whose elasticity controls the expression and fate of stem cells. The knowledge and understanding generated by studying synthetic soft materials – emulsions and foams, polymer melts and solutions, liquid crystals, etc. – also translates into understanding the behavior of biological soft materials such as cells and tissues. Better understanding of these complex biological materials will lead to the development of synthetic analogues to greatly benefit biomedical research and the development of therapeutics and diagnostics. Fundamental control of how soft matter components structure, move, and interact in a hierarchical fashion from the nano- to micro- to macroscale is critical to develop materials that respond to environmental stimuli and serve as biological mimics.

My research uses a combination of complementary techniques to characterize the behavior and the consequences of that behavior in soft matter systems from molecular to bulk time and length scales. During my Ph.D., I investigated how the dynamics of tracer particles couple to the structure and relaxations in complex, heterogeneous materials to improve the processing of composite materials and efficacy of targeted drug delivery methods. My work identified that nanoparticles diffuse orders of magnitude faster than expected when they are comparably sized to characteristic length scales in the material. For my postdoctoral research, I characterized the mechanics of dense gels of cellulose nanofibrils using bulk rheology. At high concentrations, the fibrils form physical networks whose yielding behavior bifurcates around a critical stress during aging as segmental fluctuations rebuild a percolating network. This work provides fundamental insight into the yielding of amorphous materials to aid the design of sustainable inks for additive manufacturing and of hydrogels for tissue engineering applications. Throughout my doctoral and postdoctoral experience, I identified the physics that underly material properties using static and dynamic scattering methods to drive drastic changes in the bulk material response characterized with bulk rheology. This bottom-up approach builds a comprehensive understanding of the hierarchical physical interplay that underlies soft materials.

As a faculty member, I will continue to focus on investigating and exploiting the interplay between structure, dynamics, and interparticle interactions on the nano- to macroscopic length and time scales and exploit these mechanisms to design the following:

- **soft matter systems that undergo self-sustained periodic responses under a constant environmental stimulus**
- **facile and scalable methods to separate and purify mixtures of nanoparticles, and**
- **synthetic mimics of biological tissues.**

Each of these projects focuses on critical open questions facing soft matter physics and achieving these goals will directly benefit applications ranging from energy storage, additive manufacturing, soft robotics, and tissue engineering.

To fund my research group, I will actively pursue funding resources from governmental agencies, private, non-profit groups, and industry. During my doctoral and postdoctoral research, I gained experience developing and writing successful proposals to acquire x-ray and neutron beamtime resources at multiple national labs as well as contributing to multiple funding proposals. The aim of my research overlaps with the goals of the Materials Genome Initiative and of many funding agencies, including NSF, ACS PRF, DOE, and DARPA, to which I will submit grant applications. Additionally, I aim to collaborate with potential industrial partners such as Unilever, Proctor & Gamble, and ExxonMobil due to my focus and expertise in soft matter materials and processing conditions.

Teaching Interests

I see beauty in the ability of chemical engineers to distill problems into simple, useful models. For example, biological cells are amazingly complex objects with multiple functions that interact with and respond to their surroundings. As chemical engineers, we can look at a cell and see a chemical reactor in miniature, with multiple inputs of raw materials and energy, internal reactions, control loops, and product outputs. Although the analogy of a cell as a reactor ignores much of the biological complexity, it distills the problem into a tractable and useful model that allows engineers to exploit the biological specificity of a cell to commercially produce valuable products such as insulin and monoclonal antibodies. With strong fundamentals and experience identifying and developing connections, students should be able to quickly master problems such as these even if they do not have specific training on how *E*. coli produces insulin. As an educator, I aim to focus on fundamental theories and critical thinking to inspire students to make connections between disparate fields and develop independent and creative solutions. Students should graduate as chemical engineers with the ability and self-confidence to jump into new disciplines, breakdown problems into basic building blocks, develop a solution, and justify their decisions without needing specific experience on those exact problems.

My training as a chemical engineer has prepared me to teach any of the core chemical engineering courses in the undergraduate curriculum. I am particularly interested in teaching classes relating to transport phenomena, such as heat and mass transfer or fluid dynamics, which will bring my research into the classroom. On a graduate level, I am interested in teaching advanced transport phenomena but also electives focusing on soft matter and polymer physics. My vision for a soft matter graduate elective course would focus on colloidal dynamics and structure, surface forces, and techniques to measure and characterize these phenomena on multiple length scales, with specific focus on scattering and rheological methods. For a polymer course, I would want to build a strong foundation of thermodynamic driving forces and how polymers relax and structure over different time and length scales.

Selected Publications (17 total, 8 first author)

For more information, please visit: http://www.ryanps.info

Poling-Skutvik, R.; Slim, A. H.; Narayanan, S.; Conrad, J. C.; Krishnamoorti, R. Soft interactions modify the diffusive dynamics of polymer-grafted nanoparticles in solutions of free polymer. *ACS Macro Lett.* 2019, *8*, 917-922.

Poling-Skutvik, R.; Roberts, R. C.; Slim, A. H.; Narayanan, S.; Krishnamoorti, R.; Palmer, J. C.; Conrad, J. C. Structure dominates localization of tracers within aging nanoparticle glasses. *J. Phys. Chem. Lett.* 2019, *10*, 1784-1789.

Poling-Skutvik, R.; Lee, J.; Narayanan, S.; Krishnamoorti, R.; Conrad, J. C. Tunable assembly of gold nanorods in polymer solutions to generate controlled nanostructured materials. *ACS Appl. Nano Mater.* **2018**, *1* (2), 877–885.

Poling-Skutvik, R.; Olafson, K. N.; Narayanan, S.; Stingaciu, L.; Faraone, A.; Conrad, J. C.; Krishnamoorti, R. Confined dynamics of grafted polymer chains in solutions of linear polymer. *Macromolecules* **2017**, *50* (18), 7372–7379

Poling-Skutvik, R.; Mongcopa, K. I. S.; Faraone, A.; Narayanan, S.; Conrad, J. C.; Krishnamoorti, R. Structure and dynamics of interacting nanoparticles in semidilute polymer solutions. *Macromolecules* 2016, *49* (17), 6568–6577.

Poling-Skutvik, R.; Krishnamoorti, R.; Conrad, J. C. Size-dependent dynamics of nanoparticles in unentangled polyelectrolyte solutions. *ACS Macro Lett.* **2015**, *4* (10), 1169–1173.

587613 Electrokinetically enhanced particle focusing in Non-linear microfluidic flows

<u>**Research Interests:**</u>

Particle focusing in non-linear flows: Cross-stream migration of particles suspended in fluid flows occurs due to inherent non-linearity: inertia or polymeric stresses. This physics, in the past, has been utilized in microfluidics to focus cells in biological fluids. Recent experimental studies, towards the manipulation of focusing via external fields, have drawn my attention to ask: how electrokinetics affects the particle migration in non-linear flows? Use of classical mathematical techniques (such as perturbation theory and asymptotics in conjunction with Lorentz reciprocal theorem) has helped me gain fundamental insights into the migration of charged particles and reveal the primary source behind the breaking of symmetry in weakly inertial [P5, P3] and viscoelastic flows [P4]. I am currently examining this behavior in high Reynolds number flows.

Other research interests and experience

Swimmers in complex fluids: Synthetic swimmers such as Janus particles and rods can propel themselves by generating concentration gradients along their surface. This results in a 'diffusio-osmotic' slip which consequently propels the particle autonomously. A significant portion of the potential applications lies in drug-delivery and biological fluid systems which exhibit non-Newtonian behavior. This led me to study the changes in the slip and swimming of Janus particles in complex fluids. [P6]

Process intensification in microreactors: Microchannels are known to offer small length scale for rapid heat and mass transport. Despite of this, several reaction processes are limited by slow diffusion, which causes formation of undesired products and safety hazards. I am interested in intensification methods and modeling of such systems through CFD simulations. For instance, a gently curved channel produces vortices which generates cross-section mixing and enhancement in heat and mass transport. [P2]

<u>**Teaching Interests:**</u>

Graduate Courses: Mathematical Methods in Chemical Engineering, Multiphase flows, Dynamical Systems.

During the course of my PhD at IIT Madras, I was a teaching assistant of the aforementioned courses and had numerous opportunities to give lectures. Although I enjoyed teaching a subject related to my research interests, I also learned to give lectures outside my research. Apart from theoretical lectures, I also conducted seminars (as a part of the course work) which introduced first year graduate students to computational techniques using Matlab and Mathematica.

<u>**Publications:**</u>

[P6] **A. Choudhary**, T. Renganathan and S. Pushpavanam, *"Non-Newtonian effects on the slip and mobility of a self-propelling active particle"*, (Under review process in Journal of Fluid Mechanics)

[P5] T. Krishnaveni, **A. Choudhary**, T.Renganathan and S. Pushpavanam, *"Inertial migration of a particle in flows with transverse viscosity variation"* (Under review process in Physics of Fluids)

[P4] **A. Choudhary**, Di Li, T. Renganathan, Xiangchun Xuan, S. Pushpavanam, *"Electrokinetics Enhances Cross-stream Particle Migration in Viscoelastic Flows"* (Under review process in Journal of Fluid Mechanics)

[P3] **A. Choudhary**, T. Renganathan, S. Pushpavanam, *"Inertial migration of an electrophoretic rigid sphere in a 2D Poiseuille flow"*, Journal of Fluid Mechanics vol. 874(2019) 856-890

[P2] **A. Choudhary** and S. Pushpavanam, "*Process intensification by exploiting Dean vortices in catalytic membrane microreactors*" Chemical Engineering Science Volume 174, 31 December 2017, 413-425

[P1] C. Sun, Z. Luo, **A. Choudhary**, P. Pfeifer and R. Dittmeyer, "*Influence of the Condensable Hydrocarbons on an Integrated Fischer−Tropsch Synthesis and Hydrocracking Process: Simulation and Experimental Validation*" Ind. Eng. Chem. Res.2017564513075-13085

587793 Recycle, micromixing and particle migration in a microchannel

Lab on chip devices involve flow through channels of very low characteristic length scales. The performance of these devices depends on the residence time i.e. the average time that a fluid element spends in the system. Conventionally, the residence time is increased by increasing the length of the channel or by decreasing the flowrate. Here, we propose a novel technique to increase the residence time by recycling the fluid in the microchannel using electroosmosis. Toward this end, Navier-Stokes equation for the flow field and the Poisson-Boltzmann equation for potential field are solved simultaneously using the finite difference method (FDM) and the minimum threshold value of electric field for inducing recycle flow is determined.

Recycling is then employed as means to enhance mixing of the unreacted reactants. Due to the characteristic low length scales, mixing in a microchannel is limited by diffusion. The extent of mixing can, however, be improved by inducing secondary flows in the system. In our work, we study the enhancement of mixing in pressure-driven flow using a spatially periodic electric field. The stream function formulation is used to solve the Navier-Stokes equation and the Poisson-Boltzmann equation using FDM. The enhanced mixing is quantified using Poincare maps, Shannon entropy, and a scalar species transport equation.

Particle/cell sorting is one of the microfluidic applications, where inertia in the flow is exploited for separation. The inherent nonlinearity due to the presence of inertia in the flow causes lateral migration of the particles, which then self-assemble at certain equilibrium locations. This phenomenon is known as inertial focusing. Inertial focusing can be manipulated by varying the shear gradient profile of the underlying flow. In this work, we study inertial focusing by varying the viscosity. We consider the particle migration in three cases: continuous transverse viscosity variation, a liquid-liquid stratified flow of two miscible liquids, and a liquid-liquid stratified flow of two immiscible liquids. We perform direct numerical simulations, where level set and immersed boundary methods are used to model the interfacial and the particle dynamics. The optimal operating conditions for particle separation are identified.

Keywords: *Electrokinetics, Shannon entropy, inertial focusing, liquid-liquid stratified flow, immersed boundary method, and level set method*

Publications:

1. **Krishnaveni, T**., Renganathan, T., Pushpavanam, S., 2017. *Recycle Flows in Lab-on-Chip Applications Using Electroosmotic Effects.* Eng. Chem. Res. 56, 4145–4155. https://doi.org/10.1021/acs.iecr.6b04942
2. **Krishnaveni, T.**, Renganathan, T., Picardo, J.R., Pushpavanam, S., 2017. *Numerical study of enhanced mixing in pressure-driven flows in microchannels using a spatially periodic electric field.* Rev. E 003100, 1–15.
3. **Krishnaveni, T**., Renganathan, T., Pushpavanam, S., *Inertial focusing in two dimensional flows with sharp viscosity stratification in a microchannel.* http://arxiv.org/abs/1910.09028

581843 Understanding and Designing Non-Equilibrium States at Soft/Hard Material Interfaces

Research Interests:

My career objective is to interface soft and hard materials with controlled non-equilibrium interfacial states to create material platforms that address societal challenges in energy, electronics, health and water. Interfaces between soft and hard matter ubiquitously exist in natural and man-made systems such as cartilage/bone interface in body, fluid/electrode interface in catalyst, and solid/electrolyte interphase in battery. These interfaces often carry deterministic impacts on material behavior, property and functionality by governing transport and transformation of mass and charge. Such exchanges fundamentally link to the free energy information encoded in the interfacial material states. Soft and hard materials differ profoundly in the strength and density of chemical bonds, giving rise to largely mismatched energy density at interface. *How to control this energy discontinuity stand as a core consideration that dictates the scope of soft/hard hybrid systems.*

The nonstoichiometric coordination environments at the interface create spaces for materials away from equilibrium (*e.g.*, dangling molecules, glassy materials, grain boundaries, and metastable phases). Conventional soft/hard composites have overlooked these domains due to the shortage of generalizable methodologies that can afford predictable control. Nevertheless, non-equilibrium materials encompass free energy levels common to neither soft nor hard material bulks, and thus can be exploited to design the interfacial energy landscape. Their time-dependent and stimuli-responsive features further promise otherwise unreachable hierarchy and dynamics. Therefore, *understanding and designing the compositional, spatial, structural, and temporal characters of interfacial non-equilibrium states will offer unique opportunities to build unprecedented soft/hard material platforms.*

My academic training in hard (PhD) and soft matter (postdoc) will allow me to establish a compelling interdisciplinary research program on interface science. I have accumulated expertise in a broad range of synthesis and characterization methods relevant to oxides, halides, fluidic and polymeric systems, particularly in confined environments. Based on these foundations, I propose the following future research themes:

1. **Soft/hard heterostructures with programmable hierarchy and phases**: The increasing demands of sophisticated functionalities force us to go beyond monolithic materials and pursue hybrid matters. *I propose to develop soft/hard 3D heterostructures that have controllable hierarchy and phases using liquid crystal elastomers (LCE)/halide perovskites as an exemplary soft/hard system.* To unleash the potential that non-equilibrium interfaces could bring, *I plan to locally nucleate and grow perovskites inside the open modules defined by LCE networks.* Such a unique synthesis route may lead to the following outcomes: (i) promoted or retarded crystallization kinetics of perovskites due to the nanoscale confinement of LCE networks; (ii) programmable surface reconstruction of perovskites by controlling LC isotropy; (iii) configurable size and geometry distributions of perovskites via design of the LCE network. This platform contains rich diffusion and crystallization questions, for instance, whether the nucleation and growth of inorganic in confined and/or disordered polymeric surfaces follows concepts of classical nucleation theory or falls within the broader context of “nonclassical” pathways involving formation, aggregation, and transformation of transient material states (*i.e.*, Ostwald-Lussac law). It will also bring technological implications to optoelectronics as the received hybrid is virtually a flexible optoelectronic unit that features stimuli-responsive light emission and long-term stability.

2. **Soft/hard grain-scale epitaxy**: Epitaxy means the growth of a crystalline film on top of a crystalline substrate undergoing a small lattice mismatch. It is a vital concept in solid-state semiconductor field, conventionally irrelevant to soft materials. However, the nature of crystallinity and lattice refers to structural order, which do exist in crosslinked polymer network. The crosslinking mesh can be treated as the “lattice” for polymer though it has larger degree of freedom to motion and assembly compared with the solid lattice. *The "unit cell" of crosslinked polymer networks is at the scale of grain sizes, super lattice, or*

artificially created patterns of a hard material. Based on that fact, I propose to selectively bind the crosslinking unit cell with the grain boundaries by controlling the assembly of bridging molecules, which I name as, grain-scale epitaxy. This epitaxial soft/hard interface could provide on-demand control of interfacial ion/charge transfers. For example, it could enable directional lithium ion transport from electrolyte to electrode in solid-state batteries. Combining with the soft/hard heterostructure proposed above, this design would yield a previously unseen hard/hard heterostructure that has commensurate grain arrangements.

3. **Multifunctional encapsulations for soft bioelectronics**: Exquisite control of soft/hard interfaces will have broad technological impacts in energy, electronics and bioengineering. *I propose to develop a multifunctional encapsulation platform for soft bioelectronics. The encapsulation will present the following characteristics: (i) flexible and transparent; (ii) low water permeability under deformation; (iii) anti-fatigue; (iv) interfacial adhesion to skin or organs.* Its design will entail a crosslinked polymer matrix with infusion of dense hydrophobic oil, inclusion of phase-separated droplets, and interfacial bonding molecules. The fluidic oil will serve as a dynamic and replenishable barrier to seal the free volume created during deformation. The droplets will contain precursors that undergo solidification upon mechanical perturbation or air exposure at crack tip. The stimuli-responsive solidification will locally bridge the crack and prevent its further propagation, while maintaining the bulk intact and elastic. This theme will provide transformative insights for the protection and electrode design of bioelectronics, and other emergent technologies, such as perovskite optoelectronics and soft robotics.

Teaching Interests:

After going through bachelor and PhD courses, as well as 11 courses audited during postdoc, I gained a clear vision about what expectations are for a great teacher. I have met several remarkable professors who not only taught me knowledge but also ignited my passion to subjects of thermodynamics and kinetics, intermolecular interactions, and fracture mechanics. Inspired by different aspects from each of them, I develop my teaching philosophy with emphasis on three core principles: *(i) fostering the students' interest to the subject and science in general; (ii) inspiring the students to find the core of scientific questions; (iii) cultivating students' self-learning ability and collaborative and independent problem-solving skills.* As engineers, we have plenty of practical implementations in real life to draw from to intrigue students. The history of science and scientists (*e.g.*, rubber and Goodyear) is always a fantastic aspect to introduce. Fundamental questions will help the students to see the heart of a phenomenon. During the lecture, I will ask questions like: what is the difference between the gas diffusion (or charge transfer) in a single crystal and that in a glassy material? Where does the difference come from? I will also encourage students to ask questions and nurture an active classroom. Team projects and in-class discussions will serve as versatile media to connect students.

I am interested in teaching engineering courses such as *"Statistical Thermodynamics" and "Kinetics and Phase Change"* at introductory or advanced level. In these courses, I plan to teach students topics including Brownian motion, Boltzmann distribution, thermodynamic laws, thermodynamic functions, specific heat, statistical ensembles, chemical equilibrium and kinetics, transition rate theory, atom movements and point defects, kinetics of phase change, nucleation and growth, and etc., covering a whole spectrum of solid-state and soft-matter, atomic and molecular interactions. I would like to develop two more graduate-level classes including *"Surface Chemistry" and "Material Interfaces"*. For example, in the course of "*Surface Chemistry*", I would like to deliver students historical motivations of surface chemistry; the fundamental knowledge of surface energy, bonding, diffusion, electronic orbitals, interfacial band structures; and the vastly exciting advancements of practical applications such as catalyst, battery, optoelectronics, photoelectrochemical fuel production, physical vapor deposition, chemical vapor deposition, atomic layer deposition, etc.

Selected publications (* represents equal contribution):

1. **Yanhao Yu***, C. Sun*, X. Yin, J. Li, S. Cao, C. Zhang, P. M. Voyles, X. Wang. "Metastable Intermediates in Amorphous Titanium Oxide: A Hidden Role Leading to Ultra-Stable Photoanode Protection" ***Nano Lett.***, 18, 5335-5342 (2018).

2. **Yanhao Yu**, X. Wang. "Piezotronics in Photo-Electrochemistry", ***Adv. Mater.***, 1800154, (2018).

3. **Yanhao Yu***, Z. Zhang*, Q. Liao, Z. Kang, X. Yan, Y. Zhang, X. Wang. "Enhanced Photoelectrochemical Efficiency and Stability Using a Conformal TiO_2 Film on a Black Silicon Photoanode" ***Nature Energy***, 2, 17045 (2017).

4. F. Wu*, **Yanhao Yu***, H. Yang, L. N. German, Z. Li, W. Yang, L. Huang, W. Shi, L. Wang, and X. Wang. "Simultaneous Enhancement of Charge Separation and Hole Transportation in TiO_2-$SrTiO_3$ Core-Shell Nanowire Photoelectrochemical System" ***Adv. Mater.***, 29, 1701432 (2017).

5. **Yanhao Yu**, Z. Li, Y. Wang, S. Gong, and X. Wang. "Sequential Infiltration Synthesis of Doped Polymer Films with Tunable Electrical Properties for Efficient Triboelectric Nanogenerator Development" ***Adv. Mater.***, 27, 4938-4944 (2015).

6. W. Yang*, **Yanhao Yu***, M.B. Starr, X. Yin, Z. Li, A. Kvit, S. Wang, P. Zhao, and X. Wang. "Ferroelectric Polarization-Enhanced Photoelectrochemical Water Splitting in TiO_2-$BaTiO_3$ Core-Shell Nanowire Photoanodes" ***Nano Lett.***, 15, 7574-7580 (2015).

7. **Yanhao Yu**, J. Li, D. Geng, J. Wang, L. Zhang, T. Andrew, M. Arnold, and X. Wang. "Development of Lead Iodide Perovskite Solar Cells Using Three-Dimensional Titanium Dioxide Nanowire Architectures" ***ACS Nano***, 9, 564-572 (2015).

8. **Yanhao Yu**, X. Yin, A. Kvit, and X. Wang. "Evolution of Hollow TiO_2 Nanostructures via the Kirkendall Effect Driven by Cation Exchange with Enhanced Photoelectrochemical Performance" ***Nano Lett.***, 14, 2528-2535 (2014).

581907 Quantifying Interfacial Transport Phenomena for Environmental and Biochemical Systems

Research Interests:

Interfacial properties are crucial to determine the microstructure of multiphase materials and give rise to a rich set of macroscale flow responses. My goal is to establish and lead a world-class research group specialized in quantifying interfacial transport phenomena in complex fluids, with the overarching objective to develop predictive models for innovating applications to environmental and biochemical systems. In the first few years of my independent career, my group will focus on the following areas:

1. Surfactant-driven flow in porous media: Drug delivery inside lung airways is an example of transport in porous media, in which flows are spatio-temporally non-uniform due to physiological architectures. Surfactant is involved whose non-uniform distribution generates a bulk fluid flow. My group will simulate the species transport and dewetting mechanics under the influence of fluid-structure interactions and interfacial kinetics, with the aim to devising drug delivery protocols to address obstructive lung diseases, including cystic fibrosis. Our developed tools are also of value to chemical and environmental settings, such as designing membranes for filtering highly toxic discharge.

2. Nonequilibrium colloid dynamics: A colloidal dispersion exhibits a phase transition from a suspension to a gel, where key governing factors are interfacial properties of the colloids and the imposed force field. My group will model the nonequilibrium microstructure formation of colloids induced by various transient gradients, including chemical, density, electric, and magnetic potentials. We will construct a mean-field theory for the macroscopic advective-diffusive colloid transport, with the aim to achieving two main goals: first, to tailor time-dependent phase transition for extending shelf life of therapeutics; second, to transport colloidal species into/out of hard-to-reach geometries, relevant to underground oil recovery.

3. Electrokinetics, peristalsis, and hydrophobic slip: Peristaltic pumping is distinct from traditional pressure actuation in that it generates local flow recirculations. Its coupling to surface electrokinetics and slip of a microchannel offers huge potential in designing microfluidic platforms. My group will analyze the interfacial instability of two-fluid non-Newtonian flows under electrokinetic and peristaltic perturbations. The predictive model developed will enable and optimize a host of biochemical and industrial applications. For instance, drug encapsulation by instability-induced emulsification; particle sorting and controlled release by steady recirculation cells; rapid surface coating by stable fluid films; and fiber manufacturing by co-extrusion of a composite jet.

Teaching Interests:

My joy of teaching stems from my teaching philosophy which centers around the interaction between students and the instructor. I have cultivated my teaching experiences in a diverse cultural setting such as teaching assistant for undergraduate engineering mechanics, thermodynamics, and fluid dynamics (The University of Hong Kong), instructor and teaching assistant for undergraduate and graduate fluid mechanics (Cornell), and course designer for general engineering science in university-wide teaching conferences (Cornell). With my multidisciplinary educational background, I believe that I can succeed in teaching any undergraduate level chemical and mechanical engineering courses. I am also confident to teach fluid mechanics, heat and mass transfer, engineering mathematics, and numerical methods at graduate level. I am interested in developing and teaching electives in rheology of complex fluids and methods of applied mathematics.

581961 Microbial Biofilm Processes: Multiscale Modeling, Simulation, and Visualization

My scientific research is focused on processes that involve the formation and evolution of microbial biofilms, i.e. dynamic microbial communities thriving into matrices of extracellular biopolymers. Spearheaded by mathematical modeling and computer-aided simulations, and also supplemented by microscale experiments, my research activities aim at improved mechanistic understanding of biofilm processes with a long-term vision to achieve some control over biofilms and engineer their behavior in practical applications like the confinement of organic pollutant plumes in the subsurface, the biodegradation of crude oil spills in marine waters, and ultimately the eradication of pathogenic biofilms in human tissues.

Research Achievements

- Development of a *novel* computer simulator of biofilm growth dynamics in porous media, which combined a continuum-based approach for fluid flow and mass transport with individual-based approaches for biofilm growth and detachment. *Main publication*: Kapellos GE, Alexiou TS, Payatakes AC. Hierarchical simulator of biofilm growth and dynamics in granular porous materials, Advances in Water Resources 30(6-7):1648-1667 (2007).
- Development of new theoretical models for interstitial flow and diffusion in cellular biological media (tissues, biofilms, etc) through a consistent combination of the spatial averaging method for the derivation of upscaled equations, with the particle-in-cell method for the calculation of macroscale parameters as functions of microscale system properties. *Relevant publication*: Kapellos GE, Alexiou TS, Payatakes AC. A multiscale theoretical model for fluid flow in cellular biological media, International Journal of Engineering Science 51:241-271 (2012).

Current Research Direction

Over the last few years, my research is mainly focused on the formation of microbial biofilms over oily substrates (hydrocarbons). My interest in this topic was fueled during my research visit at Texas Tech Univ. (2013-'14), where I participated in a GoMRI-funded collaborative research program that investigated the fate of crude oil spilled in the Gulf of Mexico after the Deepwater Horizon blowout. At present, through a Marie Sklodowska Curie project, I have joined forces with Profs. Patrick S. Doyle (MIT, USA) and Nicolas Kalogerakis (TU-Crete, Greece) towards understanding the mechanisms and quantifying the process rates of oil droplet biodegradation via a combination of microfluidics, biochemical analyses and computational modeling. *Relevant publication*: Kapellos GE, Paraskeva CA, Kalogerakis N, Doyle PS. Theoretical insight into the biodegradation of solitary oil microdroplets moving through a water column, Bioengineering, 5:15 (2018).

Research Interests:

- Biofilm growth and migration in porous media.
- Biofilm formation at liquid-liquid interfaces.
- Transport phenomena in cellular biological media (biofilms, tissues, etc.).
- Multiscale modeling and simulation.
- Experiments with microfluidic/microfabricated devices.

Teaching Interests:

- Transport Phenomena in Poroelastic and Biological Materials
- Computational Modeling of Biological Systems
- Biomechanics

Teaching Experience

At the University of Patras (Greece), Department of Chemical Engineering, for several years:

- Teaching assistant in the undergraduate courses Mass Transfer (with Profs. Alkiviades Payatakes and Christakis Paraskeva) and Introduction to Scientific Computing with Fortran (with Prof. Dimitrios Mataras).
- Co-supervisor of several diploma projects of undergraduate students (with Profs. Payatakes, Pavlou and Paraskeva).

581990 Design of Functional Soft Materials for a Sustainable Future

Research Interests:

A unifying fundamental element of my research will be the harnessing of anisometric interactions in soft matter to achieve new functions. I seek to use orientation-dependent interactions to modulate self-assembly, photonic properties and active transport processes. These interactions will also be employed in detection mechanisms to create a range of biologically and technologically relevant soft materials. These materials include biomimetic photonic structures, biosensors and materials for guided active transport. The work is inspired by the observation that anisometric interactions are common in biological systems, including underlying the organization of exoskeletons of beetles/butterflies, protein-protein interactions and in biological membrane organization. In contrast, relatively few forms of synthetic soft matter have been explored based on anisometric interactions, particularly for control of self-assembly and for modulating transport properties.

My graduate research at Georgia Institute of Technology primarily focused on optical and spectroscopic characterization of the order and molecular configurations of anisotropic fluids under different confining geometries, including toroidal shapes stabilized by yield stress fluids. Specifically, I discovered achiral symmetry breaking of liquid crystals leading to the formation of exotic doubly-twisted configurations when confined to cylinders and toroids. Further, I carried out theoretical and experimental studies on characterizing the order in soft materials using polarized Raman spectroscopy.

My current research as a postdoctoral researcher at Cornell University focuses on fundamental insights into coupling of strain in soft colloids including cell membranes with complex fluids. Specifically, I am interested in understanding how health of cell membranes is linked to the phenomenology that emerge from the coupling of internal organization, mechanical properties and shape in response to strain. In addition, I am also developing mechanisms to control directional propulsion in active systems by utilizing novel interactions between dispersed and continuous phases.

My expertise in liquid crystals, imaging and scattering techniques as well as my experience in nanomaterial synthesis and functionalization will position my group to undertake synergistic efforts in design, characterization and fabrication to develop these next-generation materials. These ideas are illustrated by the three initial research themes for my laboratory including 1) Anisotropic fluids as functional materials to interface with biological systems 2) Sustainable bio-materials by controlled self-assembly of cellulose nanocrystals 3) Environmental remediation using active transport. I will pursue collaborations with mathematicians, epidemiologists and chemists, leading to a convergence of ideas and tools to address challenges from a wide range of contexts. I believe this approach will also attract a host of funding opportunities. Building upon concepts in soft matter physics, my group will create a range of functional materials with programmed architectures and properties, guided by the challenge of addressing global societal challenges.

Selected Publications (15 total, 2 in preparation)

1) Liquid Crystals with Interfacial Ordering that Enhances Responsiveness to Chemical Targets.
K. Nayani, P. Rai, N. Bao, H. Yu, M. Mavrikakis, R. J. Twieg, N. L. Abbott. **Advanced Materials**, 1706707 (2018)

2) Chiral Interactions in Liquid Crystals.
K. Nayani, Y. K. Kim, N. L. Abbott. **Nature Materials**, 17 (1), 14 (2018)

3) Using Chiral Tactoids as Optical Probes to Study the Aggregation Behavior of Chromonics.
K. Nayani, J. Fu, R. Chang, J. O. Park, M. Srinivasarao. **Proceedings of National Academy of Sciences**, 114, 3826 (2017)

4) Spontaneous Emergence of Chirality in Achiral Lyotropic Chromonic Liquid Crystals Confined to Cylinders.
K. Nayani, R. Chang, J. Fu, P. W Ellis, A. Fernandez-Nieves, J. O. Park, M. Srinivasarao. **Nature Communications**, 6, 8067 (2015)

5) Curvature-Induced Twist in Homeotropic Nematic Tori.
P. W. Ellis*, **K. Nayani***, J. P. McInerney, Z. Rocklin, E. Matsumoto, M. Srinivasarao, A. Fernandez-Nieves. **Physical Review Letters**, 121, 247803 (2018) (*Equal Contribution)

Teaching Interests:

In addition to dissemination of knowledge, professors play a critical role in educating students by enabling independent critical thinking, invoking curiosity and, crucially, teaching effective communication of science. These responsibilities and roles are increasingly necessary to push back on a rising anti-science rhetoric among many populations in the US. My teaching philosophy, regardless of the particular subject matter, will be guided by methods that maximize student involvement in the learning process. The approach includes collaborative projects and group assignments that enable students to enrich their understanding through discussions and also promote development of skills in effective communication of science. Furthermore, discussion-enabled and curiosity-driven thinking will help develop inquisitiveness and deeper thinking in contexts that extend beyond science. I am comfortable teaching transport phenomena, thermodynamics, reaction kinetics, polymer physics, and colloidal/ interfacial phenomena. I intend to develop a course on scaling concepts for soft matter physics.

582076 Intelligent Systems, Active Colloids, and Advanced Responsive Materials

Research Interests:

Active colloids provide diverse and as yet untapped opportunity to engineer novel intelligent systems at microscopic and mesoscopic scales in fields ranging from targeted drug delivery, micro- to meso-scale robotics, active microfluidic devices, among others. Far from thermodynamic equilibrium, even seemingly simple active colloidal systems exhibit complex and counter-intuitive properties including negative viscosities, "turbulent" behavior at low Reynolds numbers, and highly enhanced transport. The field is ripe for new research that exploits this complexity to develop active colloidal systems that perform useful functions. To harness these properties, fundamental underpinnings must be developed to predict, tailor, and control their behaviors. I will provide theory to guide the development of active systems critical to realizing this vision.

I specialize in leveraging the methods of applied mathematics and numerical simulation, combined with highly integrated collaboration with experimentalists, to provide fundamental understanding and strategies to develop advanced functionality in soft materials and complex fluid systems. This positions me to perform groundbreaking research in intelligent systems utilizing active matter. I will use boundaries and spatial confinement to assert precise control over the behavior of active systems. For example, fluid interfaces present rich opportunity for enhanced transport and structure formation to engineer advanced functionality. Active colloids move along complicated, curvilinear trajectories near interfaces. Furthermore, they can become physically adhered to fluid interfaces, confining their motion to two-dimensional (potentially non-planar) surfaces. Once adhered, active colloids are subject to capillary and Marangoni forces, which can have a drastic impact on their behavior. The influence of hydrodynamic, capillary, and Marangoni forces can be widely tuned to produce desired behavior by changing the shape of the interface or adding surfactant to adjust the mechanical properties of the interface. These three forces are completely inherent to active systems at fluid interfaces. Introducing external fields (e.g., magnetic or electric) or anisotropy to the colloidal particles provide additional degrees of freedom. My vision is to harness this toolkit to make advanced, active, functional structures.

My background equips me to realize this vision. My past work has made me an expert in applying a variety of tools of applied mathematics and numerical analysis to engineering problems in transport phenomena and soft matter physics. These include finite element analysis, perturbation methods, boundary integral and singularity methods, and many others. However, my current work is all performed in very close collaboration with experimentalists. I have found this kind of communication between theory and experiment to be vital; experiments test the validity and assumptions of theoretical models, while theory can explain counter-intuitive observations in the lab and suggest new means of experimentation. I will continue to work in close collaboration with experimentalists to explain new observations in active matter systems as well as to conceive new directions for experimentation as suggested by theoretical findings.

Postdoctoral Advisor: Prof. Kathleen J. Stebe, University of Pennsylvania, Dept. of Chemical and Biological Engineering

Ph.D. Dissertation: *Locomotion and Drift in Viscous Flows: Numerical and Asymptotic Predictions*. Advised by Aditya S. Khair, Carnegie Mellon University, Dept. of Chemical Engineering

Teaching Interests:

In this data-driven era, intuition is of paramount importance to engineers to critically assess prediction. To develop these insights, we rely on powerful and fundamental concepts in many areas of mathematics, science, and engineering. I have a passion to teach these approaches to future generations of scientists and engineers, and I will design courses that provide effective and inclusive environments for learning. Tools of structured, active in-class learning provide means both to keep students engaged in the class content and to learn critical problem-solving skills. This style of learning departs from pure lecturing to allow students to reason through difficult problems with the help of the instructor, TAs, and peers. By using these pedagogical methods, I will instill the ability to use critical thinking and mathematical analysis to gain strong physical intuition about complex problems in engineering and to assess outcomes of machine learning and data-driven models. Providing such foundations to the next generation of scientists and engineers will equip them to push the limits in both industrial and academic environments.

I have had rich teaching experiences. During the first year of my postdoc, I had the privilege to teach an informal "mini-course" on micro-hydrodynamics and another on Green's functions and their application to partial differential equations. Combined, they comprised 14 individual lectures of 90–120 minutes. The 15–20 attendees of these lectures included members of my research group as well as other interested graduate students from across the engineering school. To aid learning, I also constructed several informal exercises for students to discuss and work through during or after class. Aside from this, I have also given guest lectures in graduate and undergraduate courses on interfacial phenomena, fluid mechanics, and mathematical methods. During my PhD, I served as a mentor for a master's student thesis project. Finally, I have provided informal one-on-one mentoring for graduate and undergraduate students throughout my academic career. I believe this past teaching experience has prepared me to teach and advise at the university level.

582108 Designer Soft Matter: From Passive to Active to Biomimetics

Research Interests:

Designing the dynamical and the structural properties of soft matter systems across different length scales is the central theme for a wide range of multidiscipline research fields from chemical engineering to condensed matter to biophysics. My research expertise lies at the interface between the physicochemical manipulation of soft matter systems (i.e., artificial microswimmers, lipid bilayers, passive colloids, (bio-)polymers, and emulsions), and the fabrication of novel and advanced functional materials with desired properties. It is my ambition to set up an independent research group that bridges chemical engineering, materials science, physics, and biology. The research focus will be to gain a fundamental understanding of the underlying governing principles involved in non-equilibrium systems using model systems:

1. **dynamic self-organization of living systems**,
2. **bottom-up biology**-how the cell membrane responds to highly localized forces from within, such as those exerted by cytoskeletal activity (i.e., microtubules, and actin filaments), and how these local forces translate into complex cell shapes,
3. **hierarchicalself-assembly of passive systems**.

I will first address these challenging problems using state-of-art soft matter model systems, namely active and passive colloids, and biomimetic systems, in combination with my expertise in design and synthesis of novel colloids, and vesicles of lipid bilayers, microfluidics, directed self-assembly, rheology, and real space characterization tools.

Building upon on this fundamental understanding and careful engineering of physiochemical properties of building blocks together with tailor-made interaction potentials between them would pave a way to developing unconventional materials with functionalities of living matter, and nature-inspired materials. I will specifically focus on engineering the responsive, self-healing materials, sculpting gels within, photonic bandgap at visible wavelengths, and transporting cargo to desired targets.

Selected Publications:

1. **R. Vutukuri**, J. Vermant, et al., *Sculpting vesicles with active particles: Less is more*, under review in Science.
2. **R. Vutukuri**, J. Vermant, et al., Light-switchable active particles: Fast and reversible fusion and fission, submitted.
3. **R. Vutukuri,** et al., *Rational design and dynamics of self-propelled colloidal bead chains: from rotators to flagella*,**Scientific Reports**, **7,** 16758 (2017).
4. **R. Vutukuri,** A. van Blaaderen, et al., *Directed self-assembly of micron-sized gold nanoplaelets into oriented flexible stacks with tunable interpolate distance*, **NanoLetters, 15**, 5617 (2015).
5. **R. Vutukuri**, A. F. Demirörs, B. Peng, P. D. J. van Oostrum, A. Imhof, and A. van Blaaderen; *Colloidal analogues of charged and uncharged polymer chains with tunable stiffness*, **Angewandte Chemie Int. Ed. 51**, 11249 (2012).
6. E. Leunissen, **H. R. Vutukuri**, and A. van Blaaderen; *Directing colloidal self- assembly with biaxial electric fields,* **Advanced Materials21**, 3116 (2009).

https://scholar.google.ch/citations?hl=en&user=ZGWfEWoAAAAJ&view_op=list_works

Funding:

Marie Skłodowska-Curie Individual Research Fellowship, European Research Council (ERC), **250 k$**.

Marie-Curie Research Fellow Mentor: Prof. dr. Jan Vermant, ETH Zürich, Switzerland

PhD Supervisor: Prof. dr. Alfons van Blaaderen, Utrecht University, The Netherlands

Teaching Interests:

I believe that teaching is an important component of an academic career. As a researcher, I am motivated not only by the pleasure of finding things out, but also explaining them to others. Since I first began to tutor classmates and junior students in high school, teaching has been an important part of my life.

I have extensive teaching experience in a range of environments. I have always enjoyed working with students. During my bachelors and master's studies, I taught and assisted several chemical engineering courses, including Chemical Reaction Engineering, Transport Phenomena, Thermodynamics, Statistical Mechanics, and Fluid Mechanics. While I was at the Utrecht University, The Netherlands, as a PhD student, I assisted in lecturing courses on Advanced Soft Condensed Matter Physics and Physics of Fluids. I have continued my teaching duties and involving in developing class projects for the Soft Materials course at ETH Zürich, Switzerland.

Courses I can teach:

My interdisciplinary research background has prepared me to teach courses in many areas. I would be enthusiastic to teach traditional and inter-disciplinary courses like Chemical Reaction Engg, Transport Phenomena, Thermodynamics, Statistical Mechanics, Fluid Mechanics, Soft Condensed Matter at the Bachelor's and Master's level programs in the department, but not limited to those courses. Additionaly, I would enjoy bringing new approaches to the teaching of existing courses.

582169 Coupled Ion Transport and Fluid Flow in Energy and Environmental Sciences

Research Interests: Transport Phenomena, Fluid Mechanics, Energy Storage, Porous Media, Electrokinetics, Interfacial Phenomena, High Performance Computing, Applied Mathematics

Teaching Interests: Transport Phenomena, Thermodynamics, Fluid Mechanics, Mathematical Methods for Engineering, Numerical Analysis, Parallel Programming

----

Many existing challenges in energy and environmental sciences require a multi-physics approach. Ion transport, fluid flow, and surface chemistry in porous media are all important in many problems. Examples include energy storage in batteries, water desalination through electrostatic deionization, and electrokinetic remediation of contaminated soil, as well as enhanced recovery techniques like low salinity water flooding and hydraulic fracturing. Although these problems are inherently multi-scale and multi-physics, they are fundamentally described by similar physiochemical processes. My future research will focus on understanding fundamental mechanisms of ion transport and fluid flow in porous media, using theoretical and computational approaches, to improve existing technologies and guide novel applications.

During my PhD at UC Santa Barbara, I was supervised by Prof. Frederic Gibou and Prof. Todd Squires, and developed efficient computational techniques for studying ion transport in porous media. These methods are based on advanced techniques such as adaptive mesh refinement and parallel computing to enable high fidelity simulations. In my PhD, I illustrated the significance of pore microstructure and surface conduction in accelerating the charging kinetics which inspired new guidelines for designing porous electrodes. As a postdoctoral researcher at MIT in Prof. Martin Bazant's lab, I am currently investigating applications of electric fields in controlling fluid instabilities and enhancing hydraulic fracturing. My work in this area has led to the discovery that electro-osmotic flows can control viscous fingering instabilities. This finding has exciting implications for active fluid manipulation under confinements. I am further investigating use of electric fields in improving the efficiency of hydraulic fracturing for shale gas harvesting.

As a new faculty, I will apply my combined knowledge of transport phenomena and mechanics as well as computational skills to challenging problems in energy and environmental sciences. In particular, my research group will focus on exploring novel applications in the area of geomechanics and two-phase electrokinetic flows for enhanced oil and gas recovery. For instance, while recent experimental observations suggest that low-salinity water injection can substantially improve secondary oil recovery, the exact cause is still debated. This is currently a very active area of research involving multi-physical processes such as ion transport, fluid flow, and surface chemistry which span several orders of magnitude in both time- and length-scales. I am also interested to further understand and optimize the performance of capacitive deionization technology for water desalination and energy storage. In particular, recent developments in "flow-electrode" design hold great promise as a fast switching and high-density grid-scale energy storage solution where power demands are intermittent. I am confident that my expertise in computational and theoretical modeling put me in a unique position to tackle these problems and help me to advance these fields in collaboration with experimentalist colleagues.

582187 Dilational Rheology of Lung Surfactant Inhibitors and Its Effects on Acute Respiratory Distress Syndrome

Research Interests:

Fluid-fluid interfaces are common in foods, cosmetics, oil/gas industries and in living systems. One such relevant example would be air/water interface in lung surfactants. Adding a curvature on top of the already complex interfacial structure makes it more complicated. It has been observed that the domain size of lung surfactant monolayer is fairly large, approximately 100-300 micron on flat interface which is comparable to the average size of human alveoli, ~150 micron. So, it is important to understand the contribution of curvature and dilatational deformation at the interface. Recent development in interfacial rheological development allows us to measure the response for shear, dilational, and extensional deformations with surface active molecules or particles. These techniques can be even more powerful if added with visualization. We developed instruments/techniques where one can measure the rheological response and see the structural change under deformation simultaneously. These techniques help us understand a wide variety of issues, ranging from how lungs get effected by diseases like Acute Respiratory Distress Syndrome to colloidal stability of the particle stabilized emulsions.

Postdoc Project: Dilational rheology of lung surfactant inhibitors and its effects on Acute Respiratory Distress Syndrome (NIH funded)

Adviser: Dr. Joseph Zasadzinski, University of Minnesota-Twin Cities

PhD Thesis: Explaining Interfacial Behavior of a Particle Laden Interface Using Microstructure

Analysis (NSF funded)

Adviser: Dr. Gordon Christopher, Texas Tech University

Research Experiences:

As an experimentalist, I developed an interfacial shear rheometry visualization system where we can see the structural change of particles while deformation is taking place. I also examined the change in stress behavior as bulk salt and surfactant concentration changes. This study allows us to design stable particle laden interface which has extensive industrial use in oil recovery and personal products. The instrument is versatile enough to study from proteins to asphaltenes. In my postdoctoral work, I am studying dilational rheology of lung surfactant inhibitors. Because of the spherical alveolar shape, it is essential to study pulmonary surfactant monolayer in curved interface as it experiences constant area compression and expansion. We developed a version of capillary pressure tensiometer for a purely spherical interface with added visualization under confocal microscope. Initial result indicates that lipase cleaved dppc, lyso-PC, makes the interface viscous and decreases combined modulus as the concentration increases well within breathing frequency.

Teaching Interests:

As graduate student, I worked as TA for Fluid Mechanics and as an Instructor for undergraduate Thermodynamics course. After my graduation, I worked as Course Instructor for Heat transfer, Fluid Mechanics and Thermodynamics I & II for five semesters including two short summer semesters. I sincerely enjoy teaching and extremely value my 2 years of teaching experience. During this time, I conducted extra problem-solving sessions and offered flexible office hours for provide additional assistance to students. I was able to inspire numerous students to join graduate school and get involved in research projects.

Future Direction:

Going forward, I want to combine my various interfacial transport knowledge in more biological and industrial systems. I am planning on continued work on Lung Surfactant formulation and structural analysis through visualization techniques. Knowledge on extensional rheology (on microfluidic channel) from our collaboration at Minnesota can be applied in coalescence and foam stability study.

Unlike flat interface where insoluble monolayer can be easily spread out, curved interface posed a strong challenge. We are developing techniques to spread insoluble surfactants on fluid/fluid or air/fluid interface in a controlled manner. These techniques can potentially open up a pathway to solve curvature effects on complex fluid/fluid interfaces. Depositing DPPC, a major insoluble component of lung surfactant, on curved interfaces and determining the effect of essential and foreign soluble proteins and lipids essentially explain the possible mechanism of working principle of lung surfactant and their inhibitors. Employing innovative design with microscopy techniques complement the understanding of rheological and transport properties with structure information concurrently. I am also extending these techniques with modelling to solve engineering problems like better filter design, coalescence, colloidal stability, enhanced oil recovery etc.

SELECTED PUBLICATIONS

1. Barman, S., Christopher, G., "Role of Capillarity and Microstructure on Interfacial Viscoelasticity of Particle-Laden Interfaces", J. Rheol. 60, 35 (2016)
2. Barman, S., Christopher, G., "Simultaneous Interfacial Rheology and Microstructure Measurement of Densely Aggregated Particle-Laden Interfaces Using a Modified Double Wall Ring Interfacial Rheometer", Langmuir, 2014, 30 (32), pp 9752–9760, DOI: 10.1021/la502329s
3. Yu-Jiun Lin, Sourav Barman, Peng He, Gordon F. Christopher, Sibani Lisa Biswal, "Combined Interfacial Shear Rheology and Microstructure Visualization of Asphaltenes at Air-Water and Oil-Water Interfaces", Journal of Rheology 62 (1), 1-10 (2018)
4. Syed E Rahmun, Nader Laal-Dehghani, Sourav Barman, Gordon F. Christopher, "Modifying interfacial interparticle forces to alter microstructure and viscoelasticity of densely packed particle laden interfaces", Journal of Colloid and Interface Science, 536, 30-41 (2019)
5. Snoeyink, C., Barman, S., Christopher, G., "Contact Angle Distribution of Particles at Fluid Interfaces", Langmuir, 2015, 31 (3), pp 891–897, DOI: 10.1021/la5040195
6. Zhang, Z., Barman, S., Christopher, G., "The role of protein content on the steady and oscillatory shear rheology of model synovial fluids", Soft Matter, 2014, 10, 5965-5973, DOI: 10.1039/C4SM00716F
7. Zhang, Z., Barman, S., Christopher, G., "Effect of Interfacial Viscoelasticity on the Bulk Linear Viscoelastic Moduli of Globular Protein Solutions," Physical Review E, 2013. 89
8. Snoeyink, C., Christopher, G., Barman, S., Wereley, S., "Sub-diffraction Limit Three-Dimensional Particle Tracking Velocimetry", ASME 2013 International Mechanical Engineering Congress and Exposition IMECE 2013 (2013)
9. S. Barman, M. Ehsan, S. Bhuiyan, Low-Pressure CNG System for Diesel Power Generation, The Online Journal on Power and Energy Engineering (OJPEE), paper no. - Vol. (1) – No. (2). Web address: http://www.infomesr.org/attachments/029_10-038.pdf

562983 Ionic and Electronic Transport Properties in Covalent Organic Framework and Nanomaterial

Research Interests: Ionic transport, Solid State batteries, Electronic transport, Optical physics, Plasmonics, Photonics

Teaching Interests: Transport Phenomenon, Material Science, Mathematics, Chemical Engineering Fundamentals, Electrochemistry, Solid State Physics

Degenerately doped semiconductor nanocrystals (NC) exhibit a localized surface plasmon resonance (LSPR) that falls in the near- to mid-IR range of the electromagnetic spectrum. Unlike metal LSPR, the metal oxide LSPR can be tuned by doping, and structural control, or by in situ electrochemical or photochemical charging. While synthetic work for this class of materials has progressed significantly, quantitative correlation between their intrinsic properties to the final LSPR properties such as extinction and near field enhancement remains poorly explored, thus prohibiting their use in applications from plasmon derived smart windows to sensors. Here, we illustrate how intrinsic NC attributes like its crystal structure, shape and size, along with band structure and surface properties affects the LSPR properties and its possible applications.

First, the interplay of NC shape and the intrinsic crystal structure on the LSPR was studied using model systems of In:CdO and Cs:WO_3, the latter of which has an intrinsic anisotropic crystal structure. For both systems, a change of shape from spherical to faceted NCs led to as anticipated higher near field enhancements around the particle. However, with Cs:WO_3, presence of an anisotropic hexagonal crystal structure leads to additional strong LSPR band-splitting into two distinct peaks with comparable intensities. This new insight demonstrated how the LSPR of semiconductor NCs can be controlled with shape, size and now crystal structure to tune the LSPR from near IR to mid-IR.

Second, plasmon-molecular vibration coupling, as a proof of concept for sensing applications, was shown using newly developed F and Sn codoped In_2O_3 NCs to couple to the C-H vibration of surface-bound oleate ligands. Electron energy loss spectroscopy was used to map the near field enhancement around these NCs responsible for coupling between the LSPR and molecular vibrations. A combined theoretical and experimental approach was employed to describe the observed plasmon-plasmon coupling, the influence of coupling strength and relative detuning between the molecular vibration and LSPR on the enhancement factor, and the observed Fano lineshape by deconvoluting the combined response of the LSPR and molecular vibration in transmission, absorption, and reflection.

Third, plasmon modulation through dynamic carrier density tuning was investigated using thin films of monodisperse ITO NCs with various doping level and sizes along with an in situ electrochemical setup. From the combination of the in-situ spectroelectrochemical analysis and optical modeling, it was found that often-neglected semiconductor properties such as band structure modification upon doping and surface chemistry strongly affect the LSPR modulation behavior. The influence of band structure and effects like Fermi level pinning by surface defect states were shown to cause a surface depletion layer that alters the LSPR properties, namely the extent of LSPR modulation, near field enhancement, and sensitivity of the LSPR to the surrounding.

582096 Flow of Soft Matter in Complex Geometries

2nd Year Postdoctoral Researcher

Research Interests: Engineering the architecture of complex materials is critical in multiscale applications. These applications range widely from drug delivery by biological fluids through deformable tissue and cell membranes to enhanced oil recovery that uses structured polymer solutions in rigid porous media reservoirs. These soft matter systems such as an injection of hydrogels from a needle to joints induce non-linear and non-equilibrium mechanical responses like stress relaxation and plastic deformation during transport or use in various geometries. Especially, when contacting reactive or reorganizable regions, these materials can be activated by changing charges, ionic strength, or photons. My goal is to connect the missing pieces between macroscopic viscoelastic flow properties and the nanostructure of the material using responsive boundary conditions. I plan to achieve this target using a combined platform of scattering techniques, rheology, and 3D-printed microfluidic environments. These techniques provide excellent spatiotemporal accuracy and are important to enhance the progress of understanding these materials.

Proposal Experience: NIST Center for Neutron Research (NCNR, USA)

Postdoctoral Project: "Revealing Elastic and Flow Properties of Structured Fluids by Microfluidics." Under supervision of Norman J. Wagner and Eric M. Furst, Chemical and Biomolecular Engineering, University of Delaware and NCNR

PhD Dissertation: "Investigation of Interfacial and Rheological Properties of Asphaltenes at Solid-Liquid and Liquid-Liquid Interfaces." Under supervision of Sibani Lisa Biswal, Chemical and Biomolecular Engineering, Rice University

Research Experience: My research has focused on the development of novel frameworks to advance soft matter characterization and application. During my Ph.D. at Rice U., I pioneered microfluidic devices that represent porous media along with confocal microscopy to engineer the interfacial and rheological properties of natural macromolecules in tunable physicochemical environments. As a postdoc at U. Delaware, this microfluidic platform has been extended further to explore structure-property relationships for injectable and suspension-stabilizable structured materials by using advanced non-linear rheological methods and both X-ray and neutron scattering techniques. This advanced platform has been extensively used for polymer solutions, hydrogels, colloidal suspensions, and biofluids. I also have collaborated with beamline scientists at NIST and TA Instruments to construct a new flow cell with highly improved temporal and spatial resolution and lowered friction to simultaneously measure rheology and small angle neutron scattering (Rheo-SANS) for all of the 1-2, 1-3, and 2-3 planes.

Future Direction: My future research lab will involve theoretical and experimental work to understand how colloidal-scale and molecular-level features of complex materials contribute to non-linear properties at responsive boundaries. I am especially interested in designing a yield-stress system that is inspired by a biological equivalent to those found in nature which also contains self-healing capabilities and can be stimulated by controllable interfaces. I will use advanced rheological methods and models (orthogonal and parallel superposition, shear and extensional startup, and thixotropic structure kinetics) along with neutron and X-ray scattering to characterize the soft matter under deformation using the newly generated flow cell. Also, I will continue to develop the research platform of recoverable interfaces with a changeable deformability that can be established by 3D-printed microfluidics using smart materials and lithography techniques. Finally, I will use active micro-rheology and atomic force microscopy to probe the local mechanical stresses near the boundaries that could drive the interaction. Therefore, my laboratory will incorporate 3D complex-responsive geometries with smart soft materials characterized by advanced rheological methods and models along with scattering techniques.

Teaching Interests: I am excited about teaching fundamental courses in the field of Chemical and Biomolecular Engineering, particularly momentum, mass, and heat transfer and kinetics. I am also interested in teaching advanced courses, including colloidal and interfacial science and rheology. At Rice University, I mentored five undergraduates and one high school student. I helped them developed research projects individually, and I found that mutual interaction and scientific communication are the keys to success. Therefore, I intend to create an environment with strong interactions both in class and in the lab.

Selected Publications:

1. **Y.J. Lin**, J. Horner, B. Illie, M. Lynch, E.M. Furst, N.J. Wagner, "Revealing Macroscopic Viscoelastic Flow Properties of Structured Yield Stress Fluids by Microfluidics", *In Preparation*, (2019)
2. **Y.J. Lin**, T. Cao, M.L. Chacon-Patino, S. Rowland, R. Rodgers, A. Yen, S.L. Biswal, "Microfluidic Study of the Deposition Dynamics of Asphaltene Sub-fractions Enriched with Island and Archipelago Motifs", *Energy Fuels*, (2019), 33 (3), 1882-1891
3. **Y.J. Lin**, A. Perrard, S.L. Biswal, R.M. Hill, S. Trabelsi, "Microfluidic Investigation of Asphaltenes-Stabilized Water-in-Oil Emulsions", *Energy Fuels*, (2018), 32 (4), 4903-4910
4. **Y.J. Lin**, S. Barman, P. He, Z. Zhang, G.F. Christopher, S.L. Biswal, "Combined Interfacial Shear Rheology and Microstructure Visualization of Asphaltenes at Air-Water and Oil-Water Interfaces", *Journal of Rheology*, (2018), 62, (1), 1-10
5. **Y.J. Lin**, P. He, M. Tavakkoli, N.T. Mathew, Y.F. Yap, J.C. Chai, A. Goharzadeh, F.M. Vargas, S.L. Biswal, "Characterizing Asphaltene Deposition in the Presence of Chemical Dispersants in Porous Media Micromodels", *Energy Fuels*, (2017), 31 (11), 11660-11668
6. **Y.J. Lin**, P. He, M. Tavakkoli, N. Mathew, Y.F. Yap, J.C. Chai, A. Goharzadeh, F.M. Vargas, S.L. Biswal, "Examining Asphaltene Solubility on Deposition in Model Porous Media", *Langmuir*, (2016), 32 (34), 8729-8734
7. Y. Zhuang, A. Goharzardeh, **Y.J. Lin**, Y.F. Yap, J.C. Chai, N. Mathew, F.M. Vargas, S.L. Biswal, "Three Dimensional Measurements of Asphaltene Deposition in a Transparent Micro-channel", *Journal Petroleum Science and Engineering*, (2016), 145, 77-82
8. C.W. Liang, C.H. Lee, **Y.J. Lin**, Y.T. Lee, C.K. Ni, "MALDI Mechanism of Dihydroxybenzoic Acid Isomers: Desorption of Neutral Matrix and Analyte.", *The Journal of Physical Chemistry B*, (2013), 117 (17), 5058-5064
9. C.W. Liang, P.J. Chang, **Y.J. Lin**, Y.T. Lee, C.K. Ni, "High Ion Yields of Carbohydrates from Frozen Solution by UV-MALDI", *Analytical Chemistry,* (2012), 84 (8), 3493-3499
10. W.S. Wen, J.K. Lai, **Y.J. Lin**, C.M. Lai, Y.C. Huang, S.S. Wang, J.S. Jan, "Effects of Copolypeptides on Amyloid Fibrillation of Hen Egg-White Lysozyme", *Biopolymers*, (2012), 97 (2), 107-116

582038 Role of Species Transport on the Stability of Interfaces

Research Interests:

Stability of and transport across interfaces are important in a diverse class of problems ranging from stability of inorganic fluid-fluid interfaces in lubricants to transport across bio-interfaces that are vital for life. In many situations, such as in foams or in the blood brain barrier – interfacial transport and stability are closely intertwined. My research seeks to establish a fundamental understanding of this class of important problems through experiments, theory and simulations.

Research Experience:

Stability of fluid-fluid interfaces in the presence of evaporation, *Department of Chemical Engineering, Stanford University* (advised by Prof. Gerald Fuller)

My dissertation research broadly explores the role of evaporation driven mass transport on the stability of systems having liquid-liquid interfaces utilizing a combination of experiments and theory. Utilizing optical measurement tools such as interferometry and shadowgraphy, we are studying two problems that clearly highlight the dual role of evaporation on the stability of such systems. The first problem concerns the stability of deleterious non-aqueous lubricant foams. Lubricants are often a mixture of various oils. When foams are generated in such lubricants, our experiments reveal that the heterogeneous evaporation of lubricant components lead to stabilizing Marangoni flows that dramatically enhances foam stability. The results from this study help the lubricant industry account for the foam stabilizing effects of lubricant blending and compensate for such effects through the addition of adequate 'antifoam' additives. Current research in this project focuses both on establishing thresholds on the adequate quantity as well as the mechanism of antifoam additives. The second problem concerns hydrodynamic instabilities in drying polymer films. Our experiments reveal that evaporation can drive Rayleigh-Taylor instabilities with the film leading to undesired heterogeneous redistribution of polymers. Theoretically we show that the onset time of the instability is bound by two power law scalings in polymer concentration dictated by the relative importance of diffusion. These results can be used to develop film drying protocols that avoid Rayleigh Taylor instabilities. Current research in this project looks at evaporation driven instabilities in phase seperated polymer solutions.

Selected Publications:

- Mossige*, **V. Chandran Suja***, G.G. Fuller et.al, Evaporation driven Rayleigh-Taylor Instability in Aqueous Polymer Solutions, *submitted* (2019). *equal contribution
- **V. Chandran Suja**, G.G. Fuller et.al, Evaporation Induced Foam Stabilization in Lubricating Oils, *PNAS* (2018).
- **V. Chandran Suja**, J. F. Frostad, G.G. Fuller Impact of Compressibility on the Control of Bubble-Pressure Tensiometers, *Langmuir* (2016).

Selected Awards:

- Universite Paris-Saclay Outgoing Mobility scholarship (2015)
- Charpak Scholarship (2014-2015)
- Chemical Engineering Outstanding TA award (2019)
- Centennial TA award (2019)

Teaching Interests:

As a formally trained chemical engineer, I am excited and qualified to teach core coursework in thermodynamics, fluid mechanics and transport phenomena. I also look forward to leveraging my interdisciplinary research experience and interests to develop and teach electives in interfacial dynamics, soft matter physics, non-Newtonian fluid dynamics, and/or applied mathematics for chemical engineers.

Teaching Experience:

- Rheology short course in Beijing , Teaching Assistant (2019)
- Fluid Mechanics (ChemEng 120A), Teaching Assistant and Guest Lecturer (2018, 2019) s

581997 Silk-Based Materials: A Sustainable Route Towards Multifunctional Material Applications

Research Interests:

Silk fibroin protein, a natural biopolymer derived from the cocoons of silkworm, has received considerable attention due to its excellent biocompatibility, controllable biodegradability, outstanding mechanical strength, and versatile aqueous processability into new material formats. These properties make silk protein-based materials suitable candidates for biomedical applications. Additionally, in the last few years, silk has been also recognized as an attractive material for emerging fields, such as environmentally adaptive materials, flexible/wearable electronics, photonics, optics, sensing, and energy conversion/storage systems. These newer utilities derive from: (1) the ability of silk to be formed into unique nanoarchitectures through hierarchical self-assembly for processing into various material formats, such as films, nanofibers, hydrogels, and sponges, with high fidelity feature resolution down to the nanoscale, and (2) the availability of silk as an abundant natural resource with eco-friendly properties that can help meet the growing demand for sustainable materials and devices.

My research interests are the synergistic integration of natural and synthetic building blocks that can create functional materials, with a focus on thin film, nanofiber, and hydrogel formation, for applications in catalysis, miniaturized energy conversion/storage, sensing and adsorption/separation systems. Specifically, by understanding the fundamental structure-property relationships of silk fibroin at the nanoscale, my research aims to leverage silk, metal organic framework (MOF), polymer, and block copolymer materials to formulate hierarchical structures with tunable functional properties, which can be transferred from molecular to the macroscopic length scales.

At the initial stage of my independent research, I plan to focus on following topics.

(1) Highly Ordered Nitrogen-Doped Graphitic Carbon Thin Films with Tailored Micropores/Mesopores for Micro-Supercapacitor Applications. This would involve the study of crystallization of silk thin films that can be transformed into ordered graphitic carbons in the form of films from ß-sheet nanocrystals.

(2) Hybrid Silk/MOF Thin Films, Nanofibers, and Aerogels for CO_2 Adsorption/Separation, Air or Water Filtration, and Chemical Sensor Applications. This would entail the precise control of the position of ordered crystalline MOFs.

(3) Stretchable Ionic Conductive Silk Hydrogels for Human Health Monitoring. I am interested in a variety of sacrificial bonds and conducting components compatible with silk matrices to fabricate tough and ionic conductive hydrogels.

Research Experience:

My Ph.D. research in the Polymer Science and Engineering Department at the University of Massachusetts Amherst (with Professor Thomas Russell and Professor Kenneth Carter) focused on the understanding of the minimum amount of topographic patterning necessary to successfully guide the self-assembly of block copolymers in 2D and 3D. During this time, I developed expertise in block copolymer lithography that can fabricate highly ordered nanostructures (hexagonal arrays or line patterns) over macroscopic length scales. In addition, I quantitatively analyzed these nanostructures using grazing incidence small angle X-ray scattering (GISAXS). These studies have increased our understanding of the morphological characteristics and lateral ordering of the self-assembled block copolymer microdomains guided by the patterned substrates for the magnetic bit-patterned media applications and the fabrication of various next-generation functional nanostructures.

During my postdoctoral research in the Department of Biomedical Engineering at Tufts University (with Professor David Kaplan), I have broadened my research experience in silk-based biomaterials. Building on my graduate research experience in polymer science and engineering, I have developed Fenton reactions that can oxidize the tyrosine residues of silk fibroin, leading to dityrosine crosslinking that enables the formation of chemically crosslinked silk hydrogels for biomedical applications. This study hinted at the use of Fenton reaction to create a silk hydrogel system with enhanced mechanical properties, while overcoming the potential concerns (e.g., immunological response) with enzymatically crosslinked hydrogels. In particular, this work has increased our understanding of the interaction of iron ions with silk fibroin that could be used to modulate the degree of dityrosine crosslinking.

Teaching Interests:

My teaching interests include core courses in chemical engineering, including fluid mechanics, heat transfer, mass transfer, thermodynamics, chemical reaction engineering, and separation engineering. In addition, with my multidisciplinary research background, I am also interested in teaching polymer science and engineering, nanomanufacturing, and materials characterization both for undergraduate and graduate levels.

582043 Interfacial Materials for Electrochemical and Biomedical Devices

Research Interests:

Interfaces play a key role in many areas including electronics, energy conversion and storage, and medical technology. This poster will first introduce the development of interfacial materials for electrochemical devices such as batteries. The significant increase in energy density of batteries must be achieved by exploring new materials and cell configurations. Lithium metal and lithiated silicon are two promising high-capacity Li-containing anodes. Unfortunately, both these anodes suffer from serious environmental corrosion during electrode preparation and battery cycling processes. The interfacial materials with precise atomic compositions and tailored microstructures realize Li-containing anodes with both environmental and electrochemical stability. The interfacial engineering approaches bring huge benefit to both the existing Li-ion batteries and next-generation Li-S batteries. In the second part, this poster will show the design of interfacial materials for transient electronics. Aside from environmental benefits, transient electronics find tremendous applications as bioresorbable implantable medical devices that obviate the need for extraction surgeries. The key challenge of bioresorbable medical devices is to develop stable interfacial materials that can provide high-performance operation yet are completely dissolvable in biofluids and are fully biocompatible. By synthesizing barrier materials with optimized physical and mechanical properties, we developed biodegradable and wireless electronics with necessary operating times for therapeutic purposes, specifically as electrical stimulators to accelerate tissue regeneration.

Teaching Interests:

Besides developing new science and technology, I am also excited about a career of teaching and mentoring. My education and research background makes me able to teach broad subjects in materials science, chemical engineering, electronics, and bioengineering. At Stanford, I was a teaching assistant for the graduate class "Materials Chemistry". During my Ph.D. and postdoc period, I have mentored two junior graduates and three undergraduates in developing research skills and knowledge, including Asian, Hispanic and Arabic students, and some of them are Ph.D. students in top universities now.

582093 Machine Learning and Data-Enabled Design and Discovery of Soft and Nano Materials

Machine Learning and Data-Enabled Design and Discovery of Soft and Nano Materials

Tarak K. Patra

Center for Nanoscale Materials, Argonne National Laboratory, Argonne, IL 60439, USA

E-mail: tpatra@anl.gov

Research Interests: High-throughput materials design, Energy storage materials, Polymer nanocomposites, Superlattice, Self-healing materials, Self-assembly, Phase transition, Free energy calculation, and Machine learning

Teaching Interests: Thermodynamics, Transport phenomena, Polymer Physics, Numerical Methods and Molecular modeling

Soft materials are highly correlated many body systems with complex structures and dynamics spanning a wide range of length and time scales. Their relaxation processes involve non-equilibrium phenomena such as vitrification, jamming, and semi-crystallization, which are highly process-dependent. A significant advancement in the understanding of these processes are vital for the successful implementation of soft materials in future electronics, medicine and energy devices. The current understanding of these process in soft materials is limited by many factors. First, atomic level interactions in many soft materials, for a wide range of physicochemical conditions, is not well-established. Second, unlike biomolecules where informatics-based design has been accelerated by the availability of large, open, and homogenous structure-property relation databases such as the protein data bank, similar data in synthetic materials tend to be sparse, heterogenous, and often outright unavailable. Third, due to a general lack of rapid, parallelizable techniques for measuring materials properties, building such databases de novo for a specific problem is often cost-prohibitive. Furthermore, many critical design problems, such as the search for highly flexible barrier materials, target materials properties well outside the range of what have been discovered so far.

The overarching goal of the proposed research program is to address these limitations by combining molecular modeling, machine leaning and evolutionary computing. By combing these tools, I plan to develop more accurate and transferable force fields for a wide range of soft materials, including polymers, small organic molecules, colloids; discover their hitherto unknown structure-property relationship; and accelerate materials design. By integrating data science and first principle calculations, my goal is to address some of the open questions of soft matters – such as the non-universal growth of relaxation time upon cooling a material, the interrelationship between ion conductivity, modulus and glass formation behavior of ionic liquids, and the universal relationship between the structural order parameters and governing forces of a molecular assembly process. Addressing these questions are crucial for many future technologies, such as safer battery with high energy density and longer durability; capsule that prevents drugs from unwanted release and degradation caused by temperature variation; membrane that pulls salts out of sea water rapidly; flexible smart phone that can be wrapped around human's wrist; and coating that prevents roads and bridges from corrosion. The vision of the proposed program is, therefore, to address above challenging problems and establish new materials design rules by building large amount of accurate homogeneous structure property data; use artificial intelligence (AI) and machine learning to understand their interrelationship; and employ evolutionary algorithms for efficient screening and discovering new structures and properties. This way, the proposed research program aims to push the frontier of materials research towards more predictive designing and phenomenological modeling.

With successful implementation of this program, I hope to study complex physical science problems and attract external funding. I am confident that this research program will lead to the development of useful design rules for flexible electronics; polyelectrolytes for energy storage application; and nanostructure materials for gas storage, water purification and many other applications. In this program, students will be trained in statistical mechanics, wide range of machine learning approaches and artificial intelligence, and their applications in chemical engineering and materials science.

Selected Publications

(Published: 15, Under review: 3, Under preparation: 4, *Correspond Author)

1. Patra T K, Chan H, Podsiadlo P, Shevchenko E V, Sankaranarayanan S KRS, Narayanan B, Ligand dynamics control structure, elasticity, and high-pressure behavior of nanoparticles supercrystals, ***Nanoscale** 11, 10655 (2019)*
2. Hung J, Patra T K, Meenakshisundaram V, Mangalara J H and Simmons D S, Universal localization transition underlying glass formation: insights from efficient molecular dynamics simulations of diverse supercooled liquids, ***Soft Matter** 15, 1223 (2019)*
3. Patra T K, Zhang F, Schulman D; Chan H, Cherukara M, Terrones M; Das S, Narayanan B, Sankaranarayanan S KRS, Defect dynamics in 2D materials probed by combining machine learning, molecular simulation and high- resolution microscopy, ***ACS Nano** 12, 8006 (2018)*
4. Cheng Y, Yang J, Hung H, Patra T K*, and Simmons D S, Design rules for highly conductive polymeric ionic liquids from molecular dynamics simulations, ***Macromolecules** 51, 6630 (2018)*
5. Patra T K, Meenakshisundaram V, Hung J, and Simmons D S, Neural network biased genetic algorithm for materials design: Evolutionary algorithms that learn, ***ACS Combinatorial Science** 19, 96 (2017)*
6. Patra T K and Singh J K, Polymer directed aggregation and dispersion of anisotropic nanoparticles, ***Soft Matter** 10, 1823 (2014)*
7. Patra T K and Singh J K, Coarse-grained molecular dynamics simulations of nanoparticle-polymer melts: Dispersion vs. Agglomeration, ***Journal of Chemical Physics** 138, 144901 (2013)*
8. Patra T K, Hens A, Singh JK, Vapor-liquid phase coexistence and transport properties of two dimensional oligomers, ***Journal of Chemical Physics** 137, 084701 (2012)*

582150 Rational Design of Smart Soft Materials

Research Interests:

Arrangement of colloidal building blocks to form structures is widely found in nature, the oldest and most profound being the internal structure of a living cell. It also has many important technological implications, ranging from the rheology and performance of paint formulations to the lightweight but strong materials on a military aircraft. There are pressing needs for a path to spatial organization of nano- and micro-scale building blocks with tunability and efficiency. I will develop an experimental program that creates functional soft matter that can be applied in energy-saving technology and in biotechnology. I propose to use my expertise in complex fluids and colloids and leverage fundamental work to advance the understanding of these systems, and identify scalable approaches that can lead to manufacturing technology. I will focus my efforts on dense suspensions, which are widely processed in myriad industries including food, personal care, energy, pharmaceutical, and construction material industries. The exploration in dense particle suspensions will follow three trajectories. The first aim of my work is to create innovative ways to form structures by an evaporative scheme. The second aim of this work is to modulate particle distribution by environmental stimuli. The third aim is to study how active colloids respond to obstacles and if patterns can be used to direct collective behaviors. The emphasis of my research will be to address the need of an alternative way to quantify and control parameters for *objects* to migrate and assemble, and their dynamic consequences in a *complex environment*.

Teaching Interests:

The role of the instructor is one that serves both as a source of knowledge and a facilitator for learning. Furthermore, I value the power of clear communication of highly technical knowledge. Therefore, I seek to create a classroom that promotes peer-learning, active learning, communication and learning beyond classroom. Learning outcomes results from a combination of stimulation and perception personal control, thus it is important to adopt course assignment to learner types. My goal for graduate education is to learn and evolve the projects with the graduate students' input, so the project can develop in a direction that the student can feel a sense of ownership. I am well-prepared teach a wide range of courses offered in the undergraduate and graduate curriculum. Due to my research interests in soft matter and especially colloid science, I believe I can be most effective in teaching Fluid Transport, Thermodynamics, Heat and Mass Transport. I am also interested in developing a course in the area of Self-Assembly and Liquid crystals.

My background:

My Ph.D. work focused on directed colloid motion in a molded soft matter field. Using nematic liquid crystals (NLCs) as a model system, I designed boundary conditions that defined the orientation of the NLC molecules, known as the "director field". NLC molecules are rod-shaped. They self-arrange in a manner roughly aligned with their neighbors as well as conform to boundary cues with the goal of minimizing distortion. In this way, they also control the assembly of the inclusions embedded in them. The inclusions can be particles, droplets, biomolecules, etc., with defined surface chemistry and independent bulk identity. By tuning the geometry of the bounding surface, I demonstrated the ability to position particles, plan trajectories, and template assembly.

As a postdoctoral scholar at the University of Delaware, I am learning what rheological signatures mean for material properties and processability. My current research effort centers on answering fundamental questions for a dense colloidal suspension, such as how friction affects maximum packing and flow behaviors. An industrial collaboration with Chemours has led me to study how components migrate and what structures are formed during drying in a multi-component system, using tools ranging from particle tracking, neutron scattering to oscillatory tests. These macroscopic characterization skills complement my Ph.D. training in microphysics nicely and provide good basis for my proposed research of exploring structural-property relationship to create smart soft materials.

Select Publications:

1. Y. Luo, T. Yao, F. Serra, D. A. Beller, and K. J. Stebe, "Deck the walls with anisotropic colloids in nematic liquid crystals", Langmuir, 2019, DOI: 10.1021/acs.langmuir.9b01811
2. Y. Luo, D. A. Beller, F. Serra, and K. J. Stebe, "Tunable colloid trajectories in nematic liquid crystals near wavy walls", Nat. Comm., 2018, 3841.
3. Y. Luo, F. Serra, and K. J. Stebe, "Experimental realization of the `lock-and-key' mechanism in liquid crystals", Soft Matter, 2016, 12, 6027-6032. (Front Cover: Soft Matter, 2016, 12, 6007-6008.)

4. Y. Luo, F. Serra, D. A. Beller, M. A. Gharbi, N. Li, S. Yang, R. D. Kamien, and K. J. Stebe, "Around the corner: Colloidal assembly and wiring in groovy nematic cells", Phys. Rev. E, 2016, 93, 032705.

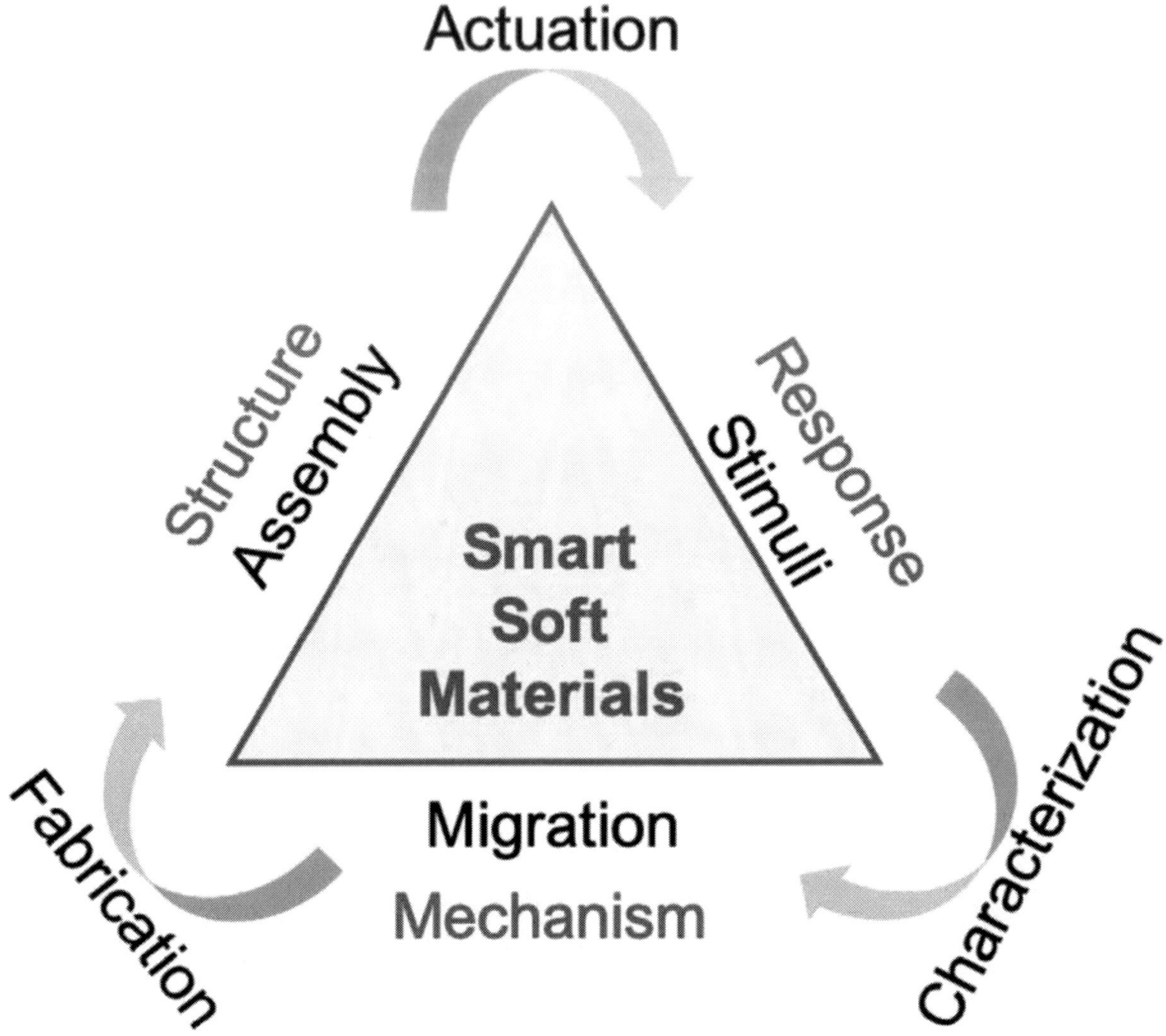

- Abstract Text

582167 Tailoring Organic Materials for Electronics and Energy: From Molecular Design to Two-Dimensional Polymerization

Research Interests:

The primary aim of my research has been to leverage the toolset of chemistry to explore and engineer organic materials for functional electronic devices and energy applications. My research covers two topics. My graduate research focused on the elucidation of structure-property relationships of contorted molecules and their applications in optoelectronics. I have shown that conjugated macrocycles and helical ribbons based on perylene diimides are new types of electronic materials for solar cells and photodetectors. With these molecules, I have managed to achieve several accomplishments that are landmarks in organic optoelectronics, including world record efficiency for non-fullerene solar cells, best narrowband organic photodetectors, etc. During my postdoctoral research, I have developed a general approach to synthesize polycrystalline two-dimensional polymer films with wafer-scale homogeneity in the ultimate limit of monolayer thickness. We call this newly developed interfacial growth method laminar assembly polymerization (LAP). The two-dimensional polymers are fully compatible with large-scale patterning and integration methods, and we demonstrate this by fabricating arrays of hybrid superlattices and electrical capacitors based on two-dimensional polymers and inorganic two-dimensional atomic crystals. The LAP process provides, for the first time, a suite of capabilities necessary for the generation and integration of molecularly designed, wafer-scale thin films, an important step toward the development of artificial solids designed at the molecular level.

In future research, I will explore design principles and synthetic approaches of two-dimensional nanoporous polymers for energy- and environment-related applications. Particularly, I am interested to develop atomically thin nanoporous polymer membranes for water desalination/purification, osmotic power generation, and selective ion-detection.

Teaching Interests:

For undergraduate teaching, topics of *General Chemistry, Thermodynamics and Kinetics*, as well as *Transport Phenomena* fit perfectly with both my interest and my background in such areas. In the lectures, I endeavor to help the students to establish an essential understanding of the big pictures of the topics, for example, interconnections between subjects, the hierarchical structures, methodologies and the development history of our course. For graduate students, my teaching interests lie in the courses of *Materials Chemistry* and *Nanoscience and Nanotechnology*. My course serves to develop their divergent thinking, catalyzing their ideas to emerge at the interfaces of different research areas. I will encourage their critical thinking and expose them to the frontier questions in the fields.

Selected Publications

[1] **Zhong, Y.**†; Chen, B.†; Park, C.; Ray, A.; Brown, S.; Mujid, F.; Lee, J.; Zhou, H.; Suh, J.; Lee, K.; Mannix, A.; Kang, K.; Sibener, S.; Muller, D.; Park, J., "Wafer-Scale Monolayer Two-Dimensional Polymers for Hybrid Superlattices" *Under Review*. (†Equal contribution)

[2] **Zhong, Y.**†; Sisto, T. J.†; Zhang, B.; Miyata, K.; Steigerwald, M. L.; Zhu, X.-Y.; Ng, F.; Nuckolls, C., "Helical Nanoribbons for Ultra-Narrowband Photodetectors." *J. Am. Chem. Soc.* **2017,** *139*, 5644. (†Equal contribution)

[3] Sisto, T. J.†; **Zhong, Y.**†; Zhang, B.; Trinh, M. T.; Miyata, K.; Zhong, X.; Zhu, X.; Steigerwald, M. L.; Ng, F.; Nuckolls, C., "Long, Atomically Precise Donor-Acceptor Cove-Edge Nanoribbons as Electron Acceptors." *J. Am. Chem. Soc.* **2017,** *139*, 5648. (†Equal contribution)

[4] Zhang, B.; Trinh, M. T.; Fowler, B.; Ball, M.; Xu, Q.; Ng, F.; Steigerwald, M. L.; Zhu, X. Y.*; Nuckolls, C.*; **Zhong, Y.***, "Rigid, Conjugated Macrocycles for High Performance Organic Photodetectors." *J. Am. Chem. Soc.* **2016,** *138,* 16426. (*Corresponding authors)

[5] Ball, M.†; **Zhong, Y.**†; Fowler, B.; Zhang, B.; Li, P.; Etkin, G.; Paley, D. W.; Decatur, J.; Dalsania, A. K.; Li, H.; Xiao, S.; Ng, F.; Steigerwald, M. L.; Nuckolls, C., "Macrocyclization in the Design of Organic n-Type Electronic Materials." *J. Am. Chem. Soc.* **2016,** *138,* 12861. (†Equal contribution)

[6] **Zhong, Y.**; Trinh, M. T.; Chen, R.; Purdum, P. E.; Khlyabich, P. P.; Sezen, M.; Oh, S.; Zhu, H.; Fowler, B.; Zhang, B.; Wang, W.; Nam, C.-Y.; Sfeir, M. Y.; Black, C.; Steigerwald, M. L.; Loo, Y.-L.; Ng, F.; Zhu, X. Y.; Nuckolls, C. "Molecular Helices as Electron Acceptors in High-Performance Bulk Heterojunction Solar Cells." *Nat. Commun.* **2015,** *6*, 8242.

[7] **Zhong, Y.**; Trinh, M. T.; Chen, R.; Wang, W.; Khlyabich, P. P.; Kumar, B.; Xu, Q.; Nam, C.-Y.; Sfeir, M. Y.; Black, C.; Steigerwald, M. L.; Loo, Y.-L.; Xiao, S.; Ng, F.; Zhu, X. Y.; Nuckolls, C. "Efficient Organic Solar Cells with Helical Perylene Diimide Electron Acceptors." *J. Am. Chem. Soc.* **2014**, *136*, 15215.

[8] **Zhong, Y.**; Kumar, B.; Oh, S.; Trinh, M. T.; Wu, Y.; Elbert, K.; Li, P.; Zhu, X.; Xiao, S.; Ng, F.; Steigerwald, M. L.; Nuckolls, C. "Helical Ribbons for Molecular Electronics." *J. Am. Chem. Soc.* **2014,** *136*, 8122.

582179 Learning from Nature: Rational Design of Multifunctional Hybrid Materials

Research Interests:

One of nature's most powerful tools to combine disparate material properties (e.g., mechanical strength, surface tension, permeability, color, catalytic properties) in a synergistic way is to integrate dissimilar components at the nanometer scale. For example, mother-of-pearl, or nacre—the iridescent material lining the inside of seashells—is made almost entirely of brittle calcium carbonate. Nevertheless, it exhibits excellent toughness and mechanical resilience because the calcium carbonate plates are arranged in a "brick-and-mortar" fashion and glued together by a tiny fraction of biopolymers. Such a meticulous integration leads to a material with properties superior to the sum of the individual properties. Recent advancements in nanotechnology are opening up new ways to approach nature-inspired organic–inorganic hybridization engineering and make novel, multi-functional nanocomposites.[1] However, much remains unknown about how nature can miniaturize hybridization synergistically, without sacrificing properties of the hybrid components, and how we can leverage this information to design nature-inspired nanocomposite hybrids. To account for these unknowns, my research will aim to understand what interaction should be controlled to maximize desired properties by inspiring by nature creatures. I will utilize that idea to solve technical problems and develop practical applications for energy, electrochemistry, electrical, mechanical, chemical, and environment related challenges.

Research Experiences:

My graduate work provided me with an in-depth tool kit for engineering hybridization at the macro- and molecular level that will be invaluable towards designing practical applications. In particular, I developed multifunctional hybrid materials to adapt mechanical flexibility and electrochemical performance synergistically, without sacrificing either properties in a given volume. Flexible batteries with high energy storage and load-bearing mechanical properties are essential for emerging wearable devices and flexible electronics which provide portability, user comfort, and flexibility of device design. Unfortunately, conventional batteries are bulky and rigid, and their components are often brittle. I developed flexible electrode materials for lithium ion batteries by adapting the brick-and-mortar structure of nacre.[2-3] Vanadium pentoxide (V_2O_5) is a promising redox-active material for flexible battery electrodes, but is limited by its mechanical rigidity, low ion and electron conductivity, and severe volumetric changes upon cycling. I addressed those problems by hybridizing V_2O_5 with an amphiphilic copolymer bearing electron- and ion-conducting blocks, poly(3-hexylthiophene)-*block*-poly(ethylene oxide) (P3HT-*b*-PEO), to form a highly flexible battery electrode. The flexible electrode mimics the brick-and-mortar structure of tough seashells, with V_2O_5 layers arranged in parallel and glued together by the copolymer. This structure significantly enhances mechanical flexibility and toughness without sacrificing the electrochemical performance of the batteries. My research revealed how the diblock copolymer creates synergy and why homopolymers do not, suggesting that tailoring intra- and inter-molecular interactions and preventing unfavorable phase separations are critical to combine multiple properties in a synergistic way.[4-6]

As another hybridization, I engineered surface-agonistic stretchable conductive coatings by mimicking the natural layer-by-layer structure of seashells.[7] Stretchable, bendable, and foldable conductive coatings which provide stable functionality while undergoing mechanical deformation are also crucial for wearable electronics and biometric sensors. MXenes are a new family of ultrathin atomic nanosheet materials that are mostly composed of layers of metals like titanium interleaved by carbon and/or nitrogen. Due to their high electrical conductivity (e.g., greater than graphene) and the variety of elements they can be composed of (e.g., compared to graphene containing only carbon), the emerging MXenes are currently booming as novel inorganic nanosystems in applications ranging energy storage and catalysis to biomedical applications. However, it is still extremely difficult to form thin MXene coatings that can withstand extreme mechanical deformation. I replicated the layer-by-layer structure of resilient seashells with negatively charged MXene and a cationic polyelectrolyte to synergistically integrate high conductivity, mechanical robustness, and mechanical flexibility into a composite coating material, and thereby address the main challenges in current MXene composites: loss of functionality under mechanical stress. I revealed the underlying mechanism of electromechanical coupling by using numerical modeling and geometric analysis, which showed reversible microscale gap formation and a corresponding change in electron conduction pathways under bending and stretching. These MXene multilayer coatings can be deposited onto nearly any surface, including fabrics and fibers, regardless of chemistry, topography, or softness, providing the basis for further development and the implementation of MXene-based conductive conformal coatings in a growing number of flexible conductive materials. Also, I developed ultrafast humidity sensors by adapting wood expansion as its moisture content changes.[8] For example, wood is constantly expanding and contracting but wood does not move equally in all directions due to the grain structure. I replicated MXene/polyelectrolyte multilayers vertically stacked to a substrate to achieve a rapid response to dynamic changes in humidity, such as for humidity respiration. I proposed the mechanism that as humidity changes, the reversible insertion of water molecules into the MXene/polyelectrolyte multilayer results in changing their interlayer distance and tunneling resistance between MXene sheets and proved by using numerical modeling and experimental measurements including *in situ* quartz crystal microbalance and *in situ* ellipsometry.

To complete the foundation for my own research program, I cultivated an electron microscopy tomography–quantitative morphometry skill set during my postdoctoral work that I will couple with my hybridization engineering tool kit. Specifically, I developed low-dose electron microscope tomography that enables investigations of beam-sensitive soft materials with heterogenous nanoscale morphologies.[9] As a model soft material, I studied polyamide membranes which are used in water desalination, purification of chemicals and pharmaceuticals, and recycling of catalysts. Polyamide membranes has been used in separation industries for over 30 years, however, it is challenging to develop the predictive design of nanometer thin polyamide membranes for two major reasons: (i) it is difficult to investigate the nonperiodic heterogenous nanostructures that emerge from synthesis and (ii) characterization of the 3D soft material structure at the nanoscale and thus quantitative understanding of their synthesis–structure–functionality relationship has been limited. To address these issues, I developed a 3D electron microscopy imaging-quantitative morphometry platform. More specifically, I took over 60 projections at various tilt angles from −60° to +60° and reconstructed a full 3D view with nanometer resolution (i.e., one 3D pixel is 6.8 Å). Next, I quantified various structural parameters governing film transport properties, such as void volume, local curvature, thickness mapping in 3D, and surface-to-volume ratio from the 3D reconstructed image as single data set. I also demonstrated that reaction conditions (i.e., reactant concentrations, reaction time) can serve as a handle on structure parameters like pore structure, local curvature, thickness, and interconnectivity. To this end, I undertook low-dose electron tomography and quantitative morphometry efforts that provide a means to map out functionally relevant, 3D features of polyamide crumples and to extract a wide variety of transport-related properties. 3D imaging and quantitative characterization of soft matter will bridge the structural organization in space and the functionality of materials.

My future research focuses concern (i) the ageing mechanisms of rechargeable batteries and the development of strategies to prolong their life, (ii) how loss of electromechanical properties of stretchable conductors could be mitigated by mimicking human elbow skin structure, where micro winkles exist and maintain flexibility over 80 years of human life. To address these questions, my research program will merge my expertise in hybridization engineering, electrochemistry, polymer, and electron microscopy tomography-quantitative morphometry. In my future research, I will seek to develop fundamental understandings of the mechano-electrochemical and mechano-electrical response mechanisms and failure modes at both the macroscopic and atomic scale. My foundation in chemical engineering and materials science leaves me well-situated to undertake this research program in interdisciplinary engineering.

Teaching Interests:

In my academic career, I have had the great fortune of being a teaching assistant on multiple occasions, for courses covering chemical engineering thermodynamics (at Texas A&M University), separation processes (at Hanyang University). I also served as laboratory instructor in a course on numerical analysis for chemical engineers (at Texas A&M University). In the process, I was able to experience first-hand how an instructor's passion could motivate and engage young, bright minds. I learned teaching know-how to provide additional support to students to tackle challenging subject matter and to improve lecture quality. For example, I will start my classes with a 5-minute overview of the previous and current lecture subjects to make sure students know exactly what they are learning and doing for the day and I will write my lecture notes by hand as class proceeds, to pace delivery of material, provide peer-teaching opportunities, and use small group collaboration. In addition to my teaching assistant experience, I developed my teaching skills through ENGR681: Preparing Future Faculty-Professional Development at Texas A&M University. In particular, I learned the importance of balance between micro- and macro-management to give students room to be autonomous and have an opportunity to improve themselves with their own knowledge.

Beyond the classroom, I have had the privilege of mentoring six undergraduates, two masters students and two PhD students in laboratories at Texas A&M University and Hanyang University. More than half of them produced peer-reviewed publications and patents under my mentorship that are rewarding experience while pursing career in academia. I am interested and qualified in teaching most of the major courses for Chemical Engineering including Thermodynamics, Separation Processes, Transport Processes, Reaction Engineering, Physical/Organic Chemistry, Polymer Chemistry, Electrochemistry, Numerical Methods, and Advanced Mathematics for chemical engineers at both undergraduate and graduate level. Additionally, I am interested in developing special interdisciplinary topic courses at the graduate level on new research topics (e.g., Quantitative Data Analysis, Polymer-Inorganic Hybrid Materials).

Selected Publications:

[1] M. Morris, **H. An,** J. L. Lutkenhaus, T. H. Epps III, "Harnessing the power of plastics: nanostructured polymer systems in lithium-ion batteries" ***ACS Energy Letters*** **2017**, 2(8), 1919-1936.

[2] **H. An,** J. Mike, K. Smith, L. Swank, Y. Lin, S. Pesek, R. Verduzco, J. L. Lutkenhaus, "Highly flexible self-assembled V_2O_5 cathodes enabled by conducting diblock copolymers" ***Scientific Reports*** **2015**, 5, 14166.

[3] **H. An,** X. Li, C. Chalker, M. Stracke, R. Verduzco, J. L. Lutkenhaus, "Conducting block copolymer binders for carbon-free hybrid vanadium pentoxide cathodes with enhanced performance" ***ACS Applied Materials & Interfaces*** **2016**, 8 (42), 28585.

[4] **H. An,** X. Li, K. A. Smith, Y. Zhang, R. Verduzco, J. L. Lutkenhaus, "Regioregularity and molecular weight effects in redox active poly(3-hexylthiophene)-*block*-poly(ethylene oxide) electrode binders" ***ACS Applied Energy Materials*** **2018**, 1(11), 5919.

[5] X. Li, **H. An,** J. Strzalka, J. L. Lutkenhaus, R. Verduzco, "Self-doped conjugated polymeric binders improve the capacity and mechanical properties of V_2O_5 cathodes" ***Polymers*** **2019,** 11(4), 589.

[6] K. Sarang, A. Miranda, **H. An,** E. Oh, R. Verduzco, J. L. Lutkenhaus, "Poly(fluorene-*alt*-naphthalene diimide) as n-type polymer electrodes for energy storage" ***ACS Applied Polymer Materials*** **2019**, 1(5), 1155.

[7] **H. An,** T. Habib, S. Sha, H. Gao, M. Radovic, M. J. Green, J. L. Lutkenhaus, "Surface-agonistic highly stretchable and bendable conductive MXene multilayers" ***Science Advances*** **2018**, 4(3), eaaq0118.

[8] **H. An,** T. Habib, S. Sha, H. Gao, A. Patel, M. Radovic, M. J. Green, J. L. Lutkenhaus, "Water sorption in MXene-polyelectrolyte multilayers for ultrafast humidity sensing" ***ACS Applied Nano Materials*** **2019** 2 (2), 948–955.

[9] X. Song, J. W. Smith, J. Kim, N. J. Zaluzec, W. Chen, **H. An,** J. M. Dennison, D. G. Cahill, M. A. Kulzick, Q. Chen, "Unraveling the morphology–function relationships of polyamide membranes using quantitative electron tomography" ***ACS Applied Materials & Interfaces*** **2019**, 11 (8), 8517–8526.

[10] C. Chalker*, **H. An*,** J. Zavala, A. Parija, S. Banerjee, J. L. Lutkenhaus, J. Batteas, "Fabrication and

electrochemical performance of structured mesoscale open shell V_2O_5 networks" ***Langmuir*** **2017**, 33(24), 5975-5981. (*co-first)

[11] **H. An*,** D. Song*, J. Lee, E. Kang, J. Jaworski, J. Kim, Y. Kang, "Promotion of strongly anchored dyes on the surface of titania by tetraethyl orthosilicate treatment for enhanced solar cell performance" ***Journal of Materials Chemistry A*** **2014**, 2, 2250. (*co-first)

[12] D. Song*, **H. An*,** J. Lee, J. Lee, H. Choi, I. Park, J. Kim, Y. Kang, "Densely packed siloxane barrier for blocking electron recombination in dye-sensitized solar cells" ***ACS Applied Materials & Interfaces*** **2014**, 6(15), 12422. (*co-first)

[13] Lee, H. Chang, **H. An,** S. Ahn, J. Shim, J. Kim, "A protective layer approach to solvatochromic sensors" ***Nature Communications*** **2013**, 4, 2461.

581402 Rational Design of Advanced Materials and Membranes for Energy, Separations and Catalytic Applications

Research Interests:

Through my academic training in different universities and institutes, I have acquired a unique perspective in understanding and developing cutting-edge materials, and I have been exposed to various research areas that span chemical engineering, materials engineering, inorganic chemistry, polymer science, membrane technology, and interface science. Since I began my PhD studies, I got inspired by Richard Feynman and his talk '*There's plenty of room at the bottom*' and believed that there are plenty of things we can do at the molecular level. My research interests focus on the rational design and synthesis of unconventional functional porous materials and membranes, providing specific fundamental insights into their molecular engineering, and exploration of their applications in energy, separations, and catalysis. I plan to harness my unique expertise in materials engineering to establish a research program centered on functional porous materials (hybrid inorganic-organic materials) and composite membranes.

Functional porous materials such as metal-organic frameworks (MOFs), covalent organic frameworks (COFs), and porous organic cages have revolutionized the field of materials science. Their "made-to-order" design leads to potentially limitless applications determined by their pore size, geometry and chemical functionality. My initial work in porous materials will address two specific aims: (1) Development and synthesis of a new class of anionic porous materials for applications in ion-conductivity and energy storage; (2) Development of hierarchal multishell core-shell nanocomposite materials for applications in separations and catalysis.

Composite membranes are often used to address the need for bulk performance or properties that cannot be achieved using single-component membranes such as polymeric membranes. Using several components in the membrane fabrication can lead to membranes with multiple properties, including optimal mechanical properties, high permeability, and superb selectivity. The control over the properties of the composite membranes makes them promising candidates that can significantly advance chemical process technology by replacing decades-old distillation and absorption-based processes. The initial work on membranes by my research group will involve: (1) Development of multicomponent membranes for separation applications; (2) Design and synthesis of a new class of metallopolymers as a promising family of functional soft materials.

The fundamental knowledge from each project will be leveraged to develop functional materials and membranes that exhibit major improvements in performance over existing state-of-the-art materials and membranes, and potentially revolutionizing the current technologies from energy efficient separation processes in the gas and petroleum industries, industrial catalytic reactions, to next-generation batteries based on earth-abundant chemistries.

In addition to the above directions, due to the interdisciplinary nature of my research work, I look forward to establishing a strong and active collaboration with fellow faculty in science and engineering to explore new opportunities at the interface of chemical engineering, chemistry, biology physics and materials science.

Research Experience

Visiting Scientist "Rational Design and Synthesis of Novel Materials and Membranes for Ionic Conductivity and Gas Separation"

Advisor: Prof. Jeffrey R. University of California, Berkeley, CA

Postdoctoral Researcher "Mixed-Matrix Membranes and Nanocomposite Materials for Post-Combustion Carbon Capture"

Advisor: Dr. David Hopkinson, National Energy Technology Laboratory, Pittsburgh, PA

Visiting Assistant Professor "Metal-organic Frameworks and Magnetic Nanocomposites for the Selective Removal of Radioactive Gases and Ions from Nuclear Reprocessing Plants" Alexandria University- Pacific Northwest National Laboratory Joint Research

PI: Praveen Thallapally, Pacific Northwest National Laboratory, Richland, WA

Assistant Professor "Functional Porous Materials Lab" Alexandria University, Egypt

PhD Dissertation: "Crystal Engineering of Functional Metal-Organic Material Platforms for Gas Storage and Separation Applications."

Advisors: Prof. Michael Zaworotko, Prof. Mohamed Eddaoudi and Prof. Brian Space, University of South Florida, Tampa, FL

Teaching Interests:

When educating and training the future generation of engineers, it is crucial to emphasize that each course is designed to improve a student's overall aptitude at solving complex problems. I passionately believe that the inclusion of state-of-the-art research findings in teaching is one of the best ways to foster an engineering mindset, as well as promote and maintain the curiosity and open-mindedness of students to novel ideas. To integrate research into the classroom, I strongly endorse the flipped classroom as a pedagogical teaching model. In this model, I present a complex problem that is related to the course material and the students need to implement this assignment during the class. During that time I provide them with immediate feedback. They work in groups, exploring solutions and preparing materials for a classroom presentation. Students can not only conquer their anxiety and confusion but even become enthusiastic about their ideas for solving problems, implementing independent research, and designing their own coursework.

I have had extensive opportunities to teach courses on diverse topics. I served as Lecturer and Assistant Professor for a number of undergraduate and graduate courses of Special Chemistry Program at Alexandria University and mentored two MS students and one PhD student. This program has an innovative curriculum and research at the interface of Chemistry, Chemical Engineering, Physics, Biology and Applied Science. I have taught the courses of Thermodynamics, Chemical Kinetics, Surface Chemistry and Catalysis, Instrumental Analysis, and Analytical Chemistry for undergraduate students and Statistical Thermodynamics and Mass Transfer in Electrochemistry for graduate students.

During my PhD, I worked as Teaching Assistant and taught General Chemistry, Organic Chemistry and Inorganic Chemistry, in addition to the Advanced Inorganic Chemistry Course for 3 years which includes intensive information on the fundamentals of Inorganic Chemistry and Materials Chemistry and the State-of-the-art research on Advanced Materials. Also, I mentored nine undergraduate students, two of them presented in the Castle Student Conference under my supervision and one got co-authored a scientific paper with me. During my postdoctoral experience, I enjoyed mentoring several undergraduates, graduates, and post-docs and seeing their success that was exciting and amazingly rewarding to me. I look forward to continuing this mentorship with my future graduate students and post-docs.

Because of my interest on developing and improving problem-solving approaches, I am most interested in teaching foundational Chemical Engineering courses such as Mass and Energy balances, Thermodynamics, Chemical Reaction Engineering, Separations, Transport Phenomena, and Reaction Kinetics. Given my research expertise, I would also be interested in developing and teaching courses related to polymer science, heterogeneous catalysis, porous materials and materials engineering.

Selected Publications: (24 total + 3 under preparation, 15 first author, 4 corresponding author, 592 citations, h-index14)

Elsaidi, S. K.; Mohamed, M. H.; Banerjee, D.; Thallapally, P. K. Flexibility in Metal-Organic Frameworks: A Fundamental Understanding. *Coordination Chemistry Reviews* **2018**, *358*, 125-152.

Elsaidi, S.; Sinnwell, M.; Banerjee, D.; Devaraj, A.; Kukkadapu, R.; Droubay, T.; Nie, Z.; Kovarik, L.; Manandhar, S.; Nandasiri, M.; McGrail, B.; Thallapally, P.; Vijayakumar, M. Reduced Magnetism in Core-Shell Magnetite@MOF Composites. *Nano Letters* **2017**, *17*, 6968-6973.

Elsaidi, S. K.; Mohamed, M. H.; Simon, C. M.; Braun, E.; Pham, T.; Forrest, K.; Xu, W.; Space, B.; Zaworotko, M. J.; Thallapally, P. K. Effect of Ring Rotation upon Gas Adsorption in SIFSIX-3- M (M = Fe, Ni) Pillared Square Grid Networks. *Chemical Science* **2017**, *8*, 2373-2380.

Zong, Z.; **Elsaidi, S. K.**; Thallapally, P. K.; Carreon, M. A. Highly Permeable AlPO-18 Membranes for N_2/CH_4 Separation. *Industrial & Engineering Chemistry Research* 2017, *56*, 4113-4118.

Mohamed, M. H.; **Elsaidi, S. K.*** Pham, T.; Forrest, K.; Schaef, H.; Hogan, A.; Wojtas, L.; Space, B.; Zaworotko, M. J.; Thallapally, P. K. Hybrid Ultramicroporous Materials for Selective Xe Adsorption and Separation. *Angewandte Chemie International Edition* **2016,** *55*, 8285

Elsaidi, S. K.; Mohamed, M. H.; Loring, J.S.; McGrail, B. P.; Thallapally, P. K. Covalent Coordination Frameworks: A New Route for Synthesis and Expansion of Functional Porous Materials. *ACS Applied Materials & Interfaces* **2016**, *8*, 28424.

Elsaidi, S. K.; Mohamed, M. H.; Wojtas, L.; Chanthapally, A.; Pham, T.; Space, B.; Vittal, J. J.; Zaworotko, M. J, Putting the squeeze on CH_4 and CO_2 through control over interpenetration in diamondoid nets. *Journal of the American Chemical Society*, 2014, *136*, 5072.

Patents: (1 published, 1 submitted, 2 under preparation)

Michael J. Zaworotko, Mona H. Mohamed, and **Sameh K. Elsaidi**, Metal-Organic Materials (MOMs) for Adsorption of Polarizable Gases and Methods of Using MOMs, US 9676807.

581356 Computational Assembly Engineering for Bio-Inspired Nanomaterials

Research Interests: My objective is to establish a world-class research program that stands at the intersection of biophysics, chemistry, computer science and data science, aimed at transformative approaches for engineering novel nanomaterials and nanodevices for medical diagnosis, drug delivery and renewable energy applications. Specifically, I am interested in exploring design rules for materials that resemble biological matter in that they are responsive to external stimuli in a controlled matter. The design rules being targeted here include not only the geometry of assembling building blocks (such as macromolecules and colloids) and the anisotropic interactions between them, but also the assembly pathways towards desirable structures and properties away from equilibrium. The fundamental challenges to the proposed approaches are that the interactions between the assembling building blocks necessarily span multiple energy and length scales. Furthermore, I am interested in nonequilibrium assembly processes occurring at the liquid-liquid interfaces and in the presence of external fields, where available theoretical predictions have been limited, yet experimental studies can be used for verification.

To address the fundamental challenges associated with the proposed approaches, I am employing multiscale modeling and large-scale computer simulation, and developing techniques borrowed from multiple disciplines. These techniques include GPU-accelerated force fields and enhanced sampling methods for molecular simulation, large deviation theory for studies of nonequilibrium steady states, and machine learning techniques for analysis of big data sets. Particularly, my near-term interest focuses on predictive design of charged copolymers for programmable nanomaterials that are used for protein delivery and energy storage applications. The outcomes of my research program would be qualitative and quantitative predictions including, but not limited to, 1) optimal polyelectrolyte architectures and compositions, and 2) efficient routes towards desirable chemical and mechanical properties of the materials, to facilitate future experimental and theoretical investigations in the field. My extensive expertise in soft matter physics, physical modeling of polyelectrolytes, proteins and liquid-liquid interfaces, and high performance computing allows me to pursue the proposed research program successfully. Furthermore, my long-time experience with algorithm and method development for large-scale open source software packages such as LAMMPS, SSAGES and HOOMD-Blue enables me develop and optimize all the codes necessary for the studies of interest.

Teaching Interests: I am interested in teaching Chemical Engineering core courses including Fluid Flow, Process Dynamics and Control, Thermodynamics and Transport. I am also developing my own courses on Polymer Physics, Intro to Machine Learning for Chemical Engineers, and Computational Nanoscience at graduate level. I would like to facilitate cooperative learning in classroom and incorporate practical aspects with fundamental problems into lecture materials. My objectives are to explain to the students why the coursework benefit them with their career, and more importantly, to inspire them to go further with the topics covered in my courses.

582599 Hierarchically Engineered Structures Using Self- and Directed- Assembly for Unique Electronic, Optical and Mechanical Properties

The properties of materials are largely determined by their formed structures. The understanding of the structure and property relationship in multilength scales is crucial in developing the next generation functional materials. The manipulation of the nanomaterial assembly could directly control their conductivity, color, and strength. My previous research experience is focused on the assembly of polymer and colloidal nanoparticles. External electric and acoustic fields are demonstrated to guide conductive polymers to grow into aligned structures or fibers with high aspect ratio. The charge transport is thus enhanced. By carefully engineering the solvent-solvent and solvent-solute interactions, a ternary system can be designed to not only achieve desired structures, but also obtain stimuli-responsive materials with on-demand tunability of properties.

Research Interests:

Structure-property relationship has always been an important research topic to understand, advance, and discover materials. The development of a broad range of fields is closely relying on well-engineered structures, including electronics, optics, catalysis, energy storage, and biomaterials. Scattering-based techniques offer a versatile platform for structural characterizations. In particular, scattering-based techniques provide quantitative analysis of samples averaged over the total volume. Moreover, they also allow time-resolved characterization and capture in-situ structural changes under varied external stimuli, such as electric field, acoustic field, and temperature gradient. In conjugation with cutting-edge imaging techniques, such as electron microscopy and confocal microscopy, a complete view of the multi-length scales structures can be unveiled and quantified. For example, besides the size and shape of crystallized polymer can be described with X-ray and neutron scattering experiments, the amount of polymer that forms crystallites can be determined. More importantly, the formation kinetics and intermediate structures can also be quantitatively compared under different external fields. This is especially important to unravel the mechanism of different physical phenomena at each stage. My research interests are to utilize the powerful scattering-based structure characterization toolbox to understand and control the structures of the assembly process in order to achieve unique functions.

Successful Proposals:

Experimental proposals that have been granted beamtimes from synchrotron- and neutron-based national labs, including 8 proposals to NIST Center for Neutron Research (NCNR), 2 proposals to Argonne National Laboratory, 2 proposals to Brookhaven National Laboratory, and 1 proposal to Oak Ridge National Laboratory (ORNL).

Postdoctoral Project:

"*Porous materials under extreme conditions*"

Under supervision of Yun Liu,

1. NIST Center for Neutron Research
2. Department of Chemical & Biomolecular Engineering, University of Delaware

PhD Dissertation:

"*Structure Engineering of Self- and Directed-Assembled Conjugated Polymers*"

Under supervision of Lilo D. Pozzo, Chemical Engineering, University of Washington

Committee members: Christine K. Luscombe, Stuart Adler, and Lih Lin

Research Experience:

My research experience is mainly focused on interdisciplinary fields involving materials and engineering. My formal training was in colloidal science with an emphasis of structural characterization using scattering techniques. In order to achieve different functions, the assembly of materials are controlled to realize specific structures. I have learned the fundamental physics behind the assembly of polymers and nanomaterials, which can be used to achieve unique functions with hierarchically engineered structures. I acquired the knowledge and invaluable experience by conducting experiments by closely working with instrument scientists in national labs. At the same time, the purpose of engineering structures is to obtain unique functions. The projects that I worked on always involved collaborations and closely related fundamental science to applications. Examples include photovoltaics, transistors, sensors, optoelectronics, batteries and biomimetic materials.

Teaching Interests:

I also had extensive teaching experience related to colloidal science and chemical engineering. I spent two semesters teaching colloids and interfacial science laboratory and two semesters teaching unit operation laboratory. Aside from those, I also mentored graduate and undergraduate students in my research group. While working as postdoctoral researcher at National Institute of Standards and Technology, I also participated in mentoring an undergraduate researcher in a collaborative research project.

Future Direction:

As a faculty, I would like to continue applying X-ray/Neutron scattering techniques as a tool to guide the designing of the material assembly in multi-length scales with unique functions. Specifically, I would like to continue pursuing the direction of achieving functional and stimuli-responsive materials with well-controlled structures. I would combine the knowledge I acquired during my Ph.D and postdoctoral research: polymer physics and liquid critical phenomena. The research will be an interdisciplinary field with a focus of answering fundamental mechanistic questions in soft materials and colloidal science. It will also seize the applications in different fields, such as optics, energy storage and thermoelectrics, which would foster extensive collaborations with groups in different fields. On the one hand, I would like to explore fundamental physics that could be responsible to assemble polymers and nanomaterials. On the other hand, I will utilize the knowledge I obtained to enhance and realize new properties.

In addition, the working experience in different national labs will be maintained and passed on to my group. The research in my group will be utilizing the world class scattering instruments and working closely with instrumental scientist in national labs to address questions in a blend of disciplines.

Selected Publications:

Yuyin Xi, Caitlyn M. Wolf, Lilo D. Pozzo, "Self-assembly of Donor-acceptor Conjugated Polymers Induced by Miscible Poor Solvents", *Soft Matter*, **15**, 1799, (2019).

Yuyin Xi, David S. Li, Greg M Newbloom, Wesley K Tatum, Matthew O'Donnell, Christine K. Luscombe, Lilo D Pozzo, "Sonocrystallization of Conjugated Polymers with Ultrasound Fields" Soft Matter, 14, 4963, (2018).

Yuyin Xi, David S. Li, Yi-Ting Lee, and Lilo D. Pozzo, "In-Situ Ultrasound sample environment for small angle scattering studies" 2018 NCNR Highlight (2018).

Yuyin Xi, Lilo D. Pozzo, "Electric Field Directed Formation of Aligned Conjugated Polymer Fibers" *Soft Matter*, **13**, 21, 3894, (2017).

Yuyin Xi, Yueh-Ling Hsieh, Ya-Hsi Hwang, Shun Li, Fan Ren, Stephen J Pearton, Erin Patrick, Mark E Law, Gwangseok Yang, Hong-Yeol Kim, Jihyun Kim, Albert G Baca, Andrew A Allerman, Carlos A Sanchez "Effect of 5 MeV proton radiation on DC performance and reliability of circular-shaped AlGaN/GaN high electron mobility transistors" *Journal of Vacuum Science & Technology B* **32,**

012201(2014).

Yuyin Xi, Lu Liu, Fan Ren, Stephen J Pearton, Jihyun Kim, Amir Dabiran, Peter P Chow, "Methane detection using Pt-gated AlGaN/GaN high electron mobility transistor based Schottky diodes" *Journal of Vacuum Science & Technology B* **31**, 032203(2013).

Yuyin Xi, Lu Liu, Ya-Hsi Hwang, Oluwadamilola Phillips, Fan Ren, Stephen J Pearton, Jihyun Kim, Chien-Hsing Hsu, Chien-Fong Lo, Jerry Wayne Johnson, "Study of hydrogen detection response time at room temperature with Pt-gated diodes fabricated on AlGaN/GaN hetero-structure" *Journal of Vacuum Science & Technology B* **31**, 032202(2013).

Jiangsheng Yu, **Yuyin Xi**, Chu-Chen Chueh, Jing-Qi Xu, Hongliang Zhong, Francis Lin, Sae Byeok Jo, Lilo D Pozzo, Weihua Tang, Alex K-Y Jen, "Boosting Performance of Inverted Organic Solar Cells by Using a Planar Coronene based Electron-transporting Layer" *Nano Energy*, **39**, 454, (2017).

581984 3D Printing of Polylactic Acid Mixed with Wood

Research Interests: Renewable Fuels, Renewable Materials

Teaching Interests: Mass and energy balances, chemical reaction engineering, biochemical and bioprocess engineering, thermodynamics

Abstract:

Polylactic acid (PLA) is a popular choice of plastic material for 3D printing. Incorporation of lignocellulosic biomass in this biodegradable plastic can increase utilization of waste plant material and manufacturing of cheaper 3D-printed biocomposites. However, mixing this hydrophobic plastic with a hydrophilic biomass material presents engineering problems in 3D printing. In this work, we show a new technology for making quality PLA-poplar hardwood filaments and their printing by fused deposition modeling (FDM) printing process. While a twin-screw extruder allows better mixing of fillers in plastics than a single-screw extruder, homogeneous biomass-filled PLA composites are difficult to produce even with a twin-screw extruder. However, with this technology, homogeneous filaments were made through a small and popular single-screw extruder and used for printing into shapes suitable for ASTM mechanical properties testing methods. Tensile, compression and impact testing, molecular weight, calorimetry, pycnometry, light and electron microscopy, infrared and NMR spectroscopy were carried out to characterize the composites. Overall, this work shows how biomass content, particle size, filament extrusion temperature and speed, and FDM parameters affect properties of 3D-printed materials made of PLA and wood.

557735 Advance Manufacturing of Materials for Flexible and Stretchable Devices

Electrochemical analysis of sweat is an ideal scheme for uninterruptedly tracing a person's physiological condition, however, developing devices for this is challenging. A wearable sweat

monitor that measures several biomarkers is an advance technology that connect smart electronic to point of care detection of biomolecules. Here, we came up with an idea where we can use advance manufacturing (3D printed as well as microfluidic liquid metal microchannels) and stretchable elastomer polydimethylsiloxane (PDMS) along with graphene as components of a flexible and stretchable plastic transistor that can analysis sweat in real-time. The liquid metals contacts encapsulated in PDMS serve as source, drain and gate electrode of a transistor. The purpose of a semiconductor is served by a thin layer of spray-painted graphene oxide on gate electrode. The 3D printed microchannels on the graphene film act as a reaction chamber for antigen–antibody interactions and provide an easy path for fast electron transport toward the electrode from the electrolyte, resulting in improved electrochemical performance of the sweat sensor. This wearable transistor sensor can also detect other biomarkers, such as, glucose, cholesterol and breast cancer which would be an economic and easy alternative to the existing senor devices.

577736 Bio-Inspired Soft Electronics

Research Interests: Soft electronics enable integration with biological systems for transformative applications in wearable health, prosthetic devices, and soft robotics. The three key functionalities of soft systems are sensing, information processing, and actuation. Biological mechanoreceptor networks collect sophisticated tactile information with minimal energy consumption. Mimicking this efficiency using neuromorphic sensor networks based on oscillating systems can provide efficient feedback mechanisms for soft electronics. Information processing in biological systems is accomplished using neurons that are connected with synapses. A key aspect of biological systems is that each neuron can be connected with up to 1000 other neurons through a 3D arrangement of synapses. Using 3D printing technologies such as electrohydrodynamic printing can enable the production of active electronic devices with a large number of interconnections. Soft actuators that can mimic the capabilities of biological muscles require the ability to spatially pattern actuation directions with a large number of degrees of freedom. 3D printing enables the design of sophisticated electrostatic actuators and the integration with sensing and information processing processing capabilities in fully integrated soft systems.

Teaching Interests: My undergrad in Nanotechnology Engineering and my PhD in Materials Science has provided me with a strong background in materials properties and materials processing, including electronic properties of materials, mechanical properties of materials, polymer physics, and rheology. My research has focused on applications of such as prosthetic devices and wearable haptics, and I would be very interested in developing new course content in cutting-edge fields such as soft electronics, emerging electronic devices, and bio-integrated electronics.

582128 Scalable Synthesis of Nano- and Micro-Scale Materials for Energy Applications

Research Interests:

In the past few decades, the exploration of nano- and micro-scale materials has led to myriad of technological advances, especially in the energy sector. These materials, particularly at the nano-scale, have unique mechanical, optical, and electrical properties that lead to exemplary performance in applications such as energy harvesting and storage, fuel production, and drinking water production using solar steam. However, the ever-growing demand for energy results in non-renewable sources of energy being widely used as the main generation source. The challenge of meeting the increasing energy demand in a sustainable and environmentally friendly way is one of the most important problems facing society. While many high-performance materials have been synthesized and reported in the literature, the synthesis process is generally expensive and low-yield, with minimal scale-up opportunities.

My research interests are in development of scalable processes for synthesizing materials at the nano- and micro-scales, particularly for use in energy storage and harvesting. Through proper process design, materials with unique morphologies (spherical, hollow, wires, columns, etc), low dimensionality (0-dimensional through 3-dimensional), and tailored composition (doped materials, solid solutions, layered compositions, etc). I aim to 1) design scalable synthesis processes for nano- and micro-scale materials, 2) investigate the effect of process parameters on fundamental material properties, 3) investigate the performance of the materials in energy storage and harvesting applications, and 4) create pilot-scale reactors for proof-of-scalability.

In this poster, I will present a variety of processes, including aerosol chemical vapor deposition, electrospray, and furnace aerosol reactors, that I have used for the synthesis of various materials, as well as characterization and performance of the synthesized materials.

My PhD work has focused on the study of aerosol reactors for the synthesis of nano- and micro-scale materials. Aerosol-based reactors provide unique opportunities to control and tailor properties of the final materials using easily controlled system parameters. For example, aerosol chemical vapor deposition (ACVD) allows for the deposition of structured metal oxide thin films by controlling reactor temperature and precursor flow rates. Additionally, doped-metal oxide thin films may be synthesized by simply introducing two precursors simultaneously. During my graduate work I have used ACVD to synthesize SnO_2, TiO_2, and Cu-doped TiO_2 films for use as battery electrodes in lithium-ion battery coin cells. Electrospray can be used to form mono-dispersed nano-droplets, which I have used to deposit a metal-organic framework coating material. Finally, furnace aerosol reactors can be used to synthesize particles ranging from the nano- to micro-scales by controlling precursor concentrations. Additionally, hollow particles can be synthesized using knowledge of the reactions taking place and tailoring system conditions. My graduate work has focused on the application of materials synthesized using these processes in energy storage, in the form of lithium-ion batteries and lithium-sulfur batteries.

My background has prepared me to work in the interdisciplinary research area between materials science and chemical engineering to develop scalable processes for material synthesis to address the pressing needs for new energy technologies.

Teaching Interests:

Teaching is about more than presenting information to the students, it is about helping the students find which aspect of the field they are most passionate about and where they can contribute to the field. Through my experience as a teaching assistant and research mentor I learned how beneficial these roles were to my own understanding of the course work or research area.

During my time as a graduate student I have been a teaching assistant for undergraduate Engineering

Analysis of Chemical Systems, Mass Transfer Operations, and Process Control Lab. Through these experiences I have covered the range of teaching new chemical engineering students the basics of mass and energy balances to helping senior students design their experiments in Process Control Lab. Additionally, in the research setting I have mentored four undergraduate students and one master's student, three of which have published papers with me. Working with students with various levels of experience in research has shown me the importance of effective communication is and has taught me how to best organize my own efforts and those of other researchers. I feel prepared to teach any of the undergraduate chemical engineering courses and would be particularly interested in teaching or developing a course on the place of chemical engineering in the energy field and renewable energy.

582530 Molecular Engineering Approach for the Development of Advanced Functional Materials: Engaging Students Through Impactful Research

Research Interests:

Polymeric materials are ubiquitous in our society due to their relatively low cost and large parameter space that can modify their physical and chemical properties for a range of purposes. As synthetic techniques improve the ability to control the design, sequence, and functionality of polymers, their breadth of applications will only continue to grow and become more specialized. Looking to the future, one of the key interests for this field is to utilize more sustainable materials to reduce the reliance on oil-sourced feed stocks, and to improve reusability to address the growing volume of plastic wastes worldwide. My research interests broadly focus on the development and optimization of polymeric systems for engineering applications. This includes: (1) sustainable polyacetal-based chemistry and reaction kinetics. (2) Dynamic bond chemistry and stimuli responsive materials. (3) The use of light as a trigger to initiate chemical changes or modify physical properties of polymeric systems. My approach to materials research emphasizes understanding and applying fundamental principles of chemistry and chemical engineering to guide molecular design for engineered systems.

My previous research experience has focused on the synthesis and engineering of stimuli-responsive polymers for applications in semiconductor manufacturing and transient technology. In particular, I have helped expand the breadth of carbonyl-based addition polymerizations to design polymers with the ability to rapidly degrade in the solid-state back to their constituent small molecule monomers. My understanding of the thermodynamic and kinetic aspects of these materials have led to me aiding chemical companies in scale-up and production of kilogram quantities of these polymers for defense applications in transient technology. I am intimately involved with several industry-collaborated projects in engineering devices and new manufacturing processes with these degradable materials.

Based on my expertise, I identified three research areas where I will have a high likelihood of success: (1) Further expanding the scope of carbonyl-based polymerization chemistries through strategic monomer and catalyst design. This includes exploring anionic and Lewis pair polymerization catalyst systems that will provide greater depth and control over monomer reactivity. (2) Combining the chemical property advantages of fluorinated polymers with the stimuli-responsive properties of polyacetals for advanced functional materials. (3) Taking advantage of dynamic acetal-based chemistries to develop materials systems that can adapt to external stresses, such as stress relaxation, self-healing, or recyclability.

The goal of this research is to advance both the fundamental understanding of polymer chemistry and physics, and applying this knowledge to better design materials systems for engineering applications. My group will utilize interdisciplinary skills to carry out significant results in both research and industry relevant areas.

Teaching Interests:

I am very enthusiastic about mentoring young researchers and teaching higher education. I first found my passion for teaching as an undergraduate teaching assistant for general chemistry under the tutelage of Dr. David Vanden Bout. I got to experience teaching using a flipped-classroom approach for class size of 300 students in a challenging subject. I enjoyed the experience so much that I spent 3 years in this position, where I became a leader with responsibilities of instructing and mentoring younger teaching assistants. I continued this passion for teaching to graduate school, where I have planned and taught over a dozen lectures for a graduate/undergraduate elective class on The Science and Engineering of Microelectronics Fabrication. The past two semesters I have also been organizing a Polymer Comunications seminar class that brings together graduate students from across departments that are interetsed in polymer research. The class brings in visiting scientists, industry representatives and graduate students speakers to discuss various aspects of polymer research. I feel very confident in teaching or implementing courses in microelectronics fabrication, kinetics and reaction engineering, polymer chemistry, and polymer science and engineering.

Mentoring young scientists and researchers has been an incredibly rewarding experience for me in my graduate career. I have had the joy of mentoring 11 undergraduate researchers in my first three years as a PhD student. In addition to a multitude of awards these mentees have won, four of these individuals were very talented and I helped aid them in pursuing their own graduate school careers at well-respected research institutions. I am truly excited about the prospect of mentoring graduate students in my future lab and aiding them to reach their career goals in science and engineering.

584437 Polymer Science in Developing Fibrous materials for Advanced Technical Applications

Research interests:

Motivation

Maintaining an unbroken period of growth over the past 40 years, the global nonwoven fabrics market accounted for $44.4 billion in 2017 and is expected to reach $98.8 billion by 2026. The market uses polymeric materials including natural fibers (cotton, jute, flax, wool), synthetic fibers (polyester, polypropylene, polyamide, rayon), special fibers (glass, carbon, nanofibers, bicomponent, superabsorbent fibers), etc in spun-bond, wet-laid, dry-laid, and e-spun technologies for end-user applications like construction, textiles, healthcare, automotive, and filtration. Due to increasing demand for nonwovens in existing markets and the new applications and opportunities that are opening up every day, the nonwovens industry continues to be adaptive, creative and relentlessly opportunistic. Rapid innovation in raw material research is driving the market away from carcinogenic glass fibers to synthetic fibers and also, rooted in the environmental impacts, from oil-based synthetics into biopolymers and natural products. In this regard, *I propose to build a polymer science program to address the needs of the nonwovens industry through the development of new products and processes*.

1. **Developing Lofty Staple Fibers**

The functionality needed to launch next-generation synthetic fabrics will likely come from composite materials. Nonwovens from wet-laid synthetic fibers suffer from lack of loftiness. This issue stems mainly from low Young's modulus in these materials. I am interested in *developing a high-yield melt-spinning process to produce staple fibers from highly filled polyester/polypropylene composites with diameters ranging 1-30 microns.* Running such a project will help the students gain theory knowledge and hands-on experience in polymer composites as well as in polymer processing and rheology. Potential funding: from the nowovens industry, Automobile Companies, Performance Fibers Companies, and MEP at NSF.

2. **Electrospinning of 3D Nonwovens for Depth Filtration**

Functional nanomaterials by electrospinning enable self-cleaning, super-hydrophilic, super-hydrophobic, and anti-microbial surfaces. Hence, not only does research on e-spun functional nonwovens train students on interfacial phenomena in fibrous composites, but also it gains industry attention for many applications, which extend to filtration, tissue engineering, scaffold constructions, wound dressings, energy conversion and storage, catalysts and enzyme carriers, protective clothing, sensors, drug delivery, cosmetics, electronic and semi-conductive materials. I am interested in *taking advantage of the fluffiness in 3D nonwovens from e-spinning for depth filtration (as opposed to surface filtration offered by 2D e-spun materials) used in purifying air*. Potential funding: from the air filter industry and MEP at NSF.

3. **Facile Route to Production of Superhydrophilic Membranes for Water Filtration**

Even though technology advances in espun nanofibers have made it practical to use nanofibrous scaffolds as a unique and breakthrough component in separation media for liquid filtration, the technique suffers from using toxic solvents and depending on many process variables. As an alternative technology, I am interested in *developing a spray system using slurries of cellulose nanofibers/chitin/PVOH in water to form superhydrophilic anti-fouling membranes for water filtration*. This will help train students in physics of polymer solutions and suspensions as well as in building and optimizing chemical processes. Potential funding: from Membrane Filtration Companies, B3 Program at USDA, and CBET at NSF.

4. **Functional Fiber Surface via Green Chemistry and Crystalline Architecture**

A current trend of adding value to nonwoven products is witnessed via enhancing function after fabric formation, not through major changes to the nonwoven process itself. Adjusting function at the end of the manufacturing process, by adding functional particles or applying topical finishes, leads to greater flexibility and more diverse surface properties. I am interested in *collaborating with the faculty focused on polymer chemistry to work on eco-friendly natural antibacterial finishes* for nonwoven fabrics based on chitosan/poly (ethylene oxide) blends. I will also work on introducing *super-hydrophobicity to the fibers by adding nano-roughness to the fiber surface*, via *adjusting polymer crystallite morphology* and by *adding nanoparticles* to the surface. Students working on these projects will become experts in polymer physical chemistry. Potential funding: from Fabric Finishing and Coating Companies, Performance Fibers Companies, and CBET at NSF.

Teaching Interests:

During 2011, I was a fulltime teacher of *Chemical Reaction Engineering, Numerical Methods, Unit Operations (+Lab), Aspen Plus Workshop, Heat Transfer (+Lab), and Fluid Mechanics (+Lab).* At UMaine, I was a TA in *Transport Phenomena*. More recently at Penn State, I taught *Polymer Rheology and Processing*. At MIT, I was also a co-lecturer in *Physical Chemistry of Polymers*. In the future, I would welcome teaching these courses, but I would also like to teach *Filtration: Process & Filter Media, Chemical Engineering Thermodynamics* and *Colloidal Science*. I have *published a textbook entitled "Numerical Methods for Process Engineers"* with Shiraz University Press and Publisher De Gruyter. I believe effective classroom teaching empowers the students' ability to think in a critical way and create their own knowledge, and this ultimately benefits the research community. I also think other legitimate purposes of teachers include the need to make a useful contribution to society by helping induct younger human beings into the society's process for thriving.

586096 Advanced Materials and Manufacturing for Electrochemical Methods in Water-Energy Nexus

The grand challenges in water – energy nexus have an opportunity to be solved by energy-efficient electrochemical methods. To realize those novel electrochemical methods, innovations with a multidisciplinary approach is needed. My research aims to leverage my background and expertise at the intersection of advanced materials and manufacturing. Using my background, *I aim to use advanced manufacturing methods to manipulate materials at different scales to create more energy efficient solutions for water and energy*. I will focus on surface and subsurfaces as the critical area for diverse applications in both the fields of water and energy. Some of my future research plans are advanced manufacturing of composite material systems for electrodes used in capacitive deionization for desalinating low-salinity waters, using hierarchical multi-material air-diffusion cathodes for in-situ hydrogen peroxide applied to water disinfection, and advanced iron-electrocoagulation to remove contaminants from wastewater and reduce fouling of membranes by removing dissolved silica. I also aim to find new material design principles for solid state energy storage to reduce defects (cracks) and self-healing properties, paving the way for reliable high-energy density next-generation batteries. Thus, I aim to build a comprehensive research program in the water-energy nexus using advanced materials, manufacturing and surface engineering.

Keywords: materials, manufacturing, surface, water, energy

- Research Interests : Chemo-mechanical coupling effect for strength of materials, sustainability, water-energy nexus, Advanced manufacturing (including additive and hybrid processes)
- Teaching Interests: Materials science for engineers, Strength of Materials, Mechanics of materials, Statics, Manufacturing processes

586603 Multiscale Framework to Engineer Non-Equilibrium Responses at Complex Materials Interfaces

Research Interests:

Materials design is at the forefront of many pioneering technological advancements. One notable direction has been to leverage unique behavior that arises at the interface of dissimilar constituents or phases, e.g., nanomaterials, complex fluids, and soft condensed materials. The additional interactions at these interfaces typically induce molecular restructuring that deviates from the characteristic order found in bulk counterparts. Intriguingly, such subtle microscopic changes tend to propagate to functionally-relevant macroscopic responses. The exploration of this idea is the central theme of my research program.

I seek to investigate fundamental molecular driving forces at materials interfaces that can be leveraged to address global challenges in energy, sustainability, and healthcare. In particular, my research group will focus on systems that are inherently out-of-equilibrium, and elucidate structure-kinetic-function relationships in order to engineer the dynamical behavior at these interfaces. My strategic approach will be to systematically study collective molecular behavior using the multiscale computational expertise that I have developed throughout my graduate and postdoctoral training. The primary benefit of such an approach is the ability to isolate individual factors through creative simulation design, which would otherwise be untenable in experimental systems. Relatedly, methodological developments will be pursued to address problem-specific challenges, as exemplified throughout my research career. My long-term goal is to establish foundational practices for materials development based on non-equilibrium strategies and computationally-guided design. To this end, I will initially focus on the following four programs:

Program 1: Plastic Crystal Electrolytes for Alkali-Metal Batteries

Program 2: Nanoporous Electrode Materials for Capacitive Deionization

Program 3: S-Layer Proteins for Programmable 2D Nanoporous Assemblies

Program 4: Virus-Inspired Nanocarriers for Drug Delivery

Research Background:

Throughout my scientific career, I have been fascinated by the ability of materials to adopt versatile macroscopic responses on the basis of microscopic changes at interfaces. This foundational paradigm has led to fundamental structure-function relationships that have rapidly accelerated technological innovation and represents an important example of fundamental science meeting engineering principles. The study of material responses at interfaces has been a consistent theme throughout my research to date. For example, my graduate work identified new means to influence collective ion and electrode charge reorganization for supercapacitor applications by engineering interfacial chemistry[1-5] and confinement[6-7] using carbon nanomaterials. My postdoctoral work identified mechanistic principles of protein self-assembly, membrane remodeling, and RNA scaffolding that collectively regulate viral morphogenesis.[8-9] An integral aspect of these past studies was the development and use of hierarchical (or multiscale) computer simulations that rigorously bridge insights from quantum mechanical (QM), molecular, and coarse-grained (CG) resolutions to connect with experiments. For instance, representing electrode polarization classically (from QM data) was critical for my work in supercapacitors.[1-7] I have also advanced new methodologies for CG modeling based on both experimental data and statistical mechanics.[8-12] Importantly, the transfer of information between scales allows these models to directly interrogate **the importance of identified microscopic behavior on collective macroscopic properties.** During my research career, I intend to utilize the versatility afforded by my training in these multiscale techniques and continue to advance frontiers in computational methods.

Successful Proposals: Ruth L. Kirschstein National Research Service Award (NIH F32)

Postdoctoral Project: "Multiscale Simulations of HIV-1 Assembly, Budding, and Maturation" under the supervision of Gregory A. Voth, Department of Chemistry, Institute for Biophysical Dynamics, James Franck Institute, The University of Chicago

PhD Dissertation: "First-Principles Investigation of Carbon-Based Nanomaterials for Supercapacitor" under the supervision of Gyeong S. Hwang, McKetta Department of Chemical Engineering, The University of Texas at Austin

Teaching Interests:

Throughout my training, my mentors have been invaluable for my growth as an independent scientist and engineer. As a faculty member, I eagerly look forward to my responsibility as a mentor for the next generation of young scientists. My core teaching philosophy can be described by the following three principles. My foremost goal is to instill a passion for learning by fostering both the curiosity and the confidence to be inquisitive without fearing failure. To this end, I have enjoyed preparing activities that enable students from local schools (from elementary to high school levels) to interact with interesting concepts from my research, e.g., seeing "self-assembly" in action. My second goal is to create tangible connections between what students learn in academic settings and real-world problems. For this reason, I firmly believe in the value of immersive, project-based learning as a supplement to lectures. Not only do these projects translate engineering skills toward problem solving, but also establish communication and collaborative skills. In fact, one of my undergraduate courses followed this model and eventually inspired me to pursue computational research. My third aim is to create an environment for my students to question underlying assumptions and to develop physical intuitions. As a teaching assistant, I have found that one of the best forums for this discourse emerges from thought-experiment challenges through rigorous debate within small student clusters. Finally, I am encouraged by the increasing ubiquity of molecular visualization tools and online educational platforms, and I will strive to translate concepts from my own research into accessible, interactive, and transparent form factors to engage future scientists.

While I am prepared to teach any core chemical engineering course, I am most interested in thermodynamics, statistical mechanics, or numerical methods. I am also looking forward to developing new courses based on my core competencies, such as a broad introduction to the growing field of computational molecular sciences.

Selected Publications:

1. Paek, E.; Pak, A. J.; Kweon, K. E.; Hwang, G. S., On the Origin of the Enhanced Supercapacitor Performance of Nitrogen-Doped Graphene. *J Phys Chem C* **2013,** *117*, 5610-5616.
2. Pak, A. J.; Paek, E.; Hwang, G. S., Tailoring the Performance of Graphene-Based Supercapacitors using Topological Defects: A Theoretical Assessment. *Carbon* **2014,** *68*, 734-741.
3. Pak, A. J.; Paek, E.; Hwang, G. S., Impact of Graphene Edges on Enhancing the Performance of Electrochemical Double Layer Capacitors. *J Phys Chem C* **2014,** *118*, 21770-21777.
4. Paek, E.; Pak, A. J.; Hwang, G. S., Large Capacitance Enhancement Induced by Metal-Doping in Graphene-based Supercapacitors: A First-Principles-Based Assessment. *ACS Appl Mater Interfaces* **2014,** *6*, 12168-12176.
5. Pak, A. J.; Hwang, G. S., On the Importance of Regulating Hydroxyl Coverage on the Basal Plane of Graphene Oxide for Supercapacitors. *ChemElectroChem* **2016,** *3*, 741-748.
6. Pak, A. J.; Hwang, G. S., Charging Rate Dependence of Ion Migration and Stagnation in Ionic-Liquid-Filled Carbon Nanopores. *J Phys Chem C* **2016,** *120*, 24560-24567.
7. Pak, A. J.; Hwang, G. S., Molecular Insights into the Complex Relationship between Capacitance and Pore Morphology in Nanoporous Carbon-based Supercapacitors. *ACS Appl Mater Interfaces* **2016,** *8*, 34659-34667.
8. Pak, A. J.; Grime, J. M. A.; Sengupta, P.; Chen, A. K.; Durumeric, A. E. P.; Srivastava, A.; Yeager, M.; Briggs, J. A. G.; Lippincott-Schwartz, J.; Voth, G. A., Immature HIV-1 Lattice Assembly Dynamics are Regulated by Scaffolding from Nucleic Acid and the Plasma Membrane. *Proc Natl Acad Sci USA* **2017,** *114*, E10056.
9. Pak, A. J.; Grime, J. M. A.; Yu, A.; Voth, G. A., Off-Pathway Assembly: A Broad-Spectrum

Mechanism of Action for Drugs That Undermine Controlled HIV-1 Viral Capsid Formation. *J Am Chem Soc* **2019,** *141*, 10214-10224.
10. Pak, A. J.; Dannenhoffer-Lafage, T.; Madsen, J. J.; Voth, G. A., Systematic Coarse-Grained Lipid Force Fields with Semiexplicit Solvation via Virtual Sites. *J Chem Theory Comput* **2019,** *15*, 2087-2100.
11. Jin, J.; Pak, A. J.; Voth, G. A., Understanding Missing Entropy in Coarse-Grained Systems: Addressing Issues of Representability and Transferability. *J Phys Chem Lett* **2019,** *10*, 4549-4557.
12. 12. Pak, A. J.; Voth, G. A., Advances in Coarse-Grained Modeling of Macromolecular Complexes. *Curr Opin Struct Biol* **2018,** *52*, 119-126.

581804 Advancing Nanomaterials for Energy and Water Applications Using Atomistic and Quantum Chemical Simulations

Two-dimensional (2D) materials, such as, graphene, transition metal dichalcogenides (TMDs) (*e.g.*, molybdenum disulfide (MoS_2)), hexagonal boron nitride (hBN), and layered metal oxyhydroxides (*e.g.*, nickel oxyhydroxide (NiOOH)), have received considerable attention due to their exceptional optoelectronic, mechanical, membrane separation, and electrocatalytic properties. However, a physical understanding of the controlled synthesis and interfacial behavior of 2D materials is still lacking. In this poster, I will discuss our research in several fundamental and applied areas pertaining to modeling thermodynamics and kinetics at the interfaces formed by 2D materials with vapor and liquid phases. Specifically, I will discuss:

1. Developing mechanistic models for the chemical vapor deposition growth of 2D materials, and the etching of 2D materials to form nanopores using density functional theory (DFT) calculations and kinetic Monte Carlo (KMC) simulations
2. Formulating force field models using a combination of lattice dynamics and DFT calculations, including using the developed force fields to understand wetting and friction at 2D material interfaces with various liquids using molecular dynamics (MD) simulations
3. Understanding (photo)electrocatalysis at 2D material-water interfaces using first-principles simulations to predict reaction overpotentials for water splitting

The theoretical and simulation-based research work discussed in this poster should inform the controlled synthesis of 2D materials and their use in various applications, including optoelectronic devices, mechanical composites, membranes for gas separation/water desalination, and electrocatalysis for splitting water molecules into oxygen and hydrogen gases.

Research Interests: The overarching goal of my research group will be the development and use of multi-scale simulation frameworks, combining quantum chemical calculations, MD simulations, and KMC simulations to advance up-and-coming **N**anomaterials for use in **E**nergy and **W**ater (NEW) applications (*e.g., (photo)electrocatalytic reduction of* CO_2*, batteries for storing energy generated through renewable means, and desalination of seawater to generate pure water*). To this end, I will seek to develop a mechanistic understanding of not only the above-mentioned applications, but also of the synthetic processes for new nanomaterials, including the incorporation of defects in them. By utilizing multi-scale simulations and realistic models for materials, my research group will link theory and simulations with experimental data using directly measurable observables (e.g., water and ion fluxes through nanopores, transmission electron microscopy images of nanoporous defects, reaction overpotentials, and ionic mobilities). The insights into various thermodynamic and kinetic processes, obtained from newly developed and validated models, will be valuable in solving the twin challenges of making available clean energy and clean water for humankind.

Teaching Interests: I have served as a teaching assistant twice during my academic career: first, for the undergraduate-level "Transport Phenomena" course at the Indian Institute of Technology (IIT) Delhi, and second, for the graduate-level "Chemical Engineering Thermodynamics" course at MIT. I have also developed and delivered a guest lecture in the "Chemical Engineering Nanotechnology" course at MIT, and two three-hour-long lectures on droplets & bubbles and fluid mechanics under the "Splash" initiative for high-school students at MIT. Finally, I have also recorded video lectures for training high-school students for the prestigious IIT Joint Entrance Examination, that are available through MIT's OpenCourseWare website (https://ocw.mit.edu/high-school/iit-jee/exam-prep/). I am passionate about teaching existing core courses on Thermodynamics and Reactor Engineering, and also about developing new, or improving existing, elective courses, on topics such as, Nanotechnology, Molecular Simulations, and Interfacial Phenomena.

581810 Catalytic Porous Materials with Tailored Nanostructures for Selective and Sustainable Chemical Processes

Research Interests:

My Ph.D. work with Prof. Wei Fan in the department of Chemical Engineering at University of Massachusetts, Amherst, focused on the synthesis of nanoporous catalytic materials such as zeolites, metal oxides and porous carbon for biomass upgrading processes. In particular, I developed a hierarchical Lewis acid zeolite by confined synthesis with mesoporous carbon, and the catalyst exhibited >3 times higher catalytic performance for sugar isomerization reactions, compared to conventional Lewis acid zeolites. In addition, I showed that Pt nanoparticles supported on Lewis acid zeolites could effectively catalyze glycerol oxidation to form lactic acid with 72% yield via one-step, whereas typical metal oxide-supported Pt catalysts did not produce lactic acid. Furthermore, I achieved biomass-derived p-xylene production with 97% yield, by developing phosphorous-containing zeolite and silica materials. In contrast, conventional Lewis or Brønsted acid catalysts led to p-xylene yield of ~70%, due to catalyst deactivation.

My current postdoctoral work with Prof. Bingjun Xu in the department of Chemical & Biomolecular Engineering at University of Delaware aims at developing multicomponent catalysts which mediate a tandem reaction. I successfully synthesized a family of zeolite-encapsulated metal nanoparticles via a novel synthetic approach. In tandem Aldol condensation and hydrogenation of furfural with acetone, the encapsulated catalyst exhibited a 4 times higher yield of target products than typical supported metal catalysts. I am also exploring new tandem catalytic system mediated by zeolite-encapsulated metal catalysts.

Before starting Ph.D., I had experience of polymer industry for several years. During this period, I had studied and designed polymer properties on a commercial scale through a variety of polymer characterization and processing techniques. This industry experience broadened my perspective towards how polymeric materials should be developed in everyday life.

My academic and industry experience in the synthesis and characterization of inorganic and organic materials as well as heterogeneous catalysis prepares me for my independent research career. I aim to establish a research program on the rational design of advanced nanoporous materials (such as zeolites, carbon, metal-organic framework (MOF), porous polymers, etc.) for energy and environmental applications. My research will be motivated by grand challenges in energy and environmental issues, in particular, shale gas conversion to chemicals, sustainable production of fuels and chemicals, emission control catalysis, CO_2 conversion, and catalysis for waste water treatment.

Selected publications: (Total: 25, 876 Citations, h-index: 17)

1. **H. J. Cho**, D. Kim, J. Li, D. Su and B. Xu, "Zeolite encapsulated Pt nanoparticles for tandem catalysis", *J. Am. Chem. Soc.*, 2018, 140 (41), 13514-13520

2. **H. J. Cho**, L. Ren, V. Vattipalli, Y.-H. Yeo, N. Gould, B. Xu, R. J. Gorte, R. Lobo, P. J. Dauenhauer, M. Tsapatsis and W. Fan, "Renewable p-xylene from 2,5-dimethylfuran and ethylene using phosphorus-containing zeolite catalysts", *ChemCatChem*, 9, 2017, 398-402

- One of the top 20 most-downloaded articles published on *ChemCatChem*, between July 2016 and June 2018
- Released in the Newspapers *Business Standard* (http://business-standard.com), *Higher Education Tribune* (http://www.highereducationtribune.com) and *Biomass Magazine* (http://biomassmagazine.com)
- Patent (No. PCT/US2017/054558), entitled "Phosphorus-containing solid catalysts and reactions catalyzed thereby, including synthesis of p-xylene"

3. K. Rieger*, **H. J. Cho***, H. Yeung, W. Fan and J. Schiffman, "Antimicrobial activity of silver ions

released from zeolites immobilized on cellulose nanofiber mats", *ACS Appl. Mater. & Interfaces*, 8 (5), 2016, 3032-3040 (***Equally contributed**)

4. P. Dornath, **H. J. Cho**, A. Paulsen, P. Dauenhauer and W. Fan, "Efficient mechano-catalytic depolymerization of crystalline cellulose by formation of branched glucan chains", *Green Chem.*, 17 (2), 2015, 769-775

- **Featured on Inside Front Cover**

5. **H. J. Cho**, C.-C. Chang and W. Fan, "Base free, one-pot synthesis of lactic acid from glycerol using a bifunctional Pt/Sn-MFI catalyst", *Green Chem.*, 16 (7), 2014, 3428-3433

- **Featured on Front Cover**
- One of the top 15 most-read articles on *Green Chem.* from May to July of 2014

6. **H. J. Cho**, P. Dornath and W. Fan, "Synthesis of hierarchical Sn-MFI as Lewis acid catalysts for isomerization of cellulosic sugars", *ACS Catal.*, 4 (6), 2014, 2029-2037

7. **H. J. Cho**, H. S. Jang, S. H. Lim, E. A. Cho, T.-H. Lim, I.-H. Oh, H.-J Kim and J. H. Jang, "Development of a novel decal transfer process for fabrication of high-performance and reliable membrane electrode assemblies for PEMFCs", *Int.J.Hydrogen Energy*, 36, 2011, 12465-12473

Teaching Interests:

I enjoy conveying my knowledge to others as well as exploring new technologies with them. I am comfortable in teaching many of the core chemical engineering courses (Chemical kinetics, Thermodynamics, Mass and Energy balances, and Heat and Mass transfer). I would also be interested in teaching more specialized courses in i) heterogeneous catalysis, ii) nanomaterials chemistry and engineering, and iii) characterization of nanostructured materials as elective courses at a graduate level.

581818 Building Autonomous Nanomachines at the Interface of Colloids and Electronics

Research Interests:

Nanotechnology is set to revolutionize our world. However, nano is currently limited to passive or, at best, stimuli-responsive materials, such as sensors and cargo release particles. But this only scratches the surface. We see the future with complex nanomachines that have computational power on-board, are capable of patching microwounds, repairing cells, performing nanosurgeries, digging nanoholes, and making autonomous decisions based on the specific situation they have encountered.

What? We aspire to develop principles and materials for a broad class of micro and nanomachines that sense, actuate, compute, and communicate in various environments that are too small or too dangerous for humans and bigger robots. We focus on three areas: (1) *Communication* – how to deliver energy and information to a remote nanoscale probe? Nano produces a lot of sensors, but we need novel photonic ideas to enable their usage in vivo. (2) *Design* – how can we create multi-functional probe with a scalable architecture? Nano makes a lot of particles, but decisively not in the form of scalable machines with computation. (3) *Applications* – what are the unique applications of these probes?

How? We are bringing nanophotonic and optoelectronic expertise into the field of chemical engineering. *For communication*, we are choosing light as it brings power and information to remote locations with ultimately the highest bandwidth possible. We have recently created a novel photonic approach allowing to transmit light up to 4.5 cm into the living tissue [1]. This will enable future machines to operate deep inside the body to perform nanosurgeries, deep phenotyping, and other functions. The work included fluorescent sensors implantation into phantom tissues and living mice, building a novel optical setup, and Monte Carlo simulations of light in random media.

For design, we are merging energy-efficient but limited in functions colloidal particles with energy-thirsty but modular in design electronic circuits. We have recently created the first cell-sized state machine capable of recording and storing the information about its environment [2]. This constitutes a modular and scalable platform that will be expanded to accommodate various elements in order to achieve applications at hand. This pursuit was based on the clean room fabrication techniques and 2D materials, and included their optoelectronic characterization.

For applications, we are using nanosensors to study plants in the approach termed plant nanobionics. We have recently created the first electromechanical sensor to directly measure plant stomatal dynamics. It successfully reveals plant diurnal cycles and can identify the onset of drought [3]. This expands our expertise into a unique direction of biosystems where complex nanomachines will be injected into. This pursuit included the development of plant epidermal electronics, studies of nanomaterial interactions with living cells, and building a mathematical model of stomatal dynamics.

Publications (total 40 with 7 in review)

[1] **V.B. Koman**, N. Bakh, F. Nguyen, D. Kozawa, M. Lee, G. Bisker, M.S. Strano, Wavelength Modulation of Fluorescent Nanosensors for High SNR Operation in Thick Tissue, *Nature Biomedical Engineering*, Submitted (2019).

[2] **V.B. Koman**, P.Liu, D. Kozawa, A.T. Liu, A.L. Cottrill, Y. Son, J.A. Lebron, M.S. Strano, Colloidal nanoelectronic state machines based on 2D materials for aerosolizable electronics, *Nature Nanotechnology*, 13, 819–827 (2018).

[3] **V. B. Koman**, T.T.S. Lew, M.H. Wong, S.Y. Kwak, J.P. Giraldo, M.S. Strano, Persistent drought monitoring using a microfluidic-printed electro-mechanical sensor of stomata in planta, *Lab on a Chip* 17(23), 4015-4024 (2017).

Teaching Interests:

In my future work, I will develop an educational program that will immerse students in the science through hands-on experience and problem-based learning. One important aspect that I plan to incorporate in the course development is the backward design principle, where one starts from the desired learning outcomes, determines the scoring system, and, finally, plans the learning experiences and instructions. Besides the clear focus on achievements of the postulated goals, such approach also brings clarity to students on what is expected from them and what they should focus on. At the same time, I will carefully look out for indicators of ersatz learning. Following my research, my teaching will be focused on materials and machines at the nanoscale, how different physical laws scale down to the nanoscale, etc. I also plan to contribute with 2D materials course that will cover the basics of different 2D materials, their fabrications methods followed by chemical, electronic, mechanical, thermal, and other exciting properties that they have to offer. Finally, I am open to teach various courses from Chemical Engineering.

I have completed Kaufman Teaching Certificate Program at MIT, mentored multiple undergrad and graduate students, and taught courses both at MIT and EPFL.

581897 Controlling Organization and Effects of Nanoparticles on Polymer-Nanoparticle Composites

Research Interests:

Synthesizing materials by design with desired properties is one of the challenges in materials science. I use molecular simulations as a tool to study the dynamics and structure of nanocomposite materials in the length and time scales that can be challenging to achieve by using experimental methods. My research focuses on two different areas: (1) self-assembly of materials at nano-scale (*e.g* self-assembly of nanomaterials functionalized with DNA), and (2) the effects of confinement on polymer dynamics (*e.g* polymer-nanoparticle composite materials and polymer thin films).

Self-assembly has offered an effective pathway to achieve the rational design of materials. One of my main research interests is to control and manipulate the organization of nanoparticles in nanocomposite materials and study how they affect the composites' properties (*e.g.* mechanical and optical). My research focuses on the study of the thermodynamic pathways that leads to self-assembly of nanomaterials and how the range and strength of interactions between nanoparticles, and shape of nanoparticles can affect the final self-assembled structures. Specifically, I'm interested in colloidal particles functionalized with DNA strands such as DNA origami, DNA-functionalized nanoparticles and Janus particles.

Polymer dynamics in confinement can be drastically different than dynamics of bulk polymers (*e.g.* the glass transition temperature and viscoelasitcity). I use molecular dynamic simulations to study the dynamics of polymer chains in different confinement geometries to understand the origins of these changes in the polymeric material properties. Specifically, I'm interested in the thin films and polymer-nanoparticle composites.

Research Experience

I have a diverse research experience. During my research career, I have used different theoretical and computational methods such as mean-field theory, molecular dynamics, Monte Carlo, and multi-particle collision dynamics to study complex systems. In most of the systems that I have studied (self-assembly of colloidal materials, properties of polymer in confinement, and freeze-casting) I have worked closely with experimental scientists. During my PhD (under supervision of Prof. Francis Starr) and postdoctoral research (under supervision of Prof. Sanat Kumar), I work and discuss closely with my experimental collaborators on daily bases. This experience has helped me to acquire necessary skills to communicate and solve scientific problems with experimentalist efficiently.

I also have experimental experience. During my work as a MS.c student, I worked on synthesizing of graphene nanoribbons from carbon nanotubes that was used in cell imaging and photothermal cancer therapy.

Below is a selective of my recent publications.

1. **Hamed Emamy**, Sanat K. Kumar, and Francis W. Starr. "Diminishing Interfacial Effects with Decreasing Nanoparticle Size in Polymer-Nanoparticle Composites." *Physical review letters* 121, no. 20 (2018): 207801.

2. **Hamed Emamy**, Oleg Gang, and Francis W. Starr. "The Stability of a Nanoparticle Diamond Lattice Linked by DNA." *Nanomaterials* 9, no. 5 (2019): 661.

3. Wenyan Liu, Miho Tagawa, Huolin L. Xin, Tong Wang, **Hamed Emamy**, Huilin Li, Kevin G. Yager, Francis W. Starr, Alexei V. Tkachenko, and Oleg Gang. "Diamond family of nanoparticle superlattices." *Science 351*, no. 6273 (2016): 582-586.

Teaching Interests:

During my PhD work at Wesleyan University, I had a privilege to teach introductory physics lab to students. During these lab sections I thought the basic physical concepts theoretically on the chalk board and then help the students trough the experiments. Interacting with students one by one helped me to be familiar with different approaches students have to the problems. I also mentored many newly joined member of the group. I helped them to be familiar with the necessary techniques to carryout their experiments. Specifically, I supervised two undergraduates with their research projects. During this time, I discovered my passion for mentoring students. Helping students to grow as a scientist and see the results of mentoring as a published work has been so fulfilling for me. Having caring mentors, I understand the importance of good education and do my best to educate the new generation of scientists. I have a wide range of teaching interests, from fundamental courses such as statistical mechanics to specialized courses such as advanced methods in molecular simulations.

581937 Engineering Redox-Active Materials from Electrocatalysis to Pseudocapacitance

Research Interests:

Redox-active materials represent the most promising prospect towards electrocatalytic systems and energy storage technologies. However, the poor understanding of redox chemistry at the molecular level limits the activity improvement and further industrial application.

I will first describe our research efforts on the synthesis of high valued chemicals and fuels from renewable electricity and CO_2 [1-5]. We focus on developing, understanding, and eventually designing new catalysts for electroreduction of CO_2 and water oxidation reaction using computational materials science, *in situ*/*ex situ* synchrotron X-ray spectroscopies, and advances in materials chemistry. Specifically, I will report a concept to engineering the redox chemistry by introducing isolated single sites and incorporating the third element to modulate the active species. Next, I will discuss the molecular insights to engineer the redox reactions for pseudocapacitance [6]. We change the dominant charge storage mechanism from surface redox reactions to ion intercalation into bulk material by atomic-level structure engineering of metal oxides.

1. **Zheng X**, Tang J, Ji Y, Wang J, Liu B, Steinrück H, Lim, K, Li Y, Toney M, Chan K, Cui Y, Theory-guided discovery of Cu-Sn alloys with high formate selectivity at low overpotentials, *Nature Catalysis*, 2019, 2, 55-61.
2. **Zheng X**, Zhang B, Sargent E H, et al. Theory-driven design of high-valence metal sites for water oxidation confirmed using *in situ* soft X-ray absorption. *Nature Chemistry*, 2018, 10, 149-154. (IF=27.9).
3. **Zheng X**, De Luna P, Sargent E H, et al. Sulfur-modulated tin sites enable highly selective electrochemical reduction of CO_2 to formate, *Joule*, 2017, 1, 794-805. (Featured article, Cell Press)
4. Zhang B*, **Zheng X***, Voznyy O*, Sargent E H, et al. Homogeneously dispersed multimetal oxygen-evolving catalysts. *Science*, 2016, 352(6283): 333-337 (* co-first author). (IF=33.6)
5. **Zheng X**, Song J P, Ling T, et al. Strongly coupled nafion molecules and ordered porous CdS networks for enhanced visible-light photoelectrochemical hydrogen evolution. *Advanced Materials*, 2016, 28, 4935-4942. (Cover, IF=18.96)
6. Ling T*, Da P*, **Zheng X***, Ge B*, Hu Z, Wu M, Du X, Hu W, Jaroniec M, Qiao S, Atomic-level structure engineering of metal oxides for high-rate oxygen intercalation pseudocapacitance, *Science Advances*, 2018, 4, eaau6261.

Teaching Interests:

I would be very interested in teaching a course on *Sustainable Energy and Catalysis*, including fuel cells, batteries, photocatalysis, carbon dioxide reduction, N_2 reduction etc. This course could be offered to both undergraduate and graduate students, and provide learning for students in the important area of sustainable energy. I would be very happy to work with my colleagues and teach a course on *Advanced X-ray Techniques for Materials and Chemistry.* The unique properties of synchrotron radiation, including continuous spectrum, high flux and brightness, and high coherence, provide unprecedented experimental tools to study materials science, bioscience, geoscience, physical and chemical sciences, among many others. The goal of the course is to get students familiar with the advanced X-ray techniques for their research in chemical engineering, materials science, chemistry, physics, etc. The course will consist of fundamentals of interaction between X-ray and matter, X-ray sources, spectroscopy, scattering, imaging, and software learning. This course could be offered to students of graduate students' levels.

581955 Engineering Nanomaterials for Biomedical Applications

Research Interests:

Engineering Nanomaterials for Biomedical Applications

Nanoscale materials often possess functionalities that go beyond conventional molecular systems. Unlike bulk materials, their properties not only depend on their composition, but may vary with the size, shape, and relative arrangements of constituents. As a chemist with extensive training in nanotechnology, I have expertise in the synthesis, characterization, and assembly of nanomaterials, as well as their interfacing with biological systems. In addition, I have had the opportunity of working in multiple different areas spanning many different length scales of "small structures", including cluster physics (molecular scale), nanomaterial design for drug delivery and cellular imaging (nanoscale), and microdroplet chemistry (micron scale). I have worked in collaboration with physicists, chemists, engineers, and medical doctors, and published 19 articles in total (11 as first or a co-first author). In my independent career, I aim to develop a research program that will explore the fundamental properties of novel nanoscale materials and then exploit them for various biomedical applications. In particular, I am interested in learning how nanomaterials interact with biological systems and utilizing this knowledge for developing improved nanomaterial-based diagnostic and therapeutic platforms.

Postdoctoral research with Prof. Chad A. Mirkin at Northwestern University

Nanomaterial-based tools for bioimaging. My research focuses on engineering nucleic acid-based probes for imaging intracellular analytes in live cells. By using aptamers, oligonucleotide sequences that bind to target analytes with high selectivity and sensitivity, it is possible to detect almost any analyte, provided the binding event can be transduced into a signaling event. I have developed a new class of signaling aptamers called "FIT-aptamers". These aptamers contain a visco-sensitive dye that is forced to intercalate or FIT between base pairs upon target binding, turning the fluorescence on due to the restricted rotation of the dye. The FIT strategy reduces false-positive signals common to all fluorophore-quencher systems, provides up to 20-fold fluorescence enhancement upon target binding, and allows target detection down to nanomolar concentrations in complex milieu such as human serum. I am now interfacing these FIT-aptamers with nanoparticles to enable transfection-reagent free cellular uptake for live cell imaging applications.

Graduate research with Prof. Richard N. Zare at Stanford University

Electroresponsive nanoparticles for drug delivery on demand. Programmable and controllable delivery of drugs is one of the main challenges in drug administration today. My thesis work was focused on developing an electroresponsive drug delivery system (DDS) as a solution to this problem. The DDS consists of drug-loaded polypyrrole nanoparticles (PPy NPs) that undergo changes in redox state upon electric stimulation, releasing their drug cargo with spatiotemporal precision. I demonstrated that various drugs ranging from small molecules to polypeptides can be released in a pulsed manner by applying <1 V. Moreover, by incorporating metallic elements into the PPy scaffold, we developed an improved DDS that can release drugs at unprecedented low voltages (50-75 mV). The widened window of operating voltage allows the release of methotrexate, an anti-cancer drug susceptible to redox degradation at higher voltages, with retained bioactivity. We next proceeded to couple PPy NPs to wireless electronics to develop a next-generation implantable DDS. In collaboration with the Arbabian Lab (Stanford Engineering), we demonstrated that drug release can be triggered wirelessly through acoustic excitation using a millimeter-sized piezoelectric transducer. Taken together, these results represent a cornerstone towards developing minimally invasive implants which can treat various chronic diseases (e.g. chronic pain/diabetes/cancer).

Reactions in aqueous microdroplets. Bulk water serves as an inert solvent for many chemical/biological reactions. We demonstrated that in micron-sized water droplets, various molecules undergo spontaneous redox reactions without any added electron donors/acceptors or applied voltage. While none of these reactions proceeds spontaneously in bulk water, redox efficiencies can reach >90% in microdroplets.

Moreover, we observed that gold nanoparticles form spontaneously from metal precursors in aqueous microdroplets and self-assemble to form nanowires, representing the first example of self-assembly without using added ligands, templates, or electric fields. The size/shape can be controlled/temporally resolved by modulating the microdroplet lifetime. Compared to bulk synthesis, the size and growth rate of AuNPs are enhanced by 2- and 100,000-fold, respectively, in microdroplets. These results demonstrate that microdroplets have a unique environment and could be potentially used as powerful microreactors for synthesizing various compounds and nanomaterials.

Teaching Interests:

I believe teaching is an integral part of learning. As a graduate student, I have taught over 450 students in both lab and lecture-based settings. Specifically, I have been a teaching assistant for various courses such as General Chemistry (Beginner and Advanced), Organic Chemistry, Analytical Chemistry, and Physical Chemistry. I have also had the opportunity to design and teach my own courses through the Stanford Splash! program to ~130 high school students. My most enjoyable teaching endeavor has been as a research mentor to 12 students including high school, undergraduate, and graduate students. Notably, all my mentees are currently pursuing science-related degrees or careers. My ideal teaching program would be focused on teaching upper-level undergraduate classes or specialized topic courses for graduate students, including Separation Science, Thermodynamics, Soft Matter and Interfacial Phenomena, Chemical Kinetics, Molecular Engineering, and Design of Drug Delivery and Diagnostic Systems.

Funding

I have assisted Prof. Zare at Stanford University and Prof. Mirkin at Northwestern University in writing multi-PI proposals resulting in >$15 million in funding from the NIH and DOE. As a graduate student, I secured self-funding (~$150,000) for two years from the Center for Molecular Analysis and Design at Stanford. I was supported for another two years by a Stanford Graduate Fellowship. As a postdoctoral fellow, I have received $10,000 in research support as an HHMI Hanna Gray Fellow finalist.

582006 An *Active* Approach to Colloidal Self-Assembly

Both as a postdoctoral researcher and Ph.D student, I have been particularly interested in understanding how colloidal active matter can be used as a tool to engineer the microscopic. This rapidly growing field focuses on characterizing the individual and collective behavior of synthetic micro-swimmers and self-propelled colloids. These *active particles* can be thought of as the synthetic analogues of swimming bacteria. However, a major benefit of these synthetic variants over their biological counterpart is the ability to systematically tailor interparticle interactions and dynamically modulate swim speed. The functionality of synthetic active particles potentially makes them the ideal tool to manipulate and self-assemble matter at the microscale. In this poster, we explore to what extent activity can be used as an extra handle to promote the self-assembly of colloidal building blocks designed to form a variety of microstructures and under what conditions the complex self-assembly landscape can be simplified and biased toward a particular target structure.

Research Interests: Using a combination of cutting-edge computer simulations and analytical theory, my research revolves around developing novel techniques to manipulate, direct, and self-assemble matter at the micro-scale. A significant thrust in my work is developing advanced computational methods that provide a consistent and faithful description of colloidal systems with the highest level of computational efficiency.

Teaching Interests: Thermodynamics, Kinetics, Statistical Mechanics, General Chemistry, Physics, Computational Methods

582021 Low-Cost, Scalable, and Rapid Platforms in Ultrathin Materials Synthesis for Water and Energy Technologies

Research Interests:

With the impending rise in global temperatures over the next century, technologies that address water scarcity or reduce emissions will be critical for mitigating the consequences of global warming. However, many proposed innovations in water and energy technologies require the ability to design novel nanomaterials and nanostructures with unprecedented structural and compositional control, such as ultralow polydispersity pore sizes in membranes to replace energy intensive distillations. With increasingly stringent requirements for precise atomic placement in these materials and the desire to deploy their associated technologies at scale, there is a need for processes that can make materials with sub-nanometer precision that can be performed cheaply and scalably. Further, the future search for nanomaterials will require both novel frameworks to create previously unsynthesizable materials and the development of platforms that can rapidly screen materials properties.

My research interests focus *on developing novel low-cost, scalable, and rapid platforms for the engineering of ultrathin materials with atomically-precise nanostructures necessary for water and energy technologies.*

I plan to establish an independent and self-sustaining research program that will build upon my past research experience in vapor-phase, layer-by-layer synthetic processes, kinetic modeling, both ex situ and in situ structure and spectroscopic characterization tools, and computer-controlled deposition and testing equipment. Processes that can make ultrathin (<50 nm) films will be essential for creating advanced membrane and catalyst materials where thinner coatings dramatically increase efficiency and performance. Additionally, because vapor-phase chemistries do not require solvents, they can be easily combined with roll-to-roll processing and paired with in situ characterization methods, making them ideal for rapid materials screening. Filling the gap in large, high quality, consistent experimental datasets provides a great opportunity for collaboration with computational researchers who focus on materials.

Because control over confinement and composition in ultrathin coatings have been identified as critical areas for improvement in catalyst and membrane design, my group will initially propose projects related to: (a) controlling pore sizes in ultrathin materials through the vapor-phase synthesis of coordination polymers and MOFs for catalysis and membranes, (b) layer-by-layer ternary materials synthesis through laser-control of temperature for gas separation and electronics applications, and (c) ultrathin organic coatings for breaking scaling relations in electrochemical O_2 and CO_2 catalysis.

Graduate Research

My Ph.D. dissertation with Professor Stacey F. Bent at Stanford University in the Department of Chemical Engineering focused on the development of one novel platform to enable new nanotechnologies, molecular layer deposition (MLD). Derived from atomic layer deposition, this process is a new vapor-phase method for creating coordination networks and polymeric films with sub-nanometer thickness and compositional control that uses surface reactions to grow films layer-by-layer. In addition to working with collaborators to create several new organic chemistries for this platform, I lead the development of an MLD-based electrochemically active manganese alkoxide catalyst film for the production of oxygen. By blending carbon species into manganese oxide, the catalyst showed improved structural stability during electrochemical cycling when compared to traditional manganese oxide. To address a major gap in understanding, I also developed a new paradigm to explain the nucleation and growth behavior of MLD-grown films, dramatically broadening the applicability of this approach. By varying monomer selection, and combining synchrotron-based X-ray characterization methods with kinetic modelling, I was able to draw correlations between the choice of monomers and the resulting film structure and growth rate. The combination of organic and inorganic moieties for the formation of previously unmade materials structures is just beginning to be realized (e.g. metal-organic frameworks), and because these materials can be quickly and easy tested through vapor-phase synthesis techniques, there is great potential to expand work and knowledge in this area.

Postdoctoral Research

I am currently in my second year of postdoctoral research with Professor Jeffrey C. Grossman at the Massachusetts Institute of Technology in the Department of Materials Science. During my time here, I have been building upon my prior work to expand the toolkit for atomic-level control of surfaces while concomitantly applying these advances to membrane architectures. In the United States, ~15% of all energy consumption is used for separations, largely through distillation. In some industries, there is a potential to significantly reduce this energy consumption through the introduction of membrane separation technologies. However, in many cases, membrane fouling prevents their widespread deployment. One promising approach that has already been shown to dramatically reduce fouling is to make membranes electrically conductive. However, current methods for making conductive membranes destroy their underlying structure, or are expensive and non-scalable to produce.

Using the insights gained during my PhD, we have come up with a completely new way to make these membranes conductive while maintaining their structure. By infiltrating traditional polymeric membranes with inorganic precursors, they can be laser annealed into a graphitic form without losing their porosity. By replacing expensive metal or graphene-based conductive coatings with cheaper alternatives, membranes can begin to leverage electric fields for new architectures and improved separation efficiency.

Summary

During the poster session, I will present a more thorough overview of my past and current research, as well as my plan for future research activities, which will continue the development of new platforms for controlling surface properties to enable advanced membrane and catalyst devices. Using my PhD and postdoctoral experience in the assembly of computer-controlled deposition and testing equipment, my future vision is to develop rapid materials screening approaches that leverage the synthetic ease of vapor-phase processes combined with in situ spectroscopic and ellipsometric characterization. By expanding the materials processing toolkit and building fundamental insight into the origins of nanostructures, it is my goal to advance the field of materials development, enabling access to unsynthesizable materials structures which will be critical for future water and energy technologies.

Teaching Interests:

My primary goal in teaching is to enable students to be successful in their future careers by inspiring them to seek understanding and knowledge, by equipping them with the tools to articulate and answer important questions, and by providing them with an inclusive and active teaching style conducive to effective learning. I plan to rely heavily on recent pedagogical advancements in my teaching practice, focusing on active learning approaches, as well as the use of computer, streaming, and video tools to both reduce the redundancy of repeated yearly classes and to broaden access to the teaching materials I produce.

In addition to serving for four quarters as a TA at the University of Washington in computer science, where I taught the introductory Computer Programming I course and the senior level Programming Languages course, I also TAed at Stanford University for the senior level Kinetics and Reactor Design course for two quarters, for which I received a Teaching Assistant Award. I delivered several guest lectures in this kinetics course, served as the departmental Head TA Mentor for two years, and was honored with Stanford's university-wide Centennial Teaching Assistant Award. I have also worked to improve teaching at Stanford and MIT by helping to run and design material for the Stanford ChemE yearly TA training and by co-organizing an Inclusive Teaching Workshop at MIT.

Because of my educational background, teaching experience, and training in pedagogy, I am confident in my ability to teach any core chemical engineering course at the undergraduate or graduate level. In particular, I am interested in courses that leverage my experience in kinetics, spectroscopy, or synchrotron-based characterization, such as a course in *Kinetics and Reactor Design*, *Surface Chemistry*, or *Spectroscopy*.

Selected Publications:

1. **David S. Bergsman**, et al. "Manganese Alkoxide Films Grown by Hybrid Atomic/Molecular Layer Deposition for Electrochemical Applications." *Under Review*
2. Richard G. Closser, Mie Lillethorup, **David S. Bergsman**, Stacey F. Bent. "Growth of a Surface-Tethered, All-Carbon Backboned Fluoropolymer by Photoactivated Molecular Layer Deposition." *ACS Appl. Mater. Interfaces*, 11 (2019) 21988
3. **David S. Bergsman**[‡], Tzu-Ling Liu[‡], et al. "Formation and Ripening of Self-Assembled Multilayers from the Vapor-Phase Deposition of Dodecanethiol on Copper Oxide," *Chem. Mater.*, 30 (2018) 5694
4. Callisto MacIsaac, ..., **David S. Bergsman**, Stacey F. Bent. "Atomic and Molecular Layer Deposition of Hybrid Mo-Thiolate Thin Films with Enhanced Catalytic Activity," *Advanced Functional Materials*, 28 (2018) 1800852
5. **David S. Bergsman**, Richard Closser, Stacey F. Bent. "Mechanistic Studies of Chain Termination and Monomer Absorption in Molecular Layer Deposition" *Chemistry of Materials*, 30 (2018) 5087
6. Richard G. Closser, **David S. Bergsman**, and Stacey F. Bent. "Molecular Layer Deposition of a Highly Stable Silicon Oxycarbide Thin Film Using an Organic Chlorosilane and Water," *ACS Applied Materials and Interfaces*, 10 (2018) 24266
7. Mie Lillethorup, **David S. Bergsman**, et al. "Photoactivated Molecular Layer Deposition through Iodo-Ene Coupling Chemistry," *Chemistry of Materials*, 29 (2017) 9897–9906
8. **David S. Bergsman**, et al. "Effect of Backbone Chemistry on the Structure of Polyurea Films Deposited by Molecular Layer Deposition," *Chemistry of Materials*, 29 (2017), 1192
9. Richard G. Closser, **David S. Bergsman**, Luis Ruelas, Fatemeh S. M. Hashemi, Stacey F. Bent. "Correcting Defects in Area Selective Molecular Layer Deposition," *Journal of Vacuum Science& Technology A*, 35 (2017) 031509 (Editor's Pick)
10. **David Bergsman**, Han Zhou, Stacey F. Bent. "Molecular Layer Deposition of Nanoscale Organic Films for Nanoelectronics Applications," *ECS Transactions*, 64 (2014), 87

581767 Nanostructured Anodic Oxides of Metals: From Corrosion Protection to Nanotechnology and Emerging Applications

Research Interests:

Anodic oxidation of metals, especially Al and Cu

Fundamentals of Al and Cu anodizing

Nanostructured materials

AAO-templates nanofabrication

Electrodeposition

CO_2 reduction reaction on CuO_x nanostructures

Corrosion protection

Teaching Interests:

Materials chemistry, physical chemistry, nanotechnology, corrosion protection, characterization techniques in corrosion, spectroscopy, electron microscopy, X-ray diffraction

Nanostructured Anodic Oxides of Metals: from Corrosion Protection to Nanotechnology and Emerging Applications

Wojciech J. Stepniowski*, Kuo-Kuang Wang, Wojciech Z. Misiolek

Loewy Institute and Department of Materials Science and Engineering, Lehigh University

5 East Packer Ave., Bethlehem, PA 18015 USA

Anodizing is a metal passivation technique, contributing mainly to aluminum alloys corrosion protection, known for over century. It allows to form a compact, dense, adherent, dielectric layer on a protected metal, hindering galvanic coupling. In industrial practice, a basic, one-step approach is applied. As an effect, a porous oxide film is obtained and consequently the pores are being chemically sealed. Corrosion performance of such films depends strongly on the applied electrolyte, voltage, temperature, but also chemicals applied for pore sealing. Numerous approaches have been developed and are patented all over the world.

A milestone research was published by Masuda and Fukuda in 1995 [1]. Till then, there was a belief that anodizing provides layers with disordered pores and the only application is corrosion protection. Masuda and Fukuda reported a totally new approach, resulting in one of the most popular, nowadays nanomaterials. After formation of the anodic aluminum oxide (AAO), the formed nanoporous oxide has to be chemically removed in a mixture of chromic and ortophosphoric acid. Then, pre-patterned aluminum surface layer is exposed. Finally, re-anodization of so pre-patterned aluminum allows to obtain, after the second step of anodization, highly-ordered anodic aluminum oxide, with hexagonally arranged, parallel and uniform pores (Fig. 1a). Since this milestone publication, AAO attracted attention of numerous researchers, becoming one of the most popular template in nanofabrication. Thanks to the application of AAO, a large variety of nano-architectures, including nanowires, nanotubes, nanodots and nanoporous arrays made of diverse materials, including metals, alloys, semiconductors, superconductors, salts, carbon materials and polymers was possible to achieve. In order to form such nanostructures, various techniques, from simple ones like electrodeposition, to advanced ones like atomic layer deposition were applied (Fig. 1 b-c) [2]. The method we have developed, cuts down costs and the number of steps of metallic nanowires fabrication: we have demonstrated that even anodizing of technical purity aluminum and subsequent stepwise voltage decrease, results in a through-hole AAO template with conductive bottoms of the pores. This enabled electrodeposition of copper from standard $CuSO_4$ bath, and allowed to form nanowires (Fig. 1 b-c).

Figure 1. Top-view of anodic aluminum oxide (AAO) formed in sulfuric acid (a) and copper nanowires formed by electrodeposition into AAO (b-c) [2].

Research on AAO triggered experiments with other metals. It was found that majority of transition metals can be successfully subjected to anodizing. For example, anodization of titanium allows to form nanotubes and nanoporous arrays made of amorphous TiO_2. Such formed TiO_2 found numerous applications, including formation of back-side illuminated dye sensitized solar cells, electrochromic devices, photocatalysis and as an additional protective layer in biomaterials. Other anodic oxides also form nanoporous or nanotubular layers. It is worth mentioning that anodic WO_3 is a promising H_2O splitting photocatalyst, while anodic ZrO_2 puts contribution in surface enhanced Raman spectroscopy.

Our recent study focuses on copper anodization. The first reason is fundamental understanding of the formation mechanism. Other oxides have fixed stoichiometry, are amorphous (except ZrO_2) and form nanoporous arrays or nanotubes (except nanowires made of ZnO). Products of copper anodic oxidation, due to the complexity of the occurring phenomena on the electrodes, are diverse: Cu_2O, CuO and $Cu(OH)_2$ are major, but often not the only products of the reactions. Additionally, the as-formed anodic films are crystalline: major components are cuprite and tenorite, while usually $Cu(OH)_2$ in an amorphous shell of the nanostructures. What is also interesting, anodization of copper results in the formation of nanowires (Fig. 2). Thus, so formed anodic films, with high-surface area 1D-nanostructures shall be perfect for catalytic applications [3]. Recently, Cu_2O and CuO are being reported as catalysts for electrochemical carbon dioxide reduction reaction (CO_2RR), thus so formed anodic films can be considered as a perfect candidate for such application. In opposite to so far reported catalysts, the anodic oxides are well-adherent to the metallic substrate and employ facile, easy to scale-up technique of fabrication. Nonetheless, still much fundamental issues of copper anodizing mechanism have to be explored and explained.

The most obvious approach in searching for the formation of nanostructures on metals, is the linear polarization of the metal in electrolyte that provides passivity at certain potential ranges. In the first attempts in the formation of copper oxides nanostructures, voltammetric scans in such solutions as KOH was done, resulting in the formation of the nanostructures (Fig. 2a). Majority of the researchers in this field also apply hydroxide solutions as the electrolytes [3]. Although, Pourbaix diagram shows passivity even at more neutral values of pH [3]. Our recent approaches utilize sodium and potassium carbonates and bicarbonates as electrolytes. Research conducted in K_2CO_3 shed some light on the phenomena occurring when copper is anodized. It was found that the diameter of the obtained nanowires increases linearly with the applied voltage [4]. Furthermore, among expected products of copper anodizing: CuO, Cu_2O and $Cu(OH)_2$, chemical compounds like Cu_4O_3 and $Cu_2CO_3(OH)_2$ were detected using X-ray diffraction (XRD). Thus, identification of malachite shows incorporation of carbonate anions into the growing oxide, what is analogous phenomenon to the incorporation of acidic anions into AAO.

Further anodization research, conducted in $NaHCO_3$ showed, for the optimized conditions, formation of close-packed nanowires with diameter down to 20 nm (Fig. 2 b-c). Also, in this, unpublished research, the nanowires are crystalline and their diameter increase linearly with the voltage.

In current research, the influence of various additives into electrolyte has been studied. After anodization in 0.01 M $KHCO_3$ at voltages ranging from 10 to 60 V as the reference, the influence of NaCl as additive was investigated. On one hand, anodizing in $KHCO_3$ without additives allowed to form nanowires. On another hand, there are reports showing that anodic polarization of copper, in NaCl-rich electrolytes, results in the formation of Cu_2O and CuO powders, depending on the operating conditions. In our research, we have studied the gap between these two opposite situations: the anodization of copper in $KHCO_3$ with various amounts of NaCl. Registered current curves showed the expected impact on the process: the greater the NaCl concentration, the greater the conductivity of the electrolyte and consequently, the greater the registered current densities. Of course, this had a significant impact on the formed nanostructures. For optimized conditions, hemi-spheres made of nanowires thicker than ones formed in $KHCO_3$ without additives were formed. Another studied additive was ethylenediaminetetraacetic acid (EDTA). The following assumption was made: water-soluble Cu^{2+} species, like $Cu(OH)_4^{2-}$ are formed during copper anodizing and contribute by re-deposition process in the formation of the nanostructures. Thus, chelating them with EDTA would prevent the re-deposition and influence the phase composition and morphology of the grown nanowires. Currently, the in-depth characterization, including optical properties of the grown nanostructures and spectra of the electrolytes after anodizing is being performed.

Figure 2. Copper oxides formed by passivation in three-electrode system in KOH (a) and two-electrode system in $NaHCO_3$ (b-c).

To sum up, anodization of metals allows to form various nanostructural oxides. Anodic aluminum oxide, consisting of hexagonally arranged nanopores, serves often as a template for further nanofabrication. In that way, for example, nanowires made from metals can be formed by electrodeposition. Another example of the nanostructures formation by anodizing are nanowires grown by copper electrochemical oxidation. Still much of the fundamental work has to be performed to reveal new anodizing regimes, improving control of the morphology and composition, such nanostructures can successfully serve as catalysts for electrochemical CO_2 reduction.

References

[1] H. Masuda, K. Fukuda, Science 268 (1995) 1466-1468.

[2] W.J. Stepniowski, M. Moneta, K. Karczewski, M. Michalska-Domanska, T. Czujko, J.M.C. Mol, J.G. Buijnsters, J. Electroanal. Chem. 809 (2018) 59-66.

[3] W.J. Stepniowski, W.Z. Misiolek, Nanomaterials 8 (2018) 379.

[4] W.J. Stepniowski, D. Paliwoda, Z. Chen, K. Landskron, W.Z. Misiolek, Mater. Lett. 252 (2019) 182-185.

581651 Electrochemical and Optoelectronic Transformations in Dynamic Semiconductor Nanomaterials

Research Interests: Next-generation energy technologies will rely on responsive semiconductor materials to rapidly convert and store energy, for applications including high-rate batteries, low-power electronics and efficient photovoltaics. These materials require facile exchange of charges and fields across interfaces and throughout a solid bulk. Solid-state transport – particularly of electrons and ions – often limits these technologies. One solution is to increase interfacial area, e.g. by thin-film processing or mesostructuring. This approach improves transport, but also reveals phenomena that only arise at nanometer scales (< 100 nm); i.e. interfacial charge storage, local focusing of light and tunable quantum phenomena. External stimuli modulate these properties through inter-related responses: for instance, electrochemical charging can induce distinct colorations depending on how ions accumulate around or within nanoparticles. Frontier technologies such as electrochromic 'smart windows' may be realized by exploiting these convoluted processes. Yet despite active research into the physics and synthesis of responsive nanomaterials, there are unresolved challenges to reach practical scales. Indeed, most devices are comprised of an ensemble of nanoparticles, and polydispersity in structure and composition is difficult to control. Moreover, techniques to observe dynamic transformations in nanomaterials are limited by spatial and temporal resolution, and convoluted by heterogeneous interfaces and phases.

I aim to connect local nanoscale properties to macroscopic behavior by linking rational synthesis of nanocrystalline ensembles to *in situ* characterization and analytic techniques. My research will confront these challenges at both the single particle and ensemble scales:

1. How do surface and bulk energetics affect the kinetics and phase behavior of nanoparticles during dynamic transformations, such as battery charging or photocarrier migration? This question delves deeper than a simple comparison of volumetric and surface energy. Metastable phases can be kinetically trapped during transformations, particularly in nanomaterials, and dynamic surface reconfiguration can lead to unexpected responses. I will manipulate particle size, shape, composition and surface chemistry to study these responses.
2. What effect does polydispersity in particle properties have on the macroscopic properties of nanomaterial ensembles? At rest, an ensemble relaxes to equilibrium. However, during dynamic transformations, slight variations in particle properties can dictate bulk kinetics and energetics. My studies will quantify and control polydispersity in ensembles to modulate bulk material properties.

Nanocrystalline ensembles can undergo rapid transformations that occur at sub-second time scales and across nanoscale interfaces, limiting the utility of many *in situ* characterizations. *In situ* optical spectroscopies can probe rapid coloration of semiconductor nanomaterials, with specificity to particular spectral features and spatial resolution approaching microns. Moreover, well-established optical spectroscopies are amenable to versatile *in situ* charging, excitation and other schemes. My group will combine these analytic strategies with precise colloidal synthesis and other techniques to investigate responsive nanomaterials for electrochemical and optoelectronic energy technologies.

Post-doctoral Research: During my post-doctoral research with Prof. Michael Chabinyc at UC Santa Barbara I have studied 2D layered hybrid perovskites for thin-film photovoltaics. These materials are tunable and robust analogues of 3D lead halide perovskites (e.g. $MAPbI_3$), and show exceptional promise for low-cost solar technologies. 2D perovskites are composed of atomically-thin layers of semiconducting lead halide sheets separated by insulating spacer molecules. The atomic layering of 2D perovskites creates similar properties to confined nanomaterials, namely quantum confinement with tunable optical and electronic properties, which extend throughout the bulk crystal. I investigated how the distribution of sheet thickness (i.e. how many lead atoms in each 2D layer) can be synthetically controlled, and how ensemble polydispersity in this layering affects bulk properties. I discovered through *in situ* optical and structural techniques that intermediate reactions skew the expected distribution of 2D layered phases. By carefully tuning reaction conditions, I revealed simple methods to control the distribution of layered phases – and decrease the concentration of detrimental defects – in thin-films. Building on these results, I then explored how accelerated reaction kinetics form colloidal dispersions of 2D perovskites. The insulating spacer molecules in these dispersions, composed of alkylammonium cations that ionically bond with lead halide sheets, dynamically interact with one another and with the solvent to direct colloidal structure. I discovered a distinction between alkylammonium cations that bridge adjacent lead iodide sheets into layered structures, and other spacer cations that act as terminating surfactants to stabilize the colloidal dispersion. This discovery provides a rational template to control the size, shape and structure of colloidal 2D perovskite nanoparticles, and demonstrates how ensemble dispersity can tune the bulk optoelectronic properties of practical devices.

Doctoral Research: My doctoral work with Prof. Delia Milliron at the University of Texas explored the electrochromic effect in metal oxide nanocrystal films. I discovered that distinct electrochemical charging processes, such as surface capacitance, ion insertion, or metal-insulator transitions, cause independent colorations in nanocrystal films. During my studies of TiO_2 nanocrystals, I resolved independent, reversible infrared and visible electrochromic responses by carefully controlling the electrochemical potential in a Li-ion electrolyte. My work showed that infrared coloration is caused by nanoscale confinement of free electrons during capacitive charging, leading to tunable plasmonic interactions with light. The small size of TiO_2 nanocrystals also allows small cations, such as Li^+, to rapidly diffuse into interstitial voids in the crystal structure, inducing a visible darkening. My studies revealed that a single nanostructured material can control infrared and visible optical coloration for next-generation 'smart window' coatings. Moreover, my measurements of independent optical responses to charging demonstrate a sensitive *in situ* probe of local nanoscale transformations. I used the infrared plasmonic response to make fundamental discoveries of charge transport in anisotropic metal oxide nanocrystals, and exploited the visible coloration of lithiated TiO_2 particles to probe ion diffusion kinetics. These studies have revealed the potential for full-spectrum *in situ* optical spectroscopies to measure dynamic transformations in nanomaterial ensembles.

Teaching Interests: My goal as a teacher is to cultivate an engaged and responsive learning environment. As a graduate teaching assistant I helped my doctoral advisor build two chemical engineering courses from : an undergraduate survey course in materials science and a graduate course on applied solid-state physics for materials science. I employed tools that I acquired from curriculum design courses during my graduate studies to help incorporate simple feedback channels throughout both classes, including peer-directed projects, low-stakes quizzes and mid-semester surveys. As a mentor, I also strive for open communication and regular feedback to ensure a productive relationship. My work frequently recalls my core chemical engineering curriculum, and I will use my research experience as a teaching device within the canonical Chemical Engineering curriculum. I have been a consistent advocate of STEM education for under-represented and resource-limited communities, and as a faculty member I will continue to seek opportunities to advance inclusion and representation in science.

AIChE 2019 Presentation: "Spanning the Atomic to the Agglomerate Dimensions in Colloidal Dispersions of 2D Lead Halide Perovskites", Session: *Synthesis and Characterization of Materials for Optical and Electronic Applications*, Nov. 14th, 8:40AM.

581554 Applications of Nanofillers in Advanced Composites

Research Interests: As a researcher, I plan to expand my work on polymer composites which includes polymer derived ceramics, using MXene in polymer composites, additive manufacturing of thermoset & ceramic composites, and molecular simulations of nanofiller-polymer interface. My research goal is to focus on combining nanoscale processing, material characterization and reactive molecular dynamics to develop a deeper understanding of nano-mechanics and how properties of nanofillers translate to large scale composites. Recently, I have been developing Data Science based algorithms to understand nanofiller-polymer interactions in thermosetting matrices with aim to extend these methods to other nanocomposites.

Teaching Interests: The foundation of my teaching philosophy is to educate and foster critical thinking and problem-solving strategies in my students and further facilitate the acquisition of life-long learning skills. As a recipient of Distinguished Teaching Fellowship at Penn State, I realized that teaching is my passion. Apart from teaching undergraduate courses, I plan to develop graduate courses on "Atomistic Scale Simulation Methods for Engineers" and "Nano-structured composite".

581328 Charge Transport in Self-Assembled Biomolecular Systems

Research Interests:

Charge transport in self-assembled molecular systems is not only essential for energy transfer in biological systems but also the fundamental basis for molecular electronics. In particular, biomolecules exhibit promising capability as building blocks for constructing hierarchical molecular systems. My research focuses on understanding the charge transport mechanisms in biomolecular systems, and leveraging this understanding to develop new integrated circuits via self-assembly of electronic functional biomolecules, inspired by neuron systems in nature. My research will directly address the fundamental scientific issues in these systems by leveraging a unique set of skills in molecular nanomaterials.

Teaching Interests:

My interdisciplinary research background provides me the fundamentals necessary to further extend my teaching abilities to additional classes. As far as topic areas, in an undergraduate curriculum, I could teach any entry-level courses. With my interdisciplinary research background, I can relate the fundamental principles learned in the course work to new field in chemical and biomolecular engineering and materials science and engineering, including Soft Materials, Surface Engineering, Electronic Materials, Advanced Materials, Energy Materials, Mechanics of Materials, Thermodynamics of Materials. I have a strong interest in teaching dedicated courses in nanomaterials and polymer science like: Nanomaterials and Nanotechnology, Nanoelectronics, Inorganic Chemistry, Polymer Physics, Physical Chemistry of Polymers, Introduction to Modeling and Simulation.

578098 Rational Synthesis and Assembly of Doped Nanomaterials for Quantum and Optoelectronic Applications

Research Interests:

The ability to engineer materials with precise atomic structure over large length scales represents one of the major challenges in chemistry and materials science. Fundamental studies to achieve this goal have led to widespread applications in medicine, chemistry, physics, and engineering. Among the many possible synthetic parameters, optically driven processes play an increasingly important role due to their ability to resonantly target nanomaterials and to create highly non-equilibrium environments, enabling the exploration of new chemistries and technologies.

I'm interested in leveraging photonic effects to demonstrate high fidelity control of nanomaterial composition and assembly across a wide range of length scales. Specifically, I'm interested in the synthesis and characterization of doped materials for quantum computing and sensing applications. During my presentation, I'll outline a research plan to study the properties of nanomaterials and to develop scalable processing methods for the integration these materials into macroscopic devices for application in quantum technologies, metamaterials, and photochemistry.

Teaching Interests: While I'm specifically interested in teaching transport processes and developing a class on light-matter interactions in nanomaterials that emphasizes modeling, I'm excited to teach the entire chemical engineering curriculum. Specifically, I'm excited to incorporate example based learning where students learn by encountering examples of phenomena before learning the equations the underscore a phenomena.

579153 Designing Nanomaterials and Electronic Devices for Multi-Level Neural Interfacing

Research Interests:

Recent developments in materials, electronics and genetics have enabled both deeper understanding of nervous system as well as enhanced therapeutic solutions in neural diseases. The goal of my research program will be developing novel nanomaterials and flexible electronic devices for improved modulation and recording of neural activity, from inside the neurons to outside of the skin. The initial research areas in my proposed program will include: 1) understanding and improving neuronal uptake and axonal transport of functional nanomaterials for non-genetic, projection-specific neuromodulation and imaging. 2) developing advanced flexible electronic skin patches for treatment of peripheral nerve disorders.

My postdoc work with Prof. Karl Deisseroth (Department of Bioengineering, Stanford University) has focused on developing a non-genetic method for projection-specific modulation of neural activity. The current optogenetics and virology technology have demonstrated that modulating the neural activity of specific projections in the brain is capable of rescuing behavioral deficits, including those relevant to depression, autism and anxiety. However, this approach requires genetic modification of neurons via viral transduction of transducers of visible light, and light penetration depth is limited by the scattering of visible light. In my approach, gold nanorods are used as near–infrared light absorbers and heat generators for neuron modulation. I found that suitably functionalized gold nanorods are rapidly internalized by neuron soma or axon terminals, are transported retrogradely/anterogradely through axons and are sufficient to modulate neural activity under near-infrared (NIR) light after axonal transport both *in vitro* and *in vivo*. This nanoparticle-based methodology demonstrates a promising approach to non-genetic, projection--specific modulation of neural activity in freely behaving mice.

My graduate work with Prof. Zhenan Bao (Department of Chemical Engineering, Stanford University) has focused on developing conjugated polymer sorted carbon nanotubes for flexible electronics applications and it consists 3 parts. In part I, I developed several approaches for selective dispersion of semiconducting carbon nanotubes by conjugated polymers: Specifically, I designed planar backbone polymers and polymers with longer side chains to increase the surface contact area and improve the dispersion yield of semiconducting carbon nanotubes. I also found that only non-polar solvents allow the dispersion of semiconducting carbon nanotubes and proposed a new mechanism to explain the selective dispersion of semiconducting nanotubes. In part II, I developed several methods to improve the characteristics of carbon nanotube transistor for flexible electronics, including: 1) using printed molecular dopants to control the threshold voltage of carbon nanotube transistors for flexible circuits with robust performances, 2) applying fluorinated polymers to remove hysteresis and improve devices' stability of carbon nanotube transistors, 3) applying N-type polymers to sort semiconducting carbon nanotubes for ambipolar transistors and circuits. In Part III, I utilized shape memory polymers to fabricate shape-controlled 3D-carbon nanotube transistors and sensors, which can wrap around any irregular surfaces, like our eyeglasses and key rings.

Teaching Interests:

During my PhD and postdoc training at Stanford, I have mentored 7 students in total, including 3 graduate students and 4 undergraduate students. Most of the students I have supervised had co-authored papers with me. I have also been very actively involved in the Stanford American Society for Engineering Education (ASEE) during my entire time at Stanford, including serving as the Vice-president of the society during 2014~2015. With another committee from ASEE, I co-developed and taught "Engineering 313: Topics in Engineering Education – Feedback and Assessment" class at spring quarter 2015 and was well received by the students. I have also been giving guest lectures to a variety of audiences, including the graduate course "CHEMENG 240: Micro and Nanoscale Fabrication Engineering", "MATSCI 384: Materials Advances for Neurotechnology" and undergraduate course "CHEM 26N: The What, Why, How and Wow's of Nanotechnology" for several years. I would be interested and capable in teaching the majority of the Chemical Engineering courses at both undergraduate and graduate levels, such as thermodynamics, kinetics, surfaces and interfaces, materials science, nanotechnology and neurotechnology.

Selective Publications:

1. **H. Wang**, L. Fenno, Y. Chen, C. Kim, C. Ramakrishnan, M. Inoue, S.Gambhir, K.Deisseroth, Projection-Specific Modulation of Neural Activity with a Non-Genetic Method, In preparation

2. J. Rivnay, **H. Wang**, L. Fenno, K. Deisseroth, G. G. Malliaras, Next-generation probes, particles, and proteins for neural interfacing, ***Science Advances***, 3, e1601649, (2017)

3. **H. Wang**, Y. Wang, B. C.-K. Tee, K.-P. Kim, J. Lopez, W. Cai, Z. Bao, Shape-Controlled, Self-Wrapped Carbon Nanotube Three-Dimensional Electronics, ***Advanced Science***, 2, 1500103, (2015)

4. **H. Wang**, Y. Li, G. Jiménez-Osés, P. Liu, Y. Fang, J. Zhang, Y.-C. Lai, S. Park, L. Chen, K. N. Houk, Z. Bao, N-type conjugated polymer-enabled selective dispersion of semiconducting carbon nanotubes for flexible CMOS-like circuits. ***Advanced Functional Materials***, 25, 1837-1844 (2015)

5. **H. Wang**, B. Hsieh, G. Jiménez-Osés, P. Liu, C.J. Tassone, Y. Diao, T. Lei, K.N. Houk, Z. Bao, Solvent effects on polymer sorting of carbon nanotubes with applications in printed electronics, ***Small***, 11, 126-133 (2015)

6. **H. Wang**, B. Cobb, A. Breemen, G. Gelinck, Z.Bao, Highly Stable Carbon Nanotube Top-Gate Transistors with Tunable Threshold Voltage, ***Advanced Materials***, 26, 4588-4593 (2014)

7. **H. Wang**, P. Wei, Y. Li, J. Han, H. R. Lee, B. D. Naab, N. Liu, C. Wang, E. Adijanto, B. C-K. Tee, S. Morishita, Q. Li, Y. Gao, Y. Cui, Z. Bao, Tuning the Threshold Voltage of Carbon Nanotube Transistors by n-type Molecular Doping for Robust and Flexible Complementary Circuits, ***Proceedings of the National Academy of Science***, 111(13), 4776-4781 (2014)

8. **H. Wang**, G. I. Koleilat, P. Liu, G. Jiménez-Osés, Y. Lai, M. Vosgueritchian, Y. Fang, S. Park, K. N. Houk, and Z Bao, High-Yield Sorting of Small-Diameter Carbon Nanotubes for Solar Cells and Transistors, ***ACS Nano***, 8, 2609-2617 (2014)

9. **H. Wang**, J. Mei, P. Liu, K. Schmidt, G.J.Oses, S. Osuna, L Fang, C.J. Tassone, A.P. Zoombelt, A.N.Sokolov, K.N.Houk, M.F.Toney, Z. Bao. Scalable and Selective Dispersion of Semiconducting Arc-Discharged Carbon Nanotubes by Dithiafulvalene/Thiophene Copolymers for Thin Film Transistors. ***ACS Nano***, 7, 2659-2668 (2013)

10. **H. Wang**, J. Luo, A. Robertson, Y. Ito, W. Yan, V. Lang, M. Zaka, F. Schaffel, M. H. Rummeli, G. A. D. Briggs, J. H. Warner. High performance field effect transistors from solution processed carbon nanotubes. ***ACS Nano***, 4, 6659-6664 (2010)

581739 Liquid Metal Encapsulation: Towards Next Generation Flexible Electronics

The frontiers of technology in the electronics industry are in bendable, lightweight, low-cost devices. Printing technologies for the fabrication of electronic circuits and devices allow for the use of flexible substrates, offer further advantages in terms of speed, cost, material usage, and environmental footprint, and can leverage the existing technology and scale of a $160 billion dollar printing industry. Printing technologies also stand to greatly transform the U.S. Department of Defense's typical operations using highly customized, low-volume parts needed in remote locations on tight timelines. My research interests are focused on the development of stable, conductive inks, and are balanced between fundamental colloidal science and leveraging this knowledge into practical impact on the advancement and development of manufacturable devices.

Metals such as the eutectic gallium- indium (EGaln) alloy that are liquid at room temperature are of particular interest for these applications. This class of materials uniquely combines the thermal and electrical conductivity of a metal with the softness and pliability of a liquid, but, in contrast to mercury, are nontoxic with low vapor pressures. However, these metal alloys are not without challenges. Trace amounts of oxygen will react with gallium and form an insulating skin layer, which can be a nuisance as it will adhere to most surfaces and leave behind undesirable residue. This oxide also renders printing from the bulk metal liquid largely unfeasible. The advancement of technologies utilizing these liquid metal materials hinges on the development of new strategies for managing these interfaces and developing them into colloidal dispersions and inks that can be easily processed.

Graphene oxide (GO) and related materials have been shown previously to function as high performance interfacial micro-barriers in emulsified systems. This work expands from previously studied organic/aqueous systems to both metal/aqueous and metal/organic particle dispersions, and demonstrates the first encapsulation of EGaln liquid metal particle cores by two-dimensional materials. Details discussed will include the role of the continuous solvent and graphene surface chemistries necessary to mediate the self-assembly of the graphene coatings. Such particles can be formulated into inks for implementation into incumbent printing technologies, and engender new technology classes of ultra-soft conductive devices and composites.

Research Interests:

Colloid science, carbon science, nanotechnology, green manufacturing, flexible electronics

Teaching Interests:

Colloid science, introduction to intellectual property, communication skills for scientists and engineers, core classes

582457 Design of Flow Reactors and Functional Nanomaterials for Energy & Photocatalytic Applications

Research Interests

My research interests are strongly related to the transformation of flow-based reactors from alternative synthetic platforms to standard and scalable systems. Such systems can precisely engineer functionalized nanoscale objects, which can act as building blocks for the development of efficient and renewable energy sources for environmentally sustainable societies. In addition, I envision that flow reactors can be used as scalable and inexpensive platforms for solar energy conversion.

During my Ph.D. at ETH Zurich, I focused on the development and application of novel segmented-flow microfluidic platforms with incorporated in-situ optical detection systems for efficient and fast screening of reaction conditions and a better understanding of the mechanisms during nanocrystal synthesis.

Currently at MIT I am broadening my scientific knowledge on multistep chemistries and single nanocrystal spectroscopy. In particular, I am currently investigating the potential of using miniature continuous stirred-tank reactors for the facile synthesis and upscaling of core-shell quantum dots and functionalized oil-field nanomaterials. In addition, my current research focuses on the development of detection modalities, which can be coupled to flow systems aiming for understanding the quantum dot photo-physics at the single particle level.

My future research activities will focus on (I) the development of autonomous and reconfigurable reactor systems for the design of nanomaterials for energy-related and optoelectronic applications, (II) strategies for the formation of metal oxide photocatalysts with morphology control, and (III) the design of photocatalytic flow reactors for solar energy conversion and environmental remediation. My long-term goal is to establish my research group as a pioneer for (a) the design and discovery of nanomaterials using autonomous and reconfigurable systems and (b) the development of flow chemistry strategies for solar energy conversion and environmental cleanup. Last, I would also like to expand my interdisciplinary knowledge into fields such as flow separation processes and design of polymeric nanoparticles for oil recovery applications.

Teaching Interests:

I believe that adapting a teaching philosophy to practice takes effort, but I hope to inspire in my students the passion for learning that has led me to remain in academia. In this direction, I have served two semesters as a teaching assistant for the Biomicofluidic Engineering course at ETH Zurich, while designing a laboratory exercise for graduate students. During the past six years I also have been fortunate to have supervised a number of graduate and undergraduate students on conducting experimental and analytical work. While conducting my Ph.D. and postdoctoral work I was the direct supervisor of nine graduate and five undergraduate students conducting their master and bachelor theses with two of them gaining the ETH medal for the best master thesis. In addition, I recently completed the Kaufman Teaching Certificate program provided by MIT to further develop my teaching skills. The program also included two microteaching sessions.

I am ready to implement this pedagogical knowledge to teach a range of courses from the chemical engineering curriculum both at the undergraduate and graduate level, including reaction engineering, transport phenomena, fluid mechanics and reactor design. In addition to current course offerings, my research interests drive me toward the development of a course related to microfluidic engineering with emphasis on fundamental concepts of transport phenomena, while focusing on principles and theory of flow chemistry for the synthesis of functional nanomaterials targeting energy applications. I feel passionate about this subject matter and I can teach state-of-the-art concepts.

581973 Multi-Sensor System for Complex Chemical Environment Detection

Research Interests:

Sensors make everything smart. Bio/chemical sensors have a wide application in industrial processes and our daily life. Sensitivity, Selectivity and Stability are three key performance indicators of a chemical sensor. While selectivity has always been the most challenging issue for most of the chemical sensors, there are increasing demand for ultra-sensitive sensors for trace chemical detection (e.g. breath sensors for health diagnosis). And stability is also extremely important for industrial applications, where reliability and long lifetime are desired. My research goal is to build multi-sensor system for complex chemical environment, which can be applied for environmental monitoring, food quality and safety detection, and health diagnosis, etc. My research interest lies in designing novel ultra-sensitive nano-structured chemical sensing materials, developing miniaturized sensors arrays, and using machine learning algorithms to process sensor data for species identification, concentration determination, baseline drift and environmental compensation.

Teaching Interests:

I'm interest in teaching core courses at both undergraduate and graduate levels, such as thermodynamics, kinetics, transportation phenomena, electrochemical engineering, etc. I'm also eager to developing courses related bio/chemical sensing technologies.

580865 Evaluation of the Influence of Ce4+/Ce3+ Redox-Couple on the Cyclyc Regeneration of NiO-Pdo/CeO2 Nanoparticles for Asphaltene Steam Decomposition

Research Interests:

The main objective of this study is to evaluate the self-regenerative property of functionalized $CeO_{2\pm}$ nanoparticles with a mass fraction of 0.89 % of NiO and 1.1 % of PdO in adsorption processes and subsequent decomposition of *n*-C_7 asphaltenes in steam gasification processes though a redox cycle.

During each regeneration cycle, the adsorption capacity and the catalytic activity of the nanoparticles were evaluated. To estimate the adsorption capacity of the nanoparticles, adsorption kinetics were studied at a fixed concentration of n-C7 asphaltenes of 10 mg·L-1 as well as adsorption isotherms at three different temperatures at 25°C, 55°C, and 75°C. To evaluate the catalytic activity, the loss of mass of the nanoparticles was evaluated at isothermal conversions with a thermogravimetric analyzer at 230°C, 240°C, and 250°C, and at non-isothermal conditions involving heating of 100°C to 600°C at a 20 °C·min-1 heating rate.

The nanoparticles showed a high affinity for adsorption of n-C7 asphaltenes, and this was maintained during all evaluated regeneration cycles, as observed in the Henry's constant that increased slightly, with changes of 21%, 26% and 31% for 25 °C, 55 °C and 75 °C. Polanyi's adsorption potential decreases by 2.6% for the same amount adsorbed from the zero cycle to the eight. In addition, the catalytic activity of the nanoparticles did not change significantly, showing that they decompose 100% of asphaltenes in all cycles at 210 °C. However, the small decrease in the adsorption capacity and catalytic activity of the nanoparticles is mainly due to the presence and change in concentration and ratio of certain elements such as oxygen and iron at the surface of the nanoparticle, as shown by XPS analyses. Thermodynamic parameters such as, change in enthalpy, entropy and Gibbs free energy, and effective activation energy were calculated in order to compare adsorptive and catalytic performance during each cycle. There is an increase of 9.3% and 2.6% in the case of entropy and enthalpy respectively, and a decrease of 0.5% at 25°C for the Gibbs free energy from cycle 0 to 8 evaluated.

Teaching Interests:

Among thermal enhanced oil recovery, in particular, cyclic steam stimulation (CSS) is a technique commonly used for heavy and extra-heavy oil production. However, this method presents a great limitation in that are economically viable the first 6-7 production cycles, due to the extra-heavy oil residue retained in the reservoir, whose decomposition temperature exceed the operation temperature of de CSS and the lightest oil components vaporization. Taking into account the self-regenerative properties of the CeO2 nanoparticles to change their oxidation state from Ce4+ to Ce3+ in a redox-cycle, and the catalytic advantages that the supporting material acquires with a high metal oxides dispersion over its surface, which allows it to decompose at low temperatures (210 °C) the asphaltenes in the presence of steam, it is proposed to evaluate their performance in catalytic regeneration cycles through continuous adsorption and subsequent decomposition of n-C7 asphaltenes in gasification processes. In this way, if the porous media is decorated with these nanoparticles that are continuously adsorbing and decomposing the heaviest oil fraction, the coke inside the reservoir will be diminished, increasing the recovery factor and the useful life of the technique.

582063 Beyond Lithium Aqueous Electrochemical Energy Storage

Beyond Lithium Aqueous Electrochemical Energy Storage

Daniel S. Charles

Department of Chemical Engineering, University of New Hampshire, Durham, NH

Research Interests:

Electrochemical energy storage has become prevalent in daily life for their use in mobile electronics, and electric vehicles. More recently, the need for large scale stationary energy storage to allow for the integration of clean and renewable alternative energy sources to the electric grid has arisen. The shift towards clean energy utilizing alternative energy sources, such as wind and solar energy, will reduce CO_2 emissions and the dependence on fossil fuels. A large-scale energy storage solution is required for load shifting and frequency regulation due to the inconsistent nature of the electrical energy generation from renewable sources. In this regard, aqueous electrochemical energy storage is a promising solution to these problems. Aqueous electrochemical energy storage devices using beyond Li-ions as charge carriers (e.g. Na, K, Mg, Zn) offer an attractive solution to this problem because of their superior safety, lower cost and excellent transport properties compared to non-aqueous devices. However, the energy density and cycle life of aqueous electrochemical energy storage devices are currently insufficient. There are two approaches for improving the energy density of aqueous electrochemical energy storage devices: one is to increase the capacity of the electrode material and the other is to increase the operating voltage. The operating voltage of aqueous electrochemical energy storage devices is limited to 1.23 V by the thermodynamically stable potential window of water, beyond which, the gas evolution reactions occur. My research has focused on gaining a fundamental understanding of and making iterative improvements to electrode materials for aqueous electrochemical energy storage to support implementation of this technology as a large-scale energy storage solution.

My research experience in energy storage has covered the entire scope from synthesis of nanomaterials to fabricating prototype pouch cells. My doctoral and postdoctoral work with Professor Xiaowei Teng at the University of Hampshire explored the structure-function relationship of nanoscale and/or disordered metal oxide electrode materials for aqueous electrochemical energy storage. We are particularly interested in the interaction of water at the interphase and within the structure of the electrode materials. My work on these materials included detailed electrochemical and structural characterization including X-ray and Neutron total scattering measurements and analysis of the Atomic Pair Distribution Function (PDF). To study the structure-function relationship, I designed an electrochemical cell for in situ synchrotron X-ray diffraction and total scattering measurements to study the structural evolution of the electrode with potential during cycling.

Two research highlights from my time at the University of New Hampshire were published in Nature Communications. The first, we reported the formation of layered Mn5O8 pseudocapacitor electrode material with a well-ordered hydroxylated interphase. This unique interphase allowed for a 3.0 V kinetically stable potential window in an aqueous sodium sulfate electrolyte by suppressing the gas evolution reactions. Secondly, we reported on the water induced structural rearrangements of the vanadium-oxygen octahedra which enhanced the stability of the layered highly disordered potassium-intercalated vanadium oxide nanosheets. The promotional effects of structural water lead to much-improved K-ion storage capacity (183 mAh g^{-1} in half cell at a scan rate of 5 mV s^{-1}, corresponding to 0.89 charge transfer per vanadium) and stability of (62.5 mAh g-1 in full cells after 5,000 cycles at 10 C) of the electrode material.

Teaching Interests:

Teaching is a part of being a faculty member that I am looking forward too. As a postdoc, I have thoroughly enjoyed mentoring incoming Ph.D. students that have joined the group. Mentorship is something that I view as very valuable, as my mentors throughout my undergraduate and graduate studies have been integral to my success thus far in my career. I would strive to pass that along and cultivate interpersonal relationships with students. During my time at the University of New Hampshire, I was funded by a Research Assistantship and a Dissertation Year Fellowship. Although I was not assigned a teaching assistant position, I did have the pleasure of covering lectures for my advisor (Kinetics, Nuclear Engineering, and Electrochemical Methods), as well as help teach the laboratory section of the Electrochemical Methods course. I was fortunate to get some teaching experience as an undergraduate student at Rochester Institute of Technology. I served as a teaching assistant for the Math Techniques for Chemical Engineers course in the Fall of 2011. The following year I was a teaching assistant for two courses in the Fall of 2012, Math Techniques for Chemical Engineers and Reaction Engineering I. Both my graduate and undergraduate degrees are in chemical engineering, I am confident in teaching any of the core curricula. In addition, I would be able to teach technical electives in the field of material science and electrochemistry.

Highlighted Publications:

Charles, D.S., Feygenson, M., Page, K., Neuefeind, J., Teng, X.W. "Structural Water Engaged Disordered Vanadium Oxide Nanosheets for High Capacity Aqueous K-ion Storage". *Nature Communications*, **2017**, 7, 15520

Shan, X., Charles, D.S., Lei, Y., Qiao, R., Wang, G., Yang, W., Feygenson, M., Su, D., Xu, W., Teng, X.W. "Bivalence Mn5O8 with hydroxylated interphase for high-voltage aqueous sodium-ion storage". *Nature Communications*, **2016**, 7, 13370 (equal contribution first author)

582180 Interfacial Engineering of 2D Nanomaterials for Biomedical and Energy Application

Research Interests:

Since the emergence of single-layer graphene in lab in 2004, a vast library of 2D nanomaterials has been produced and much has been learnt about their structure-property relationships. Their unique properties have been investigated for applications in electronics, composites, biomedical and energy storage devices. Yet, we are far from systematic production and processing of these materials with well-controlled structure and function because little is known about their interfaces at solid, liquid, and gas phase. The nanosheet interfaces are governed by their statistical nature as ensemble of nanosheets and their anisotropic morphology, manifested in their size and surface chemistry polydispersity and complicated surface interactions with other compounds and their surroundings.

In my future lab, I will lead research efforts to obtain a comprehensive understanding of 2D nanomaterials interfaces and develop methods to engineer these interfaces for applications in biomedical, energy, and catalysis solutions. I will investigate the chemical interactions of radicals, ions, small molecules, macromolecules, and other nanoparticles at the nanosheet-liquid and –gas interface in moieties such as biological media and electrolytes. Understanding and cataloging these interactions will allow me to tune the composition, size distribution, surface chemistry, and surrounding of nanosheets to exploit these anisotropic interfaces for innovative designs of chemical and biosensors, molecular membranes, and electrodes. Particularly, the physical interactions such as mechanical deformation, aggregation, penetration, uptake, and friction at the nanosheet-cell interfaces will enable me to develop novel drug-delivery and imaging platforms. At the nanosheet-gas interface, I intend to engineer highly efficient compartmental catalytic units targeting the CO2 capture and conversion.

My knowledge of nanosheets production methods, their colloidal interactions and assembly in aqueous media and the characterization platforms I have developed for studying polydispersed and anisotropic materials throughout my PhD and postdoctoral studies at TAMU and MIT will be the cornerstone of my future research as a faculty member. I will apply single-particle characterization tools, spectroscopic methods, and measurements of microscopic stresses to study interfacial energy, areal distribution, and conformation of adsorbed species on the nanosheets surface. Combining my experimental knowledge with the particle-based computational tools such as Brownian dynamics and using techniques from statistical mechanics will allow me to accurately predict and quantify the interfacial interactions of the nanosheets.

Teaching Interests:

As a new faculty, I will use my experiences as graduate teaching fellow and teaching assistant to guide the undergraduate students in gaining a fundamental understanding of the key concepts of chemical engineering. I expect all my students to apply these concepts to a wide variety of physical systems based on problem categorization. I will teach my students to tackle the smaller components of problems through defining the system in terms of proper mathematical models and then using key techniques to achieve a solution. This problem-solving algorithm enables the students to approach any engineering problem in the real-life and industrial applications. In regard to graduate education, my goal is to empower students to become independent academic researchers. I will focus my efforts on not only imparting scientific knowledge, but also teaching the methodology of scientific research and independent thinking. My ultimate goal is to produce researchers who think in a multidisciplinary fashion and can generate creative solutions for challenging and complicated situations. While my teaching experience allows me to feel comfortable with teaching any of the core courses in the chemical engineering undergraduate curriculum, my specific teaching interests include transport, numerical methods in chemical engineering, and reaction engineering. My teaching interests for elective and graduate level courses closely match my research interests, and I would like to teach or develop courses in the areas of Advanced Functional Materials, and Colloids and Interfacial Phenomena. In particular, the latter course will cover an overview of the fundamental topics in colloids including the colloidal interaction at atomic and molecular level, interfacial forces, and self-assembly of the colloidal particles.

570645 Understanding of Nanoparticle Self-Assembly Mechanisms and Its Applications to Energy Storage and Bio-Medical Applications

Research Interests:

The study of self-assemblies is a research topic in which I am very interested because it provides an efficient way to fabricate material structures (i.e., the formation of complex nanostructures, such as nanowires, branched structures, and nanoporous single crystals, as well as unique nanocluster structures, such as hetero-nanostructured materials and bottom-up nanofabrication methods, such as superlattice crystal structures) at large scales in a controlled process. One of the most interesting examples is the unique assembly of inorganic nanocrystals or colloid stability that involves control of superlattice crystal structure/cluster morphology, crystal growth via oriented attachment (OA) and spontaneous wetting-dewetting process of polymers or organic ligands and characterization of ionic structure at solid-liquid interfaces. The coupling between energetics (,e.g., van der Waals force, Electrostatic force, Steric hindrance force, and Brownian force) and dynamics (i.e., hydrodynamics) plays a crucial role. Moreover, non-DLVO forces can determine the short separation distance at equilibrium state. Therefore, the coupling in self-assembly process is expected to be correlated closely with various molecular details (e.g., surface modifications via ligand or polymer coating and solvent structures at interfaces) due mainly to the length scales of crystals (i.e., $O(1)$ nm). In addition, micelle shape, which is determined by chain length of hydrophilic chain and hydrophobic chain, can affect surface energy of particle and determine particle morphology during crystal growth in solution. In addition, the unique nature of particles would create additional complexity (e.g., the tensorial nature of the dielectric property due to the crystallinity).

Teaching Interests:

In addition to supporting extant courses in the Chemical Engineering Department, I endeavor to design graduate and undergraduate level courses in Colloid Interfacial Phenomena and Microfluidics and Thermodynamics. My intention is to guide students from fundamentals to related recent research topics. The following list provides for courses I am prepared to teach:

- **Colloid Interfacial Phenomena and Microfluidics (Graduate level):** Colloid interfacial science is the study of the behavior of particles. Scientific interest, along with experimental techniques and theoretical interpretations of interparticle forces, allows for the effective manufacture of synthetic dispersions for coating, enhanced oil recovery, the development of new fuels, environmental pollution, ceramics fabrication, corrosion phenomena, biotechnology, and separation processes. My aim is to impart a quantitative understanding grounded in basic theory and coupled to experiments on well-characterized model systems. I provide for the physical side of colloid science. Subjects range from individual forces, acting between submicron particles suspended in a liquid through the equilibrium, and dynamic properties of dispersion. Relevant forces include Brownian motion, electrostatic repulsion, attraction due to dispersion forces, attraction and repulsion caused by soluble polymers, and viscous forces arising from relative motion between particles and liquid.
- **Thermodynamics (Undergraduate level):** This course introduces the concept of thermodynamics form a chemical-engineering viewpoint. This class specifically targets an undergraduate level. It is relevant to students for studying the treatment of equilibrium thermodynamics in sufficient detail to solve a wide variety of problems. This class is designed to be conceptually-based, with examples of natural phenomenon, meant to provide students with a solid foundation of this subject. Conceptual development, numerous problems taking form as weekly assignments, and term-projects, promote deep learning and provide students the ability to apply thermodynamics to real-world engineering problems.

Research Abstract:

The study of self-assemblies have attracted increasing attention because it provide an efficient way to fabricate material structures(i.e., the formation of complex nanostructures, such as nanowires, branched structures, and nanoporous single crystals, as well as unique nanocluster structures, such as hetero-nanostructured materials and bottom-up nanofabrication methods, such as superlattice crystal structures) at large scales in a controlled process. For these reasons, interesting examples what I studied before are the self-assembly characterization of nanomaterials, simulation of nanoparticle coagulations and spontaneous wetting/dewetting process of polymers or organic ligands.

To understand kinetic details of the self-assembly process, the self-assembly of gold NPs are further analyzed by calculating the contributions of various forces involved [3-4]. Our results show that these forces, including Brownian force (F_{Br}), van der Waals force (F_{vdW}), electrostatic force (F_{elec}), and hydrodynamic force (F_D), and their interplay play a key role in NP self-assembly process. The superlattice structure can be controlled *via* solvents or ligands. For a given particle pair, the repulsive steric hindrance force (F_{Sh}) or hydration force (F_{hyd}) increases with decreasing h and balances with attractive F_{vdW} at an equilibrium separation, here measured as h_{eq}, = 2.8 nm. This prediction is validated by our experimental observations; the particles assembled into a superlattice with an average h of 2~3 nm. Meanwhile, I observed 0.7 nm of h as NPs dispersed in aqueous solution contained salts. It is demonstrated that 0.7 nm of h was attributed to the formation of strong water structure on the NP surface due to hydrated ions.

Based on these knowledge, I developed the hetero-structured metal oxide nanoparticles for lithium ion battery anode materials and gold-nanorods encapsulated by gas generating amphiphilic block copolymers under external stimulus. The synthesized Co_2MnO_4 hetero-structured nanoparticles from CoO/Mn_3O_4 core/shell nanoparticles demonstrated superior electrochemical properties with a large capacity and high stability during fast charging/discharging cycles for potential applications as advanced lithium-ion battery (LIB) electrode materials. The heterogeneous structures of $Cu_{3.8}Ni/CoO$ and $Cu_{3.8}Ni/MnO$ nanoparticles could enhance the lithium ion mobility and improve the life cycle, and these materials are therefore promising candidates as high-power-density and high-energy-density anode materials for lithium-ion batteries in diverse applications, such as electrical vehicles. Meanwhile, block copolymers coated gold nanorods showed a greater absorbance/heat generation rate under IR irradiation. This system produce ultrasound contrast enhancement effects under continuous exposure to IR light for >20 min.

These results provide insights into the self-assembly of nanoparticle and polymer, and its applications. The obtained knowledge of interactions between surface coating materials can be extended to core-shell/hetero-structured nanomaterials or polymer coated nanomaterials that can be applied to the field of lithium ion batteries and ultrasound contrast.

References:

[1] M Boles, M Engel, and D. V. Talapin. Chemical reviews 116, (2016), p11220-11289.

[2] J. M. Yuk, J. Park, P.Ercius, K. Kim, D. J. Hellebusch, M. F. Crommie, J. Y. Lee, A. Zettl,; A. P. Alivisatos, , Science 336, (2012), p61-64.

[3] W.B. Russel, W B. Russel, D. A Saville and W. R. Schowalter, in "Colloidal dispersions", ed G.K. Batchelor (Cambridge university press, New York) p3-22

[4] J. N. Israelachvili, in "Intermolecular and surface forces", Academic press, (Elsevier, Santa Barbara) p3-22

581822 Colloidal Electronic Cells As Building-Blocks for Synthetic, Tissue-like Scaffolds

Research Interests:

Carbon nanotubes (1D), graphene (2D), transition metal dichalcogenides (2D) and other low dimensional materials possess desirable mechanical and electrical properties for incorporation *into* or *onto* novel colloidal particles, granting them unique electronic and chemical functions. We have shown previously that single-walled carbon nanotubes can be printed as flexible electronic 'skins' that harvest electrical power via molecular interactions with the embedded solvent environment (through an effect called '*asymmetric chemical doping*').[1-3] The electricity so harvested have been explored for on-board energy generation for particulate devices suspended in solution[4, 5] and for *in situ* electrochemical conversion of small molecules.[6] We also demonstrated the symbiotic engraftment of various 2D materials (*e.g.*, graphene, MoS_2 and *h*BN, *etc.*) onto colloidal particles, making colloidal electronic systems with form-factors reminiscent to those of a biological cell (*c.a.* 10-100 μm in diameter),[7] with the ability to perform autonomous functions such as optical energy harvesting, chemical detection, and digital memory recording.[7-9]

These advances paved the way for us to build autonomous electronic 'cells' by wiring up synthetic 'organelles', combining the modularity that stems from micro-electronic platforms with the versatility that naturally arises in synthetic material systems. The ability to prepare these 'cellular' building-blocks at scale prompts yet further efforts to explore higher-order assemblies thereof, into hierarchical synthetic 'tissues' where complex functions emerge, via a suite of top-down and bottom-up techniques. Afterall, nature builds biological scaffolds with ordering that spans many orders of magnitude, affording remarkable control not only at the molecular (*e.g.*, protein) level, but also on the micro- (*e.g.*, organelle) as well as meso- (*e.g.*, cellular) scale. Why should artificial materials design be constrained by control and heterogeneity only at the nanoscale (and occasionally at the microscale)? In the wake of recent developments in colloidal electronics fabrication, 3D printing and guided self-assembly, today's chemical engineers are well poised to address the following problems: (i) how do individual devices assemble into higher-order structures with powerful attributes? (ii) how do simple interactions transmit information hierarchically? (iii) how can we harness and coordinate these interactions to efficiently create dynamic and complex systems?

I will present a proposed research program positioned at the intersection of *materials design*, *chemical synthesis*, and *electronic device fabrication* to help answer these questions that may bear far-reaching scientific and engineering significance. I will demonstrate how principles in chemical engineering provide us a unique toolset to build novel material architectures that address challenges in the medical, environmental, and materials sciences.

References

(* denotes equal contribution; † denotes corresponding author)

[1] SG Mahajan,* AT Liu,* and MS Strano† et. al. Sustainable power sources based on high efficiency thermopower wave devices. *Energy & Environmental Science* 9 (4), 1290-1298

[2] AT Liu,* Y Kunai,* and MS Strano† et. al. Electrical energy generation via reversible chemical doping on carbon nanotube fibers. *Advanced Materials* 28 (44), 9752-9757

[3] Y Kunai,* AT Liu,* and MS Strano† et. al. Observation of the Marcus inverted region of electron transfer from asymmetric chemical doping of pristine (n, m) single-walled carbon nanotubes. *Journal of the American Chemical Society* 139 (43), 15328-15336

[4] AT Liu and MS Strano et. al. Direct electricity generation mediated by molecular interactions with low dimensional carbon materials - a mechanistic perspective. *Advanced Energy Materials* 8 (35), 1802212

[5] AT Liu, G Zhang, and MS Strano.† Energy harvesting techniques mediated by molecular interactions with nanostructured carbon materials. *Robotic Systems and Autonomous Platforms*, 389-424

[6] AT Liu, Y Kunai, and MS Strano† et. al. Solvent induced electricity for in situ electrochemistry. *manuscript in preparation*

[7] P Liu,* AT Liu,* and MS Strano† et. al. Autoperforation of 2D materials for generating two-terminal memristive Janus particles. *Nature Materials* 17 (11), 1005

[8] VB Koman, AT Liu, and MS Strano† et. al. Colloidal nanoelectronic state machines based on 2D materials for aerosolizable electronics. *Nature Nanotechnology* 13 (9), 819-827

[9] AT Liu, and T Palacios,† MS Strano† et. al. Colloidal State Machines. *Nature Reviews Materials. in revision*

Teaching Interests:

Coming from a family of academics, I aspire to teach and conduct research in chemical engineering at a collegiate level. My undergraduate experience was in a way quite unique, in that I obtained two degrees at two different colleges (BA in Chemistry at Grinnell College and BS in Chemical Engineering at California Institute of Technology) through a transfer program. While my experiences the liberal arts college versus the engineering focused institute were different, one consistent theme throughout my academic career has been my commitment towards mentoring fellow classmates. I TAed for 5 different classes as an undergraduate and 2 more (the graduate level transport phenomena taught by Profs. Bill Deen and Martin Bazant, as well as the graduate level reaction engineering taught by Profs. Klavs Jensen and Michael Strano) since I joined MIT. My recitations were generally well-received and most student have found the extra notes I wrote after each lecture helpful. As a result, I was voted the best graduate TA for 2 separate years (2017 and 2019) in the department for the 2 classes I was involved in, and was nominated for the Goodwin Teaching Medal (awarded annually to one "conspicuously effective" TA at MIT). I have also served as a Teaching Development Fellow here at MIT and hosted regular workshops for my fellow TAs in the effort to foster a better pedagogical atmosphere across the entire institute. These experiences have further honed my teaching skills and prepared me for the future role as a teacher. In fact, being able to effectively communicate what I have learned and seeing it evoke passion among the students represent one of the most satisfying moments in recent memory.

581993 Enhancement Routes of Corrosion Resistance in Steel-Reinforced Concrete By Using Nanomaterials

Research Interests: Nanomaterials, Corrosion

Teaching Interests: Materials Engineering

Corrosion of steel in reinforced concrete has been found to be a significant factor leading to premature failure of concrete constructions. The most important causes of corrosion initiation in reinforcing steel are the ingress of chloride ions and carbon dioxide to the steel surface. After initiation of the corrosion process, the corrosion products (iron oxides and hydroxides) are usually deposited in the restricted space in the concrete around the steel. Their formation within this restricted space sets up expansive or tensile stresses, which crack and spall the concrete cover. In the tension zone, concrete will always exhibit cracks. Aggressive liquids and gasses may enter these cracks and they may cause concrete degradation. Because of such cracking, aggressive substances may reach the steel reinforcement and induce corrosion that may lead to further concrete damage and possibly structural failure.

Concrete, that is not exposed to any external influences, provides physical corrosion resistance to steel reinforcement by acting as a barrier, and chemical corrosion resistance as a result of its high pH (usually exhibits a pH between 12.5 and 13.5), which defines the range of electro-chemical potential and pH for a protective passive layer formation on the surface of steel. It is believed that this layer is an ultrathin (<10 nm) protective oxide or hydroxide film that decreases the dissolution rate of steel to negligible levels. For steel embedded in concrete, the dissolution takes place in the limited volume of water solution present in the pores of the concrete surrounding the steel.

Extensive research has been conducted over the last decades to solve the durability and corrosion problem in reinforced concrete. Several methods have been proposed to provide greater corrosion resistance of reinforced concrete system, including: (1) concrete admixtures; (2) corrosion inhibition; (3) electro-chemical methods; and (4) surface treatments (steel and concrete).

Nowadays, the application of nanomaterials to enhancing the conventional concrete properties, such as mechanical strength and durability, has received significant attention. For the purpose of concrete study, the application of nanomaterials must be at least 500 nm in size. Nano-concrete is a concrete that contains nanomaterials or a concrete with nanomaterials added in the mixture, in which the size of the nano-particles is less than 500 nm. This poster presentation will address the impact of nanomaterials on controlling the corrosion rate of reinforced concrete by considering the following methods: 1) nano-admixtures and 2) surface treatments of concrete and steel by nanomaterials.

581442 Eco Friendly Strategies for Nanocellulose Production Using a Non Commercial Enzymatic Cocktail

Research Interests: Nanotechnology, Biochemical Engineering, Sustainable Engineering,

Teaching Interests: Sustainable Engineering, Nanotechnology

Replacing non-renewable fossil sources by lignocellulosic biomass as raw material for the production of fuels and chemicals is the driving force for the development of sustainable biorefineries. However, the costs associated with the biomass processing still constitute the main economic barriers to biofuels and bioproducts competitiveness. A promising strategy to improve the biorefinery profitability is the production of enzymes on-site and high added-value co-products, such as nanocelullose, which can be obtained from the residual solid fraction of the cellulose saccharification process. These nanomaterials present excellent mechanical properties, good biocompatibility, tailorable surface chemistry, and interesting optical properties. Due to these distinguished properties, nanocellulose has attracted interest for several applications in food packaging, biomedicine, as mechanical reinforcement of matrices, among many others. The current industrial process to obtain nanocellulose employs acid hydrolysis, but the development of more environmentally-friendly strategies such as via enzymatic hydrolysis is highly desirable. In this context, the present work focused on producing home-made enzymatic cocktails and applying them to obtain nanocellulose. Firstly, cellulolytic enzymes were produced by Aspergillus niger cultivated under solid-state fermentation using wheat bran as solid substrate. Then, two strategies were evaluated to obtain nanocellulose: enzymatic hydrolysis of bleached kraft eucalyptus pulp (BKEP) followed or not by a sonication step. The BKEP fibers were pretreated in a ball mill prior to enzymatic hydrolysis. The reactions were carried out in Erlenmeyer flasks containing 50 mL of home-made enzymatic cocktail (fungal fermentation broth) and 2 % (w/v) of solids loading for 96 h and the glucose concentration and cellulose conversion were determined during the reaction. Scanning electron microscope was used to analyze the initial pulp fibers and the nanocellulose material from both strategies (enzymatic treatment and enzymatic combined with mechanic treatment) were characterized by Field Emission Scanning Electron Microscopy, showing that it was possible to obtain nanocellulose using a non-commercial enzymatic cocktail and the combination with mechanical treatment contributed to improve the dispersion of the nanocelullosic material.

581156 Interfacial Molecular Engineering of Bio-Inorganic Systems from Aqueous Molecules to Assembled High Order Functional Materials

Research Interests:

Biology heavily relies on the bottom-up self-assembly of bio-inorganic materials to create higher order functional systems that impact human health, *e.g.*, hard tissue biomineralization and joint bio-lubrication. Emerging technologies have benefitted by mimicking nature's approaches leading to potential remedies for disease and current societal challenges, *e.g.*, biosensing, drug delivery and efficient catalysis. As a future junior faculty, my research interest is on molecular engineering of active bio-inorganic interfaces for bioelectronic and biomedical applications, such as tailored biosensing and biomolecular logic devices. By utilizing local solution property gradients to direct the statistical distributions of aqueous biomolecular conformation, control over the molecular interactions, structure and transport properties of the bio-inorganic interface is possible. This research effort will include the collective interrogation of biophysical, interfacial, and transport phenomena *via* unique nanoscopic characterization tools, partially developed during my PhD, corroborated with computer simulations.

The main advantage of this molecular engineering approach is the ability to rationally engineer systems based on fundamentals sans trial-and-error. In my on-going research, I have successfully implemented molecular engineering principles to rationally design peptides with predictable assembly structures and binding energies at 2D inorganic surfaces. Key to my work has been the dual application of molecular dynamics simulations and the adaption of nanoscale energetic analysis techniques to locally interrogate how environmental selection of aqueous peptide conformation *via* temperature, pH, or electric fields impacted surface binding and assembly. While my current research focuses on the aqueous and surface adsorbed phases, my future research plan entails extending the analysis to the near surface boundary regime. The molecular engineering of the bulk deviating chemical potential and transport phenomena within this dimensionally confined layer would allow for immense control over the function of bio-inorganic systems with implications to emerging technologies, such as, directed nucleation, tailored sensing in bioelectronics, and improved tribological materials.

Teaching Interests:

My graduate training in Molecular Engineering has broadened my base beyond Chemical Engineering to include the intersections of Bioengineering, Materials Sciences, and Biochemistry. As the lines between traditional engineering fields continue to blur, specifically for emerging technologies, interdisciplinary aspects involving Nanoscience and Molecular Engineering (NME) become essential aspects in our educational programs starting at the undergraduate (UG) level to prepare students for the demands of our future workforce. The University of Washington has spearheaded the incorporation of NME into the core UG Chemical Engineering program. During my PhD, I have had the opportunity to take active part as teaching assistant (TA) in the incorporation and development of NME specific courses, as well as module developments for classical Chemical Engineering transport courses in the graduate program. As future faculty, my intent is to further refine my experiences and weave NME principles throughout my teaching so students can tackle problems currently insufficiently addressed by mere phenomenological approaches. This will include, besides theoretical classroom teaching, hands-on experimental and computational module development. My aim is to lead a lab that facilitates an environment in which undergraduates can develop their research experience and cultivate technical skills critical for industry or academia. Based on personal experience, hands-on individual research experience at the UG level greatly informs related course material and the professional interests of the student. Additionally, as I have learned from mentoring four undergraduates, this emphasis on UG research gives graduate students valuable experience leading student research, thus, better preparing them for their future research careers. Beyond the incorporation of NME principles into UG education, I plan to emphasize the critical evaluation of relevant scientific papers at the UG level, since this skill is of increasing importance as information (accurate or otherwise) becomes more readily available and disseminated *via* online sources.

574647 Meet the Faculty: Investigating the Properties of Nanoparticles Containing Fullerene-like Molecules

Research Interests: molecular simulation, biomedical engineering, nanoparticles, carbon

Teaching Interests: biomedical engineering, chemistry, process systems, fluids, thermodynamics, statics, mechanics

576064 Recombinant Peptide-Templated Palladium Nanoparticle Synthesis and Their Catalytic Activity in Coupling Reactions

The catalytic ability of synthesized palladium nanoparticles (Pd NPs) using recombinant peptides with narrow size distribution prepared in an impure environment was studied. The soluble fraction of bacterial lysates containing repeat units of a metal binding domain (Pd4) fused to the carrier protein Green Fluorescent Protein (GFPuv) was used to direct nanoparticle synthesis. Eliminating the costly purification of a peptide, highly ordered NPs on the order of 2 nm were synthesized. Indeed, the heterogeneous nature of the synthesis did not interfere with the formation of small nanoparticles. NPs can be used for cross-coupling reactions. Due to their high stability and the capability of using green solvents, impure recombinant peptide-directed Pd NPs serve as a remarkable catalyst for well-known coupling reactions commonly used in the pharmaceutical industry such as Suzuki and Stille cross-coupling. Characterization of prepared Pd NPs confirmed the synthesis of highly ordered materials which are able to catalyze the synthesis of fifteen representative compounds.

Research Interests: Protein expression, Inorganic nanoparticles, nucleation and growth of nanoparticles, catalysis, pharmaceutical applications

Teaching Interests: Physical properties of materials, Mechanical properties of materials, Advanced reactor design, Thermodynamics, Nanomaterials

562958 Plant Genome Engineering with Nanotechnology for Agricultural Applications

RESEARCH INTEREST

Food security is threatened by decreasing crop yields and increasing consumption in the light of climate change, population growth, and a shortage of arable land. To mitigate these threats, genetic engineering of plants can be employed to create crops that have higher yields and nutritional value, and are resistant to herbicides, insects, diseases, drought and heat. Despite the recent significant progress in genome editing, most plant species still remain difficult to genetically engineer. The two bottlenecks of generating engineered plants are (i) efficient biomolecule delivery into plant cells through the rigid cell wall and (ii) the regeneration of transformed tissues. The workhorse method of DNA delivery to plants, Agrobacterium, limits the range of transformable species and results in uncontrolled DNA integration, hence eliciting genetically modified organism (GMO) regulatory purview. In addition, it is currently not possible to edit plant germline cells, limiting our ability to engineer consumer-relevant crops. The development of a transformation tool that is (i) plant species-independent, (ii) non-pathogenic, (iii) non-integrating and/or DNA-free, and (iv) capable of transforming germline cells will greatly advance agricultural biotechnology. The unique and tunable physiochemical properties of nanomaterials can offer strategies for such desired plant genome engineering applications. Drawing on my interdisciplinary training, I aim to develop a multifaceted toolset using nanomaterials to provide food security for future generations.

Research Experience:

In my doctoral work, (i) I developed a nanomaterial-based gene delivery platform that efficiently delivers genes into agriculturally-relevant plants, without mechanical aid, in a non-toxic and non-integrating manner. I showed delivery of plasmid DNA to tobacco, arugula, cotton, and wheat leaves with engineered carbon nanotubes (CNTs), which resulted in strong transient expression of functional proteins. Importantly, I showed that exogenous DNA does not integrate into plant genome, relevant for avoiding GMO regulations. Next, (ii) I implemented this platform to deliver CRISPR plasmids targeting endogenous genes for knock-out. Through the transient expression of Cas9 and guide RNA in plant leaf cells, I obtained stable editing of target genes without transgene integration. My results have been patented by UC Berkeley and highlighted in C&EN, on NPR's All Things Considered, and popular mechanics, among others.

In parallel, (iii) I developed a different nanoparticle surface chemistry for in planta delivery of small interfering RNA (siRNA) for DNA-free gene knock-down applications. The status quo for siRNA applications in plants involves biological delivery of a DNA coding for the siRNA. I demonstrated 95% gene silencing efficiency via direct delivery of siRNA with nanoparticles. I further verified that CNT-based delivery provides a significant delay in intracellular siRNA degradation, suggesting nanoparticles protect the cargo from nuclease degradation in addition to enabling cell wall transport. In a separate study, (iv) I systematically investigated the effect of nanomaterial parameters on plant cell uptake and gene silencing pathways. By leveraging the facile programmability of DNA nanostructures and origami, I elucidated the underlying principles of plant nanoparticle internalization and showed that plant endogenous gene silencing mechanisms can be altered by the nanostructure shape and the siRNA attachment locus. Finally, (v) I diversified my skillset with next-gen RNA sequencing to investigate the effect of nanomaterials on the plant transcriptome. Thus, armed with tools for efficient and rational biomolecule delivery into plants, I am excited to develop versatile strategies for the improvement of plant genome engineering.

Awards:

Faculty for the Future Fellowship, Schlumberger Foundation 2016-2020 ($50k/year, international)
MIT ChemE Rising Stars 2019
WCC Merck Research Award 2019
AIChE Women's Initiative Committee Travel Award 2018
USDA National Institute of Food and Agriculture Award (Helped PI; Successful)
FFAR New Innovator Award (Helped PI; Successful)
USDA BBT Eager (Helped PI; Pending)
NSF EDGE CT (Helped PI; Pending)
AIChE 2018, Carbon Nanomaterials Graduate Student Award Session, 2nd place
AIChE 2017, Bionanotechnology Graduate Student Award Session, 3rd place

Selected Publications (4 of 12 total):

1. Demirer, G.S., Zhang, H., Matos, J. et al. High Aspect Ratio Nanomaterials Enable Delivery of Functional Genetic Material Without Transgenic DNA Integration in Mature Plants. Nature Nanotechnology (2019).
2. Demirer, G.S., Zhang, H., Goh, N. et al. Carbon Nanotube-Mediated DNA Delivery without Transgene Integration in Intact Plants. Nature Protocols (In press, 2019).
3. Demirer, G.S.*, Zhang, Hu.*, Zhang, Ho. et al. DNA Origami Nanostructure-Mediated Gene Silencing in Mature Plants. PNAS (2019).
4. Demirer, G.S., Zhang, H., Goh, N. at al. Carbon Nanocarriers Deliver siRNA to Intact Plant Cells for Efficient Gene Knockdown. ACS Nano (In review, 2019).

Future Research

The overarching goal of my lab will be to leverage chemical and biological engineering to create improved crops using new generation genome engineering tools. My lab will aim to address the critical limitations of: i) sub-optimal biomolecule delivery, (ii) inability to perform transgene or DNA-free editing, (iii) lack of tools for plant germline editing, and (iv) deficiency of plant protein interactions knowledge. Specifically, I will take three separate, initially independent, routes to address these challenges:

1) Developing transgene-free plant genome editing tools. Random integration of CRISPR DNA into plant genome causes off-target effects, endogenous gene disruption, and GMO-labeling. To avoid these complications and gain public acceptance of edited crops, researchers often go through the lengthy process of transgene segregation via breeding. However, for vegetatively propagated crops, transgene segregation in offspring is not possible. Thus, DNA-free genome editing approaches that do not cause integration will be transformative. One focus in my lab will be to develop new surface chemistries for the delivery of CRISPR ribonucleoproteins using nanoparticles that could be cleaved in planta via endogenous plant proteins. Completion of this goal will enable DNA-free genome editing for all plant species (including vegetatively propagated crops) without GMO regulation in many countries.

2) Discovering plant germline editing strategies. Another focus in my lab will be the application of gene editing technologies to plant species without available transformation methods, including wild relatives of major crops and orphan crops. Delivery of genome editing components to germline cells is a promising strategy to obtain gene-edited offspring in non-transformable species, which also has the advantage of avoiding arduous tissue regeneration. For this aim, my lab will first use the platform that I developed for CRISPR plasmid delivery. To increase the germline editing efficiency, we will explore dual methods using geminivirus replication proteins and/or injection in tandem with nanoparticle carriers. We will also test different nanoparticle surface chemistries for optimal germline cell penetration. With the completion of aim 1), my lab will explore the delivery of ribonucleoproteins into germline cells of any desired plant species for stable and heritable editing without tissue regeneration.

3) Constructing plant genetic interaction maps. Plant polyploidy and genetic redundancy makes it difficult to find effective gene targets to produce desired traits. I will dedicate a subset of my lab's efforts to understanding the extensive and complicated genetic networks controlling key agronomic traits and their interaction with environmental factors, initially focusing on stress resistance pathways, for which knowledge lags behind other traits. The main techniques for constructing plant genetic interaction maps in my lab will be stable knock-out of genes using CRISPR and transient knock-down of genes using RNAi. My lab's elucidation of genetic interactions and agronomic traits will be aided by different "omics" approaches such as proteomics, metabolomics, and mass spectrometry-based screens, which will ultimately enable a comprehensive mapping of all pathways for a better handle on creating more robust crops.

In the long-term, I aim to incorporate findings from each of these tracks to address aforementioned challenges in plant genome engineering. The multidisciplinary nature of this proposed research combines chemical engineering, chemistry, plant biology, materials science, and computation, thus providing diverse training opportunities for doctoral and post-doctoral scholars.

TEACHING INTERESTS

My capability as a teacher arises from my undergraduate and graduate education, where I grew to appreciate the inherent multidisciplinary topics of Chemical Engineering coursework. As a UC Berkeley teaching assistant, I taught both core (Thermodynamics) and elective (Nanoscience and Engineering Biotechnology) classes, and also served as a team leader and teacher in the "Bay Area Scientists in Schools" program, where my team taught hands-on classes in public elementary schools about the environment. Additionally, I mentored four undergraduate research students and two high school students during my thesis studies; three of my mentees co-authored publications with me, and two of them went on to graduate school. I've learned an important lesson in my teaching and mentorship experiences: while it is rewarding to produce good science, my biggest impact will come from producing great scientists. Correspondingly, I maintain that teaching and mentorship – both in the classroom and in the laboratory – will be my most important contribution as a professor.

My initial experiences as a teacher at UC Berkeley have prominently contributed to my educational plan. I am enthusiastic to teach both core and elective courses, and interested in developing an interactive course to discuss the latest approaches in genome engineering and nanotechnology. Among the classes in my academic career, I have gained the most from the ones that encouraged critical thinking, interactive learning, and creative problem-solving, while drawing parallels to real-life scenarios. Therefore, I strive to implement such features in the courses I teach.

583556 Harnessing the Power of Nanomaterials

Research interests:

I have focused on the synthesis and characterization of 2-dimensional (2D) semiconducting nanomaterials, specifically perovskites, during the past four years at Iowa State University (ISU). Perovskites have been improving drastically in the last couple of years to achieve efficiencies close to commercially available silicon-based solar cells. However, these materials suffer from instability upon exposure to oxygen, humidity, and light. To address the humidity issue, I have synthesized 2D perovskites by covering the structure with bulky organic molecules, which act as hydrophobic blankets and could prevent the degradation caused by moisture. To further improve the stability and efficiency, I have tried to use a combination of different elements in the fabrication of thin films, addressing several challenges pertaining to synthesis and homogeneity of the films along the way. Perhaps even more importantly, if perovskites or any other solution-processed materials are to be commercialized for solar cell applications, utilization of toxic solvents in the fabrication process must be drastically reduced, or ideally eliminated. To address this issue, I have directed efforts in our lab for developing a solvent-free technique, which requires minimal solvent utilization through melting and spreading 2D perovskite materials.

I have been fortunate to have access to the Cornell High Energy Synchrotron Source facility in order to characterize my perovskite films more carefully and with greater detail. In addition, several collaborations with research groups focusing on single crystal X-ray diffraction and nuclear magnetic resonance techniques have inspired me to delve deeper into material characterization for optoelectronic applications, i.e., our lab's expertise.

Future Work

I plan to extend my understanding of 2D materials, in particular perovskites, by working on several parallel projects in collaboration with experts in material characterization. I intend to couple my material synthesis and thin-film fabrication skills with solid-state chemistry and crystallography knowledge to enable me to design materials with enhanced efficiency and stability, and to engineer methods that would facilitate the commercialization of such films. I also plan to use the techniques I learned in previous projects to develop non-toxic 2D materials such as silicon.

To pursue my carrier as a faculty member, I would like to lead my research group to undertake the most challenging problems with regards to 2D materials for electronic and optoelectronic applications. I would like to apply similar scientific principles to 1D (e.g. nanowires) and 0D (e.g. quantum dots) materials with potential applications in the next generation of LEDs and transistors.

Teaching interests:

During my graduate studies at ISU, I have had numerous opportunities to mentor students in our research laboratory including a high school student (NSF Young Engineers & Scientists Program), a high school teacher (NSF Research Experiences for Teachers), and 6 undergraduate students through the chemical engineering department and US DOE Science Undergraduate Laboratory Internship. I also have served as a teaching assistant in several courses such as chemical engineering laboratory I, chemical engineering thermodynamics, and surface and colloid chemistry. I was recognized for my efforts in teaching by receiving the teaching excellence award (recognizes the top 10% of graduate students involved in teaching each year). Furthermore, I had the chance to teach science and engineering concepts to elementary schoolers through the NSF Trinect Fellowship, during which I learned how to communicate the most complex concepts to a lay audience. Experiences such as this will undoubtedly help me in developing effective teaching and mentoring strategies as a faculty member.

Selected publications

- **Atefe Hadi**, Bradley J. Ryan, Rainie D. Nelson, Kalyan Santra, Fang-Yi Lin, Eric W. Cochran, and

Matthew G. Panthani, *Chem. Mater.* **2019**, 31, 4990–4998

- Yujie Wang, Utkarsh Ramesh, Charles K. A. Nyamekye, Bradley J. Ryan, Rainie D. Nelson, Abdulla M. Alebri, Umar H. Hamdeh, **Atefe Hadi**, Emily A. Smith, and Matthew G. Panthani, Chem. Commun., **2019**, 55, 6102—6105
- R. D. Nelson, K. Santra, Y. Wang, **A. Hadi**, J. W. Petrich, and M. G. Panthani, Chem. Commun., **2018**, 54, 3640—3643

Patent

- Panthani, M., **Hadi**, **A.**, Milot, M., Wilsey M., Verburg A., Hamdeh, U., and Nelson, R. "Organic or Inorganic Metal Halide Perovskite via Cation Exchange." Submitted Invention Disclosure (2018)

Conference presentations

- **Hadi, A.**, Ryan, B., Nelson, R., Santra, K., and Panthani, M. "Stability enhancement in layered lead halide perovskites" poster presentation at North American Solid State Chemistry Conference, Golden, CO, (2019)
- **Hadi**, **A.**, Nelson, R., Cleveland, I., Jacoby, J., and Panthani, M. "Synthesis of Nanostructured Inorganic Perovskites for Photovoltaic Applications" oral presentation at The American Institute of Chemical Engineers, Minniapolis, MN (2017)
- Nelson, R., **Hadi**, **A.**, Verburg, A., Panthani M. "Solution-Based Synthesis of Cesium Bismuth Iodide Nanocrystals." Paper presented at the AIChE annual meeting, Minneapolis, MN (2017)
- **Hadi**, **A.**, Dubey, A., and Panthani M. "Process Engineering of Pure Phase Quasi-2D Layered CsPbI3 Perovskite Thin Films for Solar Cell Applications." Poster Presentation at the Nano@ISU meeting at Iowa State University, Ames, Iowa (2017)

583905 Next-Generation Fluorescence Bioimaging: Integrating Chemical and Optical Tools

Research Interests:

Very little happens in biological processes unless two or more biomolecules come together and interact. Understanding of biomolecular dynamics in the context of the highly complex nanoscale environments are opening up the next frontier for biological discovery. However, it remains challenging to image these biomolecular assemblies, their function and interactions in structurally and dynamically complex cellular systems. Electron microscopy provides nanometer resolution but low molecular specificity. Near-field optical microscopy and atomic force microscopy are limited to imaging cell surfaces. Fluorescence microscopy offers high-specificity while the resolution is fundamentally limited to ~200-500 nm. To fill this technical void, my research will focus on the development of (1) fluorescent probes and (2) imaging methods for non-invasive nanoscale imaging tools integrating with machine-learning to understand (3) the structure and dynamics in transcriptional, translational and epigenetic processes.

Research Experience

My predoctoral training covers fluorescence spectroscopy and microscopy, organic dye synthesis, artificial molecular switches and supramolecular chemistry. Specifically, I invented photoswitchable dyes with infinite contrast for multiplexed fluorescence imaging, cancer cell detection, dynamic tracking *in vivo* and super-resolution imaging in live cells. Additionally, I developed ultra-bright fluorescent supramolecular nanocarriers for the intracellular delivery of guest cargo. My postdoctoral research extends to the optical implementation, imaging and processing method development for spectroscopic single-molecule localization microscopy (sSMLM). This innovative imaging technique allows high-throughput single-molecule spectroscopy in biological samples non-invasively and enables parallel, multiplexed super-resolution imaging in a single far-red detection channel. Using sSMLM, I have unraveled biomolecular dynamics in nuclear pore complexes in differentiation process and post-translational modification in programmed cell apoptosis. These research experiences allow me to fully understand the practical need in specific biological questions as well as the limitations of up-to-date dyes for super-resolution microscopy. With these practical guidelines in mind and the multidisciplinary skills in hand, I propose to develop an interdisciplinary research program to develop multidimensional nanoscale imaging tools and visualize complex molecular dynamics.

Teaching Interests:

I have gained extensive teaching and mentoring experiences in addition to my research career. I instructed 5 semesters of general chemistry lab sessions independently (for which I won an outstanding teaching award) and developed a chem-bio integrated Howard Hughes Medical Institute Initiative courses with the supervision of two professors for three semesters at the University of Miami (UM). I have also mentored 2 high-schoolers, 4 undergraduates at UM, and 4 graduate students at Northwestern University. My teaching philosophy is to promote diverse interests and curiosity of students to learn knowledges and discover the unknown. I think that students are most engaged when the learning process is fun and exciting. In addition, the ability of critical thinking is the key of learning, to develop the ability to judge where is the root to absorb the knowledge and how to apply the knowledge endlessly. My teaching interests include general chemistry and instrumentation at the undergraduate level. Additionally, I plan to develop two graduate level courses offering (1) fluorescence spectroscopy and microscopy; and (2) fundamentals and applications of artificial molecular machines.

Grant-writing experience: PI: NIGMS-R03 (pending); Subcontracting PI: NSF-CHE (CSDM-B, pending); NIGMS-K99/R00; Jane Childs Coffin Memorial Fellowship; assisting in the submissions of NIGMS-R01 (pending) and NSF-EFRI ($2M funded).

Selected Publications:

(Total number of publications = 29, First/co-first authored publications = 11)

1. **Zhang,**
   Photoactivatable BODIPYs Designed to Monitor the Dynamics of Supramolecular Nanocarrier. *J. Am. Chem. Soc.*, **2015**, *137*, 4709–4719
2. **Zhang**, B. Captain and F. M. Raymo. A pH-Gated Photocage. *Adv. Optical. Mater.*, **2016**, 4, 1363-1366.
3. **Zhang,** S. Tang, E. Thapaliya, L. Sansalone, F.M. Raymo, Fluorescence Activation With Switchable Oxazines. Chem. Comm., **2018**, 54, 8799-8809. (Front Cover)
4. Tang[#], **Y. Zhang[#],** P. Dhakal[#], L. Ravelo, C. L. Anderson, K. M. Collins, F. M. Raymo, Photochemical Barcodes. *J. Am. Chem. Soc.*, **2018**, 140,4485-4488. ([#] Equal Author Contribution)
5. **Zhang**, K. Song, S. Tang, L. Ravelo, J. Cusido, C. Sun, H. F. Zhang, and F.M. Raymo, Far-Red Photoactivatable BODIPYs for the Super-Resolution Imaging of Live Cells. *J. Am. Chem. Soc.*, **2018**, 140, 12741-12745.
6. K. Song, **Y. Zhang**, G. Wang, C. Sun and H. F. Zhang, Three-dimensional biplane spectroscopic single-molecule localization microscopy. *Optica*, **2019**, 6, 709-715.

584652 Scalable Electronic Separation of Single-Walled Carbon Nanotubes (SWCNTs) via Novel Chromatographic Methods

Single walled carbon nanotubes (SWCNTs), due to their diverse and superlative electronic properties, are considered the best candidate to replace silicon as the backbone of the semiconducting and electronic industry. All of the current upstream synthesis techniques result in production of mixtures of metallic and semiconducting SWCNTs. This creates impediments to their application as transistors, which require extremely high-purity fractions of semiconducting SWCNTs. Therefore, there is a great need to develop methods to separate mixtures of metallic and semiconducting nanotubes into highly enriched fractions of single electronic types. Experimental complexity, lack of scalability, and the lack of understanding about the underlying mechanisms of the current downstream separation techniques have motivated our research group to design simple and scalable chromatographic based separation methods which rely upon conventional chromatographic materials such as ion exchange, hydrophobic, and mixed mode materials in order to achieve separations of metallic and semiconducting nanotubes. In this work, we have demonstrated that the upstream purification steps, which generally are implemented to separate non-SWCNTs impurities, confer to nanotubes ionic characteristics in addition to their inherent hydrophobic structures. The induced surface charges and inherent hydrophobic surfaces of the SWCNTs were directly exploited via traditional chromatographic materials and methods, and as a result separated fractions of metallic and semiconducting SWCNTs were obtained.

Keywords: Single walled carbon nanotubes, Separation, Chromatography

Research Interests:

- Design and development of separation techniques such as liquid chromatography, aqueous two phase polymer extraction, density gradient ultracentrifugation, and filtration in order to purify and separate various nanomaterials (such as SWCNTs), and bio-macromolecules (such as proteins)
- Investigation of macromolecular interactions via molecular modeling techniques such as docking and molecular dynamics
- Model-based design of high resolution chromatofocusing technique

Teaching Interests:

- Thermodynamics
- Transport phenomena
- Separation processes
- Chemical engineering design

587008 Deep characterization boosted next-generation high-performance energy storage systems

Research interests:

Reducing carbon emission and ending the fossil fuel economy require the implementation of clean energies. Batteries will be the most important infrastructures for energy storage and distribution to power up future smart and resilient cities, where high energy, high power, high mobility, and reliable safety is highly demanded. But today's battery technology is far from satisfactory to support transportation, buildings, homes and grid storage in smart cities. Even though a variety of new materials and novel chemistry have been found to provide ultra-high theoretical capacity in past decades, very few of them has been successfully transferred from lab to market to make an immediate impact to our daily life. One reason is because the majority of these research are limited to "proof of concept" studies in a limited lab scale. In contrast, battery is a complex system with defects in materials and inhomogeneity in manufacturing, along with interphases formed under chemical/electrochemical dynamics during usage, making it tricky from the research perspective to accurately correlate the materials level, electrode level and system level properties together to solve problems across different scales. Without proper emphasizing on statistics across multiple dimensions, experimental data often can be misinterpreted. The mismatch between academic-scale and industrial-scale makes conventional battery R&D run into its bottleneck. Breakthrough relies on *deep characterization with new methods, statistical analysis using machine learning, and novel high-precision manufacturing methods.*

My Ph.D. research on quantitative diagnosis of lithium metal anode failure is a successful attempt of deep characterization concept. I developed novel multiscale characterization methods to accurately quantify inactive lithium in lithium metal batteries (LMBs), the ideal candidate for high energy density rechargeable batteries, and identified the underlying cause of low Coulombic efficiency in LMBs is unreacted metallic Li, clearing the long-term misconception in the field that solid electrolyte interphase formation is the major cause (*Nature*, 572, 511–515, 2019). The discovery with the introduced new methods provides paradigm-shifting guidance to the battery research field towards the daily use of LMBs in future transportations (*Trends in Chemistry*, 1, 152-158, 2019).

My research will be at the core of the forefronts of materials science and hold technological prospects in multiple engineering disciplines, with a focus on sustainable energy and smart grid for future cities. To further address the existing obstacles in fundamental characterization and energy storage systems, my research goal includes two main directions: (1) Advanced characterization to enable quantification, utilizing existing tools in-depth and building new methods to answer key scientific questions and bridging the gap between academia and industry. (2) Advanced battery systems. Combining with quantitative characterization induced fundamental understanding, my research effort will also be put on the rational design of anode-free batteries, which breaks the energy limitation of current battery technology for next-generation transportation, life, and a sustainable economy.

Selected publication:

1. **Fang***, J. Li*, M. Zhang, Y. Zhang, F. Yang, J. Z. Lee, M.H. Lee, J. Alvarado, Y. Yang, L. Yang, M. Cai, J. Gu, K. Xu, and Y. S. Meng, "Quantifying inactive lithium in lithium metal batteries", ***Nature***, 2019, 572, 511–515
2. **Fang,** X. Wang, Y.S. Meng, "Key issues hindering a practical lithium-metal battery", ***Trends in Chemistry***, 2019, 1, 152-158
3. A. Wynn*, **C. Fang***, M. Zhang, H. Liu, D.M. Davies, X. Wang, D. Lau, J.Z. Lee, K.Z. Fung and Y.S. Meng, "Mitigating oxygen release in anionic-redox-active cathode materials by cationic substitution through rational design", ***Journal of Materials Chemistry A***, 2018, 6, 24651
4. M. Wood, **C. Fang**, E. J. Dufek, S. C. Nagpure, S. V. Sazhin, B. Liaw, Y. S. Meng, "Predicting calendar aging in lithium metal secondary batteries: the impacts of solid electrolyte interphase composition and stability", ***Advanced Energy Materials***, 2018, 8, 1801427

5. Singer, S. Hy, M. Zhang, D. Cela, **C. Fang**, B. Qiu, Y. Xia, Z. Liu, A. Ulvestad, N. Hua, J. Wingert, H. Liu, M. Sprung, A.V. Zozulya, E. Maxey, R. Harder, Y.S. Meng, O.G. Shpyrko, "Nucleation of dislocations and their dynamics in layered oxides cathode materials during battery charging", ***Nature Energy***, 2018, 3, 641-674

Teaching interests:

As a teacher and mentor in science and engineering, my goal is to provide students solid fundamental knowledge and technical skillsets thus to build life-long problem-solving abilities. Leading a subgroup of seven people since 2017, I have been mentoring one undergraduate student and four junior graduate students with fruitful results in both research and individual development of the students. With the background of Materials Science and Engineering working on energy storage and conversion systems, I am capable to teach undergraduate and graduate level chemical engineering courses such as Kinetics, Heat Transfer, Thermodynamics, and Mass Transfer. I am also interested to develop new classes such as Sustainable Energy and Electrochemistry for senior undergrads or graduate students.

587484 Novel Nanostructured Transition Metal Oxides for Catalysis.

Research Interests:

Transition metal oxides (TMOs) have been the subject of growing interest in the catalysis community in the past decades owing to their low cost as a great alternative for precious metals and excellent catalytic performance. Their unique electronic structure of the outer d-electrons has offered them versatile abilities for optical, electrical and catalytical property engineering. The high oxygen mobility in TMOs also gives rise to easy stoichiometry tuning, heteroatoms doping as well as modifications of electronic band structures, making property control more feasible. In addition, TMOs possess high structural tunability thus allowing for the synthesis of various nanostructures with different dimensionalities ranging from 0D to 3D. In this work, a variety of facile wet synthetic approaches, such as soft templating, hard templating and template free, will be exploited to controllably synthesize novel TMO nanomaterials including copper oxide, iron oxides and manganese oxides with different nanodimensionalities. These nanomaterials and their composites were well characterized and understood via an assortment of tools including SEM/TEM/AFM, XRD/XPS/XAFS/XRF, FTIR/UV-VIS/UV-DRS/NMR/EPR/PL, ultrafast facilities transient absorption, and BET physisorption/chemisorption etc. Temperature programmed reactions such as TPR (reduction) /TPO (oxidation) /TPD (desorption) were used to study the reductive/oxidative/desorptive behaviors of these catalysts. To investigate their catalytic performance, a combination of straight gas phase catalysis and photocatalysis will be explored including CO2 hydrogenation via reverse water gas shift route (rWGS) for iron oxides and manganese oxides, methane catalytic combustion, direct Z-scheme photocatalytic water splitting that's enabled by ultrafast interfacial charge transfer, etc. In addition, first principle computations will be highly coupled with our experimental results in order to guide the catalyst designs and mechanistic understandings. The core of this work will be hypothesis-driven, enlightening from TMO structural and computational studies, and application-oriented in a sense of seeking fascinating properties.

Teaching Interests:

The candidate obtained her bachelor degree in Chemistry in 2016 and will be awarded a doctorate degree in Chemical Engineering in 2020. She is looking for a post-doctorate position and shows strong interests in teaching but not limited to the following subjects: Thermodynamics; Kinetics and Reaction Engineering; Nanomaterial synthesis and characterization; Physical Chemistry; Heterogeneous Catalysis; Photocatalysis; Electronic Materials; Lab courses.

582155 Dynamics of Carbonaceous Nanoparticles: Climate Impact & Fire Detection

Research Interests:

Incomplete combustion of fossil or biofuels and open fires release about 8 Tg of soot per year, worldwide. This is in the same order with the annual production (12 Tg) of nanostructured carbon black across the globe and motivates the multiscale design of carbonaceous nanoparticle formation, interfacing quantum mechanics, atomistic, mesoscale and continuum models [1]. Quantum mechanics and atomistic models are used to describe the electronic structure and composition of carbonaceous nanoparticles, respectively [1]. Continuum models interface Navier-Stokes equations with chemical kinetics and particle dynamics to describe the effect of process variables (e.g., temperature, residence time, fuel composition) on particle size, structure and optical properties [1].

Mesoscale discrete element modeling (DEM) for the dynamics of flame-made carbonaceous nanoparticles by surface growth, aggregation [2] and agglomeration [3] bridged quantum, atomistic and continuum models, providing a realistic pathway for carbon black formation. So, right after inception by reactive dimerization [4], small agglomerates of physically-bonded, monodisperse carbonaceous primary particles are formed by agglomeration. Then, surface growth by acetylene pyrolysis chemically bonds carbon black primary particles into compact aggregates [2]. After the complete conversion of gaseous hydrocarbons, ramified carbon black agglomerates of physically-bonded aggregates and single primary particles are formed by agglomeration alone [3]. A DEM-derived power law for the fractal-like morphology of carbon black agglomerates [3] is coupled with a continuum moving sectional model and particle size measurements to derive accurate nucleation [4], internal and surface oxidation kinetics [5]. Similar power laws can be derived for flame-made inorganic nanomaterials accounting for the agglomerate compaction during atmospheric aging in the presence of humidity [6]. The DEM-derived carbon black size distribution and morphology found excellent agreement with data from different combustion groups, explaining several experimental observations made in the last 15 years. This novel understanding of flame-made nanoparticle dynamics during surface growth, aggregation and agglomeration is essential for the accurate estimation of their environmental impact, characterization by optical diagnostics and selective sensing by fire detectors.

For example, the climate models of the Intergovernmental Panel on Climate Change (IPCC) list CO_2, CH_4 and carbon black, as the most potent contributors to global warming based on their radiative forcing (*RF*) impact. Among them, carbon black exhibits the highest uncertainty (~ 90 %). Examining closely these models, it becomes apparent that they could significantly under-predict the direct *RF* for carbon black, largely due to their assumption of spherical carbon black morphology and use of the corresponding Mie theory for light absorption. Here it is shown that carbon black rather CH_4 is the second largest contributor to global warming after CO_2 due to the reduced light scattering and enhanced light absorption by its filamentary structure [3]. The available data on agglomerate morphology of carbon black derived from various carbon sources, optical band gaps, mobility size, and mass absorption cross-section, and the analysis of the light absorption from the Rayleigh-Debye-Gans theory for carbon black are provided to affirm this discovery [7].

Optical diagnostics are widely used to monitor and mitigate carbon black emissions. For instance, early and reliable fire detection is essential to prevent thousands of human deaths and injuries every year in the US alone and protect the environment. Modern fire (smoke) sensors detect the light scattered from carbon black emitted during open fires. The optical properties of ramified carbon black agglomerates are calculated typically by Rayleigh or Mie theories for spheres impeding their selective sensing. So, fire detectors are not selective enough to light scattering from carbon black and thus produce false alarms costing up to 1 billion £/y in United Kingdom alone. Here, the morphology and light scattering from carbonaceous nanoparticle agglomerates are measured in premixed ethylene flames [8] and simulated by coupling DEM for surface growth, aggregation and agglomeration with the Discrete Dipole Approximation (DDA) [9]. Using the Mie theory for spheres overestimates the measured carbon black mass up to 10 times. This results in 60 % larger light scattering than that measured in premixed ethylene flames. In contrast, the DEM-derived carbon black structure and light scattering is in excellent agreement with those measured here in premixed flames. Thus, the DEM-DDA will be used to design and optimize the selective sensing of fire detectors to carbon black.

In sum, the development of detailed models for nanoparticle dynamics and their thorough validation with experiments provides insight into the morphology and optical properties of carbon black during flame synthesis. This facilitates the mitigation of particulate emissions, control of their environmental impact and detection by robust fire sensors.

References

[1] Buesser, B., and Pratsinis, S.E. "Design of Nanomaterial Synthesis by Aerosol Processes" (2012) *Annu Rev Chem Biomol* **3**, 103-127.

[2] Kelesidis, G.A., Goudeli, E., and Pratsinis, S.E. "Flame synthesis of functional nanostructured materials and devices: Surface growth and aggregation" (2017) *Proc Combust Inst* **36**, 29-50.

[3] Kelesidis, G.A., Goudeli, E., and Pratsinis, S.E. "Morphology and mobility diameter of carbonaceous aerosols during agglomeration and surface growth" (2017) *Carbon* **121**, 527-535.

[4] Kholghy, M.R., Kelesidis, G.A., and Pratsinis, S.E. "Reactive polycyclic aromatic hydrocarbon dimerization drives soot nucleation" (2018) *Phys Chem Chem Phys* **20**, 10926-10938.

[5] Kelesidis, G.A., and Pratsinis, S.E. "Estimating the internal and surface oxidation of soot agglomerates" (2019) *Combust Flame*, under review.

[6] Kelesidis, G.A., Furrer, F.M., Wegner, K., and Pratsinis, S.E. "Impact of humidity on silica nanoparticle agglomerate morphology and size distribution" (2018) *Langmuir* **34**, 8532-8541.

[7] Kelesidis, G.A., Kholghy, M.R., Fan, L.-S., and Pratsinis, S.E. "Radiative Forcing in Climate Change: the Critical Role of Soot" (2019) in preparation.

[8] Kelesidis, G.A., Kholghy, M.R., Zurcher, J., Robertz, J., Allemann, M., Duric, A., and Pratsinis, S.E. "Light scattering from nanoparticle agglomerates" (2019) *Powder Technol*, doi.org/10.1016/j.powtec.2019.02.003.

[9] Kelesidis, G.A., and Pratsinis, S.E. "Soot light absorption during agglomeration and surface growth" (2019) *Proc Combust Inst* **37**, 1177-1184.

Teaching Interests:

One of my main motivations for pursuing an academic career is the joy and fulfillment I gain by teaching, interacting and mentoring students.

During my PhD studies I had the chance to be a teaching assistant at the Department of Mechanical and Process Engineering of ETH Zurich, Switzerland. I also supervised the research of four BSc students (three months long), one MSc student (three months long), as well as five shorter projects for BSc level class. Furthermore, I was responsible for the organization (scheduling, exercise and project preparation and evaluation) of the Introduction to Nanoscale Engineering (BSc level) and Micro- and Nanoparticle Technology classes (MSc level). Finally, I assisted the exercise and exam sessions of the Mass Transfer class (BSc level) and conducted experimental and numerical exercises on Brownian motion for the Laboratory Practicum in Process Engineering (MSc level).

In the future, I would be interested in teaching core BSc level courses, including "Heat and Mass Transfer", "Thermodynamics" and "Numerical Analysis", that are indispensable for every chemical engineer. Furthermore, I would also like to give MSc level lectures based on my research interests, such as "Nanoparticle Science and Engineering" and "Computational Chemistry and Physics". These classes will introduce the students to the current frontiers of academic and industrial research, addressing large challenges of our society, such as global warming, and providing valuable skills for their potential future careers, e.g. in fire detection and chemical industries.

579910 Particle Simulation in Rotating Reactors

Research Interests: Multi-phase flow simulation, Particle simulation, Energy engineering, CFD etc

Teaching Interests: CFD, Multiphase flow, Particle technology, Thermodynamics, Heat and mass transfer etc.

Abstract

Particles in the rotating flow are simulated using Eulerian-Lagrangian two-phase simulation framework. Typically the particles in the rotating flow field will deviate the flow path of the vortex due to the inertia of the particles. The mechanism is often seen in the centrifugal separator. In the research, the particles flows are simulated in the Taylor green flow and the rotatory reactors, where the particle-wall interactions, particle-particle interactions and particle-fluid interactions are considered.

581864 Development of Strategies to Combat Genomic Instability in Cell Culture Engineering, Biopharmaceutical Production, Disease and Ageing

Research Interests:

Glyco-engineering of biopharmaceuticals

Glycans are sugar polymers of outstanding pharmaceutical importance. Not only do some of the most widely prescribed drugs represent glycan biopharmaceuticals (e.g. heparin), but glycans also modify and impact the efficacy and safety of protein drugs, such as hormones or antibodies. Unlike proteins, however, glycans are synthesized in highly complex, stochastic reaction networks, which complicates biopharmaceutical manufacturing of these drugs since (i) synthesis does not follow a direct template (as protein synthesis does), and (ii) the genetic factors controlling activity of glycan-synthesizing enzymes remain largely unknown.

During my time as a postdoc I have explored novel approaches to tackle both of these confounding factors of glycan biomanufacturing. To increase the predictability of the glycan reaction network, I developed a computational model that describes the network as a Markov chain, a frequently used class of stochastic processes in engineering applications. This approach enables computational simulation of glycan biosynthesis while requiring few parameters defined by the user. Thus, this facilitates the development of cell engineering strategies to obtain a desired glycan profile during manufacturing. In addition, I utilized screening techniques, such as computational binding site predictions as well as genome-wide CRISPR/Cas9 knock-out libraries, and identified previously unrecognized regulator genes of glycan biosynthesis. In particular, we found that of one these genes (a zinc finger repressor-protein) controls biosynthesis of heparin, a blood anticoagulant of significant pharmaceutical relevance. This finding may pave the way for the engineering of cell lines capable to produce this valuable drug which can currently only be extracted from porcine intestinal mucosa. Thus, clean cell-culture based production will greatly enhance cost-efficiency and biosafety.

Cell line stabilization for biopharmaceutical manufacturing

In the biopharmaceutical industry, cell lines, such as Chinese Hamster Ovary (CHO) cells, are grown at large scales to produce various protein drugs, e.g. therapeutic antibodies, enzymes, and hormones which represent powerful new medicines for metabolic disorders, congenital ailments and high-incidence diseases like cancer or auto-immune disorders. The high prices of these drugs, however, pose a significant burden on patients and health care providers, and hamper the availability of these treatments in developing countries. A key contributing factor to high prices of these drugs is the high production cost, caused by instability of the cell lines producing them, wherein cells fail to maintain an economically viable protein titer over extended periods of time. One major underlying cause is the inherent instability of the CHO genome, such as the propensity of chromosomes to suffer frequent double-strand breaks. This can lead to the loss of genomic material, particularly to the loss of transgene copies which, consequently, results in dramatic loss of expression of the desired protein. So far, effective strategies to mitigate genomic instability in CHO cells have had only very limited success. Thus, genomic instability remains a major drain on resources and time in cell line development and production.

I have deployed a novel approach to restore the cells' ability to repair naturally occurring DNA damage and thus prevent the rapid occurrence of chromosomal instability. We analyzed whole-genome sequencing data of a multitude of commonly used CHO cell lines and identified several potentially deleterious mutations in key DNA repair genes. In particular, we found genes related to double-strand-break repair pathways to be severely affected which is consistent with the observed chromosomal instability. Using genome editing technology, I was able to revert several mutations in these genes, and the resulting cell lines gained improved double-strand repair capability and more stable product titers. We are currently conducting screens using a GFP-reporter system that allows us to more rapidly identify all DNA repair genes that, if restored, will improve DNA repair capability. This will enable us to engineer DNA-repair optimized CHO cell lines with a more stable genome and thus higher and more cost-efficient production capacities. My work on this project forms the backbone of a proposal currently under review in the Cellular and Biochemical Engineering program at NSF.

Future directions

In the future I would like to continue my work on CHO cell line stabilization but also expand the scope of this project into other areas. Genomic instability is also a growing concern in cell therapies, including induced pluripotent stem cells and CAR-T cells, potentially jeopardizing their clinical applicability. Also, genomic instability is a major driver of cancer and cell senescence, the latter being the root cause of ageing. My long-term goal as a future faculty member is to develop approaches to diagnose, mitigate, and prevent genomic instability in these contexts and realize applications in industry and medicine. I strive to maintain a dual methodology, combining both computational and experimental approaches, and would like to continue exploring biotechnological applications as an inspiration and a driver in the search for basic biological discoveries. Such aims will be highly fundable at several NIH institutions (including NCI, NIGMS, and NIAID), NSF, and many other foundations.

Teaching Interests:

As a scientist, learning is my passion. And as one of my high school teachers liked to say, the best way to learn is to teach to others. During my time as a PhD student, participation in teaching was part of my job responsibilities, and one that I greatly enjoyed. I have held classes in front of large student audiences, in both undergraduate and graduate programs. I developed new laboratory classes from scratch, including lectures, class material and exams, covering cell & developmental biology, as well as scientific programming. As a postdoc I helped design and teach at workshops on CHO cell manufacturing and CRISPR/Cas9 screens, aimed at academic and industrial audiences. Both as a PhD student and as a postdoc, I was involved in hiring and mentoring students and volunteers in the laboratory.

I can cover areas like cell & molecular biology, developmental biology, genetic engineering, cell culture engineering, and statistics.

579770 Engineering Lipid and Polymer-Based Nanomaterials for Drug, Protein and Gene Delivery

Nanoparticle-based delivery systems have been demonstrated to be advantageous for solubilizing drugs for systemic delivery, improving the stability of therapeutic agents against enzymatic degradation, prolonging blood circulation lifetime, achieving site-specific drug targeting and controlled/sustained drug release, avoiding drug resistance mechanisms and facilitating cell uptake. Cumulatively, these effects serve to enhance the bioavailability and reduce the side-effects of the bioactive molecules. Materials including (but not limited to) lipid and lipid-like molecules (lipidoids), synthetic polymers and biomacromolecules, carbon and silica based nanoparticles, and metal and metal oxide nanoparticles have been developed for drug delivery. Each delivery system has its own advantages as well as disadvantages. My work has been focused on the development of safe and efficient lipid and polymer-based nanomaterials for therapeutic agents delivery, including small molecular drugs, peptides and proteins, and nucleic acids.

Research Interests: drug delivery, nanomedicine, protein/peptide delivery, DNA/RNA delivery, biomaterials, lipids, polymer chemistry, organic and supramolecular chemistry, nanotechnology, boimedical imaging, pharmaceutical science

Teaching Interests: organic chemistry, polymer chemistry, materials science, nanotechnology, pharmaceutical science, bioengineering, molecular biology

574950 Priming the Innate Immunity Using Nanomedicine

Research Interests: The innate immune system is implicated in an enormous number of disease processes such as *malignancy* and *inflammation* that affects millions of people around the world. Take cancer as an example, most cancer patients still do not benefit from current immunotherapeutic approaches that focused on enhancing T-cell responses, due to the low response rate, immunological toxicities, and expensive price. Targeting innate immunity in cancer has the potential to enhance current T-cells based immunotherapeutic approaches by inducing multilayered immune response. However, precise modulation of innate immunity is still a CHALLENGE, as therapies can often be a double-edged sword and cause either immune suppression in the whole body or serious autoimmune/non-specific inflammation. Nanotherapeutic approaches provide the opportunity to induce a precise immune response and reduce the immunological toxicities associated with current approaches. However, the clinical translation of nanotherapeutics requires thorough understanding of both disease pathology and clinical practice.

*The **OVERARCHING GOAL** of my research is to establish a fundamental understanding of the nanomedicine design and their interaction with the immune system, with a focus on cancer and inflammatory diseases. There are two major arms in my research program: (1) using ex vivo and in vivo models to understand the interaction between nanomedicine and the tissue microenvironment at the disease foci; and (2) using this information to develop nanomedicine that precisely modulate the immune response. The **ULTIMATE GOAL** of my research is to translate nanomedicine into clinics.* My lab will leverage the power of synthetic chemistry and materials science in the design of novel nano-devices for mRNA & drug delivery. We will also apply state-of-the-art molecular biology and immunology tools (e.g. genome sequencing, multicolor flow-cytometry, fluorescence-based imaging technique) to understand the interaction between nanomedicine and the immune microenvironment at the disease foci. **Teaching Interests**: this multidisciplinary work will provide training for engineering students, immunology students and clinical fellows. Specifically, I am interested in teaching these students how to solve multi-faceted challenges by integrating knowledge across boundaries.

My knowledge in engineering, immunology, and my experience of working closely with the clinical side has placed me in a unique position to establish an interdisciplinary research program that integrates training in engineering nanoparticle platforms with an understanding of nanoparticles formulation, analytical chemistry, immunology, cancer biology, and translational research. During my doctoral work with Dr. Kannan Rangaramanujam, I developed dendrimer-based nanomedicine for central nervous system diseases, now a patented technology. Through collaborating with neuroscientist, neurosurgeons, we built the guiding principles for designing nanomedicine that efficiently crosses biological barriers and targets neuroinflammation. My postdoctoral training with Dr. Matthias Stephan at *Program in Immunology*, Fred Hutchinson Cancer Research Center built my foundation in immunology and cancer biology, leading to experience in understanding immunotherapy, gene therapy, as well as developing clinically relevant tumor models. Based on my combined doctoral and postdoctoral work, I have published 19 peer-reviewed publications, 4 international patents, and was awarded the Basic Research Award by American Brain Tumor Association, an award that provides $100K over 2 years of transitioning from postdoctoral level training to a faculty position.

Education

Ph.D., 2016, Johns Hopkins University, Materials Science and Engineering

B.S., 2010, Donghua University, Materials Science and Engineering

Selected Publications

- **Zhang F**, Parayath N, Coon M, Stephan SB, Ene CI, Holland EC, Stephan MT. Genetic programming of macrophages to perform anti-tumor functions using targeted mRNA nanocarriers. **Nature Communication**. *accepted*.
- **Zhang F**, Stephan SB, Ene CI, Smith TT, Holland E, Stephan MT. Nanoparticles that reshape the

tumor milieu create a therapeutic window for effective T cell therapy in solid malignancies. Cancer research. 2018. **Cancer Research.**2018 ***Selected Cover Image*

- **Zhang F**, Nance E, Alnasser Y, Kannan RM, Kannan S. Microglia migration and interactions with dendrimer in brain in the presence of neuroinflammation. **Journal of Neuroinflammation**. 2016 Mar 22;13(1):65.
- **Zhang F***, Mastorakos P*, Mishra M, Mangraviti A, Hwang L, Zhou J, Hanes J, Brem H, Olivi A, Tyler B, Kannan RM. PAMAM Dendrimer Biodistribution in Glioblastoma and Intrinsic Targeting of Tumor Associated Macrophages. **Biomaterials.** 2015. 52:507-16.

581892 Zwitterionic Polymeric Platforms for Biologic Drug Delivery and Tissue Engineering

Research Interests:

Biologic products as therapeutics and biocompatible substitutes for tissue engineering are becoming prominent in the new era. However, both face similar fundamental challenges: undesired immunological responses and foreign body reactions, leading to complications. Polymer-based materials possess great potentials for tackling these problems as they have diverse chemical functionalities, adjustable mechanical properties, and tunable applicational features. Many studies have been conducted to provide potential solutions, but quite often as one problem is resolved, another one emerges. Therefore, I am interested in developing bioinspired, super-hydrophilic zwitterionic materials to tackle the roots of these problems. Zwitterionic materials have demonstrated their vast capabilities for medical applications as they (1) provide a hydration layer surrounding the materials to prevent non-specific interactions with the environment, (2) have water-loving properties which are compatible with living organisms, and (3) possess highly tunable features for various applications. Over five years of my Ph.D. research experiences, I have focused on and have developed interests in designing zwitterionic, polymeric platforms for biologic drug delivery and tissue engineering by hindering the recognition by the immune system. Beyond my current work on "shielding" foreign bodies, I believe that biomaterials-based proactive immunomodulation is an emerging field. My future research interests are as follows:

- ***Non-invasive systemic delivery systems of therapeutic biologics***

  Biologic products have always been an enticing idea due to their specificity towards an explicit disease. However, due to their susceptibility to environmental changes, protein drugs are usually delivered *via* intravenous injection to minimize loss. This often requires patients to spend a considerable amount of time in the hospital for unpleasant treatments. Therefore, non-invasive administration routes for large therapeutic biologics are highly desirable. Non-invasive systemic biological drug delivery *via* the pulmonary route has always been fantasized since the 1950s due to the large surface area of the lung for absorption. Yet, delivery of large proteins (molecular weight > 40 kDa) having an average bioavailability of only 1 to 5% *via* this route remains a primary constraint. In my current studies, we have developed a super-hydrophilic polymer-based delivery platform which (1) stabilizes proteins in the lung environment, (2) facilitates proteins in crossing the barrier, and (3) significantly enhances the pharmacokinetic properties of proteins. I envision that through this experience, I would like to further develop a platform for other non-invasive systemic deliveries of large proteins, such as oral delivery and transdermal delivery.

- ***Long-term implants with adjustable mechanical properties and no foreign body reaction***

  The field of tissue engineering has a different set of challenges to tackle. Developing tissue-mimicking grafts for successful integration with the host is one of the primary objectives for regenerative medicine. Many materials have been studied and utilized in the replacement of tissues with various properties. Nonetheless, similar to transplantations, these grafts often lose their functionality within a short period as a result of eliciting foreign body reaction towards the materials. In my current work, we have tackled the two significant issues of artificial implants - inappropriate mechanical properties and induced foreign body reactions. With sophisticated design, we are able to develop hydrogels (1) with adjustable mechanical properties for a wide range of applications, and (2) with no foreign body reactions after months of *in vivo* implantation. This exciting work brings promise to long-term implantation of artificial biomaterial for tissue substitutions. For one of my future directions, I aim to explore further the interaction between the materials and the cells, and the capabilities of these tunable hydrogels for various applications such as blood vessels and cartilage substitutes, with the aid of 3D printing techniques.

- ***Materials engineering for antigen-specific immunomodulation***

One of the emerging fields that are developing alongside with new insights into immunology is the engineering of materials that can modulate the immune system. Materials are now being engineered to deliver antigens with better control of the interaction with critical immune cells in order to induce tolerance against specific antigens. Zwitterionic materials provide an interesting angle as nature-derived zwitterionic polysaccharides promote immunosuppressive CD8 T-cell responses. Integrating with our vast experience with the nature of zwitterionic materials, I would like to develop new materials that can (1) prevent undesired interactions with the host immune system, and (2) proactively induce tolerance towards proteins with high immunogenicity. With the fundamental studies of how new zwitterionic materials stimulate the immunosuppression of foreign therapeutics, these platforms can be further used for treating autoimmune diseases, such as Multiple sclerosis, Type 1 diabetes, and Rheumatoid arthritis.

Teaching Interests:

Coming from a life science background, I have gained valuable teaching experiences through teaching undergraduate core courses in Chemical Engineering, including Mass and Energy Balance, Thermodynamics, Quantum Mechanics, and Senior Capstone Project. My teaching is highly recognized throughout the department as I was awarded the "Award for Excellence in Chemical Engineering Graduate Student Teaching." One of the most exciting classes I have experienced was the Special Senior Capstone Project. In a span of three quarters, I led a team of senior undergraduates to conduct experiments, develop business plans for medical devices, and participated in the "Hollomon Health Innovation Challenge" at the University of Washington. I have sharpened my mentorship, lecture designs, and project/experiment management skills through this great opportunity.

One of the major takeaways for me when I first encountered and tried to master the core courses of Chemical Engineering was the fact that all pieces of knowledge are applicable to various fundamental science studies, ranging from engineering to life sciences. I believe in teaching Thermodynamics, as well as Mass and Energy Balance, would be both engaging and informational for students wishing to enter engineering majors. Additionally, with my unique interdisciplinary background, I would like to design an introductory course at the interface between Cheimcal Engineering and Life Sciences. With my vast research experience, I would be excited to create graduate-level courses in emerging fields of Drug Delivery Systems, Nanomedicine, and Immunomodulatory Biomaterials.

I am also a firm believer in the education of communication. One of the major challenges that knowledgeable graduate students face is lacking the ability to express an idea clearly to different target groups. I wish to design lectures or create study groups focused on helping graduate students to write proposals, articles, and popular sciences. Besides conducting my independent research, I have enjoyed collaborating with researchers from various backgrounds. I recognized that it is the diversity that brings unique insights into the lives around us. I will embrace this principle in the classroom and lab in those I teach and facilitate discussions between groups of individuals with different backgrounds by nurturing alternative points of view, supporting open mindsets, and rewarding collaborations. Unlike basic sciences, such as Biology, Math, or Chemistry, Chemical Engineering is in the realm of applied sciences with engineering principles. Therefore, I aim to connect to high school students to provide more insights on "what is Chemical Engineering," in hope to inspire next-generation chemical engineers. I wish to integrate this program with existing outreach efforts to encourage more kids to enter the STEM field and increase the diversity of perspectives for future challenges.

587794 High Diffusive Peptides for Drug Carriers in Tumor Extracellular Matrix.

Current therapies against solid cancers fail in part due to their inability to penetrate through the tumor microenvironment and reach the cancer cells at therapeutic doses. Increasing penetration promotes tumor regression and improves the therapeutic index of compounds. Tumor penetration of the therapeutics are limited by three major transport steps - the vascular, transvascular, and interstitial barriers. The hyper-permeable vasculature and dysfunctional lymphatic system of tumors result in elevated interstitial fluid pressure of the tumor extracellular matrix (ECM). This reduces the convective transport through the ECM, thus, diffusion is the primary mode of transport for therapeutic agents. Diffusivity of molecules through the ECM depends on their size and their interaction with the components of ECM. These interactions are driven by the surface physicochemical properties (i.e., charge and hydrophobicity) of the therapeutics.

To address challenges presented by current surface chemistries, we used peptide-presenting phage libraries as a high-throughput approach to screen and identify peptide coatings with physicochemical properties able to facilitating improved diffusive transport through the tumor microenvironment. We constructed a phage library, where random seven amino acid peptide sequences are displayed on the surface of T7 phage. Here, each phage expresses a different peptide sequence and each phage effectively serves as a formulation with unique surface chemistry. These libraries display random peptides on the surface of phages and effectively serve as a collection of peptide-based surface chemistries with a permutation of amino acids possessing varying physicochemical properties (i.e. $10^5 - 10^9$ different sequences). Individual clones were selected through iterative screening of these phage libraries against the selective pressure of ECM components found in solid tumors and identification through next-generation DNA sequencing and analysis, and their transport was measured by diffusion assays.

We identified a net-neutral charge, hydrophilic peptide, P4, that diffused 90 times faster compared to a negative control through *in vitro* tumor ECM, achieving approximately half of its unhindered diffusivity in PBS. Using the selected peptide as a model we validated that hydrophilicity, charge and the order of the amino acid sequence can impact diffusive transport through the tumor ECM. The uptake of the selected clone is ~200 times higher compared to the negative control in *ex vivo* BxPC3 pancreatic tumor xenograft. We also verified the ability of the peptide as surface coatings to improve diffusive transport of nanoparticles through the tumor microenvironment towards improved drug delivery. The selected peptide was conjugated onto fluorescent nanoparticles, and the diffusivity in *ex vivo* pancreatic tumor tissues was quantified by multiple particle tracking. After peptide coating, the nanoparticles exhibited ~40-fold improvement in diffusivity and greater number of mobile particles compared to particles functionalized with the gold standard— poly(ethylene) glycol or iRGD-peptide ligands. The diffusivity is of similar order as that achieved by PEGylated or iRGD-peptide-coated particles, which are generally used for improved penetration of drugs in the tumor environment.

Keywords: Tumor environment, drug delivery, nanoparticles, diffusion, peptides, phage display, particle tracking

Research Interests:

Experience: My Ph.D. research focuses on enhancing nanotherapeutic diffusion through the extracellular matrix (ECM) of tumors. The heterogeneous ECM network provides interaction-based barriers that hinder the penetration of nanoparticles through the tumor ECM and their ability to reach the cancer cells at therapeutic doses. To address challenges presented by current surface chemistries, we used a high-throughput approach to identify peptide coatings with optimized surface physicochemical properties that facilitate improved diffusive transport through the charged network of tumor ECM. Mechanistically, our study suggests the net negatively charged extracellular matrix can act as a drug delivery depot for the optimized positively charged nanoparticles based on their electrostatic interaction and reversible binding affinity. Next, I aim to use the identified peptides to validate that the improvement in penetration indeed leads to enhanced drug delivery and tumor regression. As part of another project, I am working with Dr.

Zhengrong Cui's lab at UT-Austin to optimize the surface properties of their existing nanoparticle formulation to achieve improved efficacy against pancreatic tumors.

Teaching Interests:

Areas of Interest for Post-Doc: With the firm belief of finding better alternatives to animal models, I would like to pursue my post-doctoral study on the "Organs on Chips" technology for building complex and three-dimensional models of living human organs. I am also interested in the design of a personalized disease environment for drug development. I am looking forward to working in a collaborative environment to successfully engineer the biological intricacies of the disease model on a microfluidic chip.

581824 Mechanisms of Diffusion in Associative Polymer Networks: Evidence for Chain Hopping

Research Interests: design and evolution of artificial proteins, physics of soft materials, biocatalysis with artificial enzymes, antigen-specific immunotolerance

Teaching Interests: chemical engineering thermodynamics, reactor design, mass and heat transport, polymer physics and polymer chemistry, protein design and evolution, modern methods in chemical biology, advanced organic synthesis and organocatalysis for engineers, chemical engineering research as teaching, creating active learning environments for specifically *upper* level engineering coursework

Abstract: Networks assembled by reversible association of telechelic polymers constitute a common class of soft materials. Various mechanisms of chain migration in associative networks have been proposed, yet there remains little quantitative experimental data to discriminate among them. Proposed mechanisms for chain migration include multichain aggregate diffusion as well as single chain mechanisms such as "walking" and "hopping", wherein diffusion is achieved by either partial ("walking") or complete ("hopping") disengagement of the associated chain segments. Here we provide evidence that hopping can dominate the effective diffusion of chains in associative networks due to a strong entropic penalty for bridge formation imposed by local network structure; chains become conformationally restricted upon association with two or more spatially separated binding sites. This restriction decreases the effective binding strength of chains with multiple associative domains, thereby increasing the probability that a chain will hop. For telechelic chains this manifests as binding asymmetry, wherein the first association is effectively stronger than the second. We derive a simple thermodynamic model that predicts the fraction of chains that are free to hop as a function of tunable molecular and network properties. A large set of self-diffusivity measurements on a series of model associative polymers finds good agreement with this model.

581844 Programmable Deformation and Assembly of Soft Active Materials

Research Interests:

My research interests lie in the intersection of soft matter, photonics, microfluidics and nanostructure engineering for development of smart soft materials and platforms. At highly interdisciplinary interface, my independent research aims to exploit the fascinating interplay between materials chemistry, geometry and mechanical deformation, and to create novel structured materials and actuation devices with unprecedented properties and functions, including tunable optics, flexible electronics and bio-inspired robotics. The early stage of my research program will span mainly two areas below.

1. **Smart Mechanical Metamaterials**

In the course of evolution, nature has developed strategies to create materials with exquisite adaptability and optimized efficiency. Learning from nature, people have developed stimuli-responsive engineering polymers, known as a class of smart materials. Yet, there remains extensive untapped knowledge of the ways in which to harness functional, dynamic, and adaptable materials into the diverse integrated systems. To establish a rational strategy to control instabilities of soft polymeric materials, understanding of their phase behavior and active interactions with their surrounding is critical. **My research goal is to investigate dynamic behavior/mechanism of the responsive materials in terms of structure and property triggered by mechanical strain or different external field conditions by using various microscopy techniques.** This knowledge should help resolve significant challenges of materials integration into complex and robust 3D structures. I believe this research regarding reconfigurable metamaterials can address broader technological and societal challenges, including those in the areas of renewable energy harvesting, chemical sensing, water purification, and environmentally benign microfabrication.

2. **4D Printing of Soft Materials**

Two-photon polymerization (TPP), also known as direct laser writing, has been recognized as a promising technique for the fabrication of arbitrary 3D sub-micron features with high shape and functions in photosensitive materials without photomasks. Typically, it requires an ultrashort pulsed laser to trigger non-linear two-photon absorption only in the focal spot volume, so called voxel. Unlike other 3D printing technologies, thus, it allows for the highest resolution or layer thickness with a previously unavailable freedom of design. Although TPP is a powerful and attractive technique, the major bottleneck in unleashing its potential lies in point-by-point writing with high power pulsed laser and long processing times unacceptable for practical applications. **Thus, I am interested in triplet fusion photo-polymerization which allows for use of a non-coherent, continuous wave light at ultralow excitation power as a pumping source to generate high-energy photons.** Besides photochemistry, my interests include methods **to develop high-throughput microfabrication by combining of flow lithography** using digital micromirror device. I believe this approach provides a short processing time, offering potentials of mass production for high-fidelity freeform microstructures with programmed internal heterogeneity. I am utilizing them to synthesize novel functional building blocks like anisotropic, shape-deformable colloids for (biomimetic) assembled complex structures that dynamically respond to energy input from their surroundings.

Research Experience:

For my postdoctoral research at the University of Massachusetts Amherst (Advisor: Prof. Ryan C. Hayward, Polymer Science & Engineering), I am currently working on precise control of mechanical instabilities in responsive polymer patterns. I am particularly interested in programming robustly self-folding action in rigid origami to desired states by preventing misfolding from inherent bifurcation at origin. Using mathematical prediction, I have designed a new structural transformation of multilayer polymer films that can be dynamically reconfigured to change their shapes and properties on demand. I also have developed biochemo-mechanical constructs and self-limiting assemblies of soft materials using constrained deformation by external stimuli such as pH, temperature and light.

During my PhD at Georgia Institute of Technology (Advisor: Prof. Elsa Reichmanis, Chemical Engineering), I focused on optofluidic encapsulation of triplet fusion photon upconversion. I investigated photon management in intermolecular energy transfer between metastable states and diffusion-mediated delayed fluorescence in polymer thin films and microcapsules. I had been leading the development of photo-induced interfacial polymerization and one-step emulsification of multi-phase fluids with microfluidic devices for high quantum efficiency and photochemical stability. Further, I demonstrated a photonic shell containing cholesteric liquid crystals with planar alignment to amplify spontaneous upconverted emission yield and tailor optical properties with tunable bandgaps. I believe the upconversion capsules have a wide range of applications in photonics such as photonic ink, microlaser, volumetric display, solar energy harvesting, and photocatalytic water treatment.

Early in my career at KAIST (Advisor: Prof. Seung-Man Yang, Chemical Engineering), I was involved in research on creation of nanostructures with free-defects over a large area by multi-coherent light without a photomask (i.e. holographic interference lithography, HIL). I developed a new type of hydrogel photoresist that undergoes viscoelastic deformation and produced 3D photonic crystals with tunable structural color. I also worked on hierarchical porous electrodes by using colloidal templating and HIL for photovoltaic cells and batteries.

Teaching Interests:

During my teaching career development, I enjoyed serving as a teaching assistant for undergraduate courses in "Chemical Process Safety" and "Unit Operations Lab" at Georgia Tech. Chemical Process Safety requires students to tackle real safety problems with chemical engineering knowledge. As a TA, I prepared to guide their quiz and projects with many aspects such as toxicology, release & dispersion models and case studies. For the course of UO Lab, I was assigned as a TA of the fluidized bed to give students a brief introduction and standard operation procedures to run a series of experiments. I evaluated their knowledge from their reports so I also learned how to clarify concepts to help them understand. These experiences have built my confidence and an interest in teaching. Besides, I am excited to teach a wide range of the stem courses in Chemical Engineering such as "Chemical Reaction Engineering", "Mass, Momentum and Energy Transport", and "Thermodynamics". From my multi-disciplinary research experience, I am also confident in teaching specific courses of "Nano-chemical Technology", "Photochemistry and Photophysics", "Fluidic Mechanics", "Capillarity and Wetting", and "Polymer Chemistry". I look forward to the opportunity to not only teach existing courses but also work to develop new ones that I can contribute to. Also, I have a lot of experience in mentoring undergraduate students from SURE and NSF MRSEC REU and graduate students to develop and trim their own creative ideas. Based on these teaching and mentoring experience, I have learned about the roles and responsibilities of the principal investigator how to communicate better with my students and encourage them to proceed.

Successful Proposals:

The proposal titled "Polymer Chemistry: Soft Mechanical Logic" is being considered for a three-year award starting from June 2019 by Army Research Office.

Selected Publications:

1. **Kang, J.-H.**; Kim, H.; Santangelo, C. D.; Hayward, R. C., Enabling Robust Self-Folding Origami by Pre-Biasing Vertex Buckling Direction, *Advanced Materials* **2019**, in revision.
2. Kim, H.; **Kang, J.-H.**; Zhou, Y.; Kuenstler, A.; Kim, Y.; Chen, C.; Emrick T.; Hayward, R. C., Light-driven Shape Morphing, Assembly, and Motion of Nanocomposite Gel Surfers, *Advanced Materials* **2019**, 1900932.
3. Ha, S.-J.; **Kang, J.-H.**; Choi, D.H.; Nam, S.K.; Reichmanis, E.; Moon, J.H., Upconversion-Assisted Dual-Band Luminescence Solar Concentrator Coupled for High Power Conversion Efficiency Photovoltaic Systems, *ACS Photonics* **2018**, 5, 3621.
4. **Kang, J.-H**; Kim, S.-H.; Fernandez-Nieves, A.; Reichmanis, E., Amplified Photon Upconversion by

Photonic Shell of Cholesteric Liquid Crystals, *Journal of American Chemical Society* **2017**, 139, 5708.
5. **Kang, J.-H.**; Lee, S. S.; Guerrero, J.; Fernandez-Neives, A.; Kim, S.-H.; Reichmanis, E., Ultrathin Double-Shell Capsules for High Performance Photon Upconversion, *Advanced Materials* **2017**, 29, 1606830.
6. **Kang, J.-H.**; Reichmanis, E., Low-Threshold Photon Upconversion Capsules Obtained by Photoinduced Interfacial Polymerization, *Angewandte Chemie International Edition* **2012**, 51, 11841.
7. Kim, Y.; Cho, C.-Y.; **Kang, J.-H**; Cho, Y.-S.; Moon, J. H., Synthesis of Porous Carbon Balls from Spherical Colloidal Crystal Templates, *Langmuir* **2012**, 28, 10543.
8. Kim, H.-N.*; **Kang, J.-H***; Jin, W.-M.; Moon, J. H., Surface Modification of 2D/3D SU-8 Patterns with a Swelling-Deswelling Method, *Soft Matter* **2011**, 7, 2989. (*equal contribution)
9. Shin, J.-H.; **Kang, J.-H**; Jin, W.-M.; Park, J. H.; Cho, Y.-S.; Moon, J. H., Facile Synthesis of TiO2 Inverse Opal Electrodes for Dye-Sensitized Solar Cells, *Langmuir* **2011**, 27, 856.
10. Jin, W.-M.; **Kang, J.-H**; Moon, J. H., Fabrication of 3D Copper Oxide Structure by Holographic Lithography for Photoelectrochemical Electrodes, *ACS Applied Materials and Interfaces* **2010**, 2, 2982.
11. **Kang, J.-H**; Moon, J. H.; Lee, S.-K.; Park, S.-G.; Jang, S. G., Yang, S.; Yang, S.-M., Thermoresponsive Hydrogel Photonic Crystals by Three-Dimensional Holographic Lithography, *Advanced Materials* **2008**, 20, 3061.

581857 Molecular-Based Modeling of Polymer Dynamics for Material Design and Processing

Research Interests:

My material design research group will concentrate on mesoscale models for polymer dynamics and rheology. Although the polymer industry has been established for many decades, our understanding of the fundamental relations between polymer chemical composition and the properties of the final product is not complete. While we currently have decades of atomistic and molecular studies, industry still relies on quasi-phenomenological approaches in practical applications. Recently, the slip-link model, a novel computational molecular-informed method for rheology of entangled polymers grew from a fundamental concept to a practical tool for polymer processing optimization. Building upon my experience with the slip-link model, my group will seek to convert insights from molecular models to guidelines for polymer industry. The goal will be to build a hierarchical modeling framework, starting from molecular dynamics and building up to macroscopic simulations of final products for direct manufacturing optimization. While my group will employ common tools for both molecular modeling (such as atomistic and molecular dynamics/Monte-Carlo simulations) and continuum level modeling (such as finite difference or Lagrangian methods), the focus will be on bridging these levels of description using mesoscopic/coarse-grained or single chain models. Based on my expertise, I identified the following three starting research areas:

- multi-scale simulation with the slip-link model to explore the link between processing conditions and product properties for optimization of 3D printing, fiber blowing, and injection molding
- mesoscale modeling of nanocomposites to investigate the effect of rubber reinforcement by nanoparticles, such as Carbon Black, used in consumer rubber products
- a coarse-grained model for polyelectrolyte complexation to study effects of molecular weight and salt concentration on the rheology for material design in this area of growing scientific interest

The goal of these research projects is to advance our understanding of already existing polymer materials and their applications. My group will benefit from a research-focused institutional environment and from industrial connections. Particularly, contacts with experimental rheologists and polymer characterization specialists are essential. Partnerships with the polymer industry will be facilitated by studies aimed at the optimization of manufacturing processes for cost and waste reduction. However, my ultimate goal is enabling the design of novel and eco-friendly polymeric materials directly from directly molecular basis.

Teaching Interests:

Although I hold degrees in physics with specialization in hydrodynamics and soft matter, I have been involved with chemical engineering departments throughout my research career, and I am prepared to teach core classes like Transport Phenomena, Thermodynamics, and Statistical Mechanics, both at graduate and undergraduate levels. In addition, my research was in the field of polymer physics, and I have experience teaching a Rheology or Polymer Science class.

Selected Publications:

1. M Andreev, G Rutledge, A Slip-link Model for Rheology of Entangled Polymer Melt with Crystallization. Journal of Rheology, submitted
2. M Andreev, V Prabhu, JF Douglas, MV Tirrell, JJ de Pablo, Complex Coacervation in Polyelectrolytes from a Coarse-Grained Model, Macromolecules, 51(17),6717-6723, 2018
3. S Srivastava, M Andreev, AE Levi, DJ Goldfeld, J Mao, WT Heller, V Prabhu, JJ de Pablo and MV Tirrell, Gel Phase Formation in Dilute Triblock Copolyelectrolyte Complexes, Nature Communications, 8, 14131, 2017
4. M Andreev, JD Schieber, Accessible and Quantitative Entangled Polymer Rheology Predictions, Suitable for Complex Flow Calculations, Macromolecules, 48 (5), 1606-1613, 2015
5. M Andreev, RN Khaliullin, RJA Steenbakkers, JD Schieber, Approximations of the discrete slip-

link model and their effect on nonlinear rheology predictions, Journal of Rheology, (57), 535-557, 2013

581909 Molecular–Scale Engineering of Charge–Containing Polymers and Electrochemical Interfaces

Research Interests: My research group will focus on analysis, design, development, and deployment of adaptable materials with polymer composite building blocks. I propose to study these materials both from the perspective of advancing fundamental understanding of their structure–property–function relationships and for their application in electrochemical systems. The research I propose to undertake presents multiple opportunities for first–principles design of materials and interfaces that respond to chemical, electrical, and mechanical stimuli to achieve a desired electrochemical function. My specific aim is to perform fundamental and applications–oriented studies in a 'closed–loop' approach in which discovery–driven studies are performed in tandem with macroscale characterization in realistic, applications–specific contexts to address fundamental challenges in electrochemical devices. In so doing my group's work promises to move beyond current paradigms of top–down and bottom–up syntheses to enable adaptive materials design for achieving explicit control of electrochemical functions.

The proposed research program will leverage experimental and analytical skills across three broad themes:

(1) *Understanding mass and charge transport at functional interfaces*. Specifically in this research thrust, the electrochemical activities of a chemically modified electrode will be analyzed systematically using in-situ voltammetry, spectroscopy and NMR characterizations. Such studies will be leveraged both for advancing fundamental understanding of how specific interfacial chemistries influence electrochemical behavior and for innovating new interfacial design that can enable emergent energy storage and electrochemical sensor formats.

(2) *Designing Structured Polymer networks for charge transfer processes.* Here, supramolecular polymer architectures will be designed to decouple the charge transport (ions or electrons) and physical properties (modulus or electrochemical stability window or stretchability etc). This thrust area would be impactful for developing all solid–state charge conductors having multiple desirable properties through fundamental research.

(3) *Structure–Transport relationships in interactive nanoparticle–polymer hybrids.* Here, we will evaluate the self-assembly and stimuli-responsive behavior in nanocomposites, where the nanoparticle surface undergoes favorable or unfavorable interactions with a polymer matrix. The structural changes due to various factors like temperature, ionic strength, electric field will be systematically evaluated using suite of characterization tools. The successful design and fundamental understanding of such hybrids can enable multifunctionality in a single material; for example, high electronic and ionic conductivity in semi-solid electrodes for flow batteries or simultaneous high solubility of CO_2 and ion conduction in electrolytes for electrocatalytic devices.

Overall, the proposed research builds on my past and current expertise that I developed as a graduate student and as a postdoctoral researcher. My background in the synthesis of self–healing polymer networks and polymer grafted nanoparticles with different chemistries; on physical analysis of disordered materials using x–ray/neutron scattering, NMR, and mechanical rheology, will be particularly valuable. Likewise, my expertise in surface chemistry along with state–of–the–art electrochemical analysis techniques will help me in developing theory–driven rational approaches for designing electrochemical interfaces to develop a research program that pushes the frontiers of electrochemical systems. My proposed work lies at the intersection of several fields, synthetic organic chemistry, electrochemistry, material science, and chemical engineering. I believe students at all levels (graduate and undergraduate) educated under my guidance will possess a versatile set of core skills that will make them as desirable as knowledge creators in academia and as product developers in industry.

Teaching Interests: I have prior experience of teaching three undergraduate level courses as a Teaching Assistant (TA) in Cornell University. The courses include, (1) Heat and Mass Transfer (twice); (2) Polymeric Materials and (3) Mechanics and Heat. My responsibilities in these courses included weekly recitation sessions; designing/grading homework and exam questions as well as taking office hours and review sessions for students. I received the highest honor of Outstanding Teaching Assistant Award for my services in the Chemical Engineering department.

In my PhD research group, I have mentored several junior graduate (two PhD, three M.S. and one M.Eng.) students as well as three undergraduate students. In my postdoctoral lab at Stanford University, I continue my role as a mentor by advising a PhD and an undergraduate student.

As a tenure-track Professor, I am enthusiastic to teach and develop curriculum for undergraduate and graduate courses related to unit operations, electrochemistry, and polymer science and complex fluids. Continuing from my TA experiences, I am willing to enhance the core chemical engineering course of heat and mass transfer. Further, I will specifically design two interesting courses, titled 'Electrochemistry for Chemical Engineers' and 'Soft Matter Characterization Techniques', for senior undergraduate and graduate students, which will focus on providing over–arching knowledge and expertise in these respective areas. These courses will include reviewing the latest and most important research articles in the field, practical demonstration through my labs, and provide an opportunity for students to collaborate with each other through a research proposal and class presentation. My teaching philosophy in all my courses will revolve around active learning, effective communication, and sharing of the ideas.

582015 Optimizing the Properties of Advanced Materials through Molecular Design

Research Interests:

My research interests broadly lie in polymer chemistry research. This includes: (1) using light as a trigger to catalyze the formation and modify the surface of polymers for various applications including dental materials, fuel cell membranes, and self-healing materials. (2) The use of click-chemistry in polymers and hydrogels for facile, rapid and orthogonal synthesis and functionalization of these materials. (3) Applying and developing new dynamic polymers through both associative and dissociative exchange processes. The overall research approach starts from fundamental chemical principles such as small molecule kinetics and monomer synthesis that in turn help improve macromolecular design. A deep understanding of the mechanical properties of these polymers through various rheological and spectroscopic techniques are further essential to improve the materials and target them for their desired application.

Teaching Interests:

My teaching interests lie in developing and improving core chemical engineering and chemistry classes through the use of evidence-based learning and multidisciplinary approaches to teaching through experiential learning. Specifically, I am interested in developing classes that utilize peer instruction, cooperative-based learning and inquiry based chemical engineering labs. Furthermore, my research in materials science, polymer composites for energy and healthcare applications, photochemistry, and dynamic networks has prepared me to teach electives in polymer technology and manufacturing. Finally, I am extremely in developing innovative interdisciplinary electives that combine both the arts and STEM fields through collaborating with professor other departments.

582075 Polymer Mechanochemistry in Molecular Machines and Medical Treatments

Research Interests:

Part 1. Current Research: Polymer mechanochemistry--New Mechanophore Systems in Molecular Machines and Medical Treatments (postdoc, 2018-current, advisor: Prof. Jeffrey Moore)

Project 1. Design and synthesis of polymer mechanochemical molecular machines

Mechanophores are functional molecular units that exhibit selective chemical responses to mechanical perturbations. Polymers embedded with mechanophores are responsive to forces (elongation, compression, ultrasonication) and the outcome involves various forms: changes in color and/or fluorescence, chemiluminescent light emitting, activation of latent catalyst, as well as the generation of reactive species such as free radicals, ylides, acids, and small organic molecules. These materials have demonstrated a variety of stress-responsive properties including stress sensing, damage reporting, remote light generation, and self-healing. Among these applications, releasing small molecules from polymer backbone is one of the most challenging tasks. Generally, to transduce the mechanical force experienced by the polymer matrix to the mechanophores, both ends of the mechanophores need to be attached to the polymer backbone. Therefore, after mechanophore activation the residual mechanophore moieties remain bonded to the polymer backbone. Therefore, releasing small molecules by mechanophore is very rare.

So far, small-molecule releasing is usually based on two approaches: (1) Activating a rearrangement reaction that release molecules; and (2) molecular fragmentation. Due to the nature of most rearrangement reactions, the former approach is usually limited to releasing simple species (*e.g.* HCl, SO_2). The latter fragmentation approach is an example of releasing more complex molecules; however, it generally requires specialized molecular design with limited varieties, and demands significantly higher mechanical stress for the activation process.

Herein, I'm demonstrating my research on the design and synthesize of a new mechanophore platform with molecule-releasing capability via the introduction of mechanically interlocked molecular machines. Molecular machines are molecular constructs that are controlled to perform programmed tasks from the molecular level. Most molecular machines are designed with mechanically interlocked structures, such as a rotaxane that compose of a mobile "wheel" threaded by an "axle" and trapped by bulky stopper groups. In a rotaxane molecular machine, tasks are accomplished by controlling the movement of the wheel along the axle. Therefore, we hypothesize that by guiding the directional sliding of the wheel along the axle, mechanical forces are applied to mechanophores *via* topological hindrance, allowing for activation of mechanophores on the axle and "extrude out" the generated small molecules.

Project 2: Mechanophore-Based Biomaterials for In-Vivo Ultrasound-Triggered Chemiluminescent Phototherapy

In contrast to the great success in materials science, the application of mechanophores in biomaterials was underdeveloped. Recently, our groups reported that the high-intensity focused ultrasound (HIFU), a non-invasive medical treatement tool, can remotely activate mechanophores in polymer networks. An interesting candidate—dioxetane mechanophore developed by Sijbesma and coworkers, attracted our attention for its ability to generate light from chemical energy *via* force-induced chemiluminescence. Under HIFU stimulation, blue light was clearly observed in our dioxetane containing polydimethylsiloxane (PDMS) materials at the focal spot of the ultrasound beam, providing excellent spatial and temporal control. This preliminary result opens a pathway towards biomedical applications of mechanophores such as non-invasive sono-optical treatments for optogenetics therapy. To achieve this final goal, we are aiming at two targets: (1) Improving light intensity.

In our previous experiments the light intensity did not meet the minimum threshold for optogenetics or other photo therapies. In our recent studies, we discovered that the poor solubility (max. ca. 0.5 wt%) of the energy acceptor, diphenylanthracene (DPA), was a limiting factor in the light intensity, presumably due to low energy transfer efficiency. By replacing DPA with a more soluble hexyl-functionalized DPA (HDPA), we can now incorporate up to 7.5 wt% HDPA into PDMS. As a result, the light intensity from HIFU activation was increased by 70-fold, up to 14 $\mu W/cm^2$. The increased intensity is higher than the reported threshold of some channelrhodopsins and thus is promising for optogenetics applications. (2): Designing an injectable and biocompatible hydrogel. With success on PDMS-based materials, I'm also working on adapting the system into injectable hydrogels. In our preliminary study, a dioxetane mechanophore was crosslinked into a DMSO-swollen PEG gel network, which successfully generates blue light upon HIFU sonication. We are modifying the molecular structure of dioxetane and DPA derivatives to achive water solubility and injectability without sacrificing the light intensity.

Part 2. Selected Past Research:

Project 1: Poly[n]catenanes: Synthesis of molecular interlocked chains (Ph.D. 2011-2016, postdoc 2016-2018, advisor: Prof. Stuart Rowan)

As the macromolecular version of mechanically interlocked molecules, mechanically interlocked polymers are promising candidates for the creation of sophisticated molecular machines and smart soft materials. Poly[*n*]catenanes, where the molecular chains consist solely of interlocked macrocycles, contain one of the highest concentrations of topological bonds. As a molecular analog to the chains we use in daily life, poly[n]catenane chains are predicted to exhibit high flexibility and good robustness. It is also an ideal candidate for the study of entanglement in polymer materials. The synthesis of poly[n]catenanes, however, remains a challenge for decades.

We reported the first synthetic approach toward this distinctive polymer architecture in high yield (~75%) via efficient ring closing of rationally designed metallosupramolecular polymers. Light-scattering, mass spectrometric, and nuclear magnetic resonance characterization of fractionated samples support assignment of the high–molar mass product (number-average molar mass ~21.4 kilograms per mole) to a mixture of linear poly[7–26]catenanes, branched poly[13–130]catenanes, and cyclic poly[4–7]catenanes. Increased hydrodynamic radius (in solution) and glass transition temperature (in bulk materials) were observed upon metallation with Zn^{2+}.

Selected Publications:

(1) **Q. Wu** *et al.*, Poly[n]catenanes: Synthesis of molecular interlocked chains. *Science,* 358, 1434-1439

(2) R. Wojtecki, **Q. Wu** *et al.,* Optimizing the formation of 2,6-bis(N-alkyl-benzimidazolyl)pyridine-containing [3]catenates through component design. *Chemical Science,* 4, 4440-4448

Project 2: Graphene based composite materials for electrochemical supercapacitors (M.S. 2009-2011, advisor: Prof. Gaoquan Shi)

Graphene-based materials are excellent candidates for electrochemical energy storage because of the outstanding electrical conductivity, good mechanical strength and huge specific surface area. My M.S. research was focused on the incorporation of redox-active polymers (e.g. polyaniline) and small molecules (e.g. anthraquinone) to prepare graphene-based composite materials with large capacity, good rate capability and long cycle life.

Selected Publications:

(1) **Q. Wu** *et al.*, Supercapacitors based on flexible graphene/polyaniline nanofiber composite films. *ACS Nano,* 4, 1963-1970.

(2) **Q. Wu** *et al.* High-performance supercapacitor electrodes based on graphene hydrogels modified with 2-aminoanthraquinone moieties. *Phys. Chem. Chem. Phys.* 13, 11193-11198.

(3) Y. Sun, **Q. Wu** and G. Shi, Graphene based new energy materials. *Energy Environ. Sci.* 4, 1113-1132

Teaching Interests:

1. Teaching and mentoring experience

During my PhD I've been working as instructor for the departmental MALDI-TOF instrument for three years (2013-2016) and taught at least thirty 1-hour training sessions (up to 4 new users per session). The training session covers the basic mechanism of MALDI-TOF methodology, instrument operation, data processing and sample preparation. I've mentored/co-mentored three graduate students, five undergraduate students and one high-school student. Currently, I'm supervising three undergraduate students working on three different research projects, respectively.

2.Teaching capability and philosophy

Besides solid education I received on polymers—I ranked overall #1 GPA for all polymer-related courses during every stage (B.S., M.S. and Ph.D.), another advantage of mine the multidisciplinary background—B.S. from department of chemical engineering, M.S. from department of chemistry. In my future teaching, I will add more chemical engineering and chemistry components into polymer courses. Also, I have the capability to teach some fundamental chemical engineering as well as chemistry classes. My key philosophy for good teaching is understanding my audience (students) before trying to teach them. I will find out their prior knowledge and their limits at the beginning of each semester and modify my teaching accordingly. My favorite strategy is activating their curiosity as intrinsic driving force to obtain more knowledge.

582113 Microfluidic Processes to Engineer Hydrogel Particles and Their Applications in Biomedical Engineering

Research Interests:

Combining my expertise in microfluidics and soft matter, my group's research will focus on the hydrogel microparticle synthesis, characterization, and manipulation to improve their capabilities in diagnosis and bioengineering. The overall goal is to fine-tune the functionalities of microparticles and to spatially arrange them in a specific pattern to engineer new functional soft matters. I envision that my group will resolve the current limitation of diagnosis (i.e. sensitivity, assay time, dynamic range, multiplexity) and provide a complex in vitro platform which cannot be fabricated by conventional technologies. More specifically, two major research area is outlined below.

Functional hydrogel microparticle synthesis

Hydrogel provides the solution-like environment at a molecular level, enabling the molecular transport through the material and solution-based reaction inside it. On the other hand, it also has the solid-like mechanical properties, making it easy to handle. These characteristics make hydrogel appealing in a variety of biomedical fields, such as medicine, tissue engineering, and diagnosis. As the technology evolves, there is an increasing interest to make micron-sized hydrogels while finely tuning their structure and functionalities. For instance, cell-sized hydrogels enable to make a complex microenvironment with a high spatial resolution. In bioassay, the ability to encode more than a million code in a small sensing region provides high sensitivity and multiplexity. During my Ph.D. under Dr. Patrick S. Doyle at MIT, I worked on the microfluidic lithography technique, stop flow lithography, to synthesize functional hydrogel microparticles. Among various particle synthesis techniques, microfluidic lithography enables to synthesize the uniform hydrogel microparticle with high geometric and chemical anisotropies. During my doctoral research, I synthesized the microparticles with the following functionalities: multiplexed, highly sensitive microRNA detection; spatially selective cell-adhesiveness; and upconverting emission for anti-counterfeiting. In my group, we will advance the microfluidic lithography techniques. We will utilize our knowledge of polymer synthesis and chemical reaction to provide more diverse functionalities. In addition, for the rational design of microparticles, we will conduct the dimensionless analysis to understand the mechanism of the particle synthesis process and the factors governing the bioassays. Our lab's effort will improve the capabilities of hydrogel microparticles in biomedical engineering.

Soft matter manipulation

Arranged particles can provide unlimited encoding capacities and fast decoding capabilities. Large scale particle arrays would reduce the significant amount of time and effort in the analysis. Furthermore, spatial positioning exponentially increases the encoding capacities and play an important role in both biomedical application and material science. During my doctoral research, I developed a platform, porous microwell array, which utilizes the hydrodynamic force to arrange the microparticles. This technology simultaneously fulfills the desired capabilities of particle manipulation: scalability, precision, specificity, and versatility. It arranges the large scale of particles at a precise location with high throughput. In addition, I conducted the scaling analysis to propose a dimensionless number for specific positioning of microparticles depending on their characteristics (i.e., size modulus, shape), and experimentally validated this scaling theory. The driving force, hydrodynamic force, acts on particles in the same manner regardless of their chemical compositions, fulfilling the versatility. Using this platform, I demonstrated the following applications: microenvironment fabrication for neutrophil chemotaxis, and upconverting particle arrays for anti-counterfeiting. Furthermore, I applied the same technology to arrange the following soft matters other than hydrogel microparticles: particle-in-droplet arrays for signal amplified miRNA detection; live microbe array for the study of antibiotic susceptibility and immune response. During my postdoctoral research with Dr. Daniel Irimia and Dr. Adam B. Raff at Harvard Medical School and Massachusetts General Hospital, using the same principle, I have worked on the following fields: human skin-on-a-chip for skin infection diagnosis; and neuro-spheroid arrays for the study of Alzheimer disease.

In my group, we will advance the particle manipulation technique to improve its capabilities. We will utilize other porous materials to apply the technology to smaller objects, such as Brownian colloids or nanoparticles. We will study the synergistic effect of hydrodynamic force and other driving forces to achieve higher specificity with more sorting standards. In addition, we will achieve the complete automation of device fabrication and particle manipulation process for commercialization. As demonstrated in the previous research, we will apply the same technology to various micron-sized objects, such as droplets, cells, crystals, and spheroids. I envision that our soft matter manipulation technology will pave the way to advance the current limitations in diagnosis and tissue engineering.

Teaching Interests:

I am interested to teach a wide range of core courses in Chemical Engineering. For undergraduate courses, I look forward to teaching the following courses: fluid mechanics, heat and mass transfer, thermodynamics, reaction engineering, and engineering mathematics. For graduate levels, I am prepared to teach transport phenomena and thermodynamics. Besides those core courses, I would like to also create new courses in the fields of microfluidics and colloid/interface science.

Selected publications:

Tentori, A.M., Nagarajan, M.B., **Kim, J.J.**, Zhang, W.C., Slack, F.J. and Doyle, P.S. "Quantitative and multiplex microRNA assays from unprocessed cells in isolated nanoliter well arrays", Lab on a Chip, 18(16): 2410-2424, **2018**.

Kim, J.J.*, Reátegui, E.*, Hopke, A., Jalali, F., Roushan, M., Doyle, P.S. and Irimia, D. "Large-scale Patterning of Living Colloids for Dynamic Studies of Neutrophil-Microbe Interactions", Lab on a Chip, 18(11): 1514-1520, **2018**.

Chen, L.*, **Kim, J.J.***, Doyle, P.S., " Microfluidic platform for selective microparticle parking and paired particle isolation in droplet arrays", Biomicrofluidics, 12(2): 024102, **2018**. (This paper has been selected as an Editor's Pick)

Kim, J.J.*, Chen, L.*, Doyle, P.S., "Microparticle Parking and Isolation for Highly Sensitive MicroRNA Detection", Lab on a Chip, 17(18): 3120-3128, **2017**.

Kim, J.J.*, Bong, K.W.*, Reátegui, E., Irimia, D., and Doyle, P.S., "Porous Microwells for Geometry-Selective, Large-Scale Microparticle Arrays," Nature Materials, 16(1): 139-146, **2017**. (This paper has been featured at MIT news: http://news.mit.edu/2016/better-way-to-assay-microparticle-arrays-0930)

Eral, H.B.*, Safai, E.R.*, Keshavarz, B.*, **Kim, J.J.**, Lee, J., and Doyle, P.S., "Governing Principles of Alginate Microparticle Synthesis with Centrifugal Forces," Langmuir, 32(28): 7198-7209, **2016**.

Bong, K.W.*, **Kim, J.J.***, Cho, H., Lim, E., Doyle, P.S., Irimia, D., "Synthesis of Cell-Adhesive Anisotropic Multifunctional Particles by Stop Flow Lithography and Streptavidin-Biotin Interactions," Langmuir, 31(48): 13165-13171, **2015**.

Lee, J.*, Bisso, P.W.*, Srinivas, R.L., **Kim, J.J.**, Swiston, A.J., and Doyle, P.S., "Universal Process-Inert Encoding Architecture for Polymer Microparticles", Nature Materials, 13(5): 524-529, **2014**. (This paper has been featured at MIT news: http://news.mit.edu/2014/tiny-particles-could-help-verify-goods)

* denotes equal contribution.

582173 Functional Polymers for Molecular and Materials Design

Research Interests:

Molecular shape impacts the self-assembly and physical properties of all soft materials. In polymer science and engineering, shape control can be achieved by tuning the polymer architecture, local non-covalent interactions, and/or responses to external stimuli. My research program will connect polymer chemistry and polymer physics in order to explore the impact of molecular shape in different contexts across molecular and materials design. We will use creative chemistry in order to address challenges in polymer physics, and in turn bring insights from polymer physics to enable new reactivity.

Research Experience:

Graduate Student, Department of Chemistry, California Institute of Technology, Prof. Robert H. Grubbs

Synthesis and Self-Assembly of Bottlebrush Polymers. Bottlebrush polymers represent a unique molecular architecture and a modular platform for materials design. However, the properties and self-assembly of bottlebrush polymers remain relatively unexplored, in large part due to synthetic challenges imposed by the sterically demanding architecture. My graduate research aimed to close these gaps, connecting (1) the synthesis of polymers with precisely tailored molecular architectures, (2) the study of fundamental structure-property relationships, and (3) the design of functional materials. I developed a versatile strategy to synthesize polymers with tailored architectures via grafting-through ring-opening metathesis polymerization (ROMP). One-pot copolymerization of a macromonomer and a small-molecule co-monomer provides opportunities to control the backbone sequence and therefore both the spacing and distribution of side chains. Homo- and copolymerization rate trends were identified and elucidated by complementary mechanistic and density functional theory (DFT) studies. Building on this foundation, I prepared complex molecular architectures and explored the physical consequences of varying the grafting density and graft distribution in two contexts: block polymer self-assembly and linear rheological properties. The molecular architecture strongly influences packing demands and therefore polymer conformation. Collectively, our studies represent progress toward a universal model connecting the chemistry and conformations of graft polymers.

Beckman Postdoctoral Fellow, Department of Chemical Engineering and Materials Science, University of Minnesota, Twin Cities; Prof. Frank S. Bates

Tough and Sustainable Plastics via Graft Block Polymer Design. The molecular architecture – that is, the way in which polymer chains are connected – dramatically impacts the properties of plastics. My postdoctoral research has focused on developing a modular platform for tough and sustainable plastics based on graft block polymers. Linking two or more chemically distinct blocks structures plastics at the nanoscale, improving the mechanical properties; introducing grafts amplifies shear thinning and strain hardening behavior, improving the melt processability. Precise synthetic control over the graft block architecture enables systematic studies of key structure-property relationships. Understanding design rules based on the molecular architecture — *independent* of specific choices of block and backbone functionality — allows the materials to both meet the needs of sustainability and be customized for specific applications.

Teaching Interests:

I am prepared and excited to teach a range of subjects, including thermodynamics, chemical kinetics, soft matter physics, and polymer science and engineering. My graduate training in chemistry, together with my postdoctoral experience in a chemical engineering department, provide competency in core areas of chemical engineering as well as perspective on communicating with students with different backgrounds and interests. Beyond the classroom, I am excited to engage in science outreach and research with broader impacts. Students and postdoctoral researchers in my group will benefit from interdisciplinary training in polymer science and engineering, spanning chemistry, physics, and materials design.

Selected Peer-Reviewed Publications:

(* denotes corresponding author; + denotes equal contributions)

1. Qian, Z.; Koh, Y. P.; Pallaka, M. R.; Chang, A. B.; Lin, T.-P.; Guzmán, P. E.; Grubbs, R. H.; Simon, S. L.; McKenna, G. B. Linear Rheology of a Series of Second-Generation Dendronized Wedge Polymers. *Macromolecules* **2019**, *52*, 2063.
2. Sunday, D. F.;* Chang, A. B.;* Liman, C. D.; Gann, E.; DeLongchamp, D. M.; Matsen, M. W.; Grubbs, R. H.; Soles, C. L. Evidence for Backbone Flexibility of Bottlebrush Block Copolymers Driven by Low-χ Assembly. *Macromolecules* **2018**, *51*, 7187. (*Corresponding authors.)
3. Haugan, I. N.; Maher, M. J.; Chang, A. B.; Lin, T.-P.; Grubbs, R. H.; Hillmyer, M. A.; Bates, F. S. Consequences of Grafting Density on the Linear Viscoelastic Behavior of Graft Polymers. *ACS Macro Lett.* **2018**, *7*, 525.
4. Chang, A. B.;+ Lin, T.-P.;+ Thompson, N. B.; Luo, S.-X.; Liberman-Martin, A. L.; Chen, H.-Y.; Lee, B.; Grubbs, R. H. Design, Synthesis, and Self-Assembly of Polymers with Tailored Graft Distributions. *J. Am. Chem. Soc.* **2017**, *139*, 17683.
5. Lin, T.-P.;+ Chang, A. B.;+ Luo, S.-X.; Chen, H.-Y.; Lee, B.; Grubbs, R. H. Effects of Grafting Density on Block Polymer Self-Assembly: From Linear to Bottlebrush. *ACS Nano* **2017**, *11*, 11632.
6. Chang, A. B.; Bates, C. M.; Lee, B.; Garland, C. M.; Jones, S. C.; Spencer, R. K. W.; Matsen, M. W.; Grubbs, R. H. Manipulating the ABCs of Self-Assembly via Low-χ Block Polymer Design. *Proc. Natl. Acad. Sci.* **2017**, *114*, 6462.
7. Lin, T.-P.; Chang, A. B.; Chen, H.-Y.; Liberman-Martin, A. L.; Bates, C. M.; Voegtle, M. J.; Bauer, C. A.; Grubbs, R. H. Control of Grafting Density and Distribution in Graft Polymers by Living Ring-Opening Metathesis Copolymerization. *J. Am. Chem. Soc.* **2017**, *139*, 3896.
8. McNicholas, B. J.; Blakemore, J. D.; Chang, A. B.; Bates, C. M.; Kramer, W. W.; Grubbs, R. H; Gray, H. B. Electrocatalysis of CO2 Reduction in Brush Polymer Ion Gels. *J. Am. Chem. Soc.* **2016**, *138*, 11160.
9. Bates, C. M.; Chang, A. B.; Schulze, M. W.; Momčilović, N.; Jones, S. C.; Grubbs, R. H. Brush Polymer Ion Gels. *J. Polym. Sci., Part B* **2016**, *54*, 292.
10. Bates, C. M.; Chang, A. B.; Momčilović, N.; Jones. S. C.; Grubbs, R. H. ABA Triblock Brush Polymers: Synthesis, Self-Assembly, Conductivity, and Rheological Properties. *Macromolecules* **2015**, *48*, 4967.
11. Chang, A. B.; Miyake, G. M.; Grubbs, R. H. Sequence-Controlled Polymers by Ruthenium-Mediated Ring-Opening Metathesis Polymerization. In *Sequence-Controlled Polymers: Synthesis, Self-Assembly, and Properties*; Ouchi, M., Meyer, T., Lutz, J.-F., Eds.; ACS Symposium Series 1170; American Chemical Society: Washington, DC, 2014; pp 161.

Patents:

1. Grubbs, R. H.; Lin, T.-P.; Chang, A. B.; Chen, H.-Y.; Bates, C. M. Control of Polymer Architectures by Living Ring-Opening Metathesis Copolymerization. U.S. Patent Appl. 15/914,762. March 7, 2018.
2. Grubbs, R. H.; Bates, C. M.; Chang, A. B.; McNicholas, B. J.; Jones, S. C. Brush Block Copolymer Electrolytes and Electrocatalyst Compositions. U.S. Patent Appl. 15/065,317. March 9, 2016.
3. Grubbs, R. H.; Bates, C. M.; Chang, A. B.; McNicholas, B. J.; Jones, S. C. Triblock Brush Block Copolymers. U.S. Patent Appl. 15/065,291. March 9, 2016.

Selected Awards:

- Arnold O. Beckman Postdoctoral Fellowship (2019–Present)
- McCoy Award, California Institute of Technology (2018)
- Gray-Hill Lecturer, Occidental College (2018)
- SciFinder Future Leaders Award (2017)
- American Chemical Society Excellence in Graduate Polymer Research Award (2016)

- American Chemical Society Women Chemists Committee / Merck Research Award (2016)
- National Defense Science and Engineering Graduate Fellowship (2014–2017)
- Phi Beta Kappa, Columbia University (2013)
- Rabi Science Fellowship, Columbia University (2009–2013)

Selected Successful Proposals:

- Arnold O. Beckman Postdoctoral Fellowship (2019–Present)
- National Defense Science and Engineering Graduate Fellowship (2014–2017)
- Advanced Photon Source at Argonne National Laboratory: GUP-43955 (240 hours of beamtime) (2015–2017)
- Advanced Photon Source at Argonne National Laboratory: GUP-55715 (200 hours of beamtime) (2017–2019)

580142 Towards Sustainable Polymers through Integration of Dynamic Covalent Chemistry and Advanced Processing

Postdoctoral Research Highlight:

Increasing demand for high performance plastics and elastomers and the associated environmental issues present society with enormous engineering challenges.[1] In this poster, I will highlight my postdoctoral work on promoting the sustainability of polymers, under the guidance by Prof. Bates and Prof. Ellison at University of Minnesota and in collaboration with many other researchers in the Center for Sustainable Polymers. The projects will be discussed in this poster can be summarized as follows:

Reversibly Cross-Linked Nonwoven Fibers: Sustainability Meets Melt Blowing

Melt blowing is a 60-year-old polymer processing method for producing nonwoven polymer fibers. It combines polymer melt extrusion with high-velocity hot air jet fiber drawing to produce nonwoven fibers in a single step. Due to its simplicity and high-throughput nature, melt blowing produces more than 10% of the $50 billion global nonwovens market. Semi-crystalline thermoplastic feedstocks, such as poly(butylene terephthalate) and polypropylene, dominate the melt blowing industry due to facile melt processability and product thermal/chemical resistance; other amorphous commodity thermoplastics (e.g., (meth)acrylates) are generally not employed because they lack one or both attributes. Cross-linking nonwoven products could enable more demanding applications, but cross-linking must be implemented after fiber formation. This poster describes melt blowing linear acrylic polymers containing cross-linkable functional groups, which were activated via either a cooling-induced Diels-Alder reaction[2] (e.g., furan-maleimide reaction) or a light-induced cycloaddition reaction[3] (e.g., anthracene photodimerization). The resulting fibers possessed nearly 100% gel content and exhibited enhanced thermomechanical properties with higher upper service temperatures (e.g., ~180 °C for anthracene-dimer cross-linked acrylic fibers) relative to the linear precursors. Due to the dynamic nature of the reversible cross-links at elevated temperatures, these cross-linked fibers can be recycled after use, providing new approaches to producing sustainable nonwoven products.

Reprocessable Thermoset Photopolymers: Sustainability Meets Three-Dimensional (3D) Printing

Three-dimensional (3D) printing, also referred to as additive manufacturing, has been globally recognized as a revolutionary technology which enables manufacturing materials with complex architectures that are often challenging to achieve using conventional methods. Among the existing 3D printing techniques, vat photopolymerization (e.g., stereolithography) offers superior feature resolution and contributes to nearly half of the 3D printing market. In a typical vat photopolymerization process, a liquid, reactive precursor mixture (usually comprised of acrylic monomers/oligomers, one of which being multifunctional and acting as a cross-linker) is selectively cured by a light-activated polymerization, resulting in permanently cross-linked photopolymers with robust mechanical properties and superior chemical/thermal resistance. These very characteristics that make them attractive also prohibits them from being reprocessed and thus recycled after use. Therefore, conventional 3D printed photopolymers are generally disposed as waste at the end of their product life, leading to issues ranging from economic loss to environment problems. Herein, dynamic covalent chemistries (e.g., thermoreversible Diels-Alder chemistry) were employed in vat photopolymerization to impart repairability and recyclability into 3D printed structures. The printed materials possessed nearly 100% gel content (indicative of network structures) and exhibited tailorable thermomechanical properties. Additionally, the dynamic linkages present in the network could enable effective reprocessing and reuse of these reversibly cross-linked photopolymers. For example, they could be reused as feed materials for extrusion-based processing techniques, such as fused deposition modeling (a type of 3D printing) and melt blowing (for cross-linked nonwoven production). Overall, this study demonstrates a promising route to recyclable thermoset photopolymers, providing sustainability to 3D printed materials and potentially advancing other applications such as self-healable coatings and adhesives.

References

1. Xu and Jin et al., *Macromolecules* **2018,** *51* (21), 8585-8596.
2. Jin et al., *ACS Macro Lett.* **2018,** *7* (11), 1339-1345.
3. Jin et al., *ACS Appl. Mater. Interfaces* **2019,** *11* (13), 12863-12870.

Research Interests:

Following my passion for the science and technology of polymeric materials and my commitment to a sustainable future, I am pursuing a tenure track faculty position. As a faculty member, I plan to establish an interdisciplinary research group dedicated to both fundamental and applied aspects of materials and technologies that can help address pressing global issues in energy, environment, and sustainability. More specifically, I aim to develop hierarchical polymeric materials (e.g., block polymers for photovoltaics, polymer nanocomposites for all-solid-state batteries, and covalent organic frameworks as porous membranes and absorbents) that can facilitate the development of efficient renewable energy harvesting processes and high-capacity energy storage systems as well as addressing environmental challenges, including air and water purification. As a chemical engineer who is equipped with synthetic chemistry skills and advanced characterization techniques, I will design hierarchical polymeric materials at a molecular level, direct assembly at a microscopic scale, and thereby control macroscopic material properties and performance.

Teaching Interests:

During my Ph.D. studies, I had the opportunity to teach ~30%, 10 lectures in total, of an undergraduate/graduate level course (*Introduction to Polymers*), which helped me develop the skills of a lecturer and instructor (e.g., designing exams/homework sets, holding office hours, and grading exams). Outside of the classroom, I have directly advised 7 students (including undergraduate, master, and first-year Ph.D. students) in total, through which I have honed my mentorship skills and enhanced my inter-personal skills. These experiences have sculpted my teaching and mentoring philosophy, central to which are three main concepts: *know the audience*, *teach fundamental knowledge with problem-solving skills*, and *learn through teaching*. My interdisciplinary background would allow me to teach many undergraduate/graduate classes, especially those in a standard chemical engineering curriculum (e.g., *reaction kinetics* and *transport phenomena*). As a polymer scientist, I am enthusiastic to teach topical courses in polymer science, including *polymer chemistry*, *polymer physics*, *rheology*, *polymer engineering/processing*, and *soft materials characterization*. Additionally, I am always willing to teach other subjects that might be outside my area of expertise as I believe that the best way to learn is to teach. Throughout the course, I will integrate current research challenges and build discussion points into lectures to engage students by having them use their fundamental knowledge to analyze and solve real-world problems. I believe this can help develop their critical thinking and problem-solving skills which are key to their success as the next generation of well-rounded scientists.

569234 Specialized Polymers for Integrating Advanced Synthetic and Biological Materials

Research Interests: Polymers have entered into virtually every aspect of modern society, from familiar commodity items such as disposable plastic bags to high-value technological applications in electronics, energy, and medicine. As the vibrant field of polymer science progresses to new frontiers, the incorporation of more advanced macromolecules into emerging health applications faces challenging engineering obstacles. Two prominent issues include controlling compartmentalization into polymer domains and predicting transport strategies into harsh physiological environments, which are often complex, dynamic, and responsive.

My long-term goal is to apply my expertise in *polymer science/engineering*, *molecular engineering*, and *precision drug formulation* to develop multifunctional materials with prescribed chemical attributes to address these unfulfilled needs. To achieve this, I will employ synthetic tools and ingredients to construct custom-built materials and elucidate the contributions of structure to function across hierarchical length and time scales. As an independent investigator, my research will focus on motifs of storage and delivery, extending from a convergence science approach to develop nanocarriers with controlled functionality and programmable performance. This is ultimately in pursuit of translating fundamental science into tangible applications with design principles.

The theme of my future research builds on my interdisciplinary training in chemical engineering, polymer chemistry, materials science, and the physical/life sciences. This can enable direct investigations of the spatiotemporal behavior for polymer carriers in solid- and solution-states. The core of this research enterprise will be broadly amenable to enabling analogous encapsulation/release strategies in energy management, agricultural/food production, and personal care. Moving forward, I anticipate that my laboratory infrastructure can readily host such avenues and collaborate with diverse groups in academia, industry, and national labs.

For initial projects in my group, I will leverage my broad skillset into three main aims:

(1) Accelerating Discovery and Data Science of Complex Macromolecular Architectures

(2) Surmounting Biological Barriers of Genome Editing Therapeutics Delivery

(3) Orchestrating Reconfigurable Structures in Soft Matter Assemblies

PhD Dissertation: *Tunable Polymers as Specialized Excipients for Oral Drug Delivery*, Department of Chemical Engineering and Materials Science, University of Minnesota (Advisors: Prof. Frank S. Bates and Prof. Theresa M. Reineke)

Postdoctoral Project: *Static and Dynamic Properties of Polyelectrolyte Complexes*, Pritzker School of Molecular Engineering, University of Chicago (Advisor: Prof. Matthew V. Tirrell)

Funding, Successful Proposals, Achievements, Service: As potential funding sources, the scientific challenges in these projects align with the major initiatives of national funding agencies, including NSF (NSF-DMR, including NSF's 10 Big Ideas), NIH (NIGMS, NIH Director's New Innovator Award, NIBIB), DOE (DOE-MSE), DOD (CDMRP, ARO), ACS (ACS PRF), and private foundations (Camille and Henry Dreyfus Foundation, Baxter International Foundation, HHMI, Pew Charitable Trusts). My previous record of select proposals, distinctions, and professional service are summarized:

Proposals: Oakridge National Laboratory Center for Nanophase Materials Science Research (2017); Argonne National Laboratory Advanced Photon Source (2016).

Distinctions: PMSE Future Faculty Scholar (2018); NIST-CHiMaD Postdoctoral Fellowship (2016); AIChE Pharmaceutical Discovery, Development and Manufacturing Student Award (2015); Doctoral

Dissertation Fellowship (2015); NSF Graduate Research Fellowship (2011).

Service: AIChE 08A Inhomogeneous Polymers Session Chair (2019); ACS POLY Community Engagement Team (2019); Future Faculty Workshop Attendee (2018); Joint Research Safety Initiative Founding Vice President (2018); 31 Reviewer Contributions (8 Macromolecules, 7 ACS Macro Letters; 4 Molecular Pharmaceutics)

Research Background: My research has relied on not only my technical training in chemical engineering, but also overarching aspects of chemistry, biomedical engineering, and materials science. I have led multidisciplinary projects supported by both industry and national laboratories, which I believe positions me to initiate new basic and applied research partnerships. Because of this, my perspective on translating advanced polymeric materials into end-use technologies also can facilitate ideal collaborations with existing research endeavors.

My graduate research involved the (1) controlled synthesis of polymer systems that were judiciously tailored to drugs, (2) preparation of amorphous solid dispersions, and (3) *in vitro / in vivo* dissolution experiments to assess the oral solubility advantage. This enabled the advancement of multicomponent copolymerization[1] and practical guidelines for solid dispersion formulation.[2-4] Macromolecular systems were engineered from the molecular level up to impart favorable non-covalent interactions for stabilizing otherwise-intractable anti-cholesterol, anti-seizure, and anti-cancer drugs.

My postdoctoral work aims to better understand the fundamentals of ion-containing polymers. Polyelectrolyte complexes, mixtures of oppositely-charged polymers in water, have highly tunable materials properties that are appealing for numerous bioapplications. However, harnessing these materials for clinical applications remains an enduring challenge. To this end, I am using carefully-selected building blocks to explore chemical, ionic, and architectural effects in well-defined polymer libraries. In this rich state space, monomer constituents are being systematically combined across multiple project platforms,[5-8] with time-resolved X-ray and neutron scattering, spectroscopy, and imaging investigations.

Teaching Interests: My interests include courses related to polymers (e.g., Polymer Chemistry, Polymer Physics), including topical courses I have taken like Rheology (under Prof. Chris Macosko), Scattering (under Prof. Timothy Lodge), Colloids/Interfaces (under Prof. Chris Macosko), or Biomaterials (under Prof. Chun Wang). I am also confident in teaching core courses in chemical engineering, such as Transport Phenomena or Thermodynamics.

My teaching philosophy focuses on mentoring students to become independent leaders, equipped with critical thinking and creative problem-solving skills for successful careers. In this manner, I strive to be cognizant of different ways to guide their overall education while supporting each student's individuality and valuing their diversity. At the University of Minnesota, I was a Teaching Assistant in Thermodynamics and Polymer Chemistry. Each course involved grading for ~120 undergraduate students. I also held review sessions and met with students to review unclear topics. I was awarded a faculty-nominated Chemical Engineering Outstanding TA Award and a student-nominated Council of Graduate Students Award for Graduate Teaching. I was also a Recitation Teaching Assistant in Transport Phenomena. Here, I led supplementary lectures in weekly recitation classes to ~40 students. Seeing mentored students succeed gives me immense gratification and strengthens my commitment to higher education.

Furthermore, I have led informal science education efforts described in the *Journal of Chemical Education* and *ACS Omega*.[9-10] These positive experiences taught me to tailor my science and teaching directions toward broader impacts, which I believe will help me develop competitive junior faculty application packages like the NSF CAREER award.

Outreach, Mentorship, and Diversity: Public outreach, mentorship, and commitment to diversity will be hallmarks of my research group, with a central goal of inspiring students and their families from diverse backgrounds to proactively engage with science and engineering. This is particularly important to me because I believe that the key to solving pressing global challenges ahead is investment in inclusion of an increasingly diverse workforce, especially for under-represented groups such as women, people of color,

LGBTQ+, or persons with disabilities. In my career, I want to be known for training the next generation of academic and industrial leaders to develop curiosity-driven scientific solutions for grand challenges in biotechnology, human health, and welfare.

Select Publications (10/15 total, 1 under revision, 4 submitted):

1. Ting, J. M.; Navale, T. S.; Bates, F. S.; Reineke, T. M. "Precise Compositional Control and Systematic Preparation of Multimonomeric Statistical Copolymers" *ACS Macro Lett.* **2013**, *2*, 770–774.
2. Ting, J. M.; Navale, T. S.; Jones, S. D.; Bates, F. S.; Reineke, T. M. "Deconstructing HPMCAS: Excipient Design to Tailor Polymer–drug Interactions for Oral Drug Delivery" *ACS Biomat. Sci. Eng.* **2015**, *1*, 978–990. (**ACS Editors' Choice**)
3. Ting, J. M.; Tale, S.; Purchel, A. A.; Jones, S. D.; Widanapathirana, L.; Tolstyka, Z. P.; Li, G.; Guillaudeu, S. J.; Bates, F. S.; Reineke, T. M. "High–throughput Excipient Discovery Enables Oral Delivery of Poorly Soluble Pharmaceuticals" *ACS Cent. Sci.* **2016**, *2*, 748–755.
4. Ting, J. M.; Porter III, W. W.; Mecca, J. M.; Bates, F. S.; Reineke, T. M. "Advances in Polymer Design for Enhancing Oral Drug Solubility and Delivery" *Bioconjugate Chem.* **2018**, *29*, 939–952.
5. Ting, J. M.; Wu, H.; Herzog-Arbeitman, A.; Srivastava, S.; Tirrell, M. V. "Synthesis and Assembly of Designer Styrenic Diblock Polyelectrolytes" *ACS Macro Lett.* **2018**, *7*, 726–733.
6. Wu, H.; Ting, J. M.; Werba, O.; Tirrell, M. V. "Non-equilibrium Assembly and Kinetic Pathways in Polyelectrolyte Complexes" *Chem. Phys.* **2018**, *149*, 163330.
7. Acar, H.; Ting, J. M.; Srivastava, S.; LaBelle, J. L.; Tirrell, M. V. "Molecular Engineering Solutions for Therapeutic Peptide Delivery" *Soc. Rev.* **2017**, *46*, 6553–6569.
8. Marras, A. E.; Vieregg, J. R.; Ting, J. M.; Rubien, J. D.; Tirrell, M. V. "Polyelectrolyte complexation of oligonucleotides by charged hydrophobic – neutral hydrophilic block copolymers" *Polymers* **2019**, *11*, 83.
9. Ting, J. M.; Ricarte, R. G.; Schneiderman, D. K.; Jiang, Y.; Saba, S. A.; Hillmyer, M. A.; Bates, F. S.; Reineke, T. M.; Macosko, C. W.; Lodge, T. P. "Polymer Day: Outreach Experiments for High School Students" *Chem. Educ.* **2017**, *94*, 1629–1638.
10. DeWilde, J. F.; Rangnekar, E.; Ting, J. M.; Franek, J. E.; Bates, F. S.; Hillmyer, M. A.; Blank, D. A. "Evaluating Large-Scale STEM Outreach Efficacy with a Consistent Theme: Thermodynamics for Elementary School Students" *ACS Omega* **2019**, *4*, 2661–2668.

572295 3D Printed Reconfigurable Liquid Crystal Elastomers Via Dynamic-Covalent Bonds

3D printing (3DP) of liquid crystal elastomers (LCEs) enables programming of complex, fully reversible 3D shape changes. Classical LCEs must be continuously heated to maintain their programmed actuated shape; here, we demonstrate reconfigurable 3DP LCE architectures by incorporating dynamic-covalent bonds into the LCE backbone. These 3DP LCEs can 'lock-in' on demand to maintain their complex programmed actuated state via bond exchange. Specifically, we create a printable ink containing dynamic-covalent bonds by oligomerizing liquid crystal mesogens with an allyl dithiol chain extender. During printing, the ink is crosslinked to form a LCE that undergoes reversible programmed shape-change upon heating above its nematic-to-isotropic transition temperature, T_{NI}. We leverage light-activated bond exchange to ensure the 'trigger' for bond exchange is orthogonal to the temperature change required for actuation. These reconfigurable LCEs enable architectures to be realized that are inaccessible by printing other classes of soft materials. Finally, by integrating these bond-exchangeable LCEs with standard LCEs via multimaterial 3D printing, we demonstrate a path towards hierarchical, reconfigurable, and responsive structures.

Research Interests: Liquid crystalline polymers, block copolymers, sequence defined polymers, additive manufacturing, stimuli-responsive materials, sustainable polymers

Teaching Interests: Polymer physics, additive manufacturing methods, thermodynamics, kinetics, transport

560698 Granular Composite-Enabled Multi-Scale Dynamic Responsive Materials

Research Interests:

Bio-materials for multi-scale chemical and mechanical dynamics. Reconfigurable photonic crystals and shape memory polymers.

Teaching Interests:

Materials of Chemical Engineering, Energy Transport phenomena, Material and Energy Balances, structure and properties of polymer.

Living tissues are an integrated, multi-scale architecture consisting of dense cell ensembles and extracellular matrices (ECM) that cooperatively enable the excellent mechanical properties and dynamic responsiveness. One key challenge in creating tissue-like materials is to identify cell-like building blocks that can synergistically respond to external stress with existing ECM-like polymer platforms. Here, we designed a granular material-enabled hybrid gel, including cell-like starch granules embedded in ECM-like synthetic hydrogel matrices that readily exhibited dynamic reconfigurable effects upon mechanical training. Multi-scale and *in-situ* characterizations reveal that the unique combination between microscopic (chemical bonding) and mesoscopic (physical friction) interactions from starch granules coöperatively give rise to the tissue-like properties, such as dynamic responsiveness, strain-stiffening, and self-healability. Our results suggest that granular materials, a largely ignored component for biomimetics, can be critical in enabling dynamic behaviors in artificial materials and even future adaptive and active metamaterials.

561697 Multiscale Modeling and Enhanced Sampling to Probe Peptide and Peptidomimetic Assemblies

Research Interests:

Recent progress in synthesis and fabrication techniques has given rise to a multitude of materials available for applications ranging from electronics to artificial implants. In many of these applications, synthetic materials form interfaces with biological matter, making it imperative to understand the interactions of biomolecules with inorganic substrates to improve material performance. Motivated by this, I will build a research program using the tools of computational materials science, with an emphasis on molecular dynamics (MD) simulations and enhanced sampling techniques (metadynamics, replica exchange, and umbrella sampling) to study molecular recognition patterns of polymeric substances to peptides and peptidomimetic matter. In order to probe these effects on a range of length and time scales, I will use atomistic and coarse-grained models. My future research themes will address challenges in (1) polymer membrane fouling due to the formation and build up of biofilms, (2) developing hybrid 'bioink' for additive manufacturing (3) drug delivery and therapeutics through the use of polymer-peptide conjugates

Research Experience:

The focus of my graduate research with Prof. Lisa Hall at the Ohio State University was to determine structure-property relationships in a class of charged polymers known as ionomers, from MD simulations. Ionomers contain a small fraction of ionic groups covalently bound to a non-polar backbone; these ions aggregate strongly along with counterions in a low dielectric medium. This work was in collaboration with the Sundaresan group, to aid in the development of 'smart' ionomer materials for 3D printing. To better map to typical experimental systems, I modified a coarse-grained ionomer model to include different ionic neutralization levels. Aggregate morphologies, as well as rheological properties were found to be in good agreement with experimental trends. By analyzing the long ranged ion correlations and calculating mechanical properties during uniaxial tensile deformation, I was able to provide molecular insight into an observed experimental phenomenon – increase in the mechanical strength of ionomers upon applying a static electric field. In addition, this model was employed to (1) delineate the effect of uniaxial strain on polymer backbone and ionic aggregates (2) study the structure and dynamics of ionic aggregates near a nanoparticle interface (3) examine the effect of a free interface on the adhesion of ionomer films.

In my current research with Prof. Jim Pfaendtner at the University of Washington (UW), I employ atomistic MD simulations to study biomolecules. Due to strong driving forces, these systems are difficult to investigate using traditional MD simulations, and my research has partially focused on developing new computational tools to address these challenges. With these tools, I work on three distinct projects (1) **Binding Mechanisms of Solid Binding Peptides (SBPs)** – SBPs are used in a wide range of applications, including targeted drug delivery and medical implants. This is due to the high degree of interfacial compatibility that can be achieved by precisely tailoring their sequence. By employing metadynamics, I construct free energy surfaces of these peptides as a function of different positional degrees of freedom. I calculate binding free energy of the peptide to compare with experimental dissociation constant, as a means to validate our model and system conditions. Detailed cluster analyses reveal an ensemble of peptide structures at the surface. This work is in collaboration with the Drobny group at UW. (2) **Peptide Aggregation in Solution** – Understanding biomineralization, the process by which living organisms produce hierarchically ordered structures with the aid of specific proteins, can help manufacture nanomaterials at ambient conditions. R5 and LKα14 peptides precipitate silica that is remarkably uniform in morphology; however, these peptides have completely different secondary structures in solution – R5 is random, and LKα14 is helical. This suggests that there are dominant interpeptide interactions that cause peptides, even random ones like R5 to assemble in a way that enables uniform templating of silica. By employing all atom MD simulations enhanced by metadynamics, I study the dominant driving forces of peptide aggregation, and ultimately compare the aggregates of R5 and LKα14 (3) **Coarse-Grained Peptoid Model** – Peptoids, positional isomers of peptides containing N-substituted glycine, are an excellent candidate for applications that require robust biopolymers, as they are easy to synthesize, and are stable over a wide range of conditions. Owing to their unique functionality, they have been shown to exhibit numerous self-assembled architectures, from tubes to sheets. An all atom peptoid forcefield was recently parameterized that accurately reproduced local structures from quantum mechanical calculations. To capture self-assembly properties of peptoids over long times, I have systematically developed a coarse-grained model using a bottom-up approach by explicitly parameterizing against the all atom forcefield. This work is in collaboration with the Ferguson group at the University of Chicago, and the Chen group at the Pacific Northwestern National Laboratory.

Teaching Interests:

I have served as a teaching assistant for undergraduate transport phenomena for one semester, during which I was a guest lecturer for two classes. I have also served as a teaching assistant for undergraduate process control for two semesters, during which I was a guest lecturer for three classes. Additionally, I volunteered to lecture on two grad level MD simulations classes. During my postdoc, I have given two lectures in a course titled Molecular Simulations.

Having been trained in chemical engineering throughout my career, I have the aptitude to teach any chemical engineering undergrad level course, although I am particularly interested in process control and thermodynamics. At the graduate level, I would like to teach advanced thermodynamics, encompassing statistical mechanics with an introduction to classical simulations.

583259 Exploring the Structure of Gradient Double Network Gels

Research Interests:

Gradient materials exhibit a continuous variation in any of their physical or chemical properties. For example, tendons, fibrous tissues found in our bodies, illustrate naturally occurring gradient materials; they exhibit a continuous change in modulus, which helps to achieve efficient connectivity between soft muscle tissues and hard bones. Mimicking such gradients using synthetic polymers networks with tunable stiffness will have a tremendous impact on the development of medical implants. The research program that I envision to develop will explore new methods for generating tough gradient hydrogels. The strategy will be to synthesize double network hydrogels (DN) and elastomers by incorporating degradable network strands, wherein the first network can be degraded under acidic conditions while maintaining the second network intact. Such a system will also allow for complete characterization of DN strands after their formation by selective degradation of either one of the two networks to examine the network properties of the remaining gel. After gaining fundamental insights into the structure of DN, the project will focus on the development of gradient DN (gDN), demonstrating a continuous variation in network density and modulus. A key advantage of gDN is that a single specimen can be used to investigate the effect (*e.g.*, elasticity or molecular concentration) of varying a given parameter systematically.

I am also interested in developing versatile synthetic routes for creating functional hydrogels reinforced with nanoparticles (*e.g.*, activated carbon and magnetic particles) and lignin which will have various industrial applications (*e.g.*, catalysis, water purification, and electronics). The strategy is to synthesize polymerizable resins incorporating nanoparticles and crosslinkers that can be printed through digital stereolithography (DSLA) using visible light. DSLA affords high-quality surface finish, dimensional accuracy, and a wide range of compatible materials that enable the fabrication of large-scale

3D objects. Having a strong background in polymer chemistry, polymer physics, and biology, I am confident that I can independently attain financial support from government and private agencies. The interdisciplinary nature of my research interest would draw potential funding from NIH (NIEHS, NIBIB), NSF (BMAT, CMMT, EBMS, POLY), ONR and many industries (*e.g.*, Eastman Chemical Company).

Teaching Interests:

My passion for research, teaching, and learning is what motivated me to consider a career in academia. Teaching students and witnessing how their understanding and scientific thinking develops is extremely rewarding. I think the process occurs not only when mentoring a student, but also every time a student internalizes concepts in a classroom. Beyond the transfer of knowledge, through teaching I wish to develop students' abilities to analyze information, engage in critical thinking and ensure retention of materials.

I am qualified to teach any undergraduate major courses in Chemical Engineering:

- Chemical Engineering Thermodynamics, Reaction Engineering, Reaction Kinetics
- Computer-Aided Chemical Engineering, Principles of Chemical Engineering
- Material and Energy Balances, Materials Science
- Chemistry for Engineers, Macromolecular Science and Engineering
- Biomaterials, Polymer Science and Technology

For graduate students, considering my previous experiences, I can teach the following courses:

- Polymer Chemistry, Polymer Physics, Polymer at Interfaces
- Proteins at Interfaces, Biomaterials
- Spectroscopic methods for Molecular Characterization,
- Surface-Analytical Techniques for Soft Matters

Selected Publications

- C. K. Pandiyarajan, Jan Genzer, "Thermally Activated One-Pot, Simultaneous Radical and Condensation Reactions Generate Surface-Anchored Network Layers from Common Polymers", ***Macromolecules 2019***, 52, 700–707 (Link).
- Joachim Jelken, C. K. Pandiyarajan, Jan Genzer, Nino Lomadze, Svetlana Santer, "Polymer Nano-Membranes with Reversible Tunable Pore Size", ***ACS Appl. Mater. Interfaces 2018***, 10, 30844-30851 (Link).
- C. K. Pandiyarajan, Jan Genzer, "Effect of Network Density in Surface-Anchored Poly (N-isopropyl acrylamide) Hydrogels on Adsorption of Fibrinogen", ***Langmuir 2017***, 33, 1974-1983 (Link).
- C. K. Pandiyarajan, Oswald Prucker, Jürgen Rühe, "Humidity Driven Swelling of the Surface-Attached Poly (N-alkylacrylamide) Networks", ***Macromolecules 2016***, 49, 8254-8264 (Link).
- C. K. Pandiyarajan, Michael Rubinstein, Jan Genzer, "Surface-Anchored Poly (N-isopropyl acrylamide) Orthogonal Gradient Networks", ***Macromolecules 2016***, 49, 5076-5083 (Link).
- C. K. Pandiyarajan, "Biofunctionalized Hydrogels for Tissue Engineering", ***Advanced Science News: Regenerative Medicines*** **2016** (Link).
- Ke Li, C. K. Pandiyarajan, Oswald Prucker, Jürgen Rühe, "On the Lubrication Mechanism of Surfaces Covered with Surface-Attached Hydrogels", ***Macromolecular Chemistry and Physics 2016***, 217, 526-536 (Link).
- C. K. Pandiyarajan, Oswald Prucker, Barbara Zieger, Jürgen Rühe, "Influence of the Molecular Structure of Surface-Attached poly (N-alkylacrylamide) Coatings on Blood Platelet Adhesion", ***Macromolecular Bioscience 2013***, 13, 873-884 (Link) and ***Journal Front Cover Page*** (Link).
- N. Somanathan, C. K. Pandiyarajan, W.A. Goedel, W. C. Chen, "Physico-Mechanical Studies on the Langmuir-Blodgett Films of Polythiophene Containing Mesogenic Side Chain", ***Journal of Polymer Science B: Polymer Physics, 2009***, 47, 173 (Link).

Books

- C. K. Pandiyarajan, "The Interaction of Blood Proteins and Platelets on Surface-Attached poly (N-alkyl acrylamide) Networks", Der Andere Verlag (CPI Band 23) 2014, ISBN:978-3-86247-408-0

Patents

- Slippery Coatings (Pending).
- Functionalized PBT mat for Specific Capture of Proteins from Biological Fluids (Pending)

Invited Peer Reviewer

- Soft Matter (RSC), Journal of Materials Chemistry B (RSC)
- Engineering Reports (Wiley)
- Surface & Coatings Technology (Elsevier)
- Journal of Biomaterials Science: Polymer Edition (Taylor & Francis)

582192 Engineering the Molecular Packing Autonomously

Research Interests:

One of the long-standing challenges in materials science is the control of molecular packing during processing. With the advent of new application areas (e.g. soft electronics, energy storages, bio-interfaces) for emerging polymeric materials (e.g. conjugated polymers, charged polymers, biodegradable polymers) which leverage their characteristics of mechanical softness, electronic functionality, and biological compatibility, this challenge has resurged into prominence. My research aims to utilize the basic and neoteric principles in both polymer physics and chemistry to control the polymer organization at multiple length scales, exploring the unique properties and new potentials of polymer-based electronic materials for bioelectronics. In particular, my career objectives will focus on three areas: 1) kinetically engineering the polymer packing structure for high performance bioelectronics; 2) understanding the structure-property relationships of electronic polymers under processing and deformation by advanced in-situ characterization; 3) building an automated self-learning processing platform for multifunctional thin films, and seeking their high-profile processing-morphology-property relationships and peak performance.

Short Independent Research Summary: "Developing autonomous engine for solution-processable multifunctional materials", Argonne National Lab

Postdoctoral Research Summary: "Applying polymer physics concepts to realize integrated, intrinsically stretchable transistors for skin electronics" Advisor: Zhenan Bao (Chemical Engineering), Stanford University

PhD Dissertation: "Molecular packing structures and dynamic behaviors in confined soft matters" Advisor: Xue Gi, School of Chemistry and Chemical Engineering, Nanjing University

Research Experience

My academic trainings and research experiences focus on polymer physics, fundamental understanding of molecular organization, the coupling between mechanical and electronic behaviors, and the development of innovative processing methodologies for functional materials.

I am currently an assistant scientist at Argonne National Lab since 2018. I received my PhD degree from the School of Chemistry and Chemical Engineering at Nanjing University in 2014, with a focus on the understanding of molecular packing structures and dynamic behaviors in confined soft matters. In my subsequent postdoctoral training with Prof. Zhenan Bao at Stanford University, I have applied my solid background in polymer physics to the emerging field of skin-like electronics, with the development of a new class of stretchable electronic materials and the realization of integrated, intrinsically stretchable transistors and circuits. Started from October 2018 at Argonne as an PI, I have been developing a new platform that combines artificial intelligence (AI) and robotics to leverage computational (synthetic) and experimental (e.g. morphological structures, solid properties) data to reveal mechanisms, parse the fundamental underlying laws, and AI-based feedback to guide material processing.

Beneficial from my knowledge in polymer physics gained from my PhD research, I have generated a physics concept - nanoconfinement effect - to increase polymer chain dynamics in semiconducting materials and developed a stretchy transistor that can be elongated to twice in its length with only minimal changes in its conductivity was created (Science 2017). Later on, again based on polymer physical phenomenon – phase separation, I have invented a new manufacturing process that can introduce multi-scale morphological ordering in stretchable semiconductors and boost their electrical performance by six-fold (Nature Materials 2019). This is the first time that a mass fabrication method - roll-to-roll coating - is successfully applied to fabricate active layers for stretchable electronics in meter-scale. Furthermore, my college and I co-lead a project which realized the first success in the development of the core elements for skin-like electronics—intrinsically stretchable transistor array and thus circuits, through an unprecedented scalable fabrication platform that possesses universal applicability to stretchable polymer materials, high yield and device uniformity (Nature 2018). This intrinsically stretchable transistor array and its fabrication platform developed hold the core position in the interdisciplinary area of intrinsically stretchable electronics, by bridging the material research to the electronics and application development.

Teaching Interests:

Throughout my 10 years as a student then a postdoc in Chemical Engineering Department, I have been guided and inspired by numerous great mentors and teachers, whose teaching philosophy has helped shaping that of my own. At Nanjing University, I have TAed undergraduate level classes in analytical chemical projects laboratory. At Stanford, I seized every opportunity to further enrich my learning and teaching experiences since the first day I arrived. Specifically, I actively attended many classes, such as Polymer Chemistry (CHEMENG 464), Organic semiconductors for Electronics and Photonics (MATSCI 343), Polymer Physics (CHEMENG 466). I'm constantly inspired by the professors' skills to stimulate the students' enthusiasm for learning. In addition to this effectiveness in the classroom, I equally value my mentorship ability in the lab. As a graduate student, I mentored three junior graduate students. One of them has obtained her Ph.D. with University President's Honor. As a postdoctoral scholar at Stanford, I mentored three master students, two of them successfully obtained Ph. D. offers from Stanford.

Thus, based on my extensive training in interdisciplinary fields including polymer science, chemistry, chemical engineering, electrical engineering and nanotechnology, I am interested in and capable of teaching several areas within Chemical Engineering at both undergraduate and graduate levels, including the courses in the subjects of thermodynamics, kinetics, polymer materials, nanomaterials, electronic materials & devices, semiconductor physics, etc. I welcome the opportunity to develop a new graduate elective course on microstructure evolution and characterization of electronic materials for Chemical Engineering.

Selective Publication List (24 articles in total, citation: 2104) (# indicates co-first author and * indicates corresponding author)

1. Xu, J. #; Wang, S. #; Bao, Z.*; et al., Highly stretchable polymer semiconductor films through the nanoconfinement effect. Science 2017, 355(6320): 59-64.

2. Wang, S. #; Xu J. #; Bao, Z.*; et al., Skin electronics from scalable fabrication of an intrinsically stretchable transistor array. Nature 2018, 555, 83

3. Xu, J.#; Wu, H.-C.#; Bao, Z.*; et al., Multi-scale ordering in highly stretchable polymer semiconducting films. Nature Materials 2019.

4. Wang, S. #; Oh, J. Y. #; Xu J. #; Bao, Z.*; et al., Skin-Electronics: An Emerging Paradigm, Accounts of Chemical Research 2018, 51, 1033-1045

5. Li X. #; Xu J. #; Xue, G. *; et al., Low-temperature processing of polymer nanoparticles for bioactive composites. Journal of Polymer Science Part B: Polymer Physics 2016, 54(24): 2514-2520.

6. Xu J.; Xue, G. *; et al., Sensitive characterization of the influence of substrate interfaces on supported thin films. Macromolecules 2014, 47(18): 6365-6372.

7. Xu J. #; Diao Y. #; Xue, G. *; Bao, Z. *; et al., Probing the interfacial molecular packing in TIPS-pentacene organic semiconductors by surface enhanced Raman scattering. Journal of Materials Chemistry C 2014, 2(16): 2985-2991.

8. Xu J.; Xue G. *; et al., Detection of interchain proximity and segmental motion of polymer glass. Macromolecules 2011, 44(18): 7445-7450.

9. Xu J., Chung J., Bao Z., "Stretchable Organic Semiconductor by Blending Method." U.S. Non-provisional Application. Atty. Dkt. No. 15639-000302-US-01, filed April 2017

10. Xue G., Li X., Wang X., Zhou D., Xu J., Teng C., Li L., "Preparation method of bioactive molecule

and macromolecule composite material", publication number CN 105385057, published March 2016

581885 Sustainable Approaches to Polymer Recycling

Research Interests:

Polymeric materials have become versatile and useful in the modern society. Polymers are relatively inexpensive, lightweight, and durable which can be molded into any form. The application of polymers ranges from textile fibers, shopping bags, packaging films, bottles, membranes to chips used in computers. Some commonly used polymers are polyethylene terephthalate (PET), polycarbonate (PC), polyethylene (high and low density), etc. The increase in reliance on polymeric materials generates waste, which has become a severe environmental pollution issue. Recycling of polymers is an area of significant importance for sustainable development. The post-consumer polymer wastes are being recycled or sent for energy recovery or dumped in a landfill. There are two broad approaches that are taken to recycle polymers (1) mechanical and (2) chemical recycling. In the mechanical recycling approach, used polymers are reprocessed into a new form. The disadvantage of this approach is the deterioration in the polymer properties with every cycle. The chemical recycling approach is aimed at depolymerizing the polymer to obtain its monomers. My long term goal is to develop a green approach to recycle polymers, where the polymers can be broken down to their monomers and then should be able to re-polymerize without any waste stream. The re-polymerized polymer would have the same properties as one produced from virgin material, and all solvents that are used should be reusable. Overall, my vision would be to develop a self-sufficient recycling process.

Postdoctoral Project: Fluid Phase Kinetics – Joint Industrial Project

Under the supervision of Clint P. Aichele, School of Chemical Engineering, Oklahoma State University.

Ph.D. Dissertation: "Bio-Based Active Barrier Material and Packaging Development"

Under the supervision of Maria R. Coleman and Saleh A. Jabarin, Department of Chemical Engineering, University of Toledo/ Polymer Institute, University of Toledo

Research Experience:

Throughout my academic career, I have worked on various application-oriented research. The research areas that I have worked on are the removal of dyes (adsorption), reduction of oxygen permeability through polymers, and factors affecting the rate of gas evolution from supersaturated liquids at elevated pressures. As a result of working on these projects, I have acquired experience in many research areas: water treatment, polymers, nanomaterials, emulsions, surface tensions, and gas-liquid separations. I have become proficient as an experimentalist with an ability to develop a lab that is geared towards addressing these areas of research.

Teaching Interests:

I have been a Teaching Assistant (TA) for various courses, including Process Separation, Fluid Dynamic, Reaction Kinetics, and Unit Operation Lab. For the above course, I have been well appreciated and reviewed by the students. As an instructor, my teaching philosophy would be to embrace an active classroom environment, which encourages question/discussion and solving practical problems to better understand chemical engineering concepts.

Teaching Experience:

Throughout my doctoral program, I have undertaken various TA assignments at the University of Toledo. I have assisted in handling the Unit Operations Lab for 4 semesters and given review lectures for Process Separation and Reaction Kinetics. Having received the Best Teaching Assistant award twice (University of Toledo) reflects my ability as an instructor. I have also mentored graduate and undergraduate students in their research.

Selected Publication:

1. Alden B Daniel, Sayeed A Mohammad, **Michael A Miranda,** and Clint P Aichele "Absorption and Desorption Mass Transfer Rates as a Function Pressure and Mixing in a Simple Hydrocarbon System" Journal: Chemical Engineering Research and Design (2019).
2. **Michael Angelo Miranda**, Maria Coleman, Saleh. A. Jabarin "Modification of Polyethylene terephthalate (PET) using linoleic acid for oxygen barrier improvement: Impact of processing methods" Journal of Applied Polymer Science (2017).
3. **Miranda, M.A.**, M. Coleman, and S. Jabarin. Impact of Processing Method and Loading of Active scavenger (Linoleic Acid) on Properties of Polyethylene Terephthalate. in SPE ANTEC™ Indianapolis 2016.
4. **Michael Angelo Miranda**, P. Dhandapani, M. Helen Kalavathy, and Lima Rose Miranda, "Chemically activated Ipomoea carnea as an adsorbent for the copper sorption from synthetic solutions" Adsorption (2010) 16: 75–84.

581606 Functional Materials Design Guided By Polymer Physics

Research Interests:

The functions of polymers have been expanded tremendously by the emergence of new classes of polymers, such as electron (or ion) conducting polymers (i.e., conjugated polymer or ionomer), entanglement-free bottlebrush polymers, and bond-exchanging vitrimers. Nevertheless, the semiflexible backbone of conjugated polymers and the crowded side chains of bottlebrush polymers are distinctly different from linear chain coil-like polymers. This difference from conventional polymers results in completely different chain conformations and new behaviors in glass transition, entanglements, liquid crystalline ordering, and crystallization. The lack of physically intuitive models that predicts the structure-property relationship for these new functional polymers hinders the ability to control their properties.

My research goal is thus to build upon the classic theory of polymer physics to quantitatively control or predict the properties from structure for functional polymers, thus guiding the materials design for specific applications. In the meantime, rheology and mechanical behavior can provide a powerful experimental window into the underlying polymer dynamics and verify the proposed model.

During my Ph.D. work co-advised by Professors Ralph Colby and Enrique Gomez at Penn State University, my research is centered on characterization of the fundamental properties of conjugated polymers and building their interrelationships with structure and mechanical performance. Driven by the application in stretchable electronics, complex repeat units are needed to varying the electronic structure of the conjugated backbone, but it is poorly understood how such designs impact the behavior of polymer chains in solids. We have first developed a rheological method to unambiguously determine the glass transition temperature (T_g) of conjugated polymers. Then, simplified from the group contribution theory of T_g, we have experimentally verified a new model that quantitatively predicts the T_g for various conjugated polymers with alkyl side chain. Further rheological study on the viscoelasticy of a conjugated polymer melt has led to a new phenomenon of reduced entanglements via local chain alignment in liquid crystalline phase. Both glass transition temperature and chain entanglements dictate the stretchability of polymeric material, so our work have made essential contributions to guide the design of conjugated polymer for desired mechanical performance.

My ongoing postdoctoral work, co-advised by Professors Michael Chabinyc and Chris Bates at University of California, Santa Barbara, is focused on bottlebrush polymers. Currently, we have demonstrated a versatile strategy to photo-crosslink bottlebrush polymers into a super-soft elastomer. These soft elastomers have the modulus of wet polymer gels, but are 100% solids which allows them to be used in pressure sensors. The correlation between the network modulus, which is inversely related to the pressure sensitivity, and the photo-crosslinker concentration is well described by a novel network model. This proposed model is adapted from the classic Phantom network model and has been tested across various classes of bottlebrush polymers with different backbone and side chain lengths. As a result, we can quantitatively control the network modulus of a photo-crosslinked bottlebrush elastomer.

Recent synthetic advancements have enabled facile and accurate control on the chain architecture of a bottlebrush polymer in terms of the backbone length, side chain length and grafting density, thus allowing exploration of new physical behaviors caused by this unconventional chain architecture. For instance, the glass transition temperature of a bottlebrush polymer can still approach that of a linear polymer in the high molar mass limit, despite of a lot more chain ends per unit volume. Forming liquid crystalline phase from the stretched backbone by crowded side chains in a non-mesogenic bottlebrush polymer is indeed possible but hardly controllable or predictable. On the other hand, apart from the stiffness of a conjugated backbone, the role of π-π stacking interaction in the formation of liquid crystalline phase is mostly unexplored in the field of conjugated polymer. Therefore, understanding the impact of chain architecture and molecular anisotropy on polymer dynamics could allow development of polymers with new functionality to address challenges in various fields, including stretchable electronics.

There is a clear trade-off between stretchability (or softness) and charge mobility of a conjugated polymer, since charges are known to transport faster along the conjugated and stiff backbone rather than through the electrically insulating and soft side chains. Adding more soft segments usually impedes the macroscopic charge mobility. By intelligently combining bottlebrush polymer and conjugated polymer, it's possible to break this trade-off and obtain a super-soft and semiconducting material. Nevertheless, the choices of conjugated polymer repeat unit and bottlebrush polymer architecture rely on the prior knowledge of the structure-property relationship of these emerging functional materials. Therefore, I have summarized the following questions to be addressed in my future research:

1. Can we understand how the side chain grafting density affects the glass transition temperature of a bottlebrush polymer? Do we know the role of bottlebrush backbone in the crystallization process of side chain? Can we understand how entanglement scales with chain stiffness for semiflexible conjugated polymers?
2. Can we predict the formation of liquid crystalline phase of a bottlebrush polymer from its molecular anisotropy? Can we decouple the contribution of the molecular anisotropy from that of the π-π stacking interaction for driving the liquid crystalline ordering in a conjugated polymer?
3. How do we control the phase separation between the semiconducting conjugated component and the super-soft bottlebrush component?

Teaching Interests:

Teaching is not merely passing the knowledge but to endow students with self-motivated interest and passion. A good teacher, advisor or mentor can strongly change a student's future trajectory in a positive manner or vice versa. As a Ph.D. student, I was lucky enough to have mentored two outstanding undergraduate students and one graduate student for about two years. Each had an individual project and has contributed to the published works. Among them, one undergraduate student is even motivated enough to pursue her Ph.D. in chemical engineering at University of Massachusetts, Amherst this year. In addition, teaching in the classroom, such as introduction to polymer science, rheology and polymer processing, also brings me great joy as an instructor whenever I see students' eyes light up. Also, teaching itself is of course an incredible opportunity for the lecturer to review and refresh the fundamental knowledge, which may illuminate the current research with new ideas.

581359 Mechanics and Processing of Polymeric Materials for Soft Electronics and Robotics

Research Interests:

My research interests are broadly centered around the material science and engineering of polymeric materials for soft electronics, robotics, and biomedical devices. I am interested in questions relating to the structure and mechanics of complex soft materials—from the molecular to the macroscopic scale. From a fundamental perspective, my research program will employ simulations—both molecular and continuum methods—in tandem with experimental characterization to understand mechanisms of elasticity in semiconducting polymers. Knowledge gained from such studies will be leveraged to design new materials, processing strategies, and integrated devices. Processing techniques including ink jet printing, 3D printing and replica molding will be employed to integrate functional soft devices with electronic circuitry consisting of highly deformable conducting and semiconducting polymers. Such processing capabilities will enable the development of electronically-controlled soft robotic devices that are easily integrated with both traditional hard robotic components and human users to provide new and useful functionality in automation and healthcare.

Teaching Interests:

My teaching interests encompass the entirety of the chemical engineering curriculum. I have had specific experience assisting in teaching graduate courses on applied mathematics (partial differential equations) and intermolecular & surface forces, in addition to an undergraduate class on polymeric materials—for two different semesters. I am also an expert computer programmer, and have had a lot of experience teaching groups of colleagues and mentees how to use Python for a variety of tasks related to simulation and data analysis in chemical engineering research. My favorite subjects are thermodynamics and transport. I hope to eventually teach a graduate-level course on statistical mechanics and molecular simulation methods. I am a supporter of project-based learning and my grading would be equally weighted by projects and exams.

585847 Design of Protein-Polyelectrolyte Complexes

PhD Research Summary

PI: Prof. Venkat Ganesan, University of Texas, Austin

Keywords: Proteins, Single Chain in Mean Field, Polyelectrolytes, Monte Carlo, Polymer, Electrostatics. Coacervation.

A widely used technique to stabilize proteins is to entrap those using highly charged polyelectrolytes. This methodology is used in stabilizing proteins in biosensors to increase their shelf life, extraction of proteins in food industries like in preparing whey powder and forming new biomaterials that can mimic properties of commercial non-biodegradable polymeric materials. The objective of our research is to build a generalized computational framework that can bring a comprehensive understanding of the physicochemical aspects underlying the structure and rationally guide the efforts to design and optimize protein and polyelectrolyte systems to obtain desired properties. To model this system, we have developed a coarse-grained multibody simulation framework which involves a combination of single chain in mean field (SCMF) simulation and hybrid particle field framework. To accommodate the weak proteins and polyelectrolytes, we used a hybrid of constant pH method and the existing SCMF framework in a semi-grand canonical system.

The structure and phase behaviors that manifest in the protein polyelectrolyte systems are understood as a consequence of the resulting electrostatic interaction and the polymer mediated interaction. In a mixture of low homogeneously charged proteins and polyelectrolytes, clusters are formed from direct aggregation of proteins, due to attractive interaction led by depletion of polymers. However, at higher charge, oppositely charged polyelectrolytes form a bridge between the otherwise electrostatically repelling charged proteins. Charged macromolecules are found to experience increased repulsion when the dielectric constant of the proteins is lower than that of the solvent. However, the influence of the dielectric contrast is found to diminish with an increase in the protein volume fraction and its charge. Our results demonstrated, when charge heterogeneity in protein is included, irrespective of the net charge of the proteins, charge heterogeneity can exert a significant influence on the resulting characteristic of the aggregates, in some cases leading to a transformation from polymer-bridged complexes into direct particle phase. A very crucial aspect of the system is when the side chain of the amides or the polyelectrolytes are weak acid or bases. In our research so far, we have discovered that both weak and strong polyelectrolytes have a higher tendency to form bridges for a weak protein than a strong protein of the same charge. In the broadest terms, our work will bring a comprehensive understanding of the physicochemical aspects underlying the structure and phase behavior of protein and polyelectrolyte mixtures and help in rationally engineering the proteins.

Research Experience: I am also involved in a project which aims at understanding the transport properties of water and salt ions, in the mesoscopic pore structures in the membranes formed by the self-assembly of multi-block copolymers. This project is led by a postdoctoral researcher in our group. This study is crucial for designing advanced water purification membranes for experimental collaborators. This study relies on using dissipative particle dynamics (DPD) simulations using LAMMPS. During my master's program, I investigated the stability criterion of two immiscible charged liquid layers present in biological systems like lipid bilayers. The research relied on using analytical techniques involving linear perturbation analysis.

Areas of Interest for Postdoc: As a postdoc, I want to explore the areas of computational therapeutics, molecular Biophysics and/or projects related to computer aided drug discovery, drug design. In addition to diversifying my technical skill sets, I want to focus more on collaborating closely with experimental works.

587177 Rationale Design of Polymeric Materials for Biological andEnergy Applications using Multiscale Modeling

Research Interests:

Quest for clean energy resources and the rapid developments in medical industry in the recent years have resulted in a tremendous increase in the number of studies which focus on designing sustainable materials for energy and bio-medical applications. These studies primarily rely on understanding the structure-property relationship of the materials used in these applications. Nevertheless, experimental design of materials with specific intrinsic properties relies heavily on trial and error basis and on system-specific empirical understanding of structure-property relationships. In this context, my research work focuses on using *computer simulation techniques and theoretical methods* to discern the fundamental physics of different systems involving *polymeric materials*, which are pertinent to biological systems and energy storage applications. Specifically, I am interested in understanding the *self-assembly and dynamics of charged polymers and gel-forming polymers.*

To decipher the structure-property relationship of materials, it is imperative to probe the material characteristics at both short and long time/length scales. To this end, I will utilize a suite of *multiscale simulation techniques* to investigate the physical properties of materials. While coarse-grained simulation techniques like *molecular dynamics, Monte Carlo, and single chain in mean field* will be used to obtain insights into the long-range characteristics, *atomistic simulations* will help in understanding the influence of chemistry on material characteristics. Further, theoretical methods will be used to supplement or rationalize the results obtained from computer simulations. Together, these multiscale simulations will help in guiding experiments to design materials with desired properties in a rationale fashion.

Keywords: Polymers, Biopolymers, Multiscale Simulation, Design of Materials

Funding

1. A two-year grant prepared with Dr. Ganesan during PhD, was accepted by Donors of the American Chemical Society, Petroleum Research Fund (2016).
2. Prepared a NIH grant along with Dr. Dorfman during postdoctoral work.

Publications

1. **V. Sethuraman**, Michael McGovern, David C. Morse, Kevin D. Dorfman, *Soft Matter,* 15, 5431 – 5442, **2019**
   Email: avsethura@umn.edu, bvenkat@che.utexas.edu, cmorse012@umn.edu, ddorfman@umn.edu
2. **V. Sethuraman**, Kevin D. Dorfman, *Phys. Rev. Mater.,* 3, 055601, **2019**
3. H. Lee, **V. Sethuraman**, Y. Kim, W. Lee, D-Y. Ryu and V. Ganesan, *Macromolecules*, 51 (12), 4451-4461, **2018**
4. N. Rebello, **V. Sethuraman**, G. Blachut, C. J. Ellison, C. G. Willson, V. Ganesan, *Phys. Rev. E.*, 96 (5), 052501, **2017**
5. **V. Sethuraman**, S. Mogurampelly, V. Ganesan, *Soft Matter*, 13 (42), 7793-7803, **2017**
6. **V. Sethuraman**, S. Mogurampelly, V. Ganesan, *Macromolecules*, 50 (11), 4542–4554, **2017**
7. **V. Sethuraman**, V. Ganesan, *J. Chem. Phys.*, 147 (10), 104901, **2017**
8. W. Lee, S. Park, Y. Kim, **V. Sethuraman**, N. Rebello, V. Ganesan, D-Y. Ryu, *Macromolecules*, 50 (15), 5858-5866, **2017**
9. **V. Sethuraman**, V. Ganesan, *Soft Matter*, 12, 7818-7823, **2016**
10. S. Mogurampelly, **V. Sethuraman**, V. Pryamitsyn, V. Ganesan, *J. Chem. Phys.*, 144 (15),154905, **2016**
11. **V. Sethuraman**, V. Pryamitsyn, V. Ganesan, *Macromolecules,* 49 (7), 2821-2831, **2016**.
12. **V. Sethuraman**, V. Pryamitsyn, V. Ganesan, *J Polym. Sci Part B: Polym. Phys.*, 54, 859-864, **2016**
13. **V. Sethuraman**, D. Kipp, V. Ganesan, *Macromolecules*, 48 (17), 6321-6328, **2015**

14. **V. Sethuraman**, B. H. Nguyen, V. Ganesan, *J. Chem. Phys.*, **141**, 244904, **2014**
15. **M. S. Vaidyanathan**, P. Sathyanarayana, P. K. Maiti, S. S. Visweswariah, K. G. Ayappa, *RSC Adv.*, **4**, 4930-4942, **2014**

Future Research Directions

My work for the next few years will be focused on three broad fronts using simulation methods described above: (i) Understanding the self-assembly of gel-forming polymers, with focus on methylcellulose and its variants; (ii) Studying self-assembly of charged polymers on surface, with focus on intrinsically disordered proteins; (iii) Using single ion conducting electrolytes based on multiblock copolymers for lithium ion batteries. While the first two works are targeted at understanding the physics of biopolymers used in medical or food industries, the third work aims at developing sustainable polymeric materials for energy storage applications.

Teaching Interests:

My broad areas of interest in teaching relates to statistical mechanics, thermodynamics, mathematics, transport phenomena, and computer simulations. I am also excited to teach topics pertaining to fundamentals of polymer physics, and advanced simulation techniques. In addition to the theoretical courses, I am also interested in teaching courses out of my domain of expertise in collaboration with other faculty members.

581825 Accelerating Chemical Discovery and Process Development with Theoretical Models and Machine Learning

Research Interests:

Computation and modeling have played an important role in chemical engineering. Theoretical models can be used to determine the thermodynamic limits and kinetic evolution of chemical systems, thus shedding light upon controlling and optimizing chemical processes. Machine learning models have more recently demonstrated their power in predicting complicated system behaviors and have gained fast-growing interest, particularly in molecular property prediction and optimization, materials design and synthetic path planning. With the complexity of chemical systems quickly increasing, it is desirable to further expand the scope and capability of different models and explore the synergies between them. I am excited about combining these modeling techniques to study chemical reactions and process systems, in order to make the best use of historical and new experimental data, thus enabling more effective experimental efforts and speeding up process development and optimization. Initially, I would like to develop research projects on 1) applying *transfer learning* to expediate new polymer design, 2) using reinforcement learning to identify mechanisms for complicated reaction systems, and 3) automating holistic chemical process design with the help of artificial intelligence (AI).

Research Experience:

My PhD dissertation focuses on simulation and optimization of radical polymerization processes, combining mechanistic kinetic modeling and optimization algorithms. I developed schemes to accelerate kinetic Monte Carlo (KMC) method, which is used to simulate the explicit sequence of polymer chains during polymerization process. These methods significantly reduce the computational cost of KMC simulations so that they were successfully incorporated into derivative-free optimization algorithms to design and optimize polymer sequence distributions.

During my post-doc, I worked on AI-assisted synthesis route design and evaluation. I developed a deep learning model trained on a large reaction database to recommend conditions (catalyst, reagent and solvent) for given organic reactions with a high accuracy. Without explicitly given the properties of the chemicals, the model learned the similarity of solvents/reagents/catalysts, which was used to assess compatibility of reactions and identify retrosynthetic suggestions that are amenable to one-pot synthesis development. I also used optimization techniques to minimize the chemical inventory for multiple synthetic targets.

Teaching Interests:

As a chemical engineer by training, I am capable of and passionate about teaching core chemical engineering courses in general. During the 4th year of my PhD, I participated in the Teaching Apprenticeship Program at Northwestern University and taught half of the graduate kinetics class under the supervision of Prof. Justin Notestein. I have also recently participated in the Kaufman Teaching Certificate Program at MIT to fine tune my teaching methodologies and philosophy.

Selected Publications:

1. **Gao, H.**, Struble, T. J., Coley, C. W., Green, W. H., Jensen, K. F. (2018). Using Machine Learning to Predict Suitable Conditions for Organic Reactions. *ACS central science,* 11, 1465-1476. **Highlighted by *Science* - DOI: 10.1126/science.362.6420.1260-b**
2. **Gao, H.**, Waechter, A., Konstantinov, I. A., Arturo, S. G., Broadbelt, L. J. (2018). Application and Comparison of Derivative-free Optimization Algorithms to Control and Optimize Free Radical Polymerization Simulated Using the Kinetic Monte Carlo Method. *Computers & Chemical Engineering,* 108, 268-275.
3. **Gao, H.**, Konstantinov, I. A., Arturo, S. G., Broadbelt, L. J. (2017). On the Modeling of Number and Weight Average Molecular Weight of Polymers. *Chemical Engineering Journal,* 327, 906-913

4. **Gao, H.**, Konstantinov, I. A., Arturo, S. G., Broadbelt, L. J. (2016). Acceleration of Kinetic Monte Carlo Simulations for Free Radical Copolymerization: a Hybrid Approach with Scaling of Kinetic Parameters. *AIChE Journal,* 63(9), 4013-4021. **Selected as Editor's Choice: Reaction Engineering, Kinetics and Catalysis**
5. **Gao, H.**, Oakley, L. H., Konstantinov, I. A., Arturo, S. G., Broadbelt, L. J. (2015). Acceleration of Kinetic Monte Carlo Method for the Simulations of Free Radical Copolymerization through Scaling. *Industrial & Engineering Chemistry Research*, vol. 54(48), 11975-11985. **Selected as the ACS Editor's Choice for October 26th, 2015**

581945 Systems Approach to Advanced Decision-Making in Chemical Engineering, Biomanufacturing, and Society

Research Interests:

How do we make better decisions? This fundamental question for many individuals—humans, organizations, and societies—has no easy answer. Among many approaches to advanced decision-making, process systems engineering (PSE) offers a toolset for a wide range of problems. PSE is a chemical engineering discipline that overlaps with both process engineering and systems engineering. At its core, PSE is about systems thinking, which considers a system, its interacting constituents, and competing objectives. Many decision-making problems could benefit from both the technological advances in PSE and its unique systems perspective. I plan to focus my research on three broad directions—data, complex systems, and the human factor.

Like natural resources, data can be scarce or abundant. When experiments are costly to perform, one can use the design of experiments (DoE) to generate data efficiently. While DoE is a well-studied subject, from my experience working with practitioners in the biopharmaceutical industry, most experimental designs are still heuristic-based. This gap between practice and available DoE techniques creates opportunities for the industry to adopt the more systematic, automatic, and model-based approaches. The abundance of historical data and the lack of data/metadata management, on the other hand, create challenges to reproduce experiments and retrieve process parameters for model development. The transition toward better data infrastructure requires a collective effort from chemical, systems, and software engineering.

Complex systems are my core research interest. I am particularly interested in the identification and prevention of unintended negative consequences in complex systems. The positive benefits of rapid technological developments often come with unintended negative consequences to society. It is generally difficult to predict unintended consequences. Nonetheless, chemical engineers have been practicing techniques such as process hazard analysis (PHA) to identify potential vulnerabilities in a process system. The healthcare sector also adopts similar "checklist" approaches to prevent unintended negative consequences. I plan to further expand this approach to policy-making and design of emergency response protocols.

The human factor is something we engineers tend to overlook, yet it is a crucial part of engineering systems. Two aspects of this topic are worth studying. First, when stakeholders respond to signals in an engineering system, the response can become feedback and create a closed-loop situation. It requires both control theory and game theory to study the closed-loop behavior and design mechanisms such as variable pricing that are both robust and efficient. Second, one can actively involve human decision-making in the human-in-the-loop computing configuration where algorithms proactively suggestion solutions while humans safeguard the integrity of machine-generated decisions and provide feedback.

Teaching Interests:

PSE topics (including process dynamics and control, process design, process safety, optimization), artificial intelligence (including machine learning), maths, statics (including DoE), bioprocess engineering, programming

Leveraging Scale-Appropriate Principles of Metal-Oxide Reaction Engineering and Particle Technology Science for Energy Conversion Solutions

The current and projected world energy supply and demand have neared a tipping point in terms of the ever-increasing global population and the corresponding need to find economic yet sustainable solutions for mitigating increased carbon-based emissions. The need for new energy technology solutions is indisputable in order to enable a robust transition to a sustainable future. The characteristics of these new energy technology solutions tend to be very different for developing countries as opposed to developed countries. Metal- derivatives based reaction engineering and particle science principles offer a unique opportunity and a versatile technology platform that can be readily amenable to address these energy challenges for both developing and developed worlds at varying scales, scenarios, and economies.

Research Interests

Metal oxide and their derivative-based reaction engineering principles and understanding of particle science technology form the basis of solid-based chemical looping systems. My research vision is centered on leveraging these principles at various scales to provide for energy conversion solutions. The use of solids looping systems has shown potential to provide significant efficiency and cost benefits for a wide range of technology applications including energy production, CO_2 capture, pollution control and production of chemicals like hydrogen, methanol, ethylene, ethanol, and syngas from both fossil fuel and renewable energy sources. I have contributed (15 papers) and innovated (7 patents) in development of these systems, resulting in technology solutions that have shown promising economic benefits for a wide range of metal-oxide applications.

For integrated and holistic system development, I have used theoretical and experimental methods at several different orders of magnitudes with a synergistic interaction between each. Methods at small-scales include reaction mechanism investigations using density functional theory investigation, kinetic modeling coupled with experiments based parameter estimations, synthesis and characterization of materials, reactivity and functional characterization using thermogravimetric analyzers, fixed bed reactors. (Example contributions: Patent #3, Patent #5, Patent #6, Report #1,)

The next scale of technology development focuses on reactor engineering and particle science technology-based principles, including design for multi-phase flows, synchronizing hydrodynamics, and fluid flows to demonstrate sub-pilot and pilot-scale operations. (Example contributions: Patent #4, Patent #7, Report #1,)

On a systems scale, I have used process simulations and mathematical optimization principles to help with techno-economic analysis to identify critical technology gaps and potential applicability of the proposed technology. The scale-appropriate leveraging of principles of reaction engineering and particle science technology remain indispensable to technological developments and have the potential to provide promising and holistic solutions for a variety of energy applications (Example contributions: Patent #1, Patent #4, Report #1).

Patents:

#1 **Kathe M.V.**, Wang W, Chung E, Tong A.; Fan, L.-S., "Systems and methods for partial or complete oxidation of fuels", US10022693B2, 2018 (granted)
#2 **Kathe, M.V.;** Empfield, A.; Fryer, C.; Blair, E.; Fan, L.-S. "Chemical Looping Syngas Production from Carbonaceous Fuels"; US Patent App. 16/091,253, 2019 (2015 filing)
#3 **Kathe, M.V.;** Baser, D.; Fan, L.-S.; "Chemical Looping systems for conversion of low or no-carbon fuels to hydrogen"; US Patent App. 16/091,508, 2019 (2017 filing)
#4 **Kathe, M.V.;** Wang, D.; Fan, L.-S.; "Integrated chemical looping reactor system configured with unequal reactor operating pressures"; WO 2019/027972 A1, 2019 (2018 filing)
#5 **Kathe, M.V.**; Baser, D.; Sandvik, P.; Fryer, C.; Fan, L.-S.; " Systems, methods and materials for NOx decomposition with metal-oxide materials"; US Patent App. 16/260,447, 2018 (2016 filing)
#6 **Kathe, M.V.;** Baser, D.; Fan, L.-S.; Metal oxide phase-change assisted partial oxidation of methane: Stable phase (SP) syngen process, provisional patent, 2019
#7 **Kathe M.V.**; Wang, D., Xu D.; Fan, L.-S., "High Purity Syngas Generated from Gaseous Hydrocarbons Using a Fixed (Packed) Bed Reactory Assembly with Oxygen Carrying Material", provisional patent, 2019

Selected Publications (Total of 15, h-index: 13, total citations: 742)

#1 **Kathe, M.**; Empfield, A.; Na, J.; Blair, E.; Fan, L.-S.; "Hydrogen Production from Natural Gas Using an Iron-Based Chemical Looping Technology: Thermodynamic Simulations and Process System Analysis", Applied Energy, 165, 183-201 (2016)
#2 **Kathe, M.**; Empfield, A.; Sandvik, P.; Fryer, C.; Zhang, Y.; Blair, E.; Fan, L.-S.; "Utilization of CO_2 as a partial substitute for methane feedstock in chemical looping methane–steam redox processes for syngas production", Energy & Environmental Science, 10, 1345-1349 (2017)
#3 **Kathe, M.**; Fryer, C.; Sandvik, P.; Kong, F.; Zhang, Y.; Empfield, A.; Fan, L.-S.; "Modularization strategy for syngas generation in chemical looping methane reforming systems with CO_2 as feedstock", AICHE Journal, 63 (8), 3343-3360 (2017)
#4 **Kathe, M.**; Sandvik, P.;Wang, W.; Kong, F.; Fan, L.-S.; "High-Pressure Chemical Looping Reforming Processes: System Analysis for Syngas Generation from Natural Gas and Reducing Tail Gases", Energy & Fuels, (2018) (doi: 10.1021/acs.energyfuels.8b01834)
#5 **Kathe, M.**; Sandvik, P.; Fryer, C.; Kong, F.; Zhang, Y.; Grigonis, G.; Fan, L.-S.; "Coal Refining Chemical-Looping System with CO_2 as a co-feedstock for Chemicals Synthesis", Energy & Fuels, 32 (2), 1139-1154 (2018)
#6 Nadgouda, S.; **Kathe, M.**; Fan, L.-S.; "Cold gas efficiency enhancement in a chemical looping combustion system using staged H2 separation approach", International Journal of Hydrogen Energy, 42 (8), 4751-4763 (2017)
#7 Tong, A.; Zeng, L.; **Kathe, M.**; Sridhar, D.; Fan, L.S.; "Application of the moving bed chemical looping process for high methane conversion", Energy&Fuels,27(8),4119-4128, (2013)
#8 Zeng, L.; **Kathe, M.;** Chung, E.; Fan, L.-S.; "Some Remarks on Direct Solid Fuel Combustion Using Chemical Looping Processes", Current Opinion in Chemical Engineering, 1 (3), 290-295 (2012)

Selected Reports (1 out of more than 5)

#1- Kathe, M.V., Xu, D, Hsieh, T.-L., Simpson, J., Statnick, R., Tong, A. Fan, L.-S.; "Chemical looping gasification for hydrogen enhanced syngas production with in-situ CO2 capture" OSTI 1185194 (2014)

Proposed Research Thrusts:

My proposed research thrusts have been shaped by my doctoral and postdoctoral experiences of having contributed to, innovated, and developed several promising solids looping based technologies on multiple scales and orders of magnitudes. Each of these thrusts requires leveraging scale-appropriate principles of metal-oxide reaction engineering and particle technology science. In addition to writing journal articles, patents, and book chapters, I have spent a considerable amount of effort in leading research grant writing, and that experience is invaluable in terms of formulating potential research thrusts. The following are three examples that would be representative of my proposed independent research lab work:

1. Development of a multi-functional catalytic metal-oxide system to convert fuels like natural gas, coal, biomass, and CO_2 into value-added products like liquid fuels, solid carbon, etc. (dopant screening, reactor design, experimental demonstration, and techno-economic analysis)
2. Technology development to use metal-derivatives like carbonates, nitrites, and sulfides in pollution control, carbon capture and solar thermochemical applications (fundamental material development, reaction chemistry validation, reactor design, and analysis and techno-economic analysis)
3. Rigorous validation of conventional heuristics in solids processing systems using process simulations, mathematical optimization, and surrogate modeling

Teaching Interests:

I discovered that instruction and teaching are enjoyable while serving as a teaching assistant for separation processes and transport phenomena courses during my PhD education. Following up on that experience, I choose a lecturer position over industry research in chemical engineering immediately after my PhD to explore that dimension of my career.

Chemical engineers out of college are advertised as efficient problem-solvers with competent analytical and troubleshooting skills. My focus as a lecturer has always been on imparting a fundamental understanding of specific chemical engineering principles and providing enough tools and examples to be able to apply these in independent, newer situations. The process of identifying a student's level of understanding when they begin a class and explaining the thought and problem-solving process of teaching chemical engineering principles has been incredibly fulfilling. My teaching philosophy resonates strongly with the thought that education should reinforce a student's critical thinking process related to solving or troubleshooting chemical engineering problems rather than merely serving as a shortcut problem-solving tool based on a superficial understanding of chemical engineering phenomena. I have implemented and developed this style of teaching in senior chemical engineering capstone process and product design course,

course on chemical engineering economics and separation principles course at The Ohio State University (Chemical Engineering Department) over the last 3 years. My teaching performance earned excellent reviews for being able to communicate chemical engineering principles effectively (average student instructor of evaluation (SEI) rating of 4.7/5). Though my teaching evaluations by students and department faculty have been excellent, I am continuously working to improve on some aspects of my teaching. These include design of exams for all student learning styles in the class, balance between difficulty level of exam questions and further understanding the combination of in-class aids (on-board writing vs only PowerPoints) that contribute to the most effective student learning outcomes. My teaching experiences in terms of syllabus, exams, student evaluation, and peer evaluation comments are available for review.

Based on my doctoral education, my area of expertise is reaction engineering-based traditional chemical engineering principles and particle science technology. The following are the courses that I would be very comfortable with teaching in chemical engineering as my research and past teaching endeavors have overlapped with each in some aspect. I would also be interested in any other courses that the department requires. I would be very interested in offering a technical elective on clean energy and carbon capture.

Courses that I could teach:

- Chemical Engineering Capstone Process and Product Design
- Engineering Economics and Process Design Strategy
- Principles of Separation processes
- Chemical Engineering Kinetics and Reaction Engineering
- Mass and Energy Balances
- Transport Phenomena (mass, heat and momentum transfer principles)
- Chemical Engineering Thermodynamics
- Process Control
- Unit Operations Lab
- Technical Electives Based on Energy, Carbon Capture
- Technical Electives Based on Reaction Engineering, Multi-phase flow principles

582051 Online Process Optimization of Complex Cyber-Physical Systems

Research Interests:

Online process optimization deals with driving a process towards optimal operation in the presence of varying disturbances and uncertainties. In the presence of increasing competition and increasing focus on sustainable operation, real-time optimization (RTO) is widely recognized as one of the most crucial technologies in a wide range of application areas and as an important factor in enabling new application areas such as renewable energy systems, medical and bio processes, water management, food production etc. In spite of this, traditional real-time optimization is not widely used in practice. From my four years of industrial research experience, together with three years of academic research experience in the field of process optimization, I have identified three main challenges that impede the practical application of real-time optimization in many processes, which are:

1. Lack of good models that can be used in optimization algorithms (offline model development)
2. Wrong value of model parameters used in the optimization problem (online model update)
3. Numerical robustness including computational issues

My primary research interest is to address this gap in online process optimization and develop novel decision-making methods and algorithms to tackle these challenges.

Based on the current challenges and knowledge needs, important research question that I aim to address are:

- *How can process data be used more efficiently to address the challenges with the traditional model-based optimization tools for large-scale systems?*
- *How can we handle different types of uncertainty in the real time optimization (RTO) problem?*

Building on the previous work on process optimization and control structure design, I aim to address the challenges with the traditional RTO schemes and take a new look at the possibilities and advantages of exploiting data more efficiently to address these challenges. I propose to achieve this by means of different approaches ranging from classical feedback control to more advanced machine learning and artificial intelligence (AI) tools. To this end, I make the following hypotheses that forms the basis for my research goals in the next three to five years.

<u>Hypothesis 1:</u> *Combining machine learning algorithms that uses online process data with traditional model-based RTO methods, can address the challenge of developing and updating the model (Challenge 1)*

<u>Hypothesis 2</u>: *Modified RTO algorithms that use also transient data more efficiently, can alleviate the steady-state wait time issues (Challenge 2)*

<u>Hypothesis 3</u>: *Developing systematic approaches to convert the optimization problem into a feedback control problem can alleviate computational issues associated with solving numerical optimization problems online (Challenge 3).*

<u>Hypothesis 4</u>: *A hierarchical combination of the different RTO approaches that exploits the different time scales, can handle a wider class of uncertainty in an efficient manner and provide a unified RTO paradigm that exploits the synergy between the two existing paradigms.*

By addressing the main challenges with traditional optimization approaches, the potential of process data can be better exploited in many industrial applications leading widespread application of autonomous decision-making tools in the process industries and at the same time enable new application areas.

Teaching Interests:

There is no greater joy than imparting knowledge to a generation of young minds to provide them with the tools to succeed in life and tackle the future challenges facing our society. Not only my research interests, but also my teaching interests are motivated by the need for technology to address the societal challenges. I believe that process systems engineering is an important subject area in achieving this and teaching these concepts at both undergraduate and graduate level becomes increasingly important.

My primary interest lies in teaching topics from process systems engineering (PSE), including process modelling, control, design and optimization. Having a background in mechatronics engineering with specialization in control theory, along with industrial experience in process control and PhD in chemical engineering, I have found my multidisciplinary background to be advantageous, since PSE as a field is becoming increasingly multidisciplinary. Therefore, to reflect this transformation in the PSE community, I am also interested in teaching multidisciplinary courses from a PSE context such as system identification, neural networks and optimal process control.

Teaching process systems engineering as a subject area involves a significant amount of scientific and mathematical concepts. When teaching mathematical concepts involving theorems, and formulae, it is very important that my students clearly understand the underlying concepts and how to derive these concepts. I tend to use the blackboard often when I teach. This is mainly because, I find that using a blackboard and writing the concepts is much easier for the students to follow along than when they just appear as slides in front of them. Due to varying learning rates among students, I also noticed that this gives enough time for the majority of the classroom to follow along. Since different learning style uses different parts of the brain, by involving more of the brain during learning, we remember more of what we learn. Therefore, in addition to visual and auditory teaching aides, I also advocate my students to take notes to engage in kinesthetic learning, thereby covering a broad range of learning styles in the classroom. The clarity and aesthetics of my handwriting and teaching style has also been well received by students.

I find that I myself learn a lot about the subject area from teaching, which further strengthens my research activities. I strongly believe that teaching and research are complementary.

581394 Process Control and Developpement for Energy-Efficient Applications

Research Interests:

My main research interests are in the process control and improvement. It is important to understand the involved physical and chemical mechanisms in order to enhance the transformation and transport phenomena in chemical process. Characterization techniques for monitoring, understanding and improving materials functionalities in chemical processes are, thus, instrumental. Research activities that can help solve the actual needs in sustainable development, energy efficiency and availability worldwide, as well as environmental processes are the cornerstone of the work I wish to perform.

My research activities will be focused on Process Intensification , which could be developed for both chemical and pharmaceutical applications, but for a short term program will mainly be directed on equipement enhancement and control for energy-efficient applications.

Teaching Interests:

I started my teaching experience by sharing my research projects during oral communication with colleagues. The more questions I received the more details I wanted to provide. So I went from helping interns and graduate students understand and define their project to teaching courses.

Up to now, I have been lecturer in two universities, for different courses: unit operation laboratory, elements of chemical reaction engineering, pharmaceutical engineering and design, and formulation of pharmaceutical products.

I studied in three different countries with professors from different backgrounds, which provides me with strong communication and social abilities. I believe that the way in which concepts are taught to students is as important as the way we evaluate their learning. This goes from preparing courses to teaching lessons, developing exercises or practical laboratory and providing and evaluating assignments.

582604 Uncertainty Quantification and Risk-based Decision Support Methodologies for Healthcare and Advanced Manufacturing

Research interests:

In the last decade, there has been a rapid growth in the application of optimization techniques to the solution of problems of industrial importance. In particular, the process and related industries have recently expanded the systematic use of optimization in supply-chain management, operational planning and scheduling, and optimal process control. Even pharmaceutical companies, which have been conservative in model-based decision support innovations, have turned to optimization for product design and drug delivery purposes. This global trend, primarily motivated by increasing competition, calls for further developments in optimization and alike research fields, e.g. dynamic and mixed-integer optimization, model uncertainty quantification, optimization under uncertainty and risk assessment.

In view of these considerations, my passion for applied mathematics and statistics, and my rigorous chemical engineering education, my research expertise and current research interests span the following topics:

- Dynamic modeling of chemical and pharmaceutical processes;
- Quantification of parametric uncertainty in nonlinear models;
- Risk assessment and risk-based decision making in process operations;
- Deterministic and stochastic dynamic optimization;
- Operational planning and scheduling of batch processes;
- Supply-chain and enterprise-wide optimization;

Specific examples of my research achievements include a framework for simultaneous optimization of the supply-chain and production systems of industrial gas producers [1], two new strategies for deterministic and stochastic dynamic optimization of batch operations ([2,3] and [4,5]), and a novel approach to the integration of scheduling, dynamic optimization and control in multi-unit batch processes [6]. These strategies are very flexible, and will accommodate virtually any batch processing system (reactors and bio-reactors, crystallizers, freeze-drying units, home appliances, ...) and performance metric (economic functions, indicators of process environmental impact, ...). Thus, they constitute a general, robust optimization platform for batch processes.

Additional research work, completed in the last two years, comprises an innovative method for rapid estimation of the probability distribution (PDF) of the parameters of ODE/DAE models [7,8] and a strategy for dynamic selection of the most appropriate uncertainty set in stochastic dynamic optimization problems [9], which together allow real-time optimization of systems with stochastic and time-varying features (reactors/crystallizers subject to fouling, food processing units, ...). These two approaches seek to exploit the rapid growth in the availability of process data, which makes it possible to model the underlying uncertainties that arise in real-life process applications. Therefore, they constitute the backbone of a structured approach to the analysis and utilization of big data sets.

My current research efforts are devoted to developing new optimization-driven Monte Carlo algorithms for PDF estimation [7,10] and to the application of dynamic modelling, uncertainty quantification, and deterministic/stochastic optimization to processes for purification of monoclonal antibodies. The development of the aforementioned new types of Monte Carlo algorithms is justified by their superior reliability and computational efficiency, which are essential for uncertainty quantification in complex dynamic models, such as those of most processes for purification of monoclonal antibodies.

In view of all of my research achievements, my future research plan includes the following activities:

- Development of a general platform for optimal drug delivery applicable to both standard drugs (e.g. insulin, pediatric drugs, opioids, biologics) and/or veterinary drugs. One of the several applications of this platform may be an improved artificial pancreas featuring fully automated model tuning,

more reliable insulin administration policies and automatic compensation for long-term variations in the patient's response to treatment. Other potential applications encompass pediatric drug delivery with automatic minimization of the risk for side effects, optimal administration of opioids to mitigate drug abuse and addiction problems, and ideal administration of conventional and/or broad spectrum antibiotics to animals (aminoglycosides, macrolides, penicillins, tetracyclines, ...).
- Development of general methodologies for the rational design of formulations of biologics with applications to monoclonal antibodies and other biologic drugs of interest. These same approaches may as well be applicable to conventional drugs.
- Definition of a general framework for rational design and online optimal management of pharmaceutical processes comprised of components for optimal process synthesis, dynamic optimization of upstream operations and risk-based real-time decision support for downstream processing steps. Some divisions of the pharmaceutical industry, which this framework can benefit, include generic drug manufacturing, production of biologics and/or of drugs subject to rapid degradation, and manufacturing of drugs/drug products from materials/intermediates difficult to process.
- Definition of a general framework for rational design and dynamic optimization of food processing plants with potential applications to the production of nutrition supplements (e.g. artificial meat), the fermentation of wine and liquors, the spray drying of milk and natural extracts, the lyophilization of fruits and vegetables, and so on.
- Development of accurate PDF estimation strategies applicable to high-dimensional spaces.
- Development of stochastic dynamic optimization algorithms able to account for both parametric and structural model uncertainty as well as other nonconventional sources of uncertainty (e.g. the initial conditions of the problem).
- Development of quantitative risk assessment strategies under both parametric and structural model uncertainty.

All of these activities will require use and extension of the methods and tools, described previously, as well as development of completely new ones. This tentative research plan may be partially modified based on funding availability and potential collaborations with other faculty members.

Teaching interests:

My research work and graduate studies have provided me with deep knowledge of control theory, numerical methods, dynamic and steady-state modeling, coding/parallel computing (especially in C++), and statistics. Therefore, I would be interested in teaching any course broadly related to these subjects, e.g. applied statistics, process control, numerical calculus, dynamic and steady-state modelling, and process optimization. I would also like to introduce and teach two advanced courses, whose topics lie at the boundaries of Chemical Engineering, Statistics and Computer Science:

- Applied statistics and big data analytics for chemical engineers;
- Applications of parallel computing to chemical engineering problems.

Since optimization strategies have become an essential tool for the solution of many problems of industrial relevance in the last few years, I strongly believe these interdisciplinary courses would provide new chemical engineers with useful and valuable knowledge.

References:

1. Rossi F, Manenti F, Reklaitis G. A general modular framework for the integrated optimal management of an industrial gases supply-chain and its production systems. *Computers and Chemical Engineering*. 2015;82:84-104.
2. Rossi F, Manenti F, Buzzi-Ferraris G. A novel all-in-one real-time optimization and optimal control method for batch systems: Algorithm description, implementation issues, and comparison with the existing methodologies. *Industrial and Engineering Chemistry Research*. 2014;53:15639-15655.
3. Rossi F, Copelli S, Colombo A, Pirola C, Manenti F. Online model-based optimization and control for the combined optimal operation and runaway prediction and prevention in (fed-) batch systems. *Chemical Engineering Science*. 2015;138:760-771.

4. Rossi F, Reklaitis G, Manenti F, Buzzi-Ferraris G. Multi-scenario robust online optimization and control of fed-batch systems via dynamic model-based scenario selection. *AIChE Journal*. 2016;62:3264-3284.
5. Rossi F, Manenti F, Pirola C, Mujtaba I. A robust sustainable optimization & control strategy (RSOCS) for (fed-)batch processes towards the low-cost reduction of utilities consumption. *Journal of Cleaner Production*. 2016;111:181-192.
6. Rossi F, Casas-Orozco D, Reklaitis G, Manenti F, Buzzi-Ferraris G. A computational framework for integrating campaign scheduling, dynamic optimization and optimal control in multi-unit batch processes. *Computers and Chemical Engineering*. 2017;107:184-220.
7. Rossi F, Mockus L, Reklaitis G. Rigorous Bayesian inference VS new approximate strategies for estimation of the probability distribution of the parameters of DAE models. *Computer Aided Chemical Engineering. Accepted (oral presentation at ESCAPE29, Eindhoven, NL).*
8. Rossi F, Manenti F, Buzzi-Ferraris G, Reklaitis G. A novel strategy for rapid estimation of the probability distribution of the uncertain parameters of ODE/DAE systems. *In preparation. Estimated submission at the end of 2019.*
9. Rossi F, Manenti F, Buzzi-Ferraris G, Reklaitis G. Stochastic NMPC/DRTO of batch operations: batch-to-batch dynamic identification of the optimal description of model uncertainty. *Computers and Chemical Engineering*. 2019;122:395-414.
10. Rossi F, Mockus L, Reklaitis G. A novel quasi-deterministic Monte Carlo method for estimation of probability distributions. *In preparation. Estimated submission at the end of 2019.*

581956 Review of Recent Trends in Sulfur Unit Capacity Expansion Projects

Research Interests: Avinashkumar has notable research experience includes study of iron-based Fischer-Tropsch reactions over Zeolite catalyst. He is also actively engaged in peer review of scientific journal papers.

Teaching Interests: None

Sulfur Recovery Unit, most commonly called as Claus process, is very significant gas desulfurization process unit in Refinery and Gas industries. Explorations of new natural gas fields, refining of high-sulfur crude oils, and recent crude expansion projects is needing capacity expansion of Claus unit for many companies around the world. In refineries, the Sulphur Recovery Units take acid gas from amine regeneration units and sour water strippers, converting hydrogen sulfide to elemental sulfur using the Claus process. The Claus process is hydraulically limited by mass flow rate. Reducing pressure drop across control valves, flow meters, lines, knock-out drums, and packing improves the capacity. Oxygen enrichment helps improve the capacity by removing nitrogen, this is more commonly done on all capacity expansion projects. Typical upgrades required due to oxygen enrichment are new burners, new refractory in thermal reactor, resizing of 1st condenser, instrumentation changes, and steam / condensate heat integration. Some other capacity expansion options typically considered are: tail gas compressor, replacing air blower with higher head, hydrocarbon minimization in the feed, water removal, and ammonia removal. Increased capacity related upgrades in sulfur recovery unit also need changes in tail gas treatment unit, typical changes include improvement to quench tower duty, packing area upgrades in quench and absorber towers, and increased amine circulation flow rates.

580443 Modelling of Gas Recovery from South African Shale Reservoirs

Research Interests: Oil and Gas production and processing

Teaching Interests: Chemical Engineering & Oil and Gas production and processing

Abstract

Mathematical models for a triple-porosity continuum describing the distribution of pressure, the driving force for gas flow during gas production, were developed on the basis of the continuity equation from first principles, focusing on the South African shales of the Karoo Basin. The triple-porosity continuum incorporated the matrix, the natural fracture network, and the hydraulic fracture network; the matrix and the natural fracture network systems were considered two dimensional, while the hydraulic fracture system was considered one dimensional. The three developed mathematical models are in the form of the general diffusion or heat equation, thus the numerical finite difference method (FDM) was employed in solving the triple-porosity model. MatLab tool was used to develop and execute the FDM simulation or algorithm of each system of the triple-porosity model; results for each system are presented and discussed in chapter 6 of the research report. The results generated agree with models previously developed and validated with their respective field data, substantiating the reliability of the model developed in this research report. Currently the South Africa's Barnett shale does not have field data, hence the field data from literature was used to validate the developed model. The model is thus a guide onto describing the gas flow behaviour for the South African shale gas reservoirs, and not specifically for the Barnett shale.

580444 Modeling and Simulation of Modular Refinery for Production of Fuels with Low Environmental Pollution

Research Interests: Oil and Gas production and processing

Teaching Interests: Oil and Gas production and processing

Abstract

Production and upgrading of fuels and petrochemical products, use worldwide are based on characteristics (capacity) of refinery. Although crude oil has different composition according to their origin, keys operations (units) that take place are same. Crude oil distillation, cracking and reforming constitute the keys operations in petroleum refinery for obtaining useful product. Each process has attracted more researches since the discovery of separation method of crude oil cuts. Complex operations that occur in the production of fuels, and environmental usage repercussions open doors to multiple further researches.

Process industry covers a wide range of operations that handle solid, liquid and gases as feed or products. This needs Chemical Engineering science movement to solve any problem that can take place during operations. Modeling and simulation are either combined or alone useful tools for sorting out real troubleshooting situations and/or upgrading researches. These can bring great advantages to process industry.

A good strategy to analyse real behaviours of refinery units can then provide by using modeling and simulation as tools. Due to the need of upgrading petroleum refinery process, this research is aimed to develop modeling and simulation of main units of refinery.Design data based on characteristics of crude oil, operating parameters of the involved units, and other essential data were calculated and collected,then entered into the simulation software, Aspen Plus to generate the modular refinery model. Results of complete simulation were studied and that leads to obtaining different products composition and determine conditions of petroleum products. According to this study, it is found that it is definitely possible to predict impact of sulphur on environment. Various information about the modular refinery under study were obtained using the simulation tool. The objectives of this research were accomplished.

578959 20-Plus Years after the Pioneering Work at the University of Oklahoma-Chemical Engineering: Capillary Condensation of Light Hydrocarbons in MCM-41 Type Nano-Porous Media

Research Interests: no limitations

Teaching Interests: no limitations

20-Plus years after the pioneering work at the University of Oklahoma-Chemical Engineering: capillary condensation of light hydrocarbons in MCM-41 type nano-porous media will be addressed.

578981 Detailed Stress Analysis of Floating Platforms

Research Interests: dynamic modeling and analysis

Teaching Interests: Transport phenomena

582121 Process Systems Engineering and Dynamic Process Control of Large-Scale Systems

Research Interests: My research interest lies in the field of dynamic simulation, process control, optimization and system engineering of large-scale hybrid systems. The goal is to develop complex hybrid systems and maximize overall efficiency of the systems by energy integration and exploring synergistic benefits shared within the system. Hybridizing renewable energy with fossil fuel system is the key pathway towards making renewable energy systems feasible by exploring the possible synergistic benefits. Energy integration for these large-scale plants maximizes the hybrid plant performance. Process intensification is combining different units to a single unit. This kind of operation reduces the resources and energy usage of the combined unit in comparison to the separate units. Process intensification also introduces further degrees of freedom for optimization. Optimizing the decision variables ensures plants optimum performance for single or multi-objective goals. Large scale systems have multiple inputs with disturbances creating transient effects coming into the system. Process control is essential in these systems to ensure safe and smooth output.

During my Ph.D. research, I have demonstrated design, dynamic simulation and complex control system of a highly integrated natural gas/ solar thermal hybrid power plant [1]. The study has shown that the solar-to-electric efficiency has almost doubled due to the synergistic benefits of hybridization. A detailed review on hybrid concentrated solar systems has demonstrated the state of art and the current opportunities in this research area [2]. In another article, we have demonstrated the techno-economic evaluation of solar/natural gas hybrid plant [3]. Optimizing hybrid concentrated solar system can further improvise the plant performance both in terms of economic and environmental prospects [4,5]. Using a combination of process intensification and optimization the system reduces the resources need and ensure maximum productivity. Currently as a postdoc in University of Delaware I am working on designing reactive distillation column.

Teaching Interests: My passion for teaching roots to the opportunity of mastering concepts during teaching. Teaching to a group of people further opens my horizon to different ways student think. Also delivering fundamental concepts with applied knowledge enhances my understanding in the particular area. I am passionate about teaching courses like dynamic process control, smart systems, and system engineering. The students will learn system level design and application of different software's in improvising system performance. Including machine learning in smart system course will help students to apply the strength of machine learning by predicting from data and using it for improving plant performance. Process optimization is also another important course that I am interested to teach. This course will help students to design and optimize non-linear systems and enhance the system performance by using the degrees of freedom in the model. I have TA'ed courses like renewable energy, air pollution, process safety and smart systems where I have had the experience of in class interaction with students. Solid background in my research in these areas will enable the students to view these concepts in a practical aspect.

References:

[1] Rashid, K., Safdarnejad, S. M., & Powell, K. M. (2019). Dynamic simulation, control, and performance evaluation of a synergistic solar and natural gas hybrid power plant. *Energy Conversion and Management*, *179*, 270-285

[2] Powell, K. M., Rashid, K., Ellingwood, K., Tuttle, J., & Iverson, B. D. (2017). Hybrid concentrated solar thermal power systems: A review. *Renewable and Sustainable Energy Reviews*, *80*, 215-237.

[3] Rashid, K., Safdarnejad, S. M., Ellingwood, K., & Powell, K. M. (2019). Techno-economic evaluation of different hybridization schemes for a solar thermal/gas power plant. *Energy*, *181*, 91-106.

[4] Rashid, K., Sheha, M. N., & Powell, K. M. (2018, June). Real-time optimization of a solar-natural gas hybrid power plant to enhance solar power utilization. In *2018 Annual American Control Conference*

(ACC) (pp. 3002-3007). IEEE.

[5] Rashid, K., Safdarnejad, S. M., & Powell, K. M. (2019). Process intensification of solar thermal power using hybridization, flexible heat integration, and real-time optimization. *Chemical Engineering and Processing-Process Intensification*, *139*, 155-171.

581940 CO_2 Utilization through Dry Reforming of Methane Reaction

Research Interests:

The re-use of CO2 into value-added products is one of the most desired process pathways which have the potential to reduce carbon emissions of current processes. Dry Reforming of Methane (DRM) provides such a pathway to re-use CO2 by reacting it with methane to produce a valuable and versatile intermediate, Syngas. Syngas is a mixture of CO and H2 and is a raw material to produce various petrochemicals including fuels via Fischer-Tropsch. However, the DRM process suffers from three main challenges : low H2/CO ratio of produced syngas, high energy requirement and rapid catalyst deactivation. My PhD research aims to address these challenges at different scales. The first two problems deal with the DRM process. A Life-Cycle Analysis and Process Optimization was done which resulted in formulating process flow-sheets that can have lower CO2 emissions compared to existing processes as well as solve the low H2/CO ratio problem. Specifically, it was found that a post-DRM syngas ratio adjustment step to adjust H2/CO ratio, by removing CO resulted in a process with lower carbon footprint and lower operating costs compared to existing commercial processes. At the lower scale - catalyst scale, I have tested Nickel catalysts over-coated with Alumina layers via Atomic Layer Deposition technique. The catalyst testing experiments helped understand the effectiveness of the ALD overcoat to reduce coking and sintering of catalysts. Overall, this work addresses the major challenges of the DRM process and provides some plausible solutions that can be taken at the process level and the catalyst level to help achieve DRM at large scales.

Teaching Interests:

I have been a Teaching Assistant for the following undergraduate courses: Process Dynamics and Control, Chemical Kinetics and Reactor Design.

584389 Process Systems Engineering (PSE)

Abstract: My main research focus is process systems engineering (PSE) and current application domain is continuous pharmaceutical and bio-pharmaceutical manufacturing. However, the PSE methods and tools that I developed are generic and can be applied to any manufacturing industries. Currently, pharmaceutical and bio-pharmaceutical industries are going through paradigm shift from conventional batch to continuous manufacturing process in which PSE can play a very significant role. Furthermore, there is a need of advanced modular continuous manufacturing platform technology for fast and robust development of pharmaceutical and bio-pharmaceutical products. Master formulations, process models, standardized sensing and control architecture, superior automation and novel methods and tools need to be developed to enable platform technology. This area of research and education are highly encouraged and supported by pharmaceutical and biopharmaceutical companies, academic institutions, regulatory authority (FDA) and national science foundation (NSF).

In this work, a scientific foundation for the optimal design, optimization and control of the modular continuous process involving structured organic particulate systems has been developed. The mathematical models for continuous manufacturing of active pharmaceutical ingredients and final product tablet, has been developed for virtual experimentation and applied for plant design, optimization and control. The developed continuous tablet manufacturing plant and process model has been adapted by several pharmaceutical companies. A flexible monitoring and control system incorporating different PAT techniques and tools and control strategies for continuous pharmaceutical manufacturing process has been designed and implemented to Rutgers pilot-plant. Different control strategies such as PID, PID coupled with dead time compensator, and MPC has been compared and evaluated practically for pharmaceutical process control. A new hybrid MPC-PID control strategy has been developed and implemented into the pharmaceutical tablet manufacturing plant and its performance has been found better than the other control strategies. A novel RTD based control system has been developed and implemented into the pilot-plant. Novel sensing methods have been developed for real time measurements of powder density and powder level. Feedforward control system has been integrated with feedback control system to take proactive mitigation actions on raw materials and process variability. A moving horizon based Dynamic Real Time Optimization (MH-DRTO) method has been also developed to provides the optimal operational set points for the control system. The developed control architecture has been implemented into pilot-plant.

Biosketch: I am working as an Assistant Research Professor at C-SOPS, Department of Chemical and Biochemical Engineering, Rutgers University, NJ, USA. I am the recipient of prestigious EFCE Excellence Award given in Recognition of an Outstanding PhD Thesis, from European Federation of Chemical Engineering. I received several grants as a PI/Co-PI and have well stablished research group. I have published more than 63 scientific research papers, written 12 book chapters, presented at over 100 international conferences and edited one pharmaceutical book published by Elsevier. I am actively serving as a conference session chair, Journal reviewer and guest editor.

Research Interests: Process Systems Engineering (PSE) including process control, process modelling and simulation, process optimization, machine learning, process monitoring and PAT, data science, cyber infrastructure, cyber-physical security, new methodology and software development. Continuous pharmaceutical and biopharmaceutical manufacturing.

Teaching Interests: My teaching experience that I acquired serving as a teaching assistant, and lecturer (assistant professor), makes me confidant to develop and teach any courses of chemical and biochemical engineering. The industrial invited crash courses that I have given provide me opportunity to extend my teaching skills to cover more applied courses and to interact with highly experienced scientist and operators from manufacturing companies. I demonstrated my skills to instruct hands-on exercise through several workshops that I conducted. Several training courses that I conducted for industrial and academic participants make me instrumented with additional capability of building highly skilled workforce. I developed online course modules for FDA where I learn how to develop and give online courses. I am open to teach any courses that your university will assign me. If needed, then I can develop new courses based on my research outcomes.

Grants: Extensive experience in funding proposal writing.

1. **2019 Glaxo Smith Kline (GSK)** (granted)

Development of Computational Models for Continuous Tablet Manufacturing

Role: **PI.** Source: GSK Company. Amount: **$540,000.**

2. **2019 Siemens** (granted)

Enable full implementation of "Integrated Cyber-Physical Security Framework and Novel Control Module" for continuous pharmaceutical manufacturing plant

Role: PI. Source: **Siemens Company.** Amount: **$71,800.**

3. **2018 Food & Drug Administration (FDA)** (granted)

Industry 4.0 Implementation in Continuous Pharmaceutical Manufacturing

Role: **Co-PI**. Source: FDA. Amount: **$4 Million.**

4. **2018 Food & Drug Administration (FDA)** (granted)

Advanced Bio-manufacturing

Role: **Co-PI**. Source: FDA. Amount: **$1.8 Million**

5. **2018 Glaxo Smith Cline (GSK)** (granted)

Development of Computational Models for Continuous Tablet Manufacturing

Role: **PI**. Source: GSK Company. Amount: **$539,991 (1st year);** Total: **$2.34 Million**.

6. **2018 Siemens** (granted)

Cyber-physical security in the Pharmaceutical Manufacturing Process

Role: **PI.** Source: Siemens Company. Amount: **$75,000**.

7. **2018 Diversitech Corporation** (granted)

Fluid bed drying trials

Role: **PI**. Source: **Diversitech** Company. Amount: **$1600**.

8. **2018 Rutgers Global International Travel Grant** (granted)

Role: **PI.** Amount: **$750.**

9. **2018 CNH** (granted)

Wear prediction and validation with fine particle interaction

Role: **Co-PI.** Source: CNH Company. Amount: **$118608.**

10. **2017 Glaxo Smith Cline (GSK)** (granted)

Development of tablet press feed frame RTD model and BU to CU relation

Role: **Co-PI.** Source: GSK Company. Amount: **$150000.**

11. **2017 Grant to stablish international collaboration** (granted)

Grants to Support the Initiation of International Collaboration.

Role: **Co-PI.** Source: Hamburg University of Technology. Amount: **€1610.**

12. **2016 Integra Continuous Manufacturing Systems** (granted)

Development of sensing and control strategy for direct compaction continuous pharmaceutical tablet manufacturing process. **Role: PI**. Source: Acumen Company. Amount: **$5114**.

13. **2016 Research Council Grant Award** (granted)

Development and validation of advanced model predictive feedforward/feedback control strategy for pharmaceutical process. **Role: PI**. Source: Rutgers. Amount: **$3800**.

Submitted grant proposals: 6 proposals have been submitted (total amount>$15 Million)

Publications: Scientific articles (63 published + 11 under submission/preparation); Book (1); Book chapters (12); International conference presentations (100); Citations>1303, h-index>19, i10-index >28.

Activities in AIChE 2019 conference

Session chair/co-chair

1. **Chair**: Process Monitoring & Fault Detection
2. **Co-Chair**: Applied Math for Biological and Biomedical Systems
3. **Co-chair:** Integrated Product and Process Design.

Oral presentations

1. **Singh, R.,** Ramachandran, R., Ierapetritou, M., Muzzio, F. J. **(2019)**. Industry 4.0: Advanced Bi-layer Control System for Continuous Pharmaceutical Manufacturing Pilot-plant. Oral presentation at AIChE, Orlando, FL, USA, November 10-15, 2019.
2. **Singh, R. (2019).** Cyber-Physical Security of Advanced Automated Continuous Pharmaceutical Manufacturing Pilot-Plant. Oral presentation at AIChE, Orlando, FL, USA, November 10-15, 2019.

Contact:

Email: *ravendra.singh@rutgers.edu,* ***Phone:*** *(001)848-445-4944;* ***Fax****: (001)732-445-2581*

Websites: *http://cbe.rutgers.edu/faculty; http://ravendrasingh.wixsite.com/ppse*

581891 Gas Hydrates Research: From Fundamental Science to Engineering Applications

Research Interests:

Clathrate hydrates (also known as gas hydrate or hydrates) are solid crystalline structures that trapped small gas molecules (methane, ethane, carbon dioxide, propane, etc). These solid structures form when water molecules create hydrogen bonds network around a gas molecule, thus entrapping the gas inside a cage. Hydrates typically form at high pressure and low temperature (typically higher than ice point temperature). These unique structures have significant impact on various fields including energy recovery, oil/gas transportation as well as separation technology. For example, in energy recovery, it was reported that the amount of energy as natural gas stored inside gas hydrates in nature is twice of that fossil fuels. Similarly, it was reported that separation cost using gas hydrates can be half of that membrane technology at similar capacity. However, gas hydrates technologies are not fully utilized since its nucleation and formation/dissociation kinetics are not well understood.

My research in the gas hydrate began in the area of flow assurance, specifically understanding the rheological properties as well as viscosity prediction of gas hydrate slurries. In the mid of 2000s, there was a paradigm shift in handling gas hydrates in flowlines. In order to reduce the production cost, the industry shifted from complete avoidance strategy where hydrate formation is prevented by injecting large quantities of thermodynamic inhibitors to hydrate management strategy where flowlines are being operated carefully in the presence of hydrate (avoiding hydrate plug). This shift has caused a momentum in research on gas hydrate rheological properties. In my doctoral work, I have proposed a new method in hydrate rheological measurements at high pressure. It was the first systematic measurements that decoupled various variable that affect viscosity measurements. In fact, the work on the relative viscosity model of hydrate slurries was then extended for multiscale modeling of pressure drop of multiphase flowlines. The results of this investigation showed outstanding improvement of pressure drop prediction. My research expertise has also extended to other field of gas hydrate research including the thermodynamic properties of gas hydrates systems and application of gas hydrates for engineering technology. I have performed investigatio of Vapor-Liquid-Solid-Equilibria (VLSE) of systems containing Petroleum Gas Systems. Furthermore, I have also conducted research in hydrate technology. In this project, I designed and evaluated the use of gas hydrates for desalination purpose.

As such, as a future faculty, I would to focus research in tackling the fundamentals science of hydrate nucleation, formation and kinetics. I would like to combine my knowledge and expertise in hydrate thermodynamics, kinetics and transport in promoting gas hydrate technology. An area that I would like to focus on is modeling of hydrate formation and kinetics using Raman spectroscopy and gas chromatography. The knowledge gained in this investigation can promote the hydrate drilling for energy recovery. Furthermore, I believe that gas hydrates have a lot of potential as well as impact in other fields including energy storage, transportation and separation. Therefore, another area that I would like to focus my research is using gas hydrates for water treatment as well as gas separation technology. Additionally, since I started my research in gas hydrates in the field of flow assurance and I have developed a relative viscosity model for hydrate slurries that has better prediction, I would like to continue my research in this area. In this area, I would like to focus my research on using rheological properties of gas hydrates in predicting and modeling the hydrate agglomeration phenomenon. I believe that this is the key in predicting hydrate plugging in flowlines and thus can be used by the industry as an early warning sign for hydrate plug.

Teaching Interests:

Currently, I am active in both courses and research mentorship teaching. For academic courses, I have experienced teaching both undergraduate (~150 students) and graduate (~20 students) level. Thus, I am comfortable teaching in both large and small group settings. From my teaching experience, I believe that it is necessary to use technology in teaching to promote student engagement in the class. Today's generation is heavily influenced by the technology that is constantly changing. Additionally, I believe that understanding the course materials is more important that high grades. For this reason, as future faculty, I plan to use technology in the classroom every possible way. This helps me as a faculty to obtain real-time response whether the students have understood the materials. By doing this, if the response shows that my students do not understand certain materials or concepts, I can immediately shift my focus to that materials. This will ensure that students will have better understanding of the course materials.

I have experienced teaching in several courses in chemical engineering including Thermodynamics, Separation Processes, and Chemical Reaction Engineering. Based on my experience, I am comfortable in teaching any chemical engineering courses at both graduate and undergraduate levels. However, I prefer to teach course in *Transport Phenomenon* and *Thermodynamics* at any level (undergraduate and graduate). I would like to teach this course since it is in line with my field of research. Thus, I can incorporate typical issues occurred by the industry in my classroom. This technique will provide students with a little experience working in the industry. This can certainly promote the students' interest in chemical engineering field.

581905 The Design, Synthesis and Testing of Advanced High-Performance Ionene Polymers As Gas Separation Membranes

Research Interests: The design and development of new advanced polymeric materials for **energy-saving separation technology and environmental green energy applications**. Our current projects investigate the use of versatile nature and tunable structures of ionic liquids in polymeric membrane systems for effective gas separation applications. We have focused on the design, synthesis and testing of advanced "high-performance" ionenes (polycations formed from condensation reactions) as a new generation of separation membranes with a synergistic approach that combines different transport mechanisms, especially size-sieving and solution-diffusion. Membranes prepared from these newly developed ionene polymers displayed high CO_2 separation properties, together with excellent mechanical and thermal stabilities. Profound outputs from our current work funded by the Department of Energy – Basic Energy Sciences (DOE BES) will significantly contribute toward the establishment of a strong scientific and technical base for membrane science and technology.

Teaching Interests: The faculty candidate, Dr. Kammakakam has specific plans upon the completion of his postdoc project works to develop a tenure-track research career on the energy-saving materials. Dr. Kammakakam also has a solid academic foundation and a passion for teaching graduate students, worked as a teaching assistant of Prof. Kim Tae-Hyun as a part of the Ph.D. program. From the two years of teaching experience at Incheon National University, he observed that the teaching across intellectual domains goes beyond the creativity and knowledge, and he can teach students in a way that surprises, connects and enriches them. It is his dream and goal to work as an academic faculty and develop new advanced materials for energy-saving applications.

581943 Hybrid Membranes for Challenging Energy Separations

Research Interests:

Chemical separations consume 15% of U.S. industrial energy use, while molecularly selective membranes can significantly reduce this number [1]. Numerous membrane materials have emerged to address unmet separation challenges in recent decades, among which organic-inorganic hybrid membranes show promising application prospect by combining the advantages of organic and inorganic membranes. However, the design and realization of hybrid membrane materials with targeted applications still rely greatly on trial and error due to several practical challenges: (1) identification of advanced inorganic filler, (2) interfacial problem and (3) mismatching of performance between organic and inorganic phases. Rational design of hybrid membranes requires fundamental understanding of material properties, penetrants discrepancy, and corresponding permeation results. In this regard, coupling atomistic simulation techniques and experimental approaches based on membrane theories will expedite the hybrid membrane development progress by providing insights into specific separation problems.

Prior to joining Georgia Tech, I pursued the integration of computational and experimental skills in describing gas adsorption and diffusion phenomena through porous sorbents and membranes. At Georgia Tech with the supervision of Professor William Koros, I proposed a novel olefin/paraffin separation mechanism based on fundamental understanding of membrane materials and gas molecules as well as gas transportation theory, named *conformation-controlled molecular sieving effect*. This novel mechanism *refined* the traditional molecular sieving effect, which was further experimentally verified using the in-purpose identified molecular sieve. The developed hybrid membranes incorporated with the identified molecular sieve show promising propylene/propane and *n*-butane/*iso*-butane separation performance, surpassing all reported hybrid membranes. In the following projects toward aggressive sour gas (H_2S and CO_2) sweetening, I developed novel polymeric membranes with extraordinary separation performance and explored the gas permeation fundamentals using the integrated computational and experimental skills. Surprisingly, my work suggests the counter-intuitive benefits of polymer plasticization on improving H_2S removing efficiency [2].

My future research aims to lead the developments of hybrid membranes for key energy separations assisted by computational simulations. Design of "inorganic" molecular sieves will be guided by a molecular-level understanding of permeation (adsorption + diffusion) and material structure-property relationships. Matching of "organic" polymer matrix with the molecular sieves' properties and performance will be pursued to attain optimum membrane performance. Computational simulations will be performed to provide insights to membrane performance and guide future membrane developments. Ultimately, not only successful but also frustrated organic-inorganic recipes along with their physical and chemical properties as well as experimental separation data will be organized to derive a database for several challenging energy separations. With extensive experimental experience in membranes and strong expertise in computational simulations, I am uniquely suited to undertake these challenging research efforts.

References

[1] National Academies of Sciences, Engineering, and Medicine. *A Research Agenda for Transforming Separation Science.* 2019, Washington, DC: *The National Academies Press.* Doi: https://doi.org/10.17226/25421

[2] Yang Liu, Zhongyun Liu, Gongping Liu, Wulin Qiu, Nitesh Bhuwania, Daniel Chinn, William Koros. Surprising Plasticization Benefits in Natural Gas Upgrading Using Polyimide Membranes. *Journal of Membrane Science*, 2019, submitted.

Teaching Interests:

Chemical Engineering Fundamentals, Computational Simulations, Separation Processes, Fundamentals of

Selected Publications (36 total, 14 first author, 454 citations, H-index 12):

1. Yang Liu, Zhijie Chen, Gongping Liu, Youssef Belmabkhout, Osama Shekhah, Mohamed Eddaoudi, William Koros. Conformation-Controlled Molecular Sieving Effect for Membrane-based Propylene/Propane Separation. *Advanced Materials* **2019**, *31(14)*, 1807513.
2. Yang Liu, Gongping Liu, Chen Zhang, Wulin Qiu, Shouliang Yi, Valeriya Chernikova, Zhijie Chen, Youssef Belmabkhout, Osama Shekhah, Mohamed Eddaoudi, William Koros. Enhanced CO_2/CH_4 Separation Performance of a Mixed Matrix Membrane Based on Tailored MOF-Polymer Formulations. *Advanced Science* **2018**, *5(9)*, 1800982.
3. Yang Liu, Bing Zhang, Defei Liu, Ping Sheng, Zhiping Lai. Fabrication and Molecular Transport Studies of Highly *c*-oriented AFI Membranes. *Journal of Membrane Science* **2017**, *528*, 46-54.
4. Yang Liu, Jing Liu, Y. S. Lin, Ming Chang. Effects of Water Vapor and Trace Gas Impurities in Flue Gas on CO_2/N_2 Separation using ZIF-68. *The Journal of Physical Chemistry C* **2014**, *118*, 6744-6751.
5. Yang Liu, Jing Liu, Y. S. Lin. Adsorption Equilibrium and Kinetics of Carbon Dioxide on Zeolitic Imidazolate Framework-68. *Chemical Engineering Science* **2014**, *118*, 32-40.

582007 Advanced Membranes for Sustainable Separations at the Water-Energy Nexus

Background: The overarching goal of my proposed research program is to develop advanced scalable membranes with an aim to address the growing demands of freshwater production and energy generation. The issues of water and energy tend to be inextricably linked, and the development of cost-effective separation strategies are crucial for both of these issues. Membrane-based processes like reverse osmosis and nanofiltration currently account for majority of the world's desalination operations, effectively replacing the traditional thermally driven separation operations. Development in membrane materials like polyamide thin-film composite membranes have contributed to this tremendous success of membrane technology. The emergence of new waterborne contaminants (e.g. per- and polyfluoroalkyl substances or, PFAS) as well as increased reliance on new sources of energy (e.g. shale gas), necessitate development of advanced membranes, beyond the polyamide thin-film composites. Furthermore, in the area of gas and vapor separations, energy intensive processes like distillation and amine absorption continue to be the predominant separation techniques. This continued reliance on existing techniques reflects difficulty in meeting advanced separation needs by current state-of-the-art polymeric membranes, which are limited by the so-called "*polymer performance upper bound*". My proposed research will focus on developing the next generation of high-performing liquid and gas separation membranes.

Research Interests:

A. Layer-by-layer (LbL) as a versatile platform for advanced liquid separation membranes

1. Modification of commercial membrane surfaces *via* LbL of alternately charged polyelectrolytes has been shown to be an effective method to develop fouling resistant water purification membranes. This project will seek to extend this approach beyond wastewater reuse to other important applications such as biofuel generation. I envision that such surface modified membranes can be developed as effective tools for separation of harmful inhibitory substances from useful sugar molecules during biomass pretreatment stages.

2. Shale gas recovery and extraction generate large quantities of high salinity produced water, which are challenging to purify with regular thin-film composite RO membranes. In fact, for such high-salinity feed streams, engineered osmosis processes like forward osmosis are especially attractive compared to reverse osmosis. This project will focus on a novel design of asymmetric forward osmosis membranes, wherein both the support and the selective layer will be fabricated via the LbL technique. Molecular-level engineering of multiple LbL parameters has the potential to overcome the key challenges faced in FO separations [e.g. concentration polarization (internal and external) and inadequate separation properties]. Besides inherently superior transport performances, these membranes can also provide high fouling resistances.

B. Carbon molecular sieve (CMS) membranes for challenging gas and vapor separations

1. While produced water generation during natural gas extraction provides new opportunities for advanced liquid separation membranes, removing pipeline contaminants (e.g. N_2, CO_2 etc.) from natural gas requires even more advanced approaches. Carbon molecular sieve (CMS) membranes, which are the end products of high temperature pyrolysis of polymers, are uniquely suited for such challenging separations, with the added advantage of easy scalability. This project will specifically focus on overcoming the so-called "*hyperskin*" effect in such membranes, which limits productivities in their hollow fiber formats. Engineering this feature will lead to membranes with exceptionally high productivities that would not only be useful for natural gas production, but also for other critical separations like CO_2 capture and downstream separations like olefins/paraffins.

2. The underlying porous substructure beneath the CMS selective layer plays an important role in maintaining high productivities in asymmetric forms. This project will involve creating high porosity CMS substrates, while overcoming problems with high temperature (>700 °C) pyrolysis needed to create ultra-selective CMS membranes. Specifically, current silica-based anti-collapse treatments lose

porosity above 700 °C, and my approach avoids this deficiency using nanoparticle engineering during fiber spinning. Such membranes will be "game changers" for challenging applications like olefin/paraffin separation ($\Delta < 0.2$ Å) to debottleneck energy intensive distillation processes.

Prior research experiences

My research background involves both liquid and gas separation membranes. In my PhD research at Michigan State University (Supervisor: Prof. Ilsoon Lee), I developed polyelectrolyte multilayer membranes for perchlorate ion removal and wastewater effluent treatment. For the first time, these membranes were designed for *real wastewater effluents* and their long-term performance surpassed those of the state-of-the-art commercial membranes. Currently, in my postdoctoral research at Georgia Tech (Supervisor: Prof. Bill Koros) I am working on the core fundamental aspects of CMS membranes. My work has resulted in the *identification of previously undiscovered features* in these membranes, which have significant implications for developing the next generation of high-performance gas separation membranes. These topics are covered in publications 1&2 below.

Peer-reviewed publications: (Total – 11; Google Scholar: Oishi Sanyal)

1. **Sanyal, O.;** Hicks, S.T.; Bhuwania, N.; Hays, S.; Kamath, M.J.; Karwa, S.; Swaidan, R.; Koros, W.J. "Cause and effects of hyperskin features on carbon molecular sieve (CMS) membranes," *Journal of Membrane Science*, **551**,113-122, (2018).
2. **Sanyal, O.;** Zhang, C.; Wenz, G. B.; Fu, S.; Bhuwania, N.; Xu, L.; Rungta, M.; Koros, W.J. "Next generation membranes-using tailored carbon," *Carbon*, **127**, 688-698, (2018).
3. **Sanyal, O.;** Liu, Z.; Yu, J; Meharg, B. M.; Hong J. S; Liao, W.; Lee, I. "Design of fouling-resistant clay-embedded polyelectrolyte multilayer membranes for wastewater effluent treatment," *Journal of Membrane Science*, **512**, 21-28, (2016).
4. **Sanyal, O.;** Liu, Z.; Meharg, B. M.; Liao, W.; Lee, I. "Development of polyelectrolyte multilayer membranes to reduce the COD level of electrocoagulation treated high-strength wastewater," *Journal of Membrane Science*, **496**, 259-266, (2015).
5. **Sanyal, O.;** Sommerfeld, A.N.; Lee, I. "Design of ultrathin nanostructured polyelectrolyte-based membranes with high perchlorate rejection and high permeability," *Separation and Purification Technology*, **145**, 113-119, (2015).
6. **Sanyal, O**.; Lee, I. "Recent progress in the application of layer-by-layer assembly to the preparation of nanostructured ion-rejecting water purification membranes," *Journal of Nanoscience and Nanotechnology*, **14**, 2178-2189, (2014).
7. Cao, Y.; Zhang, K.; **Sanyal, O.;** Koros, W. J. "Carbon molecular sieve membrane preparation by economical coating and pyrolysis of porous polymer hollow fibers" *Angewandte Chemie International Edition* (2019) (https://doi.org/10.1002/anie.201906653).
8. Adams, J. S.; Itta, A. K.; Zhang, C.; Wenz, G. B.; **Sanyal, O.;** Koros, W. J. "New insights into structural evolution in carbon molecular sieve membranes during pyrolysis," *Carbon*, **141**, 238-246, (2019).
9. Yu, J.; **Sanyal, O**.; Izbicki, A.P.; Lee, I. "Development of layered multi-scale porous thin films by tuning deposition time and molecular weight of polyelectrolytes," *Macromolecular Rapid Communication*, **36**, 1669-1674, (2015).
10. Yu, J.; Han, S.; Hong, J.S.; **Sanyal, O.;** Lee, I. "Synchronous generation of nano and micro-scaled hierarchical porous polyelectrolyte multilayers for superwettable surfaces," *Langmuir*, **32**, 8494–8500, (2016).
11. Gokhale, A.A.; Lu, J.; Parker, N.J.; Izbicki, A.P.; **Sanyal, O**.; Lee, I. "Conductive oxygen barrier films using supramolecular assembly of graphene embedded polyelectrolyte multilayers," *Journal of Colloid and Interface Science*,**409**, 219-226, (2013).

Teaching Interests:

I have served as a teaching assistant for core chemical engineering courses (Transport Phenomena, Mass and Energy Balance) at Michigan State University and Georgia Tech. I have also mentored 10 undergraduate students so far. Since my background is in chemical engineering, I can teach any of the core courses like Thermodynamics, Transport Phenomena, Chemical Reaction Engineering, Mass Transfer etc. Moreover, I am interested in developing a new graduate level course on membrane-based separations. Importantly, I will commit to maintaining diversity both in my research group as well as in the department.

582011 Rational Design of Nanofiltration Membranes for the Removal of Trace Organic Contaminants from Waste Water Based on Single-Molecule Dynamics and Heterogeneity

Daniel F. Kienle, Postdoctoral Researcher, University of Colorado at Boulder

Research Interests:

Research Experience

A common theme in my current and planned research is the rational design of emerging technologies based on a mechanistic understanding of their deficiencies. This approach has been specifically applied to the development of immobilized enzyme sensor for the detection of explosive chemicals. Enzyme sensor technology has incredible potential, but limited success due to the loss of enzyme function upon immobilization. Significant advances in understanding of the mechanism behind the reduced activity of immobilized enzymes has been possible through several complementary single-molecule fluorescence microscopy techniques in combination with sophisticated trajectory analysis using stochastic (non)linear state-space models. Using these methods, we have been able to quantitatively identify heterogeneity in functionality of immobilized enzymes and distinguish several important modes of failure, including inaccessible active sites, structural instability, inhibited conformational mobility, and substrate/product transport barriers, which are typically convoluted by less sophisticated measurements and analyses. Using this nuanced understanding of immobilized enzyme function, we developed several immobilization materials and methods showing enhanced performance relative to enzymes in solution.

Research Plan

As a faculty member, I will employ the same strategy and suite of techniques to address several challenges related to the removal of trace organic contaminants (TOCs), such as active pharmaceutical ingredients, organic dyes, and byproduct of chemical manufacturing, from municipal and industrial waste water via nonporous or nanoporous nanofiltration (NF) membranes. Conventional aqueous NF technology generally has low TOC rejection efficiency < 50%. Moreover, NF membranes in general are constrained by the trade-off between permeability and selectivity, which complicates optimization. To overcome this constraint, modifications to membrane materials are often made by generating pores in the membrane, incorporating porous materials, or functionalizing for selective chemical affinity; however the resulting membranes often underperform owing to the lack of a detailed description of the transport of solutes into and within the membranes, which is undoubtedly mirrors the complexity and heterogeneity of the membrane material. Current methods for assessing the membrane properties are reliant on ensemble mass transport, and microscopy based structural characterization, yielding limited mechanistic insight and ambiguity in the steps for improvement.

The single molecule tracking and non-Markovian machine learning trajectory analysis developed during my postdoc enabled the detection of 3D single-molecule diffusion in dense polymer materials with unprecedented resolution. By tracking the penetration and diffusion of many individual organic solute molecules (typically $\sim 10^5$ molecular trajectories) in NF membranes, my research will capture detailed information as to the properties that can be used to control solute adsorption, penetration, internal transport, and rejection. This will provide the feedback necessary for rationally designed modifications that will yield step-changes in nanofiltration membrane performance. The studied membrane materials and modifications will be accomplished through a combination of in-house synthesis, industrial partnerships, and academic research collaborations. If successful, these tactics will also be applied to multi-solute systems, and organic solvent nanofiltration for the purpose of optimizing solute fractionation, which is an important step in purifying chemical and pharmaceutical products.

Another critical challenge associated with NF membrane technology for water treatment is fouling by biomolecules, which reduces the effective life of the membrane. This challenge is typically addressed using surface modification with non-fouling chemicals such as hydrophilic polymer brushes. While these can significantly reduce the rate of fouling, success often occurs at the cost of permeability or selectivity. I plan to apply the methods I've developed to detect heterogeneity in single molecule conformational dynamics, adsorption, and surface diffusion of FRET labeled biomolecules on the surface of NF membrane materials. A detailed understanding of the fouling process at the molecular level, and the potential role of surface heterogeneity, as well as the changes to the small molecule transport is necessary to determine how fouling and performance can be mutually controlled. I plan to use this information to develop materials, and membrane modification strategies that will balance the reduced rate of fouling with retained membrane permeability and selectivity.

Teaching Interests:

While I am open to teaching any chemical engineering class, I have several preferences consistent with my strengths and experience. Specifically, I think that I could be an effective teacher for a computational and numerical methods course, as I have been a TA for a course in this subject area. I would also be interested in teaching transport phenomena or thermodynamics at the undergraduate or graduate level. Finally, I would like to offer an elective focused on interfacial phenomena, which is relate to my general field of research.

Publications

1. Weltz, J. S.; **Kienle, D. F.**; Schwartz, D. K.; Kaar, J. L. Dramatic Increase in Catalytic Performance of Immobilized Lipases by Their Stabilization on Polymer Brush Supports. **ACS Catal. 2019**, 4992–5001.
2. Bull, D. S.; **Kienle, D. F.**; Chaparro Sosa, A. F.; Nelson, N.; Shambojit, R.; Cha, J. N.; Schwartz, D. K.; Kaar, J. L.; Goodwin, A. P. Surface-Templated Nanobubbles Protect Proteins from Surface-Mediated Denaturation, **Phys. Chem. Lett., 2019**, 10 (11), 2641-2647
3. Morrin, G. T.; **Kienle, D. F.**; Schwartz, D. K., Standalone Interferometry-Based Calibration of Convex Lens-Induced Confinement Microscopy with Nanoscale Accuracy. ***Analyst*** **2019**, In-Press.
4. **Kienle, D. F.**; Schwartz, D. K. Complex Salt Dependence of Polymer Diffusion in Polyelectrolyte Multilayers. ***Phys. Chem. Lett.*****,2019**, *10* (5), pp 987–992
5. **Kienle, D. F.**; Falatach, R. M.; Kaar, J. L.; Schwartz, D. K., Correlating Structural and Functional Heterogeneity of Immobilized Enzymes. ***ACS Nano*** **2018**, *12*, 8091-8103.
6. Chaparro Sosa, A. F.; **Kienle, D. F.**; Falatach, R. M.; Flanagan, J.; Kaar, J. L.; Schwartz, D. K., Stabilization of Immobilized Enzymes via the Chaperone-Like Activity of Mixed Lipid Bilayers. ***ACS Appl. Mater. Interfaces*** **2018,** *10*, 19504-19513.
7. Marruecos, D. F.; **Kienle, D. F.**; Kaar, J. L.; Schwartz, D. K., Grafting Density Impacts Local Nanoscale Hydrophobicity in Poly(Ethylene Glycol) Brushes. ***ACS Macro Letters*** **2018,** *7*, 498-503.
8. ***Kienle, D. F.**; de Souza, J. V.; Kuhl, T. L., Large-Area Release and Transfer of Ultrathin, Freestanding Nanocrystalline Ceria Films. ***Thin Solid Films*** **2017,** *638*, 318-323.
9. ***Kienle, D. F.**; Kuhl, T. L., Analyzing Refractive Index Profiles of Confined Fluids by Interferometry Part II: Multilayer and Asymmetric Systems. ***Chim. Acta*** **2016,** *936*, 236-244.
10. **Kienle, D. F.**; Kuhl, T. L., Density and Phase State of a Confined Nonpolar Fluid. ***Rev. Lett.*** **2016,** *117*.
11. **Kienle, D. F.**; Kuhl, T. L., Analyzing Refractive Index Profiles of Confined Fluids by Interferometry. ***Chem.*** **2014,** *86*, 11860-11867.
12. **Kienle, D. F.**; de Souza, J. V.; Watkins, E. B.; Kuhl, T. L., Thickness and Refractive Index of Dppc and Dppe Monolayers by Multiple-Beam Interferometry. ***Bioanal. Chem.*** **2014,** *406*, 4725-4733.
13. Hardisty, M. R.; **Kienle, D. F.**; Kuhl, T. L.; Stover, S. M.; Fyhrie, D. P., Strain-Induced Optical Changes in Demineralized Bone. ***Biomed. Opt.*** **2014,** *19*.

*Corresponding Authorships

582119 Membrane Materials and Transport Studies for Sustainable Water, Energy and Life Sciences: *From Fundamentals to Applications*

Research Interests:

Overcoming the emerging global scarcity of water, energy, and other natural resources (e.g., nutrients, metals, minerals, etc.) are grand challenges faced by humanity today. These critical resources are inextricably linked, and therefore, must be considered together as new materials and technologies are developed. Membrane-based and adsorption-based separation processes promise to provide efficient and economical separation solutions which are then applicable at large scale to sustainably address the water-energy-food-ecosystems nexus. Recent advances in controlled fabrication and characterization of materials across length scales allow for creative designs of functional nanocomposite membranes with predetermined chemistry/structure and unprecedented performance. Furthermore, they provide effective tools to unravel transport mechanisms of various molecules across interfaces and/or heterophasic structures. These understandings, coupled with advanced theory and simulations, establish foundational guidelines towards exploration, rationale designs and control fabrication of novel yet resilient membrane materials that could effectively be deployed at scale.

At the interface of chemical engineering, materials science, chemistry, and physics, my research interests lie in tackling the water–energy-environment nexus by seeking robust and cost-effective solutions with the integration of materials structure, surface chemistry, module geometry, process controls and transport modeling for water remediation, chemical separations and energy storage. At its core, I am deeply interested in building a research program based upon fundamental understandings of how molecules behave near dynamical interfaces in multicomponent fluidic systems across time- and length- scales. I will leverage the advanced characterization tools to study the correlations between nanostructures and chemistry with the transport properties of chemical species across materials interfaces which, in turn, offer opportunities for better predictive models and system designs with precise control to be proposed. Gathered knowledge will then be translated into deterministically innovating and implementing scalable and selective materials and sustainable methods for effective separations, toward a future of sufficient availability of resources, reduced pollution and maintained ecological balance.

To do this, I would like to unite the gathered knowledge from my **master** degree (on porous carbon nanofibers as supercapacitor electrode materials), my **Ph.D.** (at the **University of Connecticut**, <u>on polymer-based membrane materials and transport studies for forward osmosis</u>, advised by professor Jeffrey McCutcheon) and my **postdocs** (at **Lawrence Livermore National Lab**, <u>on vertically-aligned carbon nanotube membranes</u> and **Lawrence Berkeley National Lab**, <u>on porous nanomaterials and organic-inorganic composite membranes)</u> as a toolkit in further exploring my directions. Based on the proposals that I have assembled, I foresee myself collaborating with other experimentalists, the simulation teams and industry.

Publications and Citations: https://scholar.google.com/citations?user=-9ArTS8AAAAJ&hl=en&oi=sra

Email address: nbui@lbl.gov and/or ngoc.23tuut@gmail.com

Teaching Interests:

To me, teaching and learning are deeply intertwined and have been a vital part of my life. It started quite early, since elementary school, with my love for Maths and thirst for knowledge. In college, I spent 30% of my college time on tutoring high-school students on Maths-Chemistry-Physics and volunteering to teach village children during summer breaks. In the first two years of my PhD at the University of Connecticut, I was a teaching assistant for a few professors, spending most of the time grading students' homework and exams. Observing those professors' teaching methods and how students responded both in classes and through exams helped shape my perspectives on teaching. In the last two years of my PhD and at

Lawrence Berkeley National Lab, I was given opportunities to give a few guest lectures relating my research to both graduate and undergraduate students. I have found that teaching is an effective way to reinforce established understandings while broadening my knowledge through learning from colleagues and the critical inquiries of students. As a faculty member, I am interested in teaching core chemical engineering courses such as Fluid Mechanics, Transport Phenomena, Kinetics, Thermodynamics, Fundamentals of Engineering, and other specialized courses related to fundamentals of polymers, materials and chemistry. Besides courses that align with departmental needs, I would be delighted to develop elective courses in areas related to my research (e.g. transport phenomena, membrane separations, emerging technologies for sustainable water and energy, and so on).

Looking forward to my role as a professor, I'd like to apply what has been a vital part of my growing-up, what I have observed, learnt and practiced to help shape the future of the next generations. I strive to be a role model starting with the first crucial step: being fully aware of, honoring and cultivating the diversity surrounding us. Promoting equity, inclusion and diversity is one of the core principles that I strive to continuously foster throughout my professional career. I will reinforce existing diversity and outreach programs at my school with securing funds for mentoring underrepresented students, organizing inspirational workshops, connecting with teachers/faculties at middle-/high- schools, community colleges, and historically underrepresented schools to give talks to underrepresented students in local communities, joining K-12 STEM program, and organizing Open Lab Day frequently.

557879 Biomimetic and Polymeric Membranes for Enhanced Selectivity in Desalination and Ionic Separations

Biomimetic and Polymeric Membranes for Enhanced Selectivity in Desalination and Ionic Separations

Jay R. Werber

Postdoc: University of Minnesota, Ph.D: Yale University

Google Scholar Profile jwerber@umn.edu

Research Interests: separations, membrane separations, polymer materials, biomimetic membranes, water treatment, hydrometallurgy and extractive separations, metal recycling

Teaching Interests: transport phenomena, separations, bioprocessing, biochemical engineering, polymer science

Short bio:

Jay is an environmentally-focused chemical engineer, dedicated to addressing global issues of water and resource scarcity. In particular, Jay's research work is on the design and characterization of membranes for highly selective molecular separations. Following his B.S. degree (Ch.E., Washington U., 2009), Jay worked for four years in process development at the biopharmaceutical company Genentech, engaged in protein separations using membranes and chromatography. Jay then pursued a Ph.D. at Yale in Chemical and Environmental Engineering under Prof. Menachem Elimelech. Jay's dissertation was centered on understanding the performance of biomimetic desalination membranes (depicted below), that is, membranes designed to mimic the structure and selectivity of biological cell membranes. Jay also worked on the characterization of polymeric membranes, modeling of industrial-scale membrane processes, and interactions between nanomaterials and lipid bilayers. In order to gain skills in polymer synthesis and processing, Jay is a postdoctoral associate in the polymer chemistry lab of Prof. Marc Hillmyer at the University of Minnesota. Jay seeks a faculty position in Chemical Engineering at a major research university, where he plans to continue research on membrane separations and transport in polymers.

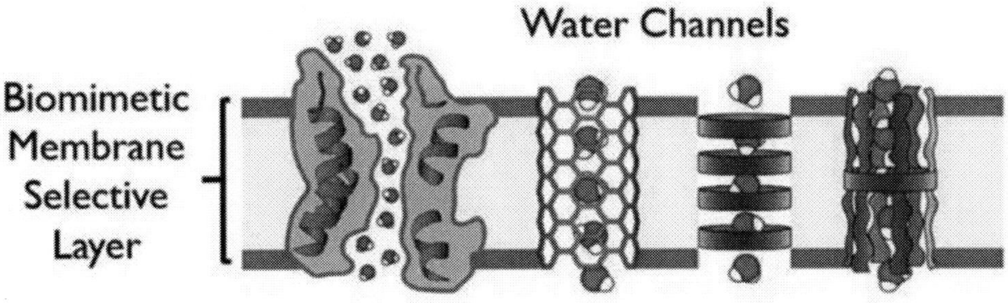

Career highlights:

- Mentor to three Ph.D. students, two undergraduates, and one high school student
- NSF Graduate Research Fellowship

- Abel Wolman Fellowship from the American Water Works Association
- Best Graduate Student Paper from the ACS Division of Environmental Chemistry
- Primary author of two successful NSF proposals (CBET 1701658, CBET 1437630) with total funding of $670,000
- Co-valedictorian of Washington U. School of Engineering

Selected publications (H-index: 11, Total citations: 1245, as of 4/2019):

1. Werber, J.R.; Porter, C.J.; Elimelech, M. A Path to Ultra-Selectivity: Support Layer Properties to Maximize Performance of Biomimetic Desalination Membranes. *Environ. Sci. Technol.* **2018**, 52, 10737-10747. DOI: 10.1021/acs.est.8b03426
2. Werber, J.R; Elimelech, M. Permeability and Selectivity Limits of Biomimetic Desalination Membranes. *Science Advances* **2018,** 4, eaar8266. DOI: 10.1126/sciadv.aar8266
3. Lu, X.; Feng, X.; Werber, J.R.; Chu, C.; Zucker, I.; Kim, J.H.; Osuji, C.O.; Elimelech, M. Enhanced Antimicrobial Activity through the Controlled Alignment of Graphene Nanosheets. *Proc. Natl. Acad. Sci. U.S.A.* **2017**, 201710996. DOI: 10.1073/pnas.1710996114
4. Werber, J.R.; Bull, S.K.; Elimelech, M. Acyl-Chloride Quenching Following Interfacial Polymerization to Modulate Permeability and Surface Charge of Desalination Membranes. *Journal of Membrane Science* **2017**, 535, 357-364. DOI: 10.1016/j.memsci.2017.04.041
5. Chen, D.*; Werber, J.R.*; Zhao, X.; Elimelech, M. A Facile Method to Quantify the Carboxyl Group Areal Density in the Active Layer of Polyamide Thin-Film Composite Membranes. *Journal of Membrane Science* **2017**, 534, 100-108. DOI: 10.1016/j.memsci.2017.04.001
6. Werber, J.R.*; Deshmukh, A.*; Elimelech, M. Can batch or semi-batch processes save energy in reverse-osmosis desalination? *Desalination* **2017**, 402, 109-122. DOI: 10.1016/j.desal.2016.09.028
7. Werber, J.R.; Deshmukh, A.; Elimelech, M. The Critical Need for Increased Selectivity, Not Increased Water Permeability, for Desalination Membranes. *Environ. Sci. Technol. Lett.* **2016**, 3, 112-120. DOI: 10.1021/acs.estlett.6b00050
8. Werber, J.R.; Osuji, C.O.; Elimelech, M. Materials for next-generation desalination and water purification membranes. *Nature Reviews Materials* **2016**, 1, 16018. DOI: 10.1038/natrevmats.2016.18

* equal contribution

579906 Biosorption of Methylene Blue Using Citric Acid Modified Arjun Bark Powder

Research Interests: Catalysis, Biofuel, Environment and Waste water Treatment

Teaching Interests: Reaction Engineering, Industrial Pollution Control Engg, Heat Transfer, Chemical Technology

ABSTRACT: In this article the adsorptive removal of methylene blue (MB) from its aqueous solution using citric acid modified arjun bark powder as an adsorbent is reported. Initially, the adsorbent was characterized by various analytical techniques (SEM, BET, FTIR and pH_{pzc}). After characterizing the adsorbent, it was used for carrying out batch investigation on the biosorption of MB. The batch experiments revealed a very high percentage removal of MB of about 100%. The batch experiment results show Temkin isotherm. The maximum monolayer adsorption capacity observed was 512.82 mg/g at natural pH of 7.5. On the other hand, results followed the pseudo-second order kinetic model better. The energy of activation in the present study was 10 KJ/mol. The adsorptive mass transfer coefficient obtained was 1.65×10^{-7} m/s. The value of diffusivity obtained was 2.27×10^{-11}. The experimental data were used to obtain the batch reactor design.

580203 Next-Generation Molecularly Selective Materials and Processes for Scalable Energy Efficient Separations

Research Interests:

My research interests are spanning broadly from chemistry, materials to chemical engineering. While primarily interested in designing materials and processes for advanced molecular separations (e.g. gas separation and water treatment), I have a particular focus on ultra-permeable and selective membranes for scalable energy-efficient gas separations. The overarching theme of my research is to develop next-generation membrane technology to mitigate the extensive energy consumption associated with conventional thermal-driven processes (e.g. cryogenic distillation). I employ a unique and synergistic approach via establishing fundamental structures by material design and manipulating material properties by process engineering. The science solutions alongside with engineering principles will create fundamentally energy efficient and cost effective membrane process outcompeting existing technologies, affording a positive impact onpresent gas separation industries.

Separation of gas mixtures at the industrial scale in US accounts for 10-15% of national energy consumption. In this context membranes can effectively reduce the enormous amount of energy required for gas separations, compared to phase-change separation processes (e.g. cryogenic distillation). Despite the promise, state-of-the-art membranes suffer from numerous challenges, predominantly relying on underdeveloped separation performance of existing materials and difficulties of scaling-up of new materials. Those issues severely hinder the growth of membrane separation technology and must be properly addressed to increase the completeness of membranes over other separation approaches. My research strives to tackle the critical hurdles facing in the community of membrane separations, aiming to minimize the energy penalty for large-scale gas separations.

Postdoc/Project Scientist Project: Metal-organic framework/polyimide mixed matrix membranes for energy efficient gas separations

Advisor: Dr. Jeffrey J. Urban, Lawrence Berkeley National Laboratory, Berkeley, CA, USA

Ph.D. Dissertation: Highly productive ester crosslinkable composite hollow fiber membranes for aggressive natural gas separations

Advisor: Prof. William J. Koros, Georgia Institute of Technology, Atlanta, GA, USA

Research Experience:

I have comprehensive research experience during my education in three distinctively different majors, working in R&D for a world-class company for nearly 5 years and current employment in a world-leading national lab for multiple projects. Take the membrane research for example: during my PhD program in Prof. William Koros group, I was exposed in a wide spectrum of advanced separation technologies, including membranes, sorbents and barrier materials. My thesis project was focusing on the industrially most preferred format of membranes – hollow fiber membranes with the highest surface to volume ratios than other competing configurations (e.g. spiral wound). I addressed two key challenges associated with hollow fiber membrane technology – low performance and high cost of materials. My primary research involves polymer synthesis and process engineering to precisely control the membrane properties to substantially enhance separation performance and reduce the membrane cost. Upon graduation, I joined Air Liquide as a membrane research scientist, working on the most advanced membrane separation technologies. As one of the few candidates, if not the only, I have unique perspectives and first-hand experience in gas separation industries. In my current position at Berkeley National Lab, I am working on MOF/polymer mixed matrix membrane technology using the world-class facilities at Molecular Foundry. I have developed a generalizable approach of making ultra-permeable membranes with unprecedented performance (far beyond Robeson upper bounds for at least three different gas pairs) and enhanced compatibility with state-of-the-art membrane manufacturing processes. Overall, I have been working on the cutting-edges of membrane science and technology over a decade with a strong research portfolio of flat-sheet membranes, hollow fiber membranes and membrane bundles.

Teaching Interests:

I have engaged with different formats of teaching practices during Teaching Assistant appointments in both undergraduate and graduate studies. In addition, I have mentored students in prestigious programs, including Science Undergraduate Laboratory Internship (**SULI**) and Berkeley Lab Undergraduate Research (**BLUR**). With a solid background in chemistry and chemical engineering, I have interests in teaching fundamentals of Chemistry and Chemical Engineering. In particular, I am interested in teaching or developing courses related to my research, such as Mass Transfer, Membrane Separations etc.

Selected Publications (**9+ first authored**):

1. Ma and J. J. Urban, "Hydrogen-bonded polyimide/metal-organic framework hybrid membranes for ultra-fast separations of multiple gas pairs", ***Adv. Funct. Mater.***, 2019, 1903243
2. Ma and W. J. Koros, "Physical aging of ester-cross-linked hollow fiber membranes for natural gas separations and mitigation thereof", ***J. Membr. Sci.***, 551 (2018) 214-221
3. Ma and J. J. Urban, "Polymers of Intrinsic Microporosity (PIMs) Gas Separation Membranes: a mini Review", ***Proc. Nat. Res. Soc.***, 2 (2018) 1-19
4. Ma et al., "Thin-skinned intrinsically defect-free asymmetric mono-esterified hollow fiber precursors for crosslinkable polyimide gas separation membranes', ***J. Membr. Sci.***, 493 (2015) 252-262
5. Ma and W. J. Koros, "Effects of hydrocarbon and water impurities on CO_2/CH_4separation performance of ester-crosslinked hollow fiber membranes",***J. Membr. Sci.***, 451 (2014) 1-9
6. Ma and W. J. Koros, "Ester-crosslinkable composite hollow fiber membranes for CO_2removal from natural gas", ***Ind. Eng. Chem. Res.***2013, *52*(31) 10495–10505 (Invited paper in honor of Prof. Enrico Drioli)
7. Ma and W. J. Koros,"High-performance ester-crosslinked hollow fiber membranes for natural gas separations", ***J. Membr. Sci.***, 428 (2013) 251–259

Selected Patents (9 total,**6 granted**):

1. Composite carbon molecular sieve membranes having anti-substructure collapse particles loaded in a core thereof, date of patent2019/1/21,patent numberUS10183258B2
2. Metallopolyimide precursor fibers for aging-resistant carbon molecular sieve hollow fiber membranes with enhanced selectivity, date of patent2018/12/4, patent number US10143973B2
3. Metallopolyimide precursor fibers for aging-resistant carbon molecular sieve hollow fiber membranes with enhanced selectivity, date of patent2018/10/30, patent numberUS10112149B2
4. Metallopolyimide precursor fibers for aging-resistant carbon molecular sieve hollow fiber membranes with enhanced selectivity, date of patent2018/10/2,patent numberUS10086337B2
5. Composite hollow fiber membranes useful for CO_2removal from natural gas, date of patent 2017/8/1, patent numberUS9718031B2
6. A direct bonding method for shortwave light auxiliary silicon slice,CN patent publication date 2009/12/23, publication numberCN100573821C

580791 Enhancement of CO_2 Absorption from Gas Streams Using Nanofluids in Hollow Fiber Gas–Liquid Membrane Contactor

Research Interests: Membrane technology, Zeolites, CO2 removal, Desalination

Teaching Interests: Chemical engineering courses

Abstract

In this work chemical absorption of CO_2 from gas streams was carried out in a polypropylene porous asymmetric hollow fibre membrane contactor (HFMC). Aqueous solutions containing 0.5, 1.0, 1.5 and 2 M monoethanolamine (MEA) were used as chemical absorbent solution.

Several stable suspensions of Fe3O4, CNT and Al2O3 nanoparticles in MEA were prepared and tested as new absorbents to investigate the effect of nanoparticles on the rate of mass transfer during CO2 absorption.

Gas mixtures were passed through the shell-side and the nanofluid flowed co-currently through the lumen side of the fibers.

The effects of different operating conditions including gas flow rate (500, 600, 700 and 800 mlit/min), liquid flow rate (50, 100, 150 and 200 mlit/min) and nano particle content (0, 0.05, 0.1 and 0.15 wt. %) on the CO_2 outlet concentrations and the separation performance were investigated.

Results showed that among operating parameters, gas flow rate and nano-particle concentration had the more significant effects on the CO2 absorption.

The results revealed also that nanofluid stability and rheological characteristics of nanoparticles in the base fluid are the key factors in nanoparticle selection for CO2 absorption in membrane contactor.

Experimental results indicated that using nanofluid in membrane contactor is very efficient in the removal of CO_2 at high gas/liquid flow ratio. The removal of CO_2 was almost complete with recovery of higher than 96%.

580802 CO_2 Removal Using Highly Permeable, Ultra Thin and Well Oriented SAPO-34 Zeolite Membrane Synthesized By a Novel Four-Step Method

Research Interests: membrane technology, zeolites, CO2 removal, Scale-up

Teaching Interests: Chemical engineering courses

Abstract:

Despite the various effective methods previously proposed to improve the synthesis of zeolite membranes, reducing production costs along with enhancing separation performance are still essential prerequisites for developing zeolite membranes in industrial scale CO_2 removal applications.

In this study a five-step method was proposed to synthesize ultra-thin, high-permeable and well-oriented SAPO-34 membranes. Using this method, zeolite crystals were oriented well, membrane thickness was controlled easily and the membrane permeability was increased significantly without scarifying the selectivity.

Based on the results, our study aimed at highlighting the advantages of the step-by-step synthesis method for the production of highly-permeable and reproducible SAPO-34 membranes.

In this study, the crystallographic preferred orientation index (CPO) was adopted to SAPO-34 zeolite membranes for the first time. Furthermore, the effects of proposed methods on performance of synthesized membranes were innovatively evaluated via CPO values.

The synthesized SAPO-34 zeolite membranes were also tested in an equimolar CO_2/CH_4 separation process as well as a membrane adsorption test.

The surprisingly high CO_2 permeance of 3.2×10^{-6} mol m^{-2} s^{-1} pa along with the high separation and adsorption selectivities of 141 and 116, respectively, were obtained for a < 2 μm thickness SAPO-34 membrane.

577792 Acoustic Cavitation Induced Chemical Functionalization of Biochar: A Feasible Strategy for Effective Removal of Heavy Metals

Research Interests: Bio-processes

- Bioenergy and Biofuels (Biodiesel and Biochar)
- Activation of Carbonaceous Graphitic Structures
- Wastewater Treatment
- CO_2 Capture
- Sono-Physics (acoustic-based bio-processes)
- Computational Fluid Dynamic (CFD) simulation
- Techno Economic Analysis (TEA)

Teaching Interests: All Chemical Engineering Courses

Abstract

The Sustainable Energy and Environment Group (SEEG) of the University of Mississippi (UM) has discovered the synergisms of activating biochar (BC) by acoustics which include *disarrangement and exfoliation of BC's graphitic structure, mineral leaching, significant increase in BC's internal surface area and porosity*, as well as *creation of new and opening the blocked mesopores*. Acoustic treatment of biochar in water with dissolved CO_2 also created *induced surface functional groups* that enhance the adsorption capacity. Accordingly, the main objective of a series of our researches is to enhance and specify the adsorption capacity of biochar through integrated structural acoustic modification with chemical functionalization. These synergisms can result in the development of advanced activated carbon (AC) for adsorption of air and water pollutants, as traditional physical and chemical activations require very high temperatures in the activation process. However, the developed acoustic activation method is conducted *at ambient temperature and pressure* for a *very short duration* (~30 sec) which requires about 1,135 kcal / kg of activated BC produced.

In order to increase the surface functionality of raw biochar, the synergism of acoustic activation was enhanced by different chemical treatments and functionalization, which include both alkaline (using KOH) and acidic (H_3PO_4 or HNO_3) treatments followed by amination. The effect of ultrasound structural modification on biochar magnetization was also assessed since magnetization usually reduces the biochar adsorption capacity significantly. The activated biochars were used to remove nickel and lead ions, which are a common problem in old water systems. The experimental results demonstrated that the metal adsorption removal of the modified ultrasonic-chemically functionalized biochars is far higher than that of raw biochar. In most cases, the modified biochars had a faster adsorption rate as well. Moreover, the synergism created by the combined effect of ultrasound and chemical functionalization resulted in a higher metal retention capacity with no leaching of adsorbed metals during adsorption with long durations. It can be concluded that the integrated acoustic and chemical functionalization provides an efficacious and economical method which led to an adsorbent with high surface functionality and adsorption capacity as the activation is conducted at low temperatures that requires much lower energy than traditional activations.

581934 Synthesis of Environmentally Friendly and Sustainable Multi-Functional Surfaces and Interfaces

Research Interests:

The synthesis of environmentally friendly and sustainable multi-functional surfaces and materials is leading the research of innovative applications in energy, water and environmental remediation. The most important biological and physicochemical process occur at the surface or the interface between two or more materials. Hence, if a material's surface can be functionalized, it could perform different tasks simultaneously such as catalysis, selective separations, sorption and adhesion, and oxidation/reduction processes.

Functional surfaces and interface science are highly dynamic scientific areas and, despite tremendous advances in high-end products to specific functionalities in electrochemistry, energy efficiency, environmental technologies, new biomolecules or nanocomposites, the successful implementation of modified surfaces or surface modifiers beyond lab scale is still difficult. Little has been done from the perspective of a robust functional material, and the vast functions that can be engineered on its surface, or for the scalability of such products.

Thousands of established technologies involving surfaces have been developed, such as adhesives or surfactants, but understanding the interfacial phenomena is sometimes controversial, even for known applications. As a chemical and environmental engineer, I am distinctively positioned to address such aspects and to design and engineer functional surfaces and interfaces as alternative for varied processes and/or the design of suitable products or modifiers.

I have worked directly or indirectly in surface chemistry and interface science since my master's studies in civil and environmental engineering, where I worked in advanced oxidation processes, and aerobic and anaerobic treatments. Then, when I was adjunct professor in chemical engineering, advised in different projects related to immobilization of catalysts, such as titanium oxide onto activated carbon and polyester for the photocatalytic degradation of wastewater of the printer cartridge reuse, and other advanced oxidation applications.

Teaching Interests:

With more than 10 years of teaching experience at the Universidad de los Andes (Colombia) in an ABET accredited department and the Instituto Alberto Merani (Colombia), I am able to design different courses and lectures covering fundamental Chemical and Environmental Engineering subjects due to the gained knowledge during the PhD and postdoctoral research. These courses may have three learning spaces:

1. Classes: discussions of theory and illustrative examples developed.
2. Independent work: lecture preparation, performing tasks and workshops, development of laboratory reports and projects.
3. Interactive experiences and design: students undertake design exercises, simulations and/or laboratory practice in this space and discuss the outcomes, learn from the others and make mistakes, which is a key to the learning process.

Other additional aspects may be included:

- Problem analysis: definition of the problem, determination of variables and their behavior, possible solutions, troubleshooting, use of tools (hands-on and computational experience)
- The ability to conduct solutions in industrial contexts with actual variables and gradations

The chemical engineering courses that I have interest and I am familiarized with are:

Separation Processes

The separation processes of mixtures are essential in the chemical, petroleum, food, biochemistry and pharmaceutical industries, among others. In this course, principles of phase equilibrium thermodynamics and transport phenomena are applied in the characterization of various separation processes using graphical and analytical methods. On this basis, skills for selection and sizing of equipment that are useful in these processes are developed.

Industrial Chemical Processes

This course aims to understand the chemical processes used in industry, particularly in the petrochemical industry. The unit is aimed to professionals from various disciplines that require basic knowledge of industrial chemical processes. This course was designed for the workers of an oil/gas company located in Colombia (Ecopetrol S. A. VRP).

Introduction to Chemical Engineering

This course aims to identify, explain and apply the basic concepts of chemical engineering and those that are common to all engineering aspects. The course also focuses into differentiate chemical engineering from other branches of engineering, and develop basic communication skills, teamwork and evaluation. It is also intended that the students identify, interpret and fulfil the rights and obligations that they have as students of the Universidad de los Andes.

Physicochemical Aspects of Polymers

This course enables students to understand concepts, basic calculations, and their application in polymer science and engineering. Specifically, the students are expected to identify the principles and methods of synthesizing polymers, their structure, physical and chemical properties, manufacturing processes, additives and associated technologies.

Fundamentals of Industrial Processes (Material and Energy Balances)

The course Fundamentals of Industrial Processes aims to prepare the student in formulating and solving material and energy balances on chemical process systems. The knowledge acquired in the course Introduction to Chemical Engineering and Thermodynamics are reviewed here, and it establishes the basis for future subjects such as Reaction Transport Phenomena, Engineering and Separation Processes. The course introduces the engineering approach used to solve problems: establishing the relationship between the balance equations, unknown process variables and relationships available. The students must select the information to find the unknown variables and solve the process properly using manual and/or computational methods.

Transport Phenomena (Mass and Energy)

This course presents the associated phenomena with mass and energy transfer in systems that are in steady or transient state. This course introduces solutions to problems involving convective molecular transport of mass and heat and applications.

Experimental Design

This course relates to design of experiments in chemical engineering, identification of controllable and uncontrollable factors of a problem. The unit develops technical knowledge and skills that permit problem solutions that incorporate a strong fundamental base in factorial design, response surface, and statistical analysis.

Environmental Chemistry

This course is designed so that the student can develop the ability to apply the concepts of thermodynamics and equilibrium to environmental systems. The course provides the student with the basic tools that predict the behavior of chemicals in the environment and in turn describes specific cases in which chemical methods are used in environmental engineering.

Industrial Chemistry

This course aims to understand the chemical processes used in industry. The unit is aimed to students of Industrial Engineering that require basic knowledge of industrial chemical processes. The goal is to prepare the student in formulating and solving material and energy balances on chemical process systems.

582149 Microporous Molecular Sieves for Membrane and Adsorptive Separations

Research Interests:

Separations take roughly half the overall energy consumption for chemical industry in the United States. Replacing traditional separation technology with membrane separation could save up to 90% of energy. The successful implementation of membrane separation strongly relies on developing suitable membrane materials and advancing of membrane fabrication techniques. Microporous molecular sieves, such as zeolites, metal-organic frameworks (MOFs), carbon molecular sieves et al., are promising membrane materials for a lot of energy and environment related separations with great industrial importance. The industrial applications of microporous molecular sieving membranes, however, are still quite limited. Two major impeding issues are: the membrane fabrication scalability and the fabrication cost. Membrane fabrication cost is a relative concept here. Less membrane area is needed for membranes with higher performance, i.e. higher permeance and selectivity. Therefore, higher performance membranes can afford to have higher fabrication cost per square meter of membrane area. A lot of efforts have also been on developing high performance membranes. Building upon my strong background on zeolite membranes and extended knowledge and experience on MOFs, hollow fiber spinning and simulated moving bed chromatography, I plan to work on following three areas for my future research.

- Cost-effective and scalable membrane fabrication method development

Hollow fiber membranes have much higher packing density than traditional disk and tubular membranes. They could significantly reduce membrane module size and overall membrane module cost. Recently, I have developed a scalable SSZ-13 zeolite hollow fiber membrane fabrication method and successfully synthesized 20 high-quality (CO_2 permeance ~5000 GPU, CO_2/C_3H_8 selectivity over 10,000) membranes together. Further development and fundamental understanding of membrane growth will be conducted towards industrialization and applicability to other types of molecular sieves.

- Advanced porous ceramic hollow fibers

Porous ceramic hollow fibers could be used as support for molecular sieving membrane fabrication or directly used for separation. Higher surface area will provide more effective area for membrane permeation is strongly preferred for membrane support. Current ceramic hollow fibers have a trade-off between surface porosity and mechanical strength. Through newly design hollow fiber structure and sophisticated hollow fiber spinning, I am targeting at a good balance between fiber mechanical strength and surface porosity. The hollow fibers could also find applications for air purification and water treatment et al. Hollow spinning and porous hollow fibers might also interest some professors working on catalysis and biology.

- Adsorptive separation with molecular sieves

Zeolites and MOFs show excellent adsorption selectivity for a lot of interesting gas and liquid mixtures. Application of zeolite/MOFs molecular sieves for pressure/temperature swing adsorption and chromatographic separation will be investigated. Special emphasis will be on transport related aspects, such as crystal size engineering, crystal immobilization, system pressure drop reduction. This work could be a good cooperation opportunity with other faculties interested in catalysis because tailored zeolite and MOFs are great catalysts, too.

Postdoctoral Projects: (1) Reactive Separations – Membrane Reactors for Olefin Production, Sponsor: Dow Chemical; (2) ZIF-8 Hollow Fiber Membranes for Propylene Separation, Sponsor: Department of Energy, Advanced Research Projects Agency – Energy (ARPA-E); (3) Hydrocarbon Separations with Simulated Moving Bed (SMB) Chromatography, Sponsor: ExxonMobil Corporation.

PhD Thesis: Ultramicroporous zeolite membranes for energy and environment related small molecule gas separation

Teaching Interests:

My teaching philosophy mainly concentrates on how to help students understand the concepts and principles with daily life examples and industrial applications. Building the connection will make the courses more attractive and engaging the students better. A filtration process as an example, we all know that water flow out slower and slower as debris clogs the sewage. Filtration process is very similar to that, how chemical engineers treat this problem and scientifically determine what to do and when? I hope my teaching could trigger students thinking more inside and outside class. More animation and videos will be used to help visualize things so that the contents are better delivered. I have solid background on chemical engineering courses and could teach Chemical Process Principles, Transport Phenomena, Kinetics and Reactor Design, Thermodynamics, Process, Product Design and Economics or their variation forms. Considering my research interests, I propose to develop an elective course for membrane separation. The course will cover various applications and many types of membrane materials, i.e. ceramic, polymer, zeolite, metal organic framework (MOF), carbon et al. I might also have small group of students to involve in projects in my labs and let them to get more in-depth learning in certain areas under lab research environment.

Teaching Assistant Experience: Equilibrium Processes, 2011; Heat Transfer, Basics, 2012; Process Dynamic Control, 2013; Materials & Energy Balance, 2014.

Mentoring experience: 3 PhD students in University of Cincinnati and 3 PhD students in Georgia Institute of Technology for fundamental knowledge, instruments, experiments and analysis.

Selected Publications:

1. **S. Yang**, Y. Chiang, S. Nair, Scalable One-Step Gel Conversion Route to High-Performance CHA Zeolite Hollow Fiber Membranes and Modules for CO2 and H2 Separations, *Energy Technol.* DOI: 10.1002/ente.201900494.
2. **S. Yang**, Y.H. Kwon, D.Y. Koh, B. Min, Y. Liu, S. Nair, Highly Selective SSZ-13 Zeolite Hollow Fiber Membranes by Ultraviolet Activation at Near-Ambient Temperature, *ChemNanoMat*, 2019, 5, 61-67.
3. **S. Yang**, A. Arvanitis, Z. Cao, X. Sun, J. Dong, Synthesis of Silicalite Membrane with an Aluminum-Containing Surface for Controlled Modification of Zeolitic Pore Entries for Enhanced Gas Separation, *Processes,* 2018, 6, 13.
4. **S Yang**, Z Cao, A Arvanitis, X Sun, Z Xu, J Dong. DDR-type zeolite membrane synthesis, modification and gas permeation studies, *J. Membr. Sci.* 2016, 505, 194-204.

5. **S. Yang**, J. Provenzano, Arvanitis, W. Jing, J. Dong, Effect of Precursor PH on DDR Zeolite Crystal Morphology in Secondary Growth Synthesis Using Sigma-1 Seeds and Reduced Organic Agents, *J. Porous Mater.* 2014, 21, 1001–1007.

582134 Biomimetic Microelectronic Systems for Deciphering Neurotransmission and Hormone Release

Research Interests:

Biomimetic microelectronic technologies have transformed biomedical research through the development of innovative devices that help to understand complex biological mechanisms, restore a lost biological function, and to treat diseases. All successful technologies that are employed in human medical applications have been designed or influenced by a clear understanding of underlying biological principles. My research to date has focused on creating biomimetic and biohybrid technologies for examining the neurotransmitter release, biosensing, drug delivery, and material synthesis. I am interested in combining my engineering, physical sciences and biology expertise toward examining the mechanism of vesicular exocytosis and its alterations in neurological disorders and diabetes. I aim to understand the molecular determinants that encode the differential rate of exocytosis of synaptic and large dense core vesicles in neuronal and endothelial cells. Specifically, I will focus on dissecting the role of fusion regulatory protein, lipid composition and co-factors in regulating the exocytosis process under both normal and diseased conditions. This will bring important insight into the vesicle fusion process and will allow us to build a coherent mechanistic model. To do so, I propose a reductionist approach comprising of biomimetic lipid membranes on a chip, neuroprosthesis and synthetic and isolated vesicles to recapitulate physiological behavior under reconstituted conditions. These studies will ultimately enable the design of optimized and targeted drugs for several neurological disorders and diabetes.

During my Ph.D., I have developed several technologies, including long wavelength hybrid silicon photodetectors, porous silicon ELISA for point of care diagnostics, hybrid ultra capacitor for energy storage in collaboration with Storedot, Israel, and controlled nanocrystals synthesis. During my research career, I have developed a number of software packages including Raman's microscopy data analysis suite, super-resolution microscopy data analysis, and single vesicle data analysis. In my postdoctoral research career, I worked on the development of methods to form lipid membranes on a chip to study the kinetics of vesicle fusion, in particular, the exquisite speed and control of neurotransmitter release from synaptic vesicles. This pore-spanning lipid bilayer system faithfully recapitulates the cellular architecture under reconstituted conditions and allows for the precision study of the single vesicle fusion kinetics. This technology has established close collaboration with researchers from Yale and UCL, London for projects, including drug screening against membrane proteins, examining membrane transport and cell signaling in conjugation with advanced microscopy techniques. Through rigorous training, I have gained valuable experience in grant writing, mentoring, and teaching and I strongly believe these experiences will allow me to excel as a principal investigator, overseeing my own laboratory and personnel.

Grant Writing Experience

NIH Pathway to Independence Award (Impact Score = 30), EGIDE Travel Grant (2012), EGIDE Travel Grant (2014).

Postdoctoral Project: "Planar asymmetric suspended lipid membranes and single molecule studies for elucidating neurotransmission"

Under the supervision of Prof. James Rothman, Nanobiology Institute, Yale University

PhD Dissertation: "Bio-Inspired Materials and Bio-Hybrid Microsystems"

Under the supervision of Prof. Csilla Gergely, CNRS, Montpellier, France.

Selected Publications:

1. **Ramakrishnan SK**, Bera M, Coleman J, Krishnakumar SS, Pincet F, and Rothman JE. "Synaptotagmin oligomers are necessary and can be sufficient to form a Ca2+-sensitive fusion

clamp." *FEBS Letters*, 2019, 593: 154-162.

2. Heo P, **Ramakrishnan SK**, Coleman J, Rothman J, Fleury JB, Pincet F. "Highly Reproducible Physiological Asymmetric Membrane with Freely Diffusing Embedded Proteins in a 3D-Printed Microfluidic Setup." *Small*, 2019, 1900725.
3. **Ramakrishnan SK**,[#] Gohlke A,[#] Li F, Coleman J, Xu W, Rothman J, Pincet F. "High throughput single vesicle fusion using free-standing membranes and automated analysis." *Langmuir*, 2018, 34 (20).
4. Coleman J, Jouannot O, [#]**Ramakrishnan SK**, [#] Zanetti N, Wang J, Salpietro V, Houlden H, Rothman J, Krishnakumar S. "PRRT2 Regulates Synaptic Fusion by Directly Moderating SNARE Complex Assembly." *Cell Reports*, 2018, 22 (3), p820-831.
5. **Ramakrishnan SK**, Martin M, Cloitre T, Agarwal V, Cusinier F, Gergely C. "Porous silicon microcavities redefine colorimetric ELISA sensitivity for ultrasensitive detection of autoimmune antibodies." *Sensors and Actuators B: Chemical*, 2018, 271.
6. **Ramakrishnan SK**[*], Zhu J, Gergely C. "Organic-Inorganic Interface simulations for new material discoveries." *Wiley Interdisciplinary Reviews: Computational Molecular Science*, 7(1), 2017.
7. Papa Z, [#]**Ramakrishnan SK,** [#] Martin M, Marquez J, Zimanyi L, Toth Z, Gergely C. "Interactions at the silicon-peptide surfaces – Evidence of peptide multilayer assembly." *Langmuir*, 2016, 32(28).
8. **Ramakrishnan SK**, Jebors S, Martin M, Cloitre T, Agarwal V, Mehdi A, Martinez J, Subra G, Gergely C. "Engineered adhesion peptides for improved silicon adsorption." *Langmuir*, 2015, 31(43).
9. **Ramakrishnan SK**, Martin M, Cloitre T, Firlej L, Gergely C. "Design rules for metal binding biomolecules: Understanding of amino acid adsorption on platinum crystallographic facets by density functional theoretical calculations." *Physical Chemistry Chemical Physics*, 2015 14;17(6):4193-8.
10. **Ramakrishnan SK**, Martin M, Cloitre T, Firlej L, Gergely C. "Molecular Mechanism of Selective Binding of Peptides to Silicon Surface." *Journal of Chemical Information and Modeling,*2014, 54(7):2117-26.
11. **Ramakrishnan SK**, Martin M, Cloitre T, Firlej L, Cuisinier FJ, Gergely C. "Insights on the facet specific adsorption of amino acids and peptides towards platinum." *Journal of Chemical Information and Modeling,*2013 53 (12), 3273-3279.

Teaching Interests:

Teaching Experience:

During my PhD, I have taught digital image and signal processing course at the graduate level. I helped many students to learn MATLAB programming for image processing. Although, my experience in teaching is limited, I enjoyed the teaching opportunities that I got, and hopefully my students are good at MATLAB and apply the knowledge for diverse scientific problems. With strong training and an academic excellence award in engineering, I am comfortable to teach all core courses in the biological and chemical engineering discipline. I am interested in developing new courses related to molecular biophysics, microfluidics techniques for biological applications, materials for biomedical applications, Neurotechnology and Neurophysiology.

Mentoring Experience:

I possess extensive experience in teaching and mentoring students to perform outstanding research. I have advised the three undergraduate students during my graduate school. During my postdoctoral career, I trained one undergraduate and a graduate student for making suspended lipid membranes. I am currently mentoring a high-school student in computer programming for efficient data analysis. Along with my research, I supervised and advised PhD students and postdocs in my lab to carry out single molecule studies and trained them in microscopy techniques. I am eager to continue an interdisciplinary training programs to students as a junior faculty member.

Teaching Philosophy:

My teaching philosophy is that keep students interested and involved during the class. It is essential to understand the material and correlates with real-world examples. By presenting concepts in the context of practical challenges, they should be able to apply the knowledge for practical applications. In my teaching experience, I have felt that demonstrating the text with examples will help the students remember even when they are out of college. Another essential tool is keeping the classroom interactive by encouraging them to ask questions during the lectures. I tend to avoid lengthy speech by interjecting short, interactive exercises and examples into the lesson. For example, 5-10-minute short conceptual explanation followed with problem-solving exercises. This facilitates the students actively engages in the class and gets instant feedback if they do not understand.

579926 FSP-Ddf Coupling Model of Lbm for the Fluid Flow and Heat Transfer in Porous Media

Research Interests: *Hydromechanics, Heat transfer*

Teaching Interests: *Computational Fluid Mechanics, Numerical Heat transfer*

Finite size particle method (FSP) and double distribution function (DDF) coupling model of lattice Boltzmann method is proposed for the fluid flow and heat transfer in porous media. In this model, one velocity distribution function is used to simulate the flow field and one temperature distribution function is employed to simulate temperature fields for both fluid and solid, where three parameters for relaxation time are used. The calculation cost is greatly reduced compared with the two temperature distribution functions. The FSP method is very convenient to construct porous media with spherical particles or other shaped particles, and it is very accurate to simulate fluid flow and heat transfer in porous media using FSP-DDF coupling model. To validate FSP-DDF coupling model, we predicted the drag coefficient of a static spherical particle in flow field to test the boundary condition of fluid and spherical particle, and the heat diffusion process around an isothermal hot sphere submerged in a cold fluid is obtained to confirm the thermal boundary condition between particle and fluid two phases. The above simulation results are in good agreement with the experimental data and simulation data in the literature. Finally, the processes of fluid flow and heat transfer of kerosene in porous media are investigated using the FSP-DDF coupling model, and the particles that make up porous media are divided into two types: constant temperature and variable temperature. Simulation results indicate that velocity fluctuation caused by porous media structure increases with the increase of Reynolds number, and the temperature and velocity vary depending on the structure of porous media and flow pattern. The effect of heat convection can be reduced by loading variable temperature particles in the flow channel under the same mass force.

582115 Novel Approaches for Prebiotic Detection and Control of Microbial Communities

Research Interests:

Prebiotics are important glycans that shape the human gut microbiota from as early as, shortly after birth, for example with human milk oligosaccharides (HMOs) found in breast milk. Given their benefits, there has been a lot of interest in replicating their function by synthesizing them. Numerous approaches have been developed for their large-scale synthesis, however, the structural analysis of glycans is challenging and is under steadily increasing demand. The range of analytical techniques currently accessible to analyze glycans is limited by a lack of suitable high-throughput techniques. I have developed two platforms for the development of high-throughput, linkage-specific screening of glycans. The first is based on a genetically engineered whole-cell biosensor and the second based on an enzymatic, paper-based assay for determination of the type of fucosylation in glycans. This part of the work provides new techniques to enable a streamlined synthesis process for HMOs with a significant reduction in analysis time of new producer strains. The platform paves the way for development of a glyco-barcode assay for rapid analysis of key glycosylation patterns not only in biosynthesis platforms but also for diagnosis or monitoring disease states in different biomarkers.

Teaching Interests:

I am very excited by the opportunity to teach classes in chemical engineering. I have a strong academic background in undergraduate and graduate level chemical engineering, biochemistry and statistics coursework and diverse teaching experiences. I am prepared to teach all core undergraduate and graduate courses and develop new electives based on my research background, particularly in synthetic biology, bioengineering and glycobiology. My teaching would extend beyond the classroom to the laboratory, engaging students in an interdisciplinary approach and developing transferrable skills to complement their technical expertise for industry or academia.

576158 Protein Condensation in Nature: Functional or Pathological?

It is well established that protein condensation is associated with numerous neurodegenerative diseases and cancer. Recently it was discovered that P-granules exhibited liquid-liquid phase separation during C. elegans germ cell development. Another example of functional protein condensation is during cell division, where within short time scales, highly dynamic network of microtubules is nucleated. Protein self-assemblies is shown to promote microtubules nucleation by enhancing the local concentration and consequently, lowering the nucleation barrier for this process. These discoveries provided new insights on the functional aspects of protein condensates. In this study, we first explore a pathological pathway for protein condensation. We demonstrate how the p53, essential tumor suppressor protein, can form anomalous liquid condensates which can initiate nucleation of pathological amyloid fibrils. We further exhibit that crowding condition mimicking inside cell environment, can alter the ability of these condensates to nucleate fibrils. To gain deeper understanding of cell mechanisms to control protein condensation, we focus on a family of carrier proteins known as importins, which control the cytoplasm-nucleus transport of cargos such as p53, spindle assembly factors (SAFs), etc. We find out that aside from regular function of importins for transport, they can regulate liquid-liquid phase separation of some cargo proteins. Surprisingly, for some of the cargo proteins, the self-association is linked with the function. To investigate this, we focus on an essential microtubule nucleator, known as TPX2, which exhibits self-association on microtubules surface in *ex vivo* conditions, and rapid liquid-liquid phase separation *in vitro*. Recent study demonstrated that the self-association of TPX2 is directly linked to its ability to nucleate new microtubules. Here, we demonstrate a molecular picture on how importins can prevent TPX2 phase separation *in vitro*. Thus, controlling the nucleation ability of TPX2 in *ex vivo* conditions. Our data will provide details on how protein condensation can act both as pathological and functional pathways and demonstrate one possible route on cell mechanism to tune these processes.

Research Interests:

Biological liquid liquid phase separation, Biophysical characterization of protein condensates, Mechanisms underlying development of neurodegenerative diseases from protein condensates, Preventing toxic protein aggregation as a pathway for cell survival (exploring new therapeutics)

Teaching Interests:

Protein biophysics, Thermodynamics, Applied mathematics to chemical engineering, fluid dynamics, separation processes in chemical engineering

581361 Thermodynamic Behavior and Thermophysical Properties of Unusual Fluids: Heavy Oils

Research Interests:

My research interest focuses on the prediction of fluid phase behavior and fluid properties for process design and simulation applications. Accurate properties are essential for cost-effective design, simulation, optimization and operation of industrial processes. For instance, transport properties are required for sizing of equipment and defining power requirements; and, thermodynamics properties, including phase equilibria, determine the feasibility of a given process. The goal of this poster is to present a glimpse of my research approach and relevant results from my doctoral, postdoctoral and industrial work.

To date, the subject of my research has been heavy oils and their mixtures with solvents. Modeling heavy oil properties and phase behaviour is challenging as these fluids consist of a wide variety of components ranging from light alkanes to the heaviest and most aromatic components found in nature. In this variety, both chemical family and molecular size are responsible for heavy oil high viscosity (as high as $1x10^6$ cP at room conditions) and its complex phase behaviour; especially partition into multiple phases (even solid phases) upon dilution. The research challenge addressed in my Ph.D. dissertation was to understand the governing physics to construct predictive models for the viscosity and thermal conductivity of heavy oils. Several experimental techniques for the measurement of viscosity and thermal conductivity of heavy oils at high temperatures and pressures (up to 200°C and 10 MPa, respectively) were proposed, developed, and validated; and, the formulated models encompass the whole phase diagram, including the critical and supercritical regions.

My postdoctoral research is aimed at interpreting and modeling the phase behaviour of diluted heavy oils. The main challenge of this project is predicting the formation of a "solid-like" phase rich in asphaltenes. Asphaltenes are a group of heavier, highly aromatic and polar components that self-associate and are insoluble in paraffinic solvents. Cubic Equations of State are not suitable for for asphaltene precipitation. SAFT Equations of State can be used to model asphaltene precipitation but require extensive tuning. I developed a Regular Solution based model that successfully predicts not only the conditions at which the asphaltene rich phase appears but also the mass of material that partition to this phase. Compared to equations of state, the proposed model is simpler, faster and predictive. I currently hold an industrial postdoctoral fellowship position at CNOOC Petroleum. My research is focused on understanding how changes of oil composition induced by thermal cracking affect its thermodynamic behavior. Additionally, I am implementing my Regular Solution based approach on one of the company's process models. This postdoctoral fellowship has been awarded by MITACS and CNOOC Petroleum North America.

Future Research Plans:

Building on my expertise on data collection, modelling and computational aspects of phase equilibria calculations, I plan to extend my research to the formulation of models for electrolytes. Electrolytes have multiple applications in chemistry, geology, biochemistry, and chemical engineering. Understanding electrolyte thermodynamics is very important in the development of industrial processes such as carbon capture and utilization, removal of carbon dioxide from flue gases, desalination of water, bioprocesses, and pharmaceuticals. However, accurate data and phase behaviour/property models are scarce. This is the gap that my future research program will address.

Teaching Interests:

My teaching experience comes from three years as a teaching assistance (TA) and one term as a course instructor at the University of Calgary (Canada). As a TA, I was responsible for conducting laboratory demonstrations such as heat transfer, process control, mass transfer and fluid dynamics. As an instructor of the graduate level course Phase Behaviour of Reservoir Fluids, I oversaw the design of the course including the outline, assignments, supporting material, exercises, and exams. My teaching interests would focus thermodynamics, transport phenomena (momentum, mass and heat transfer), numerical methods, fluid dynamics and petroleum chemistry. Additionally, I am eager to design and teach a course focused on the application and computational aspects of thermodynamic models (including equations of state) for industrial process calculations.

Selected Publications:

Ramos-Pallares, F.; Schoeggl, F.F.; Taylor, S.D.; Yarranton, H.W. "Prediction of Thermal Conductivity for Characterized Oils and Their Fractions Using an Expanded Fluid Based Model", *Fuel*, 2018, 234, 66-80.

Ramos-Pallares, F.; Schoeggl, F.F.; Taylor, S.D.; Yarranton, H.W. "Expanded Fluid- Based Thermal Conductivity Model for Hydrocarbons and Crude Oils", *Fuel*, 2018, 224, 68-64.

Ramos-Pallares, F.; Lin, H.; Yarranton, H. W.; Taylor, S.D. "Prediction of the Liquid Viscosity of Characterized Crude Oils Using the Generalized Walther Model", *SPE Journal,* 2017, 22, 1487-1505.

Ramos-Pallares, F.; Taylor, S.D.; Satyro, M.A.; Marriott, R.A.; Yarranton, H.W. "Prediction of Viscosity of Characterized Oils and Their Fractions Using the Expanded Fluid Model", *Energy and Fuels*, 2016, 30, 7134-7157.

Ramos-Pallares, F.; Schoeggl, F.F.; Taylor, S.D.; Satyro, M.A.; Yarranton, H.W. "Predicting the Viscosity of Hydrocarbon Mixtures and Diluted Heavy Oils Using the Expanded Fluid Model", *Energy and Fuels*, 2016, 30, 3575-3595.

Ramos-Pallares, F.; Lin, H.; Taylor, S.D.; Satyro, M.A.; Yarranton, H.W. "Predicting the Viscosity of Characterized Crude Oils", *Proc. 16th American Institute of Chemical Engineers Annual Meeting*, San Francisco, CA, 2016.

586713 Enhancing nano-biomaterials and chemical methods for detection, isolation and separation of chemical and biological contaminants from infected samples

Research Interests:

Improvement of water treatment techniques designed to recover wastewater and provide drinking water is critical for protecting public health and environment and has gained many attentions in the last decades. Efficient detection and removal of chemical contaminants in water-based solutions such as heavy metals, cyanides, and dyes is necessary to recover water for human consumption and agricultural applications. Membrane-based separation techniques are considered as promising candidates to remove chemical contaminants form wastewater. Membrane processes are able to produce high quality water, isolate relatively small organic pollutants, and integrate with other separation systems. Compared to traditional treatment techniques, membranes can offer a simple, easy-to-operate, and low-maintenance separation process. The recovery of minor but valuable components from a mainstream using membranes can be done without substantial additional energy costs. Besides chemical impurities, pathogens (as microscopic biological organisms capable of causing disease) include viruses (comprising DNA or RNA with a protein coating), bacteria, protozoa and toxins released by algae should also be reduced from the wastewater or deactivated before the reuse in the agriculture. Pathogenic contamination of water cause disease outbreaks and contribute to background disease rates around the world. Regarding that, it is needed to develop synthetic biological interfaces that enable sensitive and rapid capture and enrichment of microorganisms from infected water-based solutions.

Keywords: Pathogen detection; bio interface; surface chemistry; wastewater treatment; membrane

PhD Dissertation: "*Bio-functionalized polymer interfaces for sensitive and selective detection and isolation of microbial pathogens from infected solutions*"

Under supervision of Dr. Ryan R. Hansen, Tim Taylor Department of Chemical

Engineering, Kansas State University, USA

Research Experience:

Enhancing nano-biomaterials and chemical methods for environmentally infected samples has been the main target of my research. This includes developing treatment methods to isolate chemical and biological contaminants from water-based solutions. I've been working in Department of Chemical Engineering at Kansas State University (with the collaboration of Division of Biology) on the design of nanostructured biointerfaces that detect and capture bacterial contaminants from solutions with high sensitivity using inexpensive materials. These interfaces can also be used in tissue engineering and to detect and isolate other biomolecules and cancerous cells. In particular, my research focused on systematic investigation of physical and chemical surface parameters that influence microbial pathogens capture over the lectin-functionalized polymer interfaces. Project included micro/nanofabrication, Surface physico-chemical characterization (FTIR, 1H-NMR AFM, SE), and Bacteria culture/growth and development. Findings applied to construct surfaces with significantly enhanced bacteria detection sensitivity. The range of detection sensitivities achieved for the proposed biointerfaces also offer potential use in a variety of applications including diagnosis of UTI, blood infections, and detection of bacterial targets in water-applications where lectin-based capture has traditionally been limited. As another direction, I've also studied the membrane-based separation techniques and adsorption process to isolate and recovery of chemical and heavy metal contaminants from environmental samples. Here I particularly focused on application and optimization of Electrodialysis (ED) method for the efficient removal of ionic pollutants from water and waste-water streams. My research targeted various type of contaminants in waste-water solutions released from pharmaceutical companies, absorption chillers and leachate such as sodium sulphate and lithium. With collaboration of Civil and Environmental Engineering Department, I'm currently working on studying membrane biofouling processes in wastewater treatment systems. Here we focus on understanding the fundamental mechanisms that drive biofilm formation over these membrane surfaces and developing chemical and physical treatment methods to alleviate biofouling. The long-term goal of this work is to understand the key mechanisms of membrane biofouling and cell-surface interactions that occur in anaerobic membrane bioreactors (AnMBR) systems. During these years, I was also able to face the challenges of a corresponding author's role for publishing 3 research papers. I believe my analytical skills on functional materials such as surface/material characterization, environmental separation techniques, and statistical analysis, would nicely help me to continue my career as post-doc.

Research projects:

- Studied the biofilm formation and early colonizing microbes over anaerobic membrane bioreactor

(AnMBR) substrates (*Collaborated with civil and environmental engineering*)

- Developed a nanostructured lectin-based biointerface relevance to isolation of bacteria targets from infected clinical and environmental samples (*Collaborated with division of biology)*

- Used micro/nano fabrications to generate reactive surfaces of an azlactone-based biofunctional block co-polymer for biomedical/environmental applications

- Investigated membrane separation techniques to remove chemical/ionic contaminants from water and waste-water samples

Future Direction:

I see these interdisciplinary projects as the starting point for my long-term research on enhancing materials and methods for environmental detection, isolation and separation. During my academic career, I've obtained experiences related to chemical engineering, biology and environmental science and am highly motivated to broaden my knowledge on improving nano-biomaterials and bio interfaces for bioengineering and biomedicine purposes. This include investigating the detection and isolation of microbial pathogens and cancerous cells from infected samples such as blood. Besides, considering the universal need for reliable access to safe water for public health and safety and the fact that unsafe water is responsible for more human deaths, my interest is also to work on designing environmental separation processes targeted chemicals and ionic contaminants existing in wastewater.

Selected Publications:

1- M. Masigol, N. Fattahi, N. Barua, B. S. Lokitz, S. T. Retterer, T. G. Platt, R. R. Hansen, "Identification of Critical Surface Parameters Driving Lectin-Mediated Capture of Bacteria from Solution"*ACS Biomacromolecules*, 20 (2019) 2852-2863

2- M. Masigol, N. Barua, B. S. Lokitz, R. R. Hansen, " Fabricating Reactive Surfaces with Brush-like and Crosslinked Films of Azlactone-Functionalized Block Co-Polymers", *Journal of Visualized Experiments*, 136 (2018) e57562.

3- M. Masigol, N. Barua, B. S. Lokitz, S. Retterer, R. R. Hansen, "Chemical Co-patterning Strategies Using Azlactone-Based Block Copolymers", *Journal of Vacuum Science and Technology B*, 35 (2017) 06GJ01.

4- M. Masigol, W. Poonlapdecha, L. E. Erickson,R. R. Hansen, K. Tuitemwong and P.

Tuitemwong, "Nanomaterials for foodborne pathogen detection and isolation. (*in preparation*)

5- M. Masigol, A. Moheb, A. Mehrabani-Zeinabad, "An Experimental Investigation into Batch Electrodialysis Process for Removal of Sodium Sulfate from Magnesium Stearate Aqueous Slurry, *Desalination*,300 (2012) 12-18.

6- M. Masigol, A. Moheb, A. Mehrabani-Zeinabad, "Comprehensive Study on Interactive Effects of Operational Parameters by Using Response Surface Method for Sodium Sulfate Removal from Magnesium Stearate Aqueous Slurry via Electrodialysis Process", *Desalination and Water Treatment*, 56 (2016) 14145-14157.

7- A. Ehsanian Mofrad, A. Moheb, **M. Masigol***, M. Sadeghi, F. Radmanesh, "Electrochemical Properties of PES/PVP Blend Heterogeneous Cation-Exchange Membranes by Using Design of Experiment Method", *Journal of Colloid and Interface Science*, 532 (2018) 546-556.

****Corresponding Author***

8- N. Parsa, A. Moheb, A. Mehrabani-Zeinabad, **M. Masigol**, "Recovery of Lithium Ions from Sodium Contaminated Lithium Bromide Solution by Using Electrodialysis Process", *Chemical Engineering Research and Design*, 98 (2015) 81–88.

9- M. Bazrgar Bajestani, A. Moheb, **M. Masigol***, "Hydrothermal synthesis of high capacity spinel ion-sieve composite adsorbents with improved stability for selective removal of lithium from aqueous solutions, "*ACS Industrial and Engineering Chemistry Research*", 58 (2019)

12207-12215.

****Corresponding Author***

10- N. Parsa, G, Khajouei, **M. Masigol***, A. Moheb, H. Hasheminejad, "Application of Electrodialysis Process for Reduction of Electrical Conductivity and COD of Water Contaminated by Composting Leachate", *Civil Engineering Journal*, 4 (2018) 1034-1045.

****Corresponding Author***

587637 Thermocatalytic depolymerization of lignin and hydrodeoxygenation of lignin-derived monomers

In recent years, lignin, the third fraction of lignocellulosic biomass, has evolved as one of the most prominent renewable resources for value-added aromatic chemicals. Lignin constitutes around 10-30% of the total lignocellulosic biomass. It is made up of aromatic monomers (coniferyl, sinapyl, and p-coumaryl), which brands it as the most abundant renewable source of aromatic chemicals. The removal of aromatic entities from lignin is a big challenge to date, mainly due to its heterogeneous nature and tendency of the polymer to repolymerize a more condensed structure. While working in a similar area in the last few years, we have narrowed down the specific needs and targets in this area of interest. The early and recent advancement has always given a strong focus on the development and screening of efficient catalysts for lignin depolymerization. Although the catalyst is a crucial component for the lignin depolymerization to monomeric products, significant attention is also required on the catalyst-to-lignin interactions and collision possibility of inter-unit linkages with the catalyst. An appropriate solvent might regulate the collision possibility of inter-unit linkages with active site of the catalyst. The use of a suitable solvent can also reduce the formation of char and restrain the repolymerization reactions. Therefore, Hansen solubility parameters (HSP) were utilized for selecting a suitable solvent for the lignin polymer. A mixture of water and THF was observed to have higher thermodynamic compatibility for lignin and guaiacol with a RED ~ 1. The beneficial effect of water and THF mixture was also observed in the depolymerization reactions of Kraft lignin using HZSM as a catalyst and NaOH as co-catalyst. A higher lignin conversion of 77% and 79% selectivity towards guaiacols with negligle char could be attained at the optimal reaction conditions from statistical analysis - Response Surface Methodology (RSM). The water-THF ratio was also found as the major independent variable influencing the conversion and selectivity of guaiacols. The higher selectivity of guaiacols directly from Kraft lignin gave us the motivation to perform the hydrodeoxygenation studies on the lignin-derived monomer model compounds. This upgradation step is necessary for the production of
deoxygenated compounds for their utilization as a fuel, fuel additives, or chemical intermediates. Towards this direction, different studies based on inexpensive metal (s) catalysts, particularly Ni-Mo/ZrO2 and Ni-Co/Al2O3, were carried out on hydrodeoxygenation of lignin model compound "guaiacol". For Ni-Mo/ZrO2, the presence of the optimum proportion of Ni/Mo4+ at the equimolar ratio of Ni and Mo is responsible for the higher HDO activity of the catalyst. The activation Caryl-O bonds are supposed to occur on oxophilic Mo4+ sites. Whereas, the hydride transfer from neighboring metallic Ni sites further results in hydrogenolysis of the Caryl-O bond giving phenol as the major product with 75% selectivity and complete guaiacol conversion. For the Ni-Co/ Al2O3 catalyst, a total metal loading of 9.2 mmolg-1cat and Ni/Co mole ratio of 1:2 indicated a higher proportion of NiCo2O4 spinel structure in the oxide catalyst. The spinel structure subsequently transformed to Ni-Co alloy after reduction in H2 at 573 K. The Ni-Co alloyed domains provide a different adsorption site than Ni and Co specific for direct methoxylation of guaiacol resulting in higher selectivity for benzene (35.1%) at complete guaiacol conversion.

Research Interest
My research interest focuses on the development of new catalysts and processes for the production of value-added fuels/chemicals from lignocellulosic biomass. A particular emphasis is also on the physicochemical characterization of biomass feedstock and catalyst with the help of various spectroscopic and non-spectroscopic characterization techniques (BET, powder-XRD, H2-TPR, UV–vis, FTIR, pyridine-FTIR, and NMR). The characterization information is used to improve the efficiency and selectivity of the catalyst and exploration of the reaction pathways. Recent examples of research topics include the catalytic depolymerization of lignin and the upgrading of lignin-derived monomers.

Teaching Interest
I am interested in teaching heterogeneous catalysis and developing a course on advanced instrumentation techniques in chemical engineering, which focus on the underlying working principles of various spectroscopic, analytical and separation techniques for undergraduate and graduate students. My research background in heterogeneous catalysis and bio-polymers has given me enough experience in different analytical methods in reaction engineering. I can teach handling and usage of equipment such as Gas

Chromatography, X-ray Diffraction, Fourier Transform Infrared Spectroscopy, UV-vis, Raman Spectroscopy, Chemisorption Analyser, Surface Area and porosity system, Gel permeation chromatography, Zeta potential, and particle size analyzer, Scanning electron Microscopy. I also worked as a teaching assistant in various theoretical and laboratory courses for undergraduate students at the Indian Institute of Technology, Hyderabad. During my Ph.D. tenure, I also got the opportunity of mentoring M.Tech students in our research group.

569152 Custom-Built Polymers Promote Stabilization, Delivery, and Bioavailability in Precision Drug Formulation Strategies

Polymer excipients are inert but crucial ingredients in countless medicines to treat chronic, infectious, and inheritable diseases across the biomedical landscape. These enhancers serve numerous roles, such as the encapsulation of therapeutic cargos, stabilization for prolonged shelf life, protection against toxic side-effects, and control over spatiotemporal release. As early pipeline small molecule drugs and biopharmaceuticals with clinically intractable physiochemical attributes defy conventional controlled delivery strategies, renewed attention to excipients as designer biomaterials can reposition compounds with poor bioperformance to more successful and low-cost formulations. Here, I will describe a general high-throughput synthesis/screening campaign to rapidly explore the physical and chemical state spaces of multicomponent copolymers. As an example, combinatorial libraries comprising over 60 controlled polymers were constructed in three parallel runs to solubilize antiseizure and antiandrogen drugs in solid dispersions for oral drug administration. Narrow windows in chemical composition demonstrated drug supersaturation maintenance at an order of magnitude above equilibrium levels under nonsink in vitro dissolution conditions, supported by solution-state NOESY and DOSY NMR characterization. When administered to rat models, a 23-fold increase in the area-under-the-curve over the neat drug was observed, alongside superior bioavailability metrics over the use of commercial excipients. This integrated approach of combining core principles in engineering, chemistry, and biology illustrates the practical utility and broad potential of tunable polymer excipients toward the precision drug formulation of challenging therapeutic agents. I will conclude with ongoing efforts to adopt this paradigm toward the advancement of customized polyelectrolyte excipients. Advanced synchrotron scattering experiments provide insight on the encapsulation, size, and shape of designer nanocomplexes with peptides, nucleic acids, and proteins. This work can provide the foundation for nonviral delivery of CRISPR/Cas9 components in emerging gene editing frontiers.

569247 Engineering Biomaterials to Recapitulate the Stem Cell Microenvironment

Despite the tremendous potential of stem cells to repair and replace damaged or diseased tissue, little progress has been made in translating stem cell-based therapies from the laboratory to the clinic. Several critical engineering challenges remain that have limited the clinical use of stem cells, including difficulties producing the large number of stem cells needed for therapeutic applications, the efficient and reproducible differentiation of stem cells into mature cell types of interest, delivery of cells to target tissues during transplantation, and maintenance of the viability and function of the cells post-transplantation. Materials science approaches stand to overcome these significant hurdles. By designing synthetic systems to mimic the native stem cell microenvironment, we can exert greater control over cell fate decisions, such as self-renewal and differentiation. In this talk, I will focus on my work using materials-based strategies to engineer platforms to expand neural progenitor cells (NPCs). NPCs are capable of differentiating into the major neural lineages of the central nervous system, and thus represent a promising cell source for therapies to treat numerous debilitating conditions, including spinal cord injury and neurodegenerative diseases. However, recent studies have highlighted the difficulties in maintaining the regenerative potential of cultured neural progenitor cells (NPCs) for use in clinical trials. I hypothesized that culturing NPCs within 3D biomimetic hydrogels would allow for significant stem cell expansion *in vitro*, while better maintaining the regenerative capacity of the cultured cells. While it has been well established that 2D material parameters, such as substrate stiffness, influence the differentiation of NPCs, previous 3D studies have not clearly determined which parameters are key to maintaining NPC stemness and regenerative capacity. I therefore designed a family of engineered elastin-like protein (ELP) hydrogels with simultaneous tuning of stiffness, degradability, and adhesive ligand concentration to probe the contributions of these parameters toward maintenance of NPC phenotype. NPCs were encapsulated in gels with stiffness spanning a physiologically-relevant range (E~0.5-50 kPa) and high, medium, or low degradability. Expression of the NPC stemness markers Nestin and Sox2 was positively correlated with matrix degradability, but no correlation was observed with matrix stiffness. Accordingly, NPCs cultured within high degradability gels exhibited increased proliferation and better maintained their differentiation capacity compared with cells cultured in low degradability gels. In contrast to previous results with 3D cultured mesenchymal stem cells, biochemical inhibition of cytoskeletal contractility and culture in hydrogels without cell-adhesive ligands did not decrease NPC stemness in high degradability gels, suggesting a mechanism independent of tension generation. However, blocking degradation by NPC-secreted proteases through biochemical inhibitors or shRNA-mediated knockdown resulted in a marked reduction in NPC stemness. Confocal microscopy revealed enhanced cell spreading and cell-cell contact in high degradability gels. Inhibiting cadherin-mediated contact with function blocking peptides decreased NPC stemness. Independently blocking matrix degradation and inhibiting cadherin contact both decreased ß-catenin signaling, providing a mechanism by which matrix remodeling regulates NPC stemness. Culturing NPCs in two additional hydrogel systems (proteolytically-degradable poly(ethylene glycol) and physically-remodelable alginate) demonstrated the generality of our results, indicating that NPC stemness maintenance required matrix remodeling but was insensitive to matrix stiffness. These results have identified matrix remodeling as a previously unknown requirement for NPC stemness maintenance in 3D hydrogels, suggesting that biomaterials that exploit matrix remodeling may increase the regenerative potential of NPC therapies.

569484 Liquid Infused-Elastomers As a Multi-Functional Material in Implantable Bioelectronics

Implantable electronics are a class of medical devices of unrealized potential for monitoring, diagnosis, and treatment of health and disease. Although some devices have been commercialized and are issued as the standard-of-care (e.g. pacemakers, cochlear implants), others like those for neuroprosthetics cannot make clinical impact until issues regarding life-time efficacy are addressed. Diminishing efficacy that occurs in long-term implantations is thought to be due to several factors including significant surgical trauma, an aggressive foreign body response, and poor material compatibility with the biological milieu (e.g. metal corrosion, biofouling, dielectric degradation). Here, we investigated if liquid-infused elastomers incorporated into bioelectronic devices can address these challenges. Such materials consist of an elastomer network (silicone) swollen and infused with water-immiscible liquids (oil). The elastomer hosts an oil layer yielding a slippery, hydrophobic surface out of phase from a surrounding aqueous environment.

After subcutaneous implantation in mice and histological analysis of tissue, oil-infused elastomers appear to be just as biocompatible as elastomer-only, a material widely used in countless medical devices. Extracted implants were analyzed for biofouling and revealed that oil-coated samples led to a substantial reduction in the number of adhered cells. When applied as coatings to electrophysiology microwires, oil-infused elastomers reduced friction and reduced insertion forces when pierced into agarose hydrogels. To examine the impact of this friction-reducing coating on surgical trauma, coated wires were inserted into explanted mouse brains. Elastomer-only coatings resulted in tissue tearing on the insertion edge, blood vessel shearing and extravascular blood cells, while tissue pierced with oil-infused elastomers did not show signs of each of these. In summary, we demonstrate oil-infused elastomers are biocompatible, reduce biofouling and reduce insertion trauma. These results warrant future studies to examine their potential for improving chronic implantation of bioelectronics.

578268 Rapid Actuation and Tunable Control of DNA-Based Mechanisms

Precise robotic motion is ever-present within our cells in dynamic proteins like ATP synthase and kinesin. Functioning much like macroscopic machines, these proteins have multiple components and defined motion paths. Engineers across many disciplines use inorganic- and bio-materials to create nano- and micro-robots with similar functionality. Structural DNA nanotechnology enables researchers to design and build DNA-based structures with nanometer-scale spatial control by employing well understood DNA base pairing and nucleic acid synthesis methods. Our early work contributed to a library of DNA devices with controllable motion pathways, primarily using DNA handles and strand invasion to bind or displace reconfigurable components with timescales of minutes or longer. Here, we present two strategies for actuating DNA-based mechanical devices in near real time. First, we demonstrate an approach using a simple modification to existing devices that adds a network of weak binding sites on complementary components for environmentally controlled rapid actuation. A network of weak DNA handles is activated by increasing cation concentration, which raises the avidity of the network to join the two components. Likewise, reconfiguration is quickly reversible in decreased salt conditions. Second, we present an innovative method for real-time actuation using a magnetic stimulus. Micron-scale rigid arms couple magnetic beads to nanoscale devices for direct manipulation through an externally applied magnetic field. This level of spatiotemporal control over DNA devices can serve as a foundation for real-time manipulation of molecular systems.

564181 Biomimetic Production of Polymer Fibers from Reversibly Cross-Linked Polysaccharide Networks in Water

Polymer fibers are important structural components in textiles, composite materials, and numerous materials for biomedical applications. The demand for tough, biocompatible materials produced by simple processing techniques has led to significant interest in materials inspired by natural fibers such as spider silks.[1,2] Traditional methods of fiber production typically require spinning processes that suffer from high energy costs and are dependent on organic solvents, amongst other physical limitations due to burdensome protocols and highly viscous polymer solutions.[3] This work describes a biomimetic technique for fabricating polymer fibers that is both operationally simple and performed at room temperature in aqueous solutions. Similar to spider silks, these fibers are produced by pultrusion, rather than electrospinning or extrusion, and can be drawn to > 1m in length from a viscous liquid. The liquid-phase precursor consists of a network of polysaccharides and small, branched polymers reversibly bound by dynamic covalent cross-links that rapidly break and reform, allowing for fiber formation. The effects of the chemical structure, molecular weight, and extent of branching of the cross-linker on the material properties of the bulk liquid and resulting polymer fibers are investigated. Mechanical testing and microscopy studies of the fibers provide insight into the origin of these properties, in particular, high ductility. Overall, a molecular understanding of how the polysaccharide network translates to macroscopic properties provides a strategy for efficient production and rational design of fibrous biomaterials. Unlike most polymer fibers, the described structural properties directly result from a dynamic cross-linked network. This design expands the potential of these materials from being used for their physical architectures, such as in porous scaffolds for tissue growth,[4] to being used for their chemical properties, such as in stimulus-responsive drug delivery systems.

References:

1. Rising, A.; Johansson, J. *Chem. Biol.* **2015**, *11*, 309–315.
2. Wu, Y. et al *Natl. Acad. Sci.* **2017**, *114*, 8163–8168.
3. I. C.; Fang, D.; Hsiao, B. S.; Okamoto, A.; Chu, B. *Biomacromolecules* **2004**, *5*, 1428–1436.
4. Ji, Y. et al *Biomaterials* **2006**, *27*, 3782–3792.

564442 Designer Biomaterials and Integrated Platforms Toward Precision Medicine

Precision or personalized medicine is an emerging approach that matches patients with the most appropriate treatment based on the precise molecular features of an individual patient's disease. Despite huge advances in translational research, diagnosis and treatment continue to pose profound challenges in addressing significant health problems, including cancer and infectious diseases. In this talk, I will discuss two research projects that demonstrate molecular engineering efforts toward precision medicine. One focuses on new development in diagnosis: bioanalytical sensing of plasma markers predictive of impending sepsis based on a magneto-electrochemical approach. The other utilizes a novel programmable multi-therapy release strategy for orthopedic implants, enabling both bacterial eradication at the bone-hardware interface and accelerated bone tissue regeneration. These studies highlight the potential of integrating materials science, molecular engineering, and clinical science approaches to develop advanced molecular diagnostics and innovative therapies for precision medicine.

References

[1] Min J, Nothing M, Coble B, Zheng H, Park J, Im H, Weber G, Castro CM, Swirski FK, Weissleder R*, Lee H*, Integrated biosensor for rapid and point-of-care sepsis diagnosis, ACS Nano, 12(4), 3378-3384 (2018).

[2] Min J†, Im H†, Allen M, McFarland PJ, Degani I, Yu H, Normandin E, Pathania D, Patel JM, Castro CM, Weissleder R*, Lee H*, Computational optics enables breast cancer profiling in point-of-care settings, ACS Nano, 12(9), 9081-9090 (2018).

[3] Min J, Choi KY, Dreaden EC, Padera RF, Braatz RD, Spector M, Hammond PT, Designer dual therapy implant coatings eradicate biofilms and accelerate bone tissue repair, ACS Nano, 10(4), 4441-4450 (2016).

[4] Min J, Braatz RD, Hammond PT, Tunable staged release of therapeutics from layer-by-layer coating with clay interlayer barrier, Biomaterials, 35(8), 2507-17 (2014).

569796 Mechanophore-Based Biomaterials for In Vivo Ultrasound-Triggered Light Generation

Mechanical forces are constructively generated in various biopolymer systems such as muscle development. In synthetic polymers, however, mechanical forces have been considered destructive until the development of mechanophores, the molecular units that selectively exhibit chemical responses to mechanical perturbations. To date, mechanophores have endowed polymers with a variety of stress-responsive properties including color changing, chemiluminescence and small molecule releasing, etc.

In contrast to the great success in materials science, the application of mechanophores in biomaterials was underdeveloped. Recently, our groups reported that a widely used non-invasive medical tool, high-intensity focused ultrasound (HIFU), is applicable to the activation of mechanophores. An interesting candidate—dioxetane mechanophore developed by Sijbesma and coworkers, attracted our attention for its ability to generate light from chemical energy *via* force-induced chemiluminescence. Under HIFU stimulation, blue light was clearly observed in our dioxetane containing polydimethylsiloxane (PDMS) materials at the focal spot of the ultrasound beam, providing excellent spatial and temporal control. This preliminary result opens a pathway towards biomedical applications of mechanophores such as non-invasive sono-optical treatments for optogenetics therapy. To achieve this final goal, we have two aims from the materials science perspective:

Aim 1: Improving light intensity.

In our previous experiments the light intensity did not meet the minimum threshold for optogenetics or other photo therapies. In our recent studies, we discovered that the poor solubility (max. ca. 0.5 wt%) of the energy acceptor, diphenylanthracene (DPA), was a limiting factor in the light intensity, presumably due to low energy transfer efficiency. By replacing DPA with a more soluble hexyl-functionalized DPA (HDPA), we can now incorporate up to 7.5 wt% HDPA into PDMS. As a result, the light intensity from HIFU activation was increased by 70-fold, up to 14 $\mu W/cm^2$. The increased intensity is higher than the reported threshold of many channelrhodopsins and thus is promising for optogenetics applications. For practical applications, however, small molecular energy accepters may diffuse out of the polymer network. We are chemically attaching DPA moieties besides the dioxetane mechanophore (within efficient energy transfer distance) and then crosslinking into the polymer network. With this chemical modification, we expect simultaneously immobilization of energy acceptor and maximization of light intensity.

Aim 2: Designing an injectable and biocompatible hydrogel.

In our previous design, none of the three components (polymer backbone, dioxetane mechanophore and DPA energy acceptor) are water soluble. For biomedical applications, we adapt the injectable hydrogel system reported by Hubbell and coworkers based on thia-Michael reaction between terminated multi-arm polyethylene glycol thiol (PEG-thiol) and diacrylate crosslinkers. In our preliminary study, a dioxetane mechanophore was crosslinked into a DMSO-swollen PEG gel network, which successfully generates blue light upon HIFU sonication. We are modifying the molecular structure of dioxetane and DPA derivatives to improve their water solubility without sacrificing the light intensity.

561354 Present and Future of Bioinspired Phenolic Coatings for Surface and Interfacial Engineering

Polydopamine, a representative phenolic coating introduced by Messersmith Lab in 2007, is a biomimetic coating formed on most surfaces when solutions of dopamine are exposed to oxygen under mildly alkaline conditions. Polydopamine is not only effective in controlling the surface energy of the targeted (i.e., coated) material, but also enables post-modification using nucleophiles such as amines and thiols by forming stable covalent conjugates with residual quinones on the coating. As a result of its accessibility and versatility, polydopamine has been adopted as a powerful and widespread functionalization tool in biomedical, energy, and industrial applications. However, it has been generally accepted that the wear-resistance of the polydopamine is relatively low. For example, mild sonication or even slight physical stress can severely damage the coating. This limits polydopamine-related applications. We truly believe that our next mission is to achieve mechanically improved polydopamine. In this talk, I would like to share our recent progress on this by 'partial graphitization' of polydopamine through blue diode laser annealing. Here, ‘partial graphitization' means that the functional properties of polydopamine still remain intact while the mechanical properties of the polydopamine are increased by having certain portions of the graphitized polydopamine. By adjusting the laser power, we can precisely control the level of graphitization as confirmed by Raman spectroscopy.We believe that our unique approach to overcome the weak abrasion resistance of polydopamine would once again "revolutionize" the polydopamine field by enabling applications where mechanically strong properties are required.

AUTHOR INDEX

Abiose, Kabir....249
Achinivu, Ezinne....9
Achouri, Esma Ines....507
Aditham, Abhishek....435
Afzal, Shaik....518
Agrawal, Ankit....355
Alexander, Symone....15
An, Hyosung....369
Anderson, Daniel G....563
Andreev, Marat....463
Appel, Eric A....51
Ardagh, M. Alexander....137
Ardoña, Herdeline Ann M....42
Avraamidou, Styliani....260
Baccile, Joshua....24
Back, Seoin....161
Badino, A. C....430
Bai, Yuchen....310
Bailey-Hytholt, Christina M....68
Ball, Madelyn R....144
Bandodkar, Amay J....92
Bara, Jason E....524
Barlaz, David....149
Barman, Sourav....353
Bartel, Christopher J....228
Bates, Frank S....559
Batista, Victor S....445
Beitle, Robert....434
Belarbi, Zineb....237
Belcher, Donald....120
Bergsman, David S....404
Beyzavi, M. Hassan....434
Bhagia, Samarthya....380
Bilatto, Stanley....430
Bilodeau, Camille....229
Blankschtein, Daniel....393
Block, David E....81
Bose, Suman....99, 563
Bowal, Kimberly....433
Bozlar, Michael....311
Brace, Emma C....66
Bregante, Daniel T....190
Bui, Ngoc....531
Buzzi-Ferraris, Guido....508
Cao, Liang....174
Capellades, Gerard....118
Carter, Emily A....393
Castro, Carlos E....562
Chabinyc, Michael L....492
Chandler, Bert D....312
Chang, Alice B....473
Charles, Daniel S....421
Chen, Chia-Chin....247
Chen, Christopher....86
Chen, Wei Yin....292, 540
Chen, Yinuo....252
Chen, Yujia....549
Chidsey, Christopher E. D....231
Chisholm, Nicholas G....348
Chiu, Liang-Yuan....252
Cho, Hong Je....394
Choi, Jaewon....360
Chortos, Alex....382
Choudhary, A....337
Choudhury, Snehashis....465
Chowdhury, Ratul....207
Chu, Crystal K....27, 563
Chu, Henry C. W....343
Chung, Peter....72
Clark, Ezra L....140
Colby, Ralph H....492
Contreras-Naranjo, Jose C....95
Cooper, Nathanial J....307
Corral, Daniel....249
Crane, Matthew....414
Creighton, Megan A....417
Cui, Yi....362
Dahlman, Clayton J....410
Dai, Yifan....252
Daramola, Damilola A....200
Das, Debasish....327
Das, Debayan....331
Das, Saikat....304
Davidson, Emily....481
Davis, Kirsten....59
Davis, Mark E....142
Demirer, Gozde Sultan....435
Diao, Weijian....171
Ding, Fangyuan....46
Dong, Yachao....226
Dorfman, Kevin D....496
Dos Santos, Antonio C. F....4
Dutta, Palash....124
Dutta, Sarit....322
Eagan, Nathaniel....139
Elbashir, Nimir....518
El-Halwagi, Mahmoud....518
Elsaidi, Sameh....373
El-Zaatari, Bassil....467
Emamy, Hamed....398
Enam, Fatima....550
Engler, Anthony....385
Etemad, Arash....429
Etemad, David....515
Falatach, Rebecca....529
Fang, Chengcheng....443
Fang, Yin....482
Fardous, Roa'....119
Farinas, Cristiane Sanchez....430
Feaster, Jeremy T....251

AUTHOR INDEX

Fenton, Owen S. 84
Frey, Douglas D. 442
Ganesan, Venkat 496
Gang, Oleg 398
Gao, Hanyu 498
Gao, Tao 277
Garcia-Segura, Sergi 70
Garg, Abhinav 209
Gauthier, Joseph 193
Gerhardt, Michael R. 268
Gharahcheshmeh, M. Heydari 258
Ghosh, Debadyuti 457
Gibbons, William T. 153
Gomez, Enrique D. 492
Grabow, Lars C. 312
Grzetic, Douglas 215
Gunukula, Sampath 275
Hadi, Atefe 438
Hahn, Christopher 249
Halmes, Abigail 565
Han, Endao 482
Harris, Leonard Alfredo 104
He, Yulian 445
Hensley, Alyssa 146
Hernandez, Hector Lopez 51
Hernandez, Sebastian 541
Horner, Jeffrey S. 64
Howard, Michael P. 205
Hu, Shu 445
Huang, Jianzhi 252
Huang, Kai 49
Huh, Yun Suk 39
Ioneva, Mariana 514
Jackson, Nicholas 37
Jacob, Alan Ranjit 334
Jaeger, Heinrich M. 482
Jamali, Sina 56
Jaramillo, Thomas F. 249
Jeon, Tea-Joon 39
Jiang, Shaoyi 455
Jiang, Yuanwen 482
Jimenez, Juan 179
Jin, Kailong 476
Jocz, Jennifer 166
Joo, Yong Lak 261
Jorgenson, Tyler D. 431
Joseph, Alby 563
Josephson, Tyler R. 129
Jouny, Matthew 183
Jung, Haesung 278
Jung, Sungyoon 189
Kaar, Joel L. 529
Kacica, Clayton 383
Kammakakam, Irshad 524
Kang, Ji-Hwan 460
Kang, Yijin 290
Kannan, P. Chinnayan 485
Kannan, Rohit 211
Kapellos, George E. 344
Karre, Avinashkumar 511
Kathe, Mandar 501
Kelesidis, Georgios A. 446
Khandavalli, Sunilkumar 325
Kienle, Daniel 529
Kim, Gun 565
Kim, Jae Jung 471
Kim, Oleg 40
Kim, Tae-Il 63
Koman, Volodymyr 396
Korshoj, Lee 60
Koshy, David 249
Kraft, Markus 433
Krishnamoorthy, Dinesh 505
Krishnaveni, T. 339
Kulik, Heather J. 203
Kumar, Arkadeep 389
Kumar, K. B. Sravan 312
Kumar, Manish 207
Kumar, Sanat K. 398
Kwon, Kiyoon 63
Landry, Markita 435
Langer, Robert 84, 563
Lanzarini-Lopes, Mariana 70
Larrazábal, Gastón O. 177
Lee, Dong Un (Daniel) 249
Lee, Eunseon 39
Lee, Jaewon 425
Lee, Kyueui 566
Li, Bo 413
Li, Gengnan 299
Li, Jie 36
Li, Jinxing 88
Li, King 565
Li, Linqing 86
Li, Yamin 452
Li, Yuguang (Chris) 240
Lignos, Ioannis 418
Liguori, Simona 281
Lin, Xiaojie 54
Lin, Yiliang 482
Lin, Yu-Jiun 356
Liu, Albert Tianxiang 427
Liu, Chung-Chiun 252
Liu, Dehua 310
Liu, Fang 132
Liu, Mingjie 214
Liu, Yang 525
Liu, Yixin 419
Lopez, Jeffrey 271
Lu, Xu 273

AUTHOR INDEX

Lu, Yingda 293
Lundberg, Clair 565
Luo, Le 252
Luo, Yimin 365
Luo, Yu 500
Lusardi, Marcella 142
Ma, Canghai 536
Ma, Hanyu 175
Madl, Christopher M. 115, 560
Majid, Ahmad Abdul 522
Majumdar, Saptarshi 557
Malliaras, George 561
Mallory, Stewart 403
Manenti, Flavio 508
Maranas, Costas 207
Marras, Alexander E. 111, 562
Martin, Jacob W 433
Masigol, Mohammadali 554
Masuku, Cornelius Mduduzi 126
Matos, Juliana 435
Mattern, Daniell 292, 540
McCloskey, Bryan D. 231
Medina, Oscar 420
Meza-Morales, Paul 158
Miller, Konrad 81
Min, Jouha 3, 564
Miranda, Michael 490
Mirfendereski, Seyed Mojtaba 538, 539
Mirzadeh, Mohammad 352
Misiolek, Wojciech 407
Mohanty, Rashmi 457
Molaei, Mehdi 320
Mondal, Dilip 535
Mondal, Kunal 381
Monje-Galvan, Viviana 198
Mooney, David J. 19
Moore, Jeffrey 565
Morse, David 496
Mosleh, Imann 434
Murthy, Niren 36
Narayanan, Niju 11
Naserifar, Saber 216
Nayani, Karthik 346
Nazari, Behzad 387
Nelson, Diane L. 110
Nguyen, Trung 376
Ngwanza Sr., Mbinzi Kita Deddy 513
Nielander, Adam 155
Ning, Shang 565
Nkazi, Diakanua 512, 513
Oelze, Michael 565
Oh, Mi-Wha 39
Oh, Seo Yeong 39
Omar, Ahmad 459
Pak, Alexander J. 390
Panthani, Matthew G. 438
Park, Byungchun 231
Parviz, Dorsa 423
Pascazio, Laura 433
Patra, Tarak 363
Paul, Colin D. 108
Peterson, Joseph 323
Pfaendtner, Jim 483
Pfefferle, Lisa 442, 445
Phillip, Jude M. 1
Pogaku, Ravindra 295
Poling-Skutvik, Ryan 335
Pushpavanam, Subramanium 337, 339
Qwabe Sr., Thembinkosi 512
Rahimian, Saeed Khaleghi 244
Raikwar, Deepak 557
Rajan, Ananth Govind 393
Ramakrishnan, Sathish 546
Ramakrishnan, Srini 231
Ramasubramanian, V. 187
Ramos-Pallares, Francisco 552
Rapp, Peter 459
Rashid, Khalid 516
Reineke, Theresa M. 559
Reklaitis, G. V. Rex 508
Renganathan, T. 339
Renganathan, Thiruvengadam 337
Resasco, Joaquin 151
Rezaei, Payam 442
Rogers, John A. 92, 362
Root, Samuel E. 494
Rossi, Francesco 508
Roy, Shyamal 535
Rutz, Alexandra 33, 561
Saadat, Amir 21
Safari, Mohammad 551
Sahoo, Jugal Kishore 7
Sajjadi, Baharak 292, 540
Samanta, Devleena 401
Samanta, Rituparna 495
Sampath, Janani 483
Sankarasubramanian, Shrihari 241
Sanyal, Oishi 527
Schimmenti, Roberto 135
Schroeder, Charles M. 413
Schroeder, Thomas B. H. 101
Schwartz, Daniel K. 529
Sengupta, Rajarshi 332
Sethuraman, Vaidyanathan 496
Seufitelli, Gabriel 185
Shahriari, Dena 82
Shao, Baosli 549
Shapiro, Jenna M. 77
Shee, Debaprasad 557
Shen, Zhizhang 224

AUTHOR INDEX

Shih, Arthur J. 133
Siddiquee, Muhammad....301
Silverman, Bradley 459
Sim, Seunghyun....25
Singh, Ravendra....519
Sosa, Andres Chaparro 529
Spahn, Philipp 450
Squinca, Paula....430
Starr, Francis W. 398
Staskawicz, Brian....435
Stepniowski, Wojciech....407
Stevens, Michaela Burke 234
Strano, Michael 393, 427
Suja, Vineeth Chandran....358
Sun, Dan 449
Sun, Qiji....549
Svinterikos, Efstratios....317
Tang, Jing....29
Tanner, Eden E L....116
Taylor, Lauren 122
Tekin, Halil....97
Thengane, Sonal....314
Tian, Bozhi....482
Tian, Ruichao 549
Ting, Jeffrey M....478, 559
Tirrell, David....24
Tirrell, David A. 459
Tirrell, Matthew V. 559
Tolbert, Blanton....252
Tricker, Andrew W....194
Tsao, Caroline 455
Umlauf, Benjamin....113
Vashisth, Aniruddh....412
Vutukuri, Hanumantha Rao 350
Wang, Hua....19
Wang, Huiliang 415
Wang, Jin....482
Wang, Kuo-Kuang....407
Wang, Lei....163
Wang, Shiyan 329
Wang, Shuyan 549
Wang, Yuan....252
Wang, Zhen-Gang....459
Waymouth, Robert M....231
Weissleder, Ralph 3
Weltz, James S....529
Werber, Jay R. 533
Westerhoff, Paul....70
Whittaker, Todd....312
Winter, Lea....181
Wu, Qiong....468, 565
Wu, Xiaowei 252
Xi, Yuyin....377
Xiao, Xianghui....482
Xiao, Yang 168
Xie, Renxuan....492
Xu, Alex....106
Xu, Bingjun....256
Xu, Jie....487
Xu, Wei....252
Yan, Yushan 256
Yang, Chongyin....253
Yang, Ming 195
Yang, Shaowei 544
Yang, Shuren 549
Yang, Xuan 256
Yang, Zhongyue....203
Yao, Jiwei....252
Yin, Xi 285
Yu, Yanhao 340
Zamani, Somayeh....261
Zanganeh, Steven 17
Zargar, Amin....44
Zerze, Gul H. 6
Zhang, Fan....453
Zhang, Huan 435
Zhang, Huichun....252
Zhang, Wenlin 222
Zhang, Xueqiang 154
Zhang, Yang 440
Zhao, Jie....362
Zhao, Xuebing 310
Zhao, Zhenghang....220
Zhao, Zongmin 75
Zheng, Xueli....400
Zhong, Yu 367
Zolghadr, Ali....264